Y0-CDD-869

PRENTICE-HALL VOCATIONAL AGRICULTURE SERIES

BEEF PRODUCTION. Diggins and Bundy

CROP PRODUCTION. Delorit, Greub, and Ahlgren

DAIRY PRODUCTION. Diggins and Bundy

EXPLORING AGRICULTURE. Evans and Donahue

JUDGING LIVESTOCK, DAIRY CATTLE, POULTRY, AND CROPS. Youtz and Carlson

LEADERSHIP TRAINING AND PARLIAMENTARY PROCEDURE FOR FFA. Gray and Jackson

LIVESTOCK AND POULTRY PRODUCTION. Bundy, Diggins, and Christensen

MODERN FARM POWER. Promsberger, Bishop, and Priebe

PROFITABLE FARM MARKETING. Snowden and Donahoo

PROFITABLE SOIL MANAGEMENT. Knuti, Korpi, and Hide

SWINE PRODUCTION. Bundy and Diggins

USING ELECTRICITY. Hamilton

FOURTH EDITION

LIVESTOCK AND POULTRY PRODUCTION

CLARENCE E. BUNDY

Former Chairman
Department of Agricultural Education
Iowa State University
Ames, Iowa

RONALD V. DIGGINS

Former Vocational Agriculture Instructor
Eagle Grove, Iowa

VIRGIL W. CHRISTENSEN

Chairman, Agriculture and Natural Resources Department
Hawkeye Institute of Technology
Waterloo, Iowa

PRENTICE-HALL, INC., ENGLEWOOD CLIFFS, NEW JERSEY

LIVESTOCK AND POULTRY PRODUCTION, Fourth Edition.

Clarence E. Bundy, Ronald V. Diggins, and Virgil W. Christensen

© 1975, 1968, 1961, 1954 by Prentice-Hall, Inc., Englewood Cliffs, N. J. 07632. All rights reserved. No part of this book may be reproduced in any form or by any means without permission in writing from the publisher. Printed in the United States of America.

ISBN 0-13-538579-2

Cover photograph by Grant Heilman

10 9 8 7 6 5

PRENTICE-HALL INTERNATIONAL, INC., London
PRENTICE-HALL OF AUSTRALIA, PTY. LTD., Sidney
PRENTICE-HALL OF CANADA, LTD., Toronto
PRENTICE-HALL OF INDIA PRIVATE LTD., New Delhi
PRENTICE-HALL OF JAPAN, INC., Tokyo

ABOUT THIS BOOK

The Fourth Edition of *Livestock and Poultry Production* is even more informative than the first three editions, the wide acceptance of which has been most gratifying. We present in one book the most complete and up-to-date information on this broad subject.

We have added six chapters devoted to the selection, nutrition, breeding, and training of horses. Technology and practices in swine, beef and dairy cattle, horse, sheep, and poultry production change rapidly, and during recent years significant advances have been made. It is our intent to inform the reader of these changes and assist in improving the production of livestock and poultry on the farm.

Results of research conducted by a large number of agricultural experiment stations and by the United States Department of Agriculture were reviewed. Recent publications of many universities and state agricultural colleges have been used in assembling the information contained in this book. Livestock and poultry marketing information was obtained from meat packing and other processing industries. The purebred and other livestock associations contributed the latest data concerning type and production of animals.

Each type of livestock and fowl is treated separately. The subject matter is organized and systematically indexed so that the reader can quickly locate needed information. The basic principles and practices needed in solving production, management, and marketing problems are made clear without the necessity of reading through a maze of technical and semi-technical material. The descriptive material is supplemented by several hundred illustrations, graphs, and tables.

In the preparation of this manuscript, we have drawn heavily upon our experiences during the past 44 years in farming, vocational and technical agriculture, and agricultural education work. We have worked closely with hundreds of livestock producers and educators in the agriculture field, and considerable technology and many of the recommended practices were obtained directly from these sources.

We are indebted to many individuals, institutions, and organizations for illustrative and other materials included in the text. We are especially indebted to U.S.D.A., university, and experiment station personnel. Staff members in nearly 30 universities assisted us. The secretaries of the various livestock and breed associations were very helpful in providing subject matter and illustrative materials. A large number of commercial businesses and industries provided very valuable materials, as did a number of publishers of farm magazines. Appropriate credit lines appear throughout the text.

Members of the animal science, dairy science, poultry science, agricultural engineering, and dairy and food industries staffs at Iowa State University provided valuable assistance in the preparation of the manuscript. We are also grateful to the large number of vocational and technical agriculture instructors and the many reviewers who made helpful suggestions.

CLARENCE E. BUNDY

RONALD V. DIGGINS

VIRGIL W. CHRISTENSEN

CONTENTS

LIVESTOCK FEEDING AND NUTRITION

PORK PRODUCTION

BEEF PRODUCTION

DAIRY PRODUCTION

DUAL-PURPOSE CATTLE

CATTLE DISEASES AND PARASITES

HORSE PRODUCTION

SHEEP PRODUCTION

POULTRY PRODUCTION

INHERITANCE AND REPRODUCTION

LIVESTOCK FEEDING AND NUTRITION

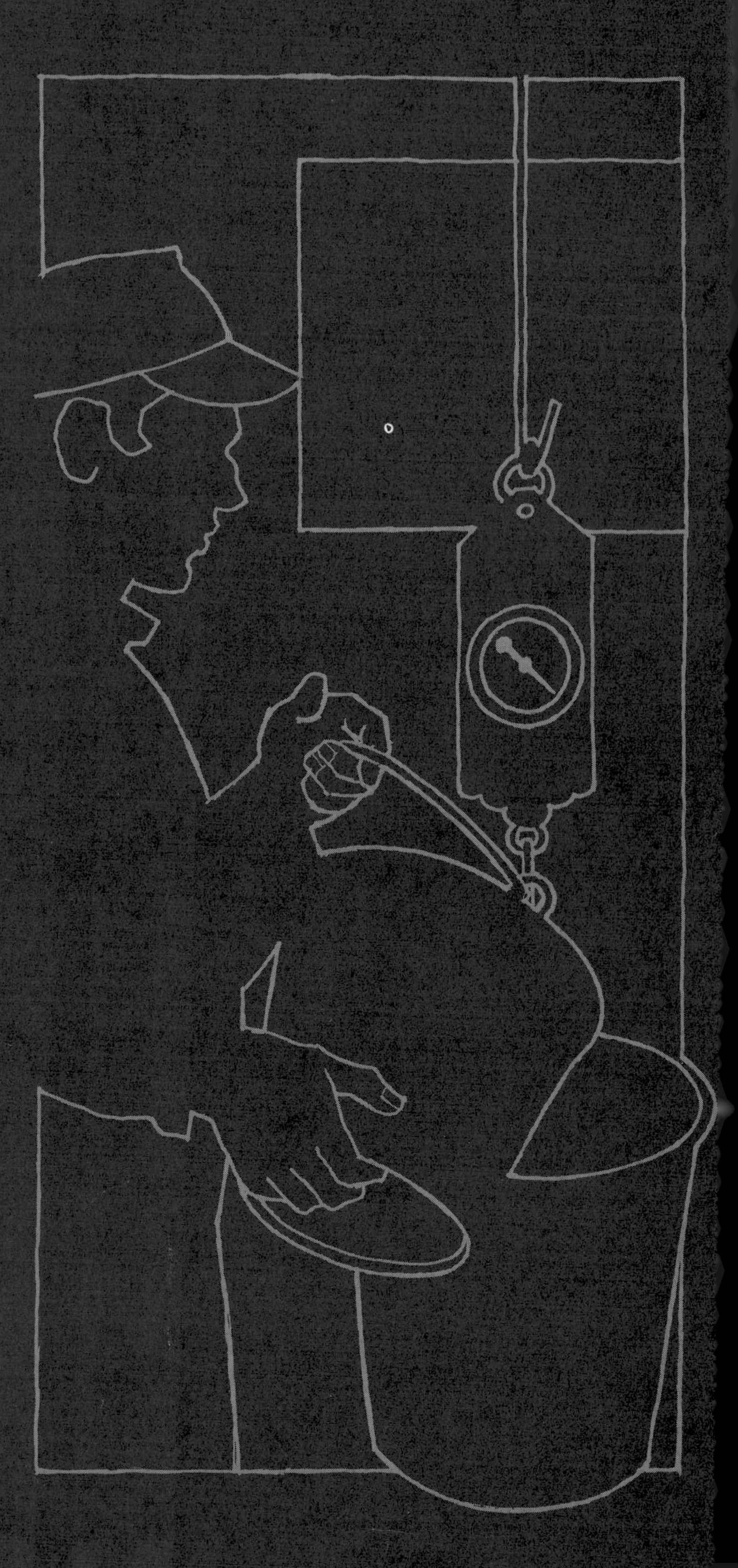

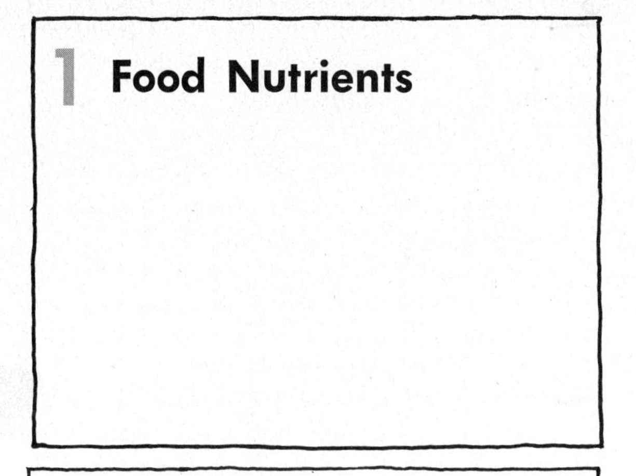

1 Food Nutrients

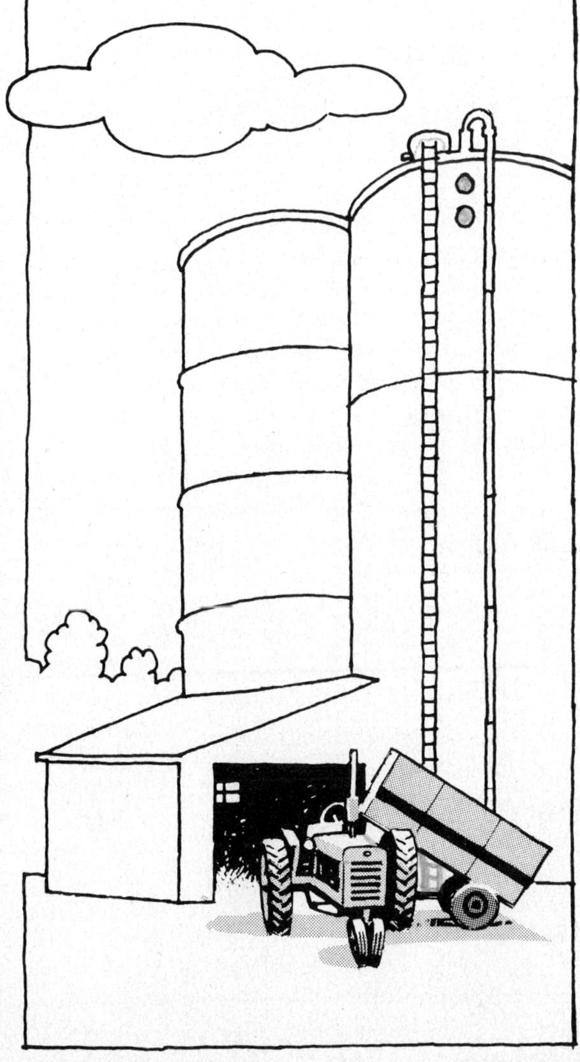

Livestock on the farm means more meat, eggs, and milk on the tables of American homes. It means a healthier population and, consequently, a stronger America. To the farmer it means a more fertile land. Livestock converts grains and hays into valuable food products, and the residues are incorporated into the land to make it richer and more productive. The farmer's returns for his time and effort depend largely upon his knowledge of the science of animal husbandry and the application of science to the livestock enterprise.

To become a good livestock producer, one must first know what foods animals require in order to grow, reproduce, and furnish man with the meat, eggs, milk, and wool he desires. Understanding certain terms will make it easier to discuss these requirements.

NUTRIENTS

The term *nutrient* means a single class of foods or group of like foods that aid in the support of life and make it possible for animals to produce efficiently the things we expect of them.

Classes of Nutrients

Nutrients are divided into five classes: carbohydrates, fats, proteins, minerals, and vitamins. Water may be classified as a sixth nutrient, but because of the various functions of water in the animal body, it will be discussed separately. Each of these nutrient classes is important to the animal's nutrition, and unless it is supplied in the feed, the healthiest and most productive livestock cannot be expected.

Carbohydrates. Carbohydrates are divided into two groups: *nitrogen-free extract* (N. F. E.) and *fiber.* Nitrogen-free extract is made up of the more easily digested and more completely digestible carbohydrate sugars and starches found in such cereal grains as corn, sorghum, wheat, barley, oats, and rye, all of which contain 60 percent or more starch. The more fibrous parts of the plants are made up largely of cellulose and lignin. These substances are much less digestible than the sugars and starches and are found primarily in hays, silage, and fodders. The more mature the plants are when made into hay, silage, or fodder, the more fiber they contain. Ruminant animals, such as cattle, sheep, and goats, can digest much larger quantities of fiber than can poultry or simple-stomach animals such as swine.

The carbohydrates, N. F. E. and fiber combined, are very important in livestock feeding, as they make up about 75 percent of the total nutrients found in plants. Both N. F. E. and fiber are composed of the same chemical elements (carbon, hydrogen, and oxygen), but each has a different combination of them.

FIGURE 1-1. This grass pasture on the Kaydell Angus Ranch, Watsonville, California, will provide most of the essential nutrients required by these replacement Angus heifers. (*Courtesy* American Angus Association)

Fats. Fats and oils are much alike except that fats are solid at ordinary temperature and oils are liquid. For the purpose of livestock feeding, both are referred to as fats. Fats are made up of the same chemical elements as carbohydrates but in different combination. Fats are easily digested by animals, and while the fat content of most seeds and plants is considerably less than that of carbohydrates, the fats are an important nutrient in livestock rations.

Proteins. Proteins are composed of a group of acids known as *amino acids*. Twenty-five or more amino acids have been identified, but it is not definitely known how many are needed by the animal body. Like fats and carbohydrates, amino acids are made up of hydrogen, carbon and oxygen, but they also contain about 16 percent nitrogen. They combine to form a large number of proteins. During the process of digestion the proteins are broken down into the amino acids, which are carried by the bloodstream to the various parts of the body and deposited as needed. All animals are apparently able to manufacture some amino acids from others if supplied in large enough quantities. It has been pretty well established that at least ten amino acids are required in the ration of simple-stomached animals, such as swine, and eleven in the rations of poultry, especially baby chicks.

It appears that the quantity of amino acid is more important than the number of amino acids in the diet of ruminant animals, such as cattle, sheep, and goats, since ruminants can manufacture all the acids they require from any one (or more) if given a large enough quantity. In this process, the nitrogen which is needed to combine with the carbohydrates in producing the amino acids may be partly provided by urea, a high nitrogen compound.

While ten amino acids are known to be essential in the rations of simple-stomached animals and eleven for poultry, the amount of each varies with the kind of animal and its maturity. Some are needed in larger quantities during the growing period, while others are required in larger quantities for reproduction. No single source of protein will supply all the amino acids in their proper balance for any animal at all stages of growth and production. Therefore, the kinds and amounts of protein-rich feed must be varied throughout these stages for best results. The following table lists the amino acids considered essential in poultry rations at some time during the growth and reproduction periods.

AMINO ACIDS ESSENTIAL FOR POULTRY †

Arginine	Phenylalanine
Isoleucine	Lysine
Histidine	Tryptophan
Leucine	Valine
Methionine	Glycine
Threonine	

While a great many feeds contain proteins and several are noted for their high protein content, one must select a combination of protein-rich feeds which will meet the requirements for each class of livestock according to the stage of growth and production. Some high protein feeds have little value as swine and poultry feeds when fed alone, because of their imbalance or lack of essential amino acids. Recommended rations, containing a proper combination of protein feeds, will be listed in the chapters dealing with each class of farm livestock.

Minerals. Minerals are generally divided into two groups, major minerals and trace minerals. The major minerals are salt, calcium, and phosphorus. They are needed in the largest quantity and are most likely to be lacking in the feed supplied. The trace minerals are those needed in very small amounts, though essential to the health of the animal. They include iron, copper, manganese, iodine, cobalt, sulphur, magnesium, zinc, potassium, and boron.

Vitamins. There are several vitamins necessary for proper nutrition of animals, but, as with the amino acids, the requirements in the ration vary with the class of animals. Simple-stomached animals and poultry require vitamin A and most of the B-complex group, which includes thiamine, riboflavin, niacin, pyridoxine, pantothenic acid, choline, pyracin, tara-amino benzoic acid, inosital, folic acid, and biotin. Vitamins B_{12}, C, D, E, K, and some unidentified vitamins are also required by all livestock. Vitamin C is apparently produced in the digestive system of all kinds of farm animals and no consideration need be given it in the ration. Vitamin K is produced in sufficient quantity in the digestive system of all farm animals except poultry, and must be included in the poultry ration. Vitamin K has been used as a treatment in some livestock ailments, particularly calf scours.

Vitamin B_{12} is produced by ruminant animals but must be included in the ration of swine and poultry. Vitamin D, known as the *sunshine vitamin,* is supplied by sunlight to animals when sufficiently exposed. However, because of long winters in much of the country, when sunlight may be at a minimum, or when animals are kept in houses for long periods of time, this vitamin may be deficient and an ample supply should be included in the ration for all classes of farm livestock.

Vitamin E seems to be required in the ration by all classes of livestock. However, calves, small lambs, and poultry seem to be more affected by a deficiency than older cattle, sheep, and swine.

† Two amino acids that are not essential in poultry rations may be used to replace a part of methionine and phenylalanine. Cystine may replace up to half the methionine requirements and tyrosine may replace 7/16 of the phenylalanine.

FIGURE 1-2. Clean fresh water is very essential to all segments of production in the beef industry. Have your water supply tested regularly to determine its purity. (*Courtesy* Hawkeye Steel Products, Inc.)

Water. The animal body is 56 to 70 percent water. Water helps liquify the nutrients fermented in the digestive system during the digestive process. Water is also important in controlling body temperature. Livestock must have an abundance of fresh, clean water if they are to be in the best health.

Feed Additives

Within the last few decades a number of drugs and drug-like substances have been introduced into the field of livestock nutrition. These products include a number of antibiotics, hormones or hormone-like substances, arsenic compounds, detergents, tranquilizers, rumen organisms, and a product known by the trade name of Tapazole. While beneficial effects, measured by increased growth, finishing ability, and feed savings, have resulted when many of these feed additives have been properly used, they cannot be considered food nutrients. A food nutrient becomes a part of the body cells and is necessary for the proper function of these cells. The feed additives do not become a part of the animal body.

Antibiotics. The antibiotics that have been used as feed additives include Aureomycin (chlortetracycline), Terramycin (oxytetracycline), penicillin, bacitracin, hygromycin B, tylosin, neomycin, Aureo 250, and streptomycin. The class and age of livestock and the conditions under which they are produced determine the effectiveness of the antibiotics and the kinds of antibiotic that will give the best results. These conditions and recommendations are given in the sections dealing with each kind of livestock.

Hormones. Certain glands in the body secrete substances called hormones. These hormones have much to do with regulating or stimulating growth, finishing, reproduction, and other body functions. It has been learned through experiments that some man-made products, when fed to animals, have the same effect as the natural hormone. Stilbestrol, for example, closely resembles the natural female hormore estrogen. When it is implanted in pellet form under the skin, a faster rate of growth is obtained. A new hormore-like substance known as melengestrol acetate (MGA) has stimulated live weight gains and improved efficiency in feedlot heifers.

Iodinated Casein. Iodinated casein is a product similar to the natural hormone thyroxine; it has induced greater milk production, especially in brood sows, and to some extent in other livestock. Recommendations as to the use of hormones will be given in the chapters dealing with each class of livestock.

Arsenicals. Although arsenic is a poison if improperly used, it has been found to have growth-stimulating ability on some classes of livestock, especially little pigs, chicks, and poults. (See chapters on the livestock classes.) The arsenic compounds used as feed additives include arsenilic acid, sodium arsanilate, and 3-nitro-4 hydroxyphenyl arsonic acid, usually referred to as 3-nitro.

Detergents, or Surfactants. Detergents, or surfactants, such as the common household detergents, have been fed experimentally to several classes of farm livestock. Some have shown beneficial results while other failed to show any value in their use. Experiments with poultry indicate possible advantages in including detergents in the ration.

Tranquilizers. Tranquilizers are drugs that have a calming effect upon livestock and may have a growth-stimulating effect through their action on hormore-secreting glands in the body. They have reduced shrink in cattle when given prior to shipping.

Rumen Organisms. It is pointed out in Chapter 3 that digestion in ruminant animals is largely dependent upon bacteria and other organisms found in the rumen. Some evidence exists that when these organisms are supplied, especially to young dairy calves that are removed from their mothers within a few days, beneficial results occur. Commercial products containing rumen organisms have been produced as feed additives for cattle and sheep. There is no definite proof that such products have any beneficial results.

FUNCTIONS OF NUTRIENTS AND FEED ADDITIVES

The various nutrients and feed additives have many complicated functions in the animal body. It is not the purpose of this book to discuss in detail the numerous and varied functions of each of the nutrients, but rather to discuss the primary functions of each nutrient so that the reader will gain a knowledge of the importance of each in the animal body.

Carbohydrates

Carbohydrates furnish heat and energy and provide materials necessary for maintenance and growth. The animal requires more total carbohydrates than all of the other nutrients combined. Since most feeds are much higher in this than in other nutrients it is not difficult to provide adequate carbohydrates in the ration.

Fats

Fats are a concentrated nutrient. They provide 2.25 times as much heat and energy as do carbohydrates. One pound of fat will equal 2¼ pounds of carbohydrates in producing heat, energy, and finishing value. An excess amount of fat may produce harmful effects. Liquid fats, such as peanut oil or soybean oil, may produce soft lard when fed to swine. The feeding of an excess amount of seeds high in oil, such as peanuts or soybeans, will have the same effect. When fed to dairy cattle, some fats may produce liquid butterfat. Cod-liver oil will actually decrease milk production when included in the dairy cow ration. However, relatively large quantities of animal fats have been successfully included in the rations of beef cattle. Most animals require less than 3 percent fat. Since animals on a high fat ration will satisfy their energy requirements more quickly, they are inclined to consume less total feed and so the protein and vitamin level will need to be increased or deficiencies in these important nutrients will occur.

Proteins

Proteins, or rather the amino acids of which they are made, are essential in livestock feeding because they help to form the greater part of the muscles, internal organs, skin, hair, wool, feathers, hoofs, and horns. Eggs and milk both contain protein; therefore, animals producing them or those suckling young must have enough protein for their own body needs and extra amounts for the products they produce.

Minerals

Minerals are needed in nearly all parts of the body but are used primarily in the bones and teeth. Minerals make up an important part of the blood. Even the heart depends upon mineral balance to maintain its regular beating. Many disease conditions in livestock, such as anemia, digestive trouble, and some types of paralysis, can be traced to a lack of one or more of the needed minerals.

Vitamins

The purpose or function of many vitamins is not clear, but we do know that serious results occur when feeds are given which do not provide the necessary amounts. If vitamin A is lacking, animals fail to reproduce, eyesight is impaired, and growth slows down. If we do not provide the B-complex group for swine and poultry, appetite fails, skin disorders appear, and disease may become a problem. B_{12} improves protein in grains and other feeds. A lack of vitamin E may cause a failure of the reproductive system and muscular disorders. Vitamin D is essential for proper utilization of minerals, calcium, and phosphorus, while vitamin K gives the blood its ability to clot, preventing excessive bleeding from injuries.

FIGURE 1-3. Leslie Keller, manager of Pioneer Feed Yards, Oakley, Kansas, checks the quality of feed in one of his feed bunks. (*Courtesy* Butler Manufacturing Company)

Feed Additives

The function of the feed additives is still not completely understood. The growth-stimulating effects of the antibiotics and the arsenic compounds are thought to be due to their ability to control undesirable bacteria in the digestive tract. They have given good results even under poor sanitary conditions. These products are effective in the prevention and treatment of internal parasites and diseases. Healthy animals do not need to utilize energy to ward off infectious diseases and parasites and are bound to do well when supplied with adequate nutrients.

Growth, finishing, and milk production are regulated by hormones secreted by body glands. An increase of these hormones over the amount normally provided by the glands apparently gives increased growth and milk production to some classes of animals when properly fed.

Some livestock nutrition experts believe that detergents improve digestion and assimilation of fats by breaking up the fat particles. They may also have functions similar to those of the arsenicals and antibiotics.

DIGESTIBLE NUTRIENTS

The term *digestible nutrients* refers to that portion of any feed that is digestible.

FIGURE 1-4. (*Above*) **The variety of the crop, the soil fertility, and the weather are all factors which will determine the quality of corn being produced on this Iowa farm.** (*Courtesy* **American Angus Association**) **FIGURE 1-5.** (*Below*) **A balance of all the essential nutrients needed to meet the requirements of cattle during gestation will help assure strong, healthy calves like these.** (*Courtesy* **TUCO, Division of Upjohn Company**)

All feeds contain a certain percentage of material that is not digestible in the animal body. The proteins and carbohydrates of various feeds differ widely in digestibility, depending largely upon the class of livestock. Ruminants digest a larger portion of the fiber than simple-stomached animals and poultry. However, cattle do not digest whole grains as thoroughly as sheep, goats, and other classes of farm animals. Whole grains are not thoroughly masticated by cattle and large quantities pass through the digestive system unbroken. The amount of digestible nutrients in various feeds, based on averages for cattle and sheep, is given in the tables in Chapter 2. For practical purposes it may be assumed that while the concentrates and very low-fiber feeds would have nearly equal digestibility for swine and poultry, high-fiber feeds would be much less digestible. Therefore, only the highest-quality legume roughages should be fed to these animals.

BALANCED RATIONS

A ration is the amount of feed which is given to an animal during a twenty-four-hour period. A balanced ration is one which supplies in the correct proportion all the food nutrients necessary during that time period and is determined by the kind of animal and the purpose for which the animal is kept. The needs of the dairy cow differ from those of the beef cow, much as the needs of the brood sow and the growing pig differ.

SUMMARY

In summarizing the various food nutrients and their functions in the animal body, let us compare the animal with the modern farm tractor. The tractor has a chassis, or a framework, which might be termed the skeleton. Just as the chassis of the tractor is made from certain kinds of iron and steel, so the skeleton of an animal is made up largely of minerals. The moving parts of the tractor and the engine proper are comparable to the muscles and the internal organs of the animal. Minerals are needed in the manufacture of these parts just as proteins are needed in the animal body.

Before the engine can move, it must be energized. Gasoline or fuel oil furnishes energy, as do the carbohydrates and fats fed to livestock.

The tractor needs oil and grease to lessen wear, prevent breakdown, and make the parts run smoothly. The amount is small, but without it the machine runs down. The vitamins and feed additives keep the animal body running. Water is used, as in an engine, to regulate the temperature.

In order for the machine to continue to run, a mechanic must be on the job replacing worn parts. In the animal, the blood can be thought of as a mechanic carrying materials to various parts so that they may continue to function.

If any parts become worn and are not replaced or if the wrong kind of fuel is fed to the motor, the machine slows down, fails to run smoothly, or stops. If livestock is not fed the kind of material necessary for proper body function, it, too, slows down or stops production.

QUESTIONS

1 What is meant by the term *nutrient*?

2 List the classes of nutrients.

3 What is the main function of each nutrient class?

4 How do fats and carbohydrates differ in their ability to produce energy and fatty tissue?

5 What are the major minerals?

6 Why are trace minerals so called?

7 Why is an abundance of clean, fresh water important to livestock production?

8 Why have feed additives become important in certain livestock feeds?

9 What is meant by *digestible nutrient*?

10 Why does the amount of fiber reduce the digestibility of a feed?

11 To what part of a feed does the term *nitrogen-free extract* refer?

12 What is the difference between a ration and a balanced ration?

REFERENCES

Iowa State University Agricultural Staff, *Midwest Farm Hand Book,* Ames, Iowa, Iowa State University Press, 1973.

Morrison, Frank B., *Feeds and Feeding,* 23rd edition, Ithaca, New York, The Morrison Publishing Company, 1967.

62 Questions about Stilbestrol Answered, Ames, Iowa, Iowa State University Bulletin, 1963, p. 133.

2 Composition and Classification of Feeds

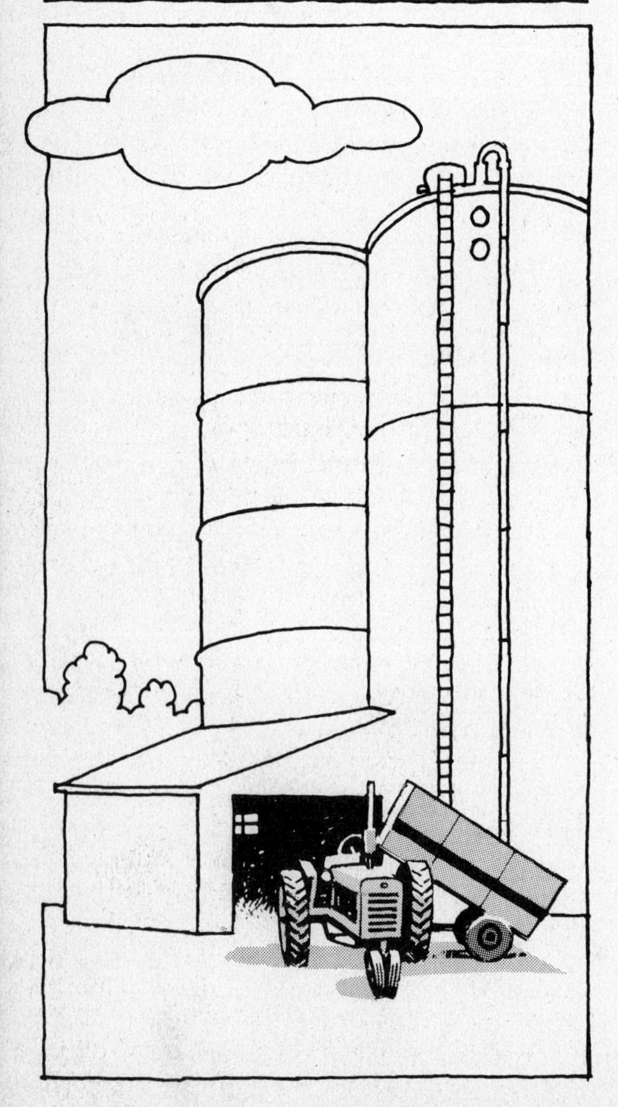

Livestock feeds are classified according to the amount of a specific nutrient *or* the amount of total digestible nutrients they provide. There are two general classes—roughages and concentrates.

ROUGHAGES

Feeds containing relatively large amounts of fiber or nondigestible material are called *roughages*. This group of feeds includes hays, silage, fodder, and other similar feeds.

Legume Roughages. Legumes are plants that have the ability to use nitrogen taken from the air. Legumes are higher in nitrogen than are other plants, and since nitrogen is essential in the manufacture of proteins, legumes are higher in protein than are other roughages. The common legumes are alfalfa, all the clovers, trefoil, soybeans, lespedeza, various kinds of peas, beans, and a large number of other less widely grown plants.

Nonlegume Roughages. Nonlegume roughages as a class are lower in protein. When they are fed as the only roughage, they seldom provide enough protein to meet the needs even of ruminants. The use of nonlegume roughages generally calls for an increased amount in the protein concentrates fed to balance the ration. The common nonlegume roughages are brome grass, orchard grass, fescue, timothy, bluegrass, prairie grass, corn fodder, corn silage, and fodder or silage made from the various sorghums.

CONCENTRATES

Concentrates are feeds which have a comparatively high digestibility. They are relatively low in fiber and include all grains and many by-products of grains and animals, such as wheat middlings, tankage, and soybean oil meal.

Protein Concentrates

Protein concentrates may be classed as a subdivision of concentrates. The exact percentage of protein which a feed must contain before it may be considered a protein concentrate has not been clearly defined, but since most grains and roughages contain less than 20 percent protein, we can consider any feed which contains 20 percent or more a protein concentrate. Such feeds are mixed with farm-produced grains and roughages to increase the protein content of the ration. Protein concentrates are generally classified as either animal proteins or vegetable proteins.

Animal Proteins. Proteins derived from animals or animal by-products, such as tankage, meat scraps, fish meal, dried skim milk, or dried buttermilk, are called animal proteins.

Animal proteins probably contain amino acids not found in other proteins, and they also contain growth-producing factors, not fully understood at this time, which make them especially valuable in the feeding of swine and poultry. Cattle and sheep, as ruminants, are able to manufacture some amino acids that simple-stomached animals must get from animal or vegetable proteins. For this reason, and because of their expense, animal proteins are only considered important in the feeding of rations to very young cattle and sheep.

Vegetable Proteins. Vegetable proteins are found in plants or in the by-products of plants. The chief vegetable protein concentrates are soybean oil meal, linseed oil meal, cottonseed meal, and peanut oil meal.

The plant or vegetable proteins, as a class, are not nearly as complete nor as high in quality as are the animal proteins, but they provide the essential amino acids in the feeding of ruminants. In swine or poultry rations, one or more animal proteins are generally combined with vegetable proteins to give the amino acid balance essential for best results.

SOURCES OF CARBOHYDRATES

Grains

Of the common feeds which are fed to livestock, the grains provide the best source of energy and have the highest growing and finishing value. Among the grains used extensively for feed are corn, oats, sorghum, and barley. Rye and wheat are used to some extent in certain areas of the United States.

Corn. Yellow corn is one of the best feed grains for all kinds of livestock. It ranks high in total digestible nutrients (about 80 percent for No. 2 grade), low in fiber, and higher in fat than any cereal grain except oats. Corn is very palatable to all kinds of livestock. However, it has deficiencies, especially of proteins and minerals, which must be corrected by the use of other feeds.

Oats. As a livestock feed, oats are very popular. They have about the same general nutritional value as corn, and the same deficiences. They contain more fiber and are less digestible. (Oats average about 70 percent digestible content compared with 80 percent for corn.) Their protein content is a little higher than that of corn, but both are lacking in quality.

The fiber content is largely in the oat hulls, and therefore the percentage of hulls determines the food value of the oats. Oats grown under certain weather conditions, and certain varieties of oats, may contain a very large proportion of hulls; oats which are grown without hulls or those which have had the hulls removed by mechanical means have a food value equal to, or higher than, that of corn.

Barley. Barley is a very good substitute for corn and can be fed successfully to all kinds of livestock. Like corn and oats, it is high in nitrogen-free extract or carbohydrates. It is less digestible than corn and has more fiber, but has the advantage of containing more protein.

Barley is higher in food value per hundred pounds than oats, and in general is considered better than oats as a finishing feed. In parts of the United States a disease known as *scab of barley* has been serious. Swine will not eat barley infected with scab unless there is no other feed available, and if they are forced to consume it digestive troubles often develop. However, barley infected with scab can be utilized by cattle.

Wheat. Wheat resembles the other cereal grains in nutritional value, having a higher carbohydrate content and a higher protein content than corn. Its food value varies considerably, depending upon the variety and the locality in which it is grown.

Wheat is a very good substitute for corn, and if the price is not too different, it can be used to replace a large part of the corn fed to all kinds of livestock.

Rye. Because rye has about the same food value as other grains, it would seem a good substitute for them. However, rye is not palatable to livestock and, when fed in large quantities, may cause digestive trouble and a reduction in feed intake. It should be mixed with other grains and fed only as a part of the ration.

Grain Sorghums. Grain sorghums resemble corn in all respects except that they are lower in fat. The sorghums make very good livestock feed and are well liked by all kinds of stock. Sorghum grain is rather hard, and grinding or rolling is necessary if it is to be fully digested.

Molasses. Molasses has a fairly high nitrogen-free extract content and furnishes a readily digestible supply of carbohydrates. Livestock find it very palatable. It is used as a supplement to other feeds and is mixed with the regular ration because it induces the livestock to consume more feed and so helps speed up the finishing process, particularly of beef cattle.

Poor or unpalatable roughages, such as corncobs, can be made more appetizing by the use of molasses.

Roughages as a Source of Carbohydrates

Roughages provide a good source of carbohydrates, and good-quality roughages, such as those produced by legumes or corn silage, will often provide enough carbohydrates by themselves for stock cattle or matured sheep. However, in feeder livestock the digestibility of many roughages is too low to produce finished animals unless the ration is supplemented with grains.

The kind of roughage, the method of harvesting, and the degree of maturity at the time of cutting are very important influences on food value. Legumes and grasses cut near full maturity may not have more than half the value per ton of those cut earlier. As an example, alfalfa cut when only a few flowers are showing will contain from 15 to 20 percent more digestible nutrients and from 25 to 30 percent more proteins than will alfalfa cut after it reaches full bloom.

Grain crops cut for silage should be reasonably mature for highest food value. Corn silage should be cut just as the lower leaves begin to dry out.

SOURCES OF FATS

Grains and Oil Seed Crops

Almost all livestock feeds contain some fats. Such feeds as soybeans and peanuts are especially high in fat, containing from 18 to 47 percent while the common feed grains such as corn, oats, sorghum, and barley range from 1.9 to 5.5 percent.

Animal Fats and Oils

Animal fats, such as lard and tallow, and the vegetable oils have been used both experimentally and practically to increase considerably the energy value and the fat content of livestock rations.

Protein Concentrates

Both animal and vegetable protein concentrates contain fats. The amount depends upon the method of processing but ranges from 1.5 to 15 percent. The animal protein concentrates are generally considerably higher in fat content than the vegetable protein feeds, especially when the oil is removed from the seeds by the solvent process.

SOURCES OF ANIMAL PROTEINS

Animal protein feeds come from animal by-products, dairy products, and dairy by-products.

Animal By-products

In the processing of meat and fish for human consumption, many by-products containing valuable animal proteins are produced. These by-products are then made into livestock feeds.

In the manufacture of all animal products that go into livestock feed, the material is heated to prevent the transmission of disease.

Tankage and Meat Scraps. Tankage and meat scraps are very much alike in composition. Both are by-products of packing houses or rendering plants. The digestible protein content will vary considerably, depending upon the contents of the feed. If the product contains a large percentage of hair or gristle, the total protein content may be high, but the digestible protein content may be low.

Since it is impossible for the user to determine the amount of nondigestible protein a feed contains, it is wise to buy only from a company known to be reliable.

Meat and Bone Scraps. In combinations of meat and bone scraps, the percentage of bone is higher. Since bone is lower than meat scraps in protein, the total protein content of the feed is generally lower.

Blood Meal. Blood meal is made from the blood collected at packing houses. In the manufacturing process, it is heated and dried. Blood meal is high in total protein, but its quality and digestibility are lower than those of good-grade tankage or meat scraps.

Fish Meal. Fish meal is manufactured either from the waste material of the fish industry or from fish not considered good for human consumption.

Fish meal is one of the best feeds for swine and poultry because it contains an abundance of high-quality protein. The comparatively high cost of fish meal, however, limits its use to mainly starter rations for young animals and poultry.

Dairy Products and By-products

Some of the best protein feeds for livestock are manufactured from products of the dairy industry.

Skim Milk and Buttermilk. Both of these products contain over 90 percent water, but the dry matter is about one third protein. Nearly all the protein is digestible and these products furnish one of the best sources of protein for swine, poultry, and very young ruminants.

Dried Skim Milk and Buttermilk. A heating process drives off most of the water in skim milk and buttermilk, leaving dry milk solids. These solid milk products furnish a high-protein food of good quality.

SOURCES OF VEGETABLE PROTEINS

All seeds, roughages, and grain by-products furnish various amounts of proteins, but the seed by-products and legume roughages are the most important sources of good-quality vegetable proteins.

Seed By-products

Seed by-products offer fairly good protein for livestock feeding, but most of them should be used in combination with other protein sources.

Soybean Oil Meal. Soybean oil meal is the residue left from soybeans after the oil has been removed. The residue is ground or formed into pellets and marketed as livestock feed.

Soybean oil meal varies in protein content from 41 to 46 percent, depending upon the method of manufacture. The protein is of very high quality, and may be used to furnish a major part of the protein concentrate in rations for most livestock.

Soybeans. Soybeans have the highest protein content of any seeds commonly used for feed. Soybeans will average about 37 percent total protein, but the quality is low. The protein in soybeans is improved by the heating process used in the manufacture of soybean oil meal. This explains the high-quality protein found in the meal compared to that in the raw soybeans.

If their price is low compared to the price of other forms of vegetable proteins, ground soybeans may be used to furnish a part of the protein in rations for cattle and older swine.

Cottonseed Meal and Cake. When cotton is processed, the seed is removed from the lint and the oil is extracted from the seed. The seed residue is manufactured into meal or cake and sold as a protein concentrate.

The protein content of cottonseed meal varies from 38 to 47 percent, depending upon the variety of cotton from which the seed is taken and upon the method of processing. Cottonseed meal is a good source of protein for ruminants, but it should be used only in limited amounts as a protein concentrate for nonruminants.

Linseed Meal. When flax seed is processed, the oil is removed from the seed and the residue is made into livestock feed. The oil is known as *linseed oil,* and the meal as *linseed oil meal.* Linseed oil meal is one of the most widely used protein feeds. Its protein content ranges from 31 to 36 percent, and it may be successfully used as the only protein feed for cattle and sheep. For swine and poultry, linseed meal should be used as only a part of the protein concentrate in the ration.

Peanut Oil Meal. Peanut oil meal is the residue of the peanut from which the oil has been removed. Peanut oil meal can be used as a protein feed for all classes of livestock but it is generally preferred as a swine or a dairy cattle feed. When fed as the only protein, it may be too laxative. Best results are obtained when peanut oil meal is used in combination with other proteins.

Corn Gluten Meal. Corn gluten meal is a by-product of corn used in the manufacture of corn oil and corn starch. The protein content ranges from 40 to 44 percent. Corn gluten meal may be used as a part of the protein in livestock rations. The quality of protein is not high, and best results are obtained when this meal is fed in combination with other proteins.

Legume Roughages. Legumes, when fed as dry roughages, silage, or pasture, furnish an excellent source of high-quality protein. Under some conditions legumes will furnish all the needed protein. Stock cattle may be successfully wintered on legume forage without any other protein feeds being provided. Bred sows can farrow large healthy litters when good legume pasture is their only feed for most of the gestation period. Legumes furnish a cheap source of protein, and when legume forage is used in the ration, the amount of protein concentrates may be reduced without affecting the growth or production of the livestock.

Urea

Urea cannot be classed as either an animal or vegetable protein. In fact it is not a protein at all, but rather a chemical compound containing 40 to 45 percent nitrogen. It has been pointed out that the amino acids contain nitrogen as well as carbon, hydrogen, and oxygen. The bacteria in the digestive system of ruminants can combine the nitrogen from urea with the carbon, hydrogen, and oxygen found in the carbohydrates and form amino acids. Therefore a limited amount of urea may be used as a protein substitute in the feeding of ruminants.

Biuret. When urea is heated, a major product, biuret, is produced. Nutritionists have been impressed with its low toxicity and adaptability as a source of protein. Biuret has little or no value as a protein when fed to simple-stomached animals, and should only be fed to ruminants.

SOURCES OF VITAMINS

Livestock, when fed a well balanced ration consisting of grains, protein concentrates, and good forage, generally get enough of the needed vitamins. However, under certain conditions which will be discussed in later chapters, it may be necessary to select feeds high in essential vitamins. Vitamins most likely to be deficient in the feed are A or carotene, the B-complex group, D, and E.

Vitamin A

This vitamin is available in several forms, and livestock should consume ample amounts of it in one or more of these forms, as discussed in the following paragraphs.

Vitamin A in Forage. Vitamin A is found in plants and in plant products as carotene, which is converted into vitamin A in the digestive system of animals. Green plants have an abundance of carotene, and there is seldom any need to give any special attention to the vitamin A content of a ration for livestock on pasture.

When forage crops are harvested, a large part of the carotene may be lost unless special care is taken to preserve it. Roughages that retain their green color and most of their leaves when cured will still be high in carotene content. Very little carotene is lost when roughages are converted into silage. Legume and grass crops harvested during the natural growing period and in good weather have a much higher carotene content than do mature or badly weathered forages. Forages lose much of their carotene content when stored for a long time.

Vitamin A in Grains. Carotene is found in yellow corn and is an important source of vitamin A. Legume forages, however, contain a much higher percentage of this vitamin than does corn.

Commercial Sources of Vitamin A. The fish liver oils are one of the best sources of concentrated amounts of vitamin A. Cod-liver oil is often used as a vitamin A supplement for small pigs and chickens not on pasture. For feed-mixing purposes vitamin A may also be purchased in a supplement containing a mixture of several other vitamins.

Vitamin B-Complex Group

Most of the vitamins in this important group may be obtained from the same sources as vitamin A. B_{12}, however, is found only in a limited number of feeds, and for this reason, we will discuss it separately.

Vitamin B-Complex Group in Forages. With the exception of vitamin B_{12}, forages provide an excellent sources of the B-complex group of vitamins.

Vitamin B-Complex Group in Animal Products. Animal products, especially milk or milk products, furnish substantial amounts of the vitamin B-complex.

Vitamin B-Complex Group in Grains. Grain crops provide most of the B vitamins, but they are not rich enough in these essential vitamins to be relied upon entirely. Unless a good forage is fed, some other source of the B group should be included in the ration.

Vitamin B Group in Brewers' Yeast and Distillers' Solubles. When there is a need for more of the B vitamins than those supplied by the ration being fed, small amounts of yeast or distillers' solubles are often used to boost the B content of the ration.

Vitamin B_{12} in Animal Products. With the exception of commercial sources, B_{12} is found only in animal products, such as milk, tankage, meat scraps, fish meal and similar feeds.

Ruminants manufacture B_{12} during the process of digestion, and therefore, with the exception of very young ruminants, it does not have to be supplied in their feed. Swine following cattle are supplied some B_{12} through the feces of the cattle.

B_{12} is an important ingredient in animal proteins because it is necessary for growth in poultry and swine. When B_{12} is supplied in other forms, the amount of animal protein fed swine and poultry can be considerably reduced without affecting their growth rate.

Commercial Sources of Vitamin B_{12}. Many laboratories and feed companies manufacture concentrated forms of vitamin B_{12} which may be purchased for feed-mixing purposes.

Vitamin D

Vitamin D is often called the *sunshine vitamin,* and animals exposed to direct sunlight seldom suffer from vitamin D deficiency. However, animals that are kept inside, especially young pigs and chickens, will require vitamin D in their ration.

Vitamin D in Forage Crops. Forage crops, especially sun-cured legume hays, are very good sources of vitamin D. The D vitamin is absorbed by the legume during curing.

Commercial Sources of Vitamin D. Concentrated amounts of vitamin D may be secured in fish liver oils and irradiated yeast, which are the best supplements to a ration low in vitamin D.

Cod-liver oil, when fed to ruminants over a long period of time, has shown injurious results. This may be due to a destruction of vitamin E caused by the fish liver oil. When it becomes necessary to add a vitamin D supplement to the ration of ruminants such as dairy calves being housed indoors, a vitamin concentrate such as irradiated yeast is recommended.

Vitamin E

Vitamin E is essential for fertility. Bulls and other male animals used heavily in breeding programs may benefit from extra amounts of vitamin E.

Vitamin E in Farm Feeds. Vitamin E is found in nearly all forages, feed grains, and protein concentrates. Animals fed a reasonable well balanced ration seldom need additional amounts of this vitamin.

Wheat Germ Oil as a Source of Vitamin E. When amounts of vitamin E greater than those supplied by the ordinary ration are desired, wheat germ oil is usually considered a good source. Wheat germ oil is manufactured commercially and can be purchased from most feed dealers.

Other more concentrated commercial sources of vitamin E are available when large quantities are needed, as in the treatment of stiff lamb disease. (See Chapter 33.)

Vitamin K

Alfalfa leaf meal is high in vitamin K. A product produced commercially called

Menadione also replaces the K vitamin needs of poultry. Except in the treatment of certain livestock disorders, poultry seems to be the only kind of farm animal for which special consideration need be given this vitamin in the ration.

Commercial Vitamin Premixes

Vitamin premixes are available for all kinds of livestock. They are often combined with antibiotics in what is termed an antibiotic-vitamin premix. One or more vitamin supplements for any kind of livestock can be purchased to provide all the vitamins likely to be deficient in the feed. When this supplement is properly mixed with the rest of the feed, the vitamins will be present in the necessary amounts.

SOURCES OF MINERALS

Several minerals are needed in well balanced livestock rations. Most of the essential minerals are present in common livestock feeds, but the amounts are generally not great enough to provide for the mineral needs of farm livestock. Therefore, mineral supplements are generally recommended.

Minerals in Farm-produced Feeds

All farm-grown feeds contain minerals, but the legume forages are the best source. Legumes are high in both phosphorus and calcium. The mineral content of all farm-grown feeds depends upon the soil on which they are grown. Legumes grown on well limed and phosphated soils have a higher

TABLE 2-1 AVERAGE COMPOSITION OF COMMON CONCENTRATES HIGH IN ENERGY AND FAT-PRODUCING VALUE (Expressed in percent)

Feed	*Total Dry Matter*	*Total Digestible Nutrients*	*Total Protein*	*Digestible Protein*	*N-free Extracts or Carbohydrates*	*Fat*	*Mineral Matter*	*Fiber*	*ENE*
Barley	89.4	77.7	12.7	10.0	66.6	1.9	2.8	5.4	77
Beet pulp (dried)	91.2	69.7	8.8	4.1	58.7	0.6	3.5	19.6	64
Corn (No. 2)	85.0	80.1	8.7	6.7	69.2	3.9	1.2	2.0	81
Ground ear corn	86.1	73.2	7.4	5.3	66.2	3.2	1.3	8.0	73
Feterita (grain)	89.4	79.8	12.2	9.5	70.1	3.2	1.7	2.2	79
Kafir (grain)	89.8	81.6	10.9	8.8	72.7	2.9	1.6	1.7	79
Milo (grain)	89.0	79.4	10.9	8.5	70.7	3.0	2.1	2.3	79
Blackstrap cane molasses	73.4	53.7	3.0	0	61.7	0	8.6	0	50
Beet molasses	80.5	60.8	8.4	4.4	62.0	0	10.1	0	57
Oats (hulled)	90.4	91.9	16.2	14.6	63.7	6.1	2.2	2.2	90
Oats	90.2	70.1	12.0	9.4	58.6	4.6	4.0	11.0	66
Rye	89.5	76.5	12.6	10.0	70.9	1.7	1.9	2.4	78
Wheat	89.5	80.0	13.2	11.1	69.9	1.9	1.9	2.6	80
Wheat bran	90.1	66.9	16.9	13.3	53.1	4.5	6.1	10.0	63
Wheat middlings	90.1	79.2	17.5	15.4	60.0	4.5	3.8	4.3	78

TABLE 2-2 AVERAGE COMPOSITION OF COMMON PROTEIN CONCENTRATES
(Expressed in percent)

Feed	*Total Dry Matter*	*Total Digestible Nutrients*	*Total Protein*	*Digestible Protein*	*N-free Extracts or Carbohydrates*	*Fat*	*Mineral Matter*	*Fiber*	*ENE*
Blood meal	91.6	60.4	82.2	60.4	0.9	1.9	5.7	0.9	52
Buttermilk (dried)	92.0	83.1	31.8	28.6	43.6	6.1	10.0	0.5	84
Corn gluten meal	91.6	79.7	43.2	36.7	38.9	2.2	3.5	3.8	79
Cottonseed meal	92.7	72.6	43.3	35.9	27.4	5.1	6.0	11.0	71
Fish meal	92.0	70.8	60.9	53.6	5.4	6.9	18.3	0.9	75
Linseed meal, solvent process	91.0	70.3	36.6	30.7	38.3	1.0	5.8	9.3	75
Meat scraps	94.2	66.7	54.9	45.0	2.5	9.4	24.9	2.5	65
Meat & bone scraps	93.7	65.3	49.7	40.8	3.1	10.6	28.1	2.2	63
Milk (dried skim)	94.0	80.7	34.7	31.2	50.3	1.2	7.8	0.2	79
Peanut oil meal, solvent process	93.0	77.3	52.3	47.6	26.3	1.6	5.9	6.9	77
Soybean seed	90.0	87.6	37.9	33.7	24.5	18.0	4.6	5.0	88
Soybean oil meal, solvent process	90.4	78.1	45.7	42.0	31.4	1.3	6.1	5.9	78
Tankage (high grade)	92.8	65.8	59.4	50.5	2.6	7.5	21.4	1.9	65

content of these important minerals than do those grown on acid and low-phosphorus soils.

The trace mineral content of grains and roughages depends largely upon the types of soil on which they are grown. The presence or absence of trace minerals in the soil may not necessarily affect the yield of the crop, but it does affect the mineral value of the feed.

Minerals in Protein Concentrates

Animal protein feeds, especially meat and bone scraps, tankage, and fish meal, are excellent sources of minerals. The mineral content in these feeds varies from 15 to 25 percent, and the necessity for other mineral supplements is reduced when such feeds are used in the ration. Since all protein feeds are relatively high in price, the amount fed is seldom enough to meet the mineral needs of the animal.

Vegetable protein meals are much lower in mineral content than are the animal proteins, but higher than are farm grains.

Chief Sources of Mineral Supplements

High-quality ground limestone is one of the best sources of calcium for mixing mineral supplements. Limestone is cheap and available in most areas.

Steamed bone meal, which is manufactured by cooking clean bones under steam pressure, is a common phosphorus supplement. It is also a very good source of calcium.

Salt is needed by all livestock and is easily obtainable either in loose or block

TABLE 2-3 AVERAGE COMPOSITION OF COMMON DRY ROUGHAGES
(Expressed in percent)

Feed	*Total Dry Matter*	*Total Digestible Nutrients*	*Total Protein*	*Digestible Protein*	*N-free Extracts or Carbohydrates*	*Fat*	*Mineral Matter*	*Fiber*	*ENE*
Alfalfa hay (average)	90.5	50.7	15.3	10.9	36.7	1.9	8.0	28.6	40
Alfalfa hay (high grade)	90.5	52.7	17.5	12.8	39.5	2.4	8.4	22.7	41
Alfalfa meal (dehydrated)	92.7	54.4	17.7	12.4	38.4	2.5	10.1	24.0	42
Birdsfoot trefoil	91.2	55.0	14.2	9.8	41.9	2.1	6.0	27.0	42
Bluegrass hay	89.4	54.8	8.2	4.8	42.1	2.8	6.5	29.8	44
Brome grass hay	88.8	49.3	10.4	5.3	39.9	2.1	8.2	28.2	35
Clover hay—red (good)	88.3	51.8	12.0	7.2	40.3	2.5	6.4	27.1	44
Clover hay—red (second cutting)	88.1	54.1	13.4	8.4	40.4	2.9	6.9	24.5	45
Corncobs (ground)	90.4	45.7	2.3	0	54.0	0.4	1.6	32.1	31
Corn fodder (medium water, well eared)	82.6	53.9	6.8	3.3	46.7	2.1	5.2	21.8	50
Corn stover (ears removed)	90.6	51.9	5.9	2.1	46.5	1.6	5.8	30.8	41
Fescue hay	89.2	52.7	7.0	3.7	43.2	1.9	6.8	30.3	41
Hegari (fodder)	86.3	52.4	6.1	3.2	52.8	1.7	7.5	18.2	49
Kafir (fodder)	90.0	53.6	8.7	4.5	44.2	2.6	9.0	25.5	49
Lespedeza hay (annual)	90.0	45.1	13.1	5.6	42.2	2.5	5.3	26.9	42
Lespedeza hay (perennial)	89.0	41.4	13.2	4.4	42.7	1.7	4.9	26.5	38
Oat hay	88.1	47.3	8.2	4.9	42.2	2.7	6.9	28.1	32
Oat straw	89.7	44.7	4.1	0.7	41.0	2.2	6.3	36.1	31
Orchard grass hay	88.7	47.4	7.5	3.5	42.7	2.4	6.8	30.4	34
Prairie hay, Western (good)	91.3	45.1	6.0	2.0	44.0	3.0	8.6	29.7	32
Reed canary grass hay	91.1	45.1	7.7	4.8	44.3	2.3	7.6	29.2	32
Sorghum (fodder sweet)	88.8	52.4	6.2	3.3	48.1	2.4	7.1	25.0	49
Soybean hay (seed well developed)	88.0	52.5	15.2	10.8	35.2	4.7	6.2	26.7	43
Sudan grass hay	89.6	50.0	11.2	6.3	41.3	1.5	9.5	26.1	40
Timothy hay (good)	89.0	50.8	7.5	4.1	44.4	2.4	4.7	30.0	38

form. Mineralized salt, which contains a mixture of trace minerals, is also available.

Trace minerals can best be secured by purchasing what is commonly known as *trace mineral mixture.* This mixture contains the trace minerals in proper proportion, and can be mixed with limestone, salt, and bone meal for a complete mineral mixture.

SOURCES OF FEED ADDITIVES

Experiments have shown that many feed additives are very important for promoting the health and growth in certain types of livestock. No farm-produced feeds contain these additives; therefore, we must rely upon commercial sources for supplementing rations with these important substances.

When we mix our own feeds, an antibiotic-vitamin supplement may be purchased and mixed in the ration according to the manufacturer's directions. There is evidence that a mixture of the known beneficial antibiotics may be better than a single antibiotic.

Iodinated casein and arsenic compounds may be purchased by feed mixers as a premix and incorporated with the ration. Detergents are available in nearly every household, and tranquilizers are available in commercial feeds.

NUTRIENT VALUES OF LIVESTOCK FEEDS

Grains and roughages vary considerably in nutrient value, depending upon their variety, the locality and type of soil in which they are grown, their maturity, the method of harvesting, and the length of time in storage The feed manufacturing process partly determines their value also. For these reasons average compositions of various feeds generally have to be used when planning livestock rations. Tables 2-1 through 2-5 give the average composition of commonly used livestock feeds.

It has been pointed out in Chapter 1 that the digestibility of the feeds as given in Tables 2-1 to 2-5 are averages for cattle

TABLE 2-4 AVERAGE COMPOSITION OF PASTURE LEGUMES AND GRASSES (Expressed in percent)

Feed	*Total Dry Matter*	*Total Digestible Nutrients*	*Total Protein*	*Digestible Protein*	*N-free Extracts or Carbohydrates*	*Fat*	*Mineral Matter*	*Fiber*	*ENE*
Alfalfa	24.4	14.8	4.6	3.5	10.0	0.9	2.2	6.7	12
Bluegrass	30.2	20.7	5.5	4.1	13.4	1.2	2.5	7.6	17
Brome grass	25.0	18.3	5.1	3.9	10.7	1.0	2.4	5.8	15
Clover (Ladino)	16.6	12.4	4.1	3.3	7.5	0.8	1.7	2.5	10
Clover (red)	25.0	16.8	4.0	2.8	11.2	0.9	2.1	6.8	15
Clover (alsike)	22.0	15.7	4.1	3.2	10.4	0.9	1.9	4.7	14
Clover (sweet)	20.8	12.8	4.1	3.2	9.2	0.7	1.9	4.9	10
Fescue	30.5	18.8	3.0	1.6	14.0	1.0	2.4	10.1	15
Lespedeza (annual)	25.0	12.7	4.1	2.0	9.2	0.5	3.2	8.0	10
Rape	16.3	12.8	2.9	2.4	8.0	0.6	2.2	2.6	10
Timothy	23.9	15.4	4.7	3.5	11.1	0.9	2.6	4.6	12
Trefoil	22.7	13.3	3.4	4.5	9.5	0.8	2.3	5.6	10
Sudan grass	21.6	14.3	3.3	2.4	10.2	0.6	1.9	5.6	11

and sheep. For practical livestock feeding the grains, grain products, and protein concentrates may be assumed to have about the same digestibility for swine and poultry when fed in a well-balanced ration and not more than the amount recommended. High-quality legume pasture, silage, or hay would have nearly equal value for swine and poultry as for cattle and sheep. However, such low-quality roughages as straw, corncobs, and corn stover should be avoided as poultry feeds. Such feeds are sometimes used in swine rations to furnish bulk and limit the feed intake when self-feeding is practiced. For example, the operator may wish to self-feed brood sows but finds that it is necessary to add ground cobs to the grain and supplement mixture to prevent overeating, which might result in too much condition.

SUMMARY

Feeds are divided into two groups: roughages and concentrates. Roughages are those feeds relatively high in fiber, such as legume and grass hays, fodders, and silages. Concentrates are the low-fiber feeds and include the grains, dairy products, and feeds manufactured from seed, dairy, and animal by-products.

Protein concentrates are feeds high in protein; they are classified as vegetable or animal proteins depending upon the material from which they are made.

The grains are higher in energy value than are other common feeds; however, roughages, although lower in nitrogen-free extract and higher in fiber, are generally a source of low-cost carbohydrates.

Tankage, meat scraps, meat and bone scraps, fish meal, milk, and milk products are all good sources of animal protein.

The vegetable proteins are supplied in soybean oil meal, linseed meal, cottonseed meal, and other feeds derived from plants. Legumes when used as feed will replace much of the protein available from other sources.

TABLE 2-5 AVERAGE COMPOSITION OF COMMON KINDS OF SILAGE AND ROOTS (Expressed in percent)

Feed	*Total Dry Matter*	*Total Digestible Nutrients*	*Total Protein*	*Digestible Protein*	*N-free Extracts or Carbohydrates*	*Fat*	*Mineral Matter*	*Fiber*	*ENE*
Alfalfa	36.0	21.3	6.0	4.1	13.7	1.4	3.2	11.7	15
Corn	27.6	18.3	2.3	1.2	16.2	0.8	1.6	6.7	18
Corn stalk silage	23.7	14.0	1.6	0.6	12.0	0.7	1.6	7.8	11
Grass-legume mixture	33.3	19.1	5.2	2.9	14.2	1.3	3.8	8.8	17
Grain sorghum	30.0	17.1	2.6	1.4	18.6	0.7	2.1	6.0	15
Beet roots (common)	13.0	10.1	1.6	1.2	8.9	0.1	1.5	0.9	8
Beet roots (sugar)	16.4	13.7	1.6	1.2	12.6	0.1	1.1	1.0	9
Mongels roots	9.2	7.0	1.3	0.9	6.0	0.1	1.0	0.8	5
Pea-vine silage	24.5	14.0	3.2	1.9	11.0	0.8	2.2	7.3	12

Tables 2-1 through 2-5 adapted by special permission of the Morrison Publishing Company, Ithaca, New York, from *Feeds and Feeding*, 23rd edition, by F. B. Morrison.

Older animals receiving rations consisting of legumes, proteins, and grains seldom need added vitamins, except possibly B_{12}. However, vitamin concentrates are often essential for best results in feeding young animals. The vitamin concentrates are found in such feeds as fish liver oil (vitamin A and D) and wheat germ oil (vitamin E). The B-complex group may be purchased in a vitamin premix available in most localities.

Feed additives have an important place in livestock feeding, and are usually provided in a premix or in commercially mixed feeds.

QUESTIONS

1. What are the two main classes of feeds?
2. How do roughages and concentrates differ?
3. List several common roughages.
4. Why are legumes higher in protein than are other roughages?
5. Name some common legumes.
6. Name some common nonlegume roughages.
7. What are some of the common feeds known as concentrates?
8. What are protein concentrates?
9. What is the chief difference between animal and vegetable proteins?
10. List the feeds commonly fed for their carbohydrate content.
11. List some of the common animal protein feeds.
12. What are some popular vegetable protein feeds?
13. What vitamins are most likely to be lacking in livestock rations?
14. Explain how vitamins may be added to a ration when the ordinary foods fail to provide essential amounts.
15. What materials may be used in mixing a mineral supplement for livestock?
16. Give two common methods of supplying feed additives in the ration.

REFERENCES

Diggins, R. V. and Bundy, C. E., *Dairy Production,* 3rd ed., Prentice-Hall, Inc., Englewood Cliffs, New Jersey, 1969.

Ensminger, M. E., *The Stockman's Hand Book,* Danville, Illinois, The Interstate Printers and Publishers, 1962.

Morrison, Frank B., *Feeds and Feeding,* 23rd edition, Ithaca, New York, The Morrison Publishing Company, 1967.

3 Digestion in Ruminants and Simple-Stomached Animals

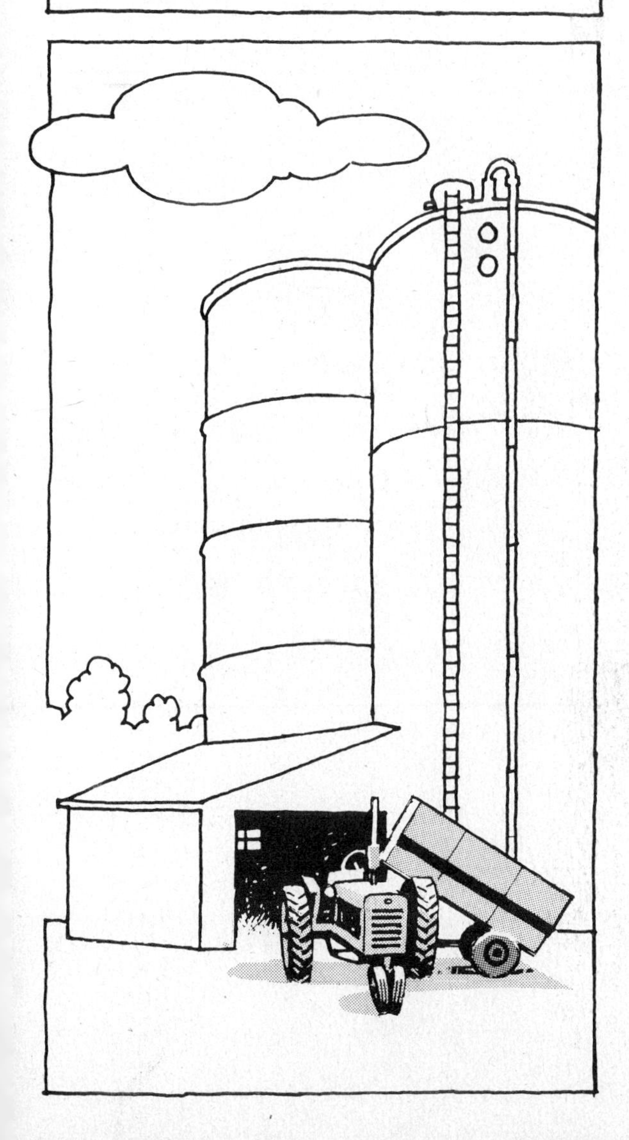

Some knowledge of the differences in digestive systems is necessary to understand livestock feeding and the reasons why some feeds are more valuable for one class of animals than for another.

DIGESTION IN RUMINANTS

To understand the food requirements of ruminants, we should know how they digest and utilize feed. *Ruminants* are animals that have four compartments to their stomachs; *simple-stomached animals* have a single-compartment stomach. Examples of ruminants among common farm animals are cattle, sheep, and goats. Hogs are the most common example of simple-stomached animals; poultry, although not classed as simple-stomached animals, have a similar digestive system.

Because of their digestive systems, ruminants do not present as many feeding problems as animals with simple stomachs. Ruminants manufacture many of the vitamins and some protein amino acids that have to be fed to simple-stomached animals. They are also able to digest large quantities of roughage.

The Four Compartments

The compartments of the ruminant stomach are known as the *rumen*, *reticulum*, *omasum*, and *abomasum*.

Rumen. The rumen is the first and by far the largest compartment of the stomach. It serves as a storage area for large quantities of feed, especially roughages, and has a capacity, depending upon the size of the animal, of 40 to 60 gallons in mature cattle.

While eating, ruminants chew the food just enough to make swallowing possible. After they have consumed their feed, regurgitation, or chewing the cud, takes place. The food is brought up from the rumen and the chewing is completed. It is then returned to the rumen for further bacterial action.

In the rumen, the food is worked upon by millions of bacteria and other microorganisms which transform low-quality proteins and even some nitrogen compounds into essential protein amino acids. They also manufacture many needed vitamins, including the vitamin B-complex group. Some of these proteins and vitamins are used by the bacteria, but when the bacteria die and are digested by the animal, these nutrients are utilized by the animal body.

The presence of the rumen, and the bacterial action that takes place in it, explains why ruminants can digest large amounts of roughage and convert it into human foods; for this reason, too, the B-com-

FIGURE 3-1.(*Above*) The stomach of a mature ruminant animal. Note the size of the rumen compared to the other three stomach compartments. FIGURE 3-2.(*Below*) Here we see a comparison of the mature (*left*) and the young (*right*) ruminant stomachs.

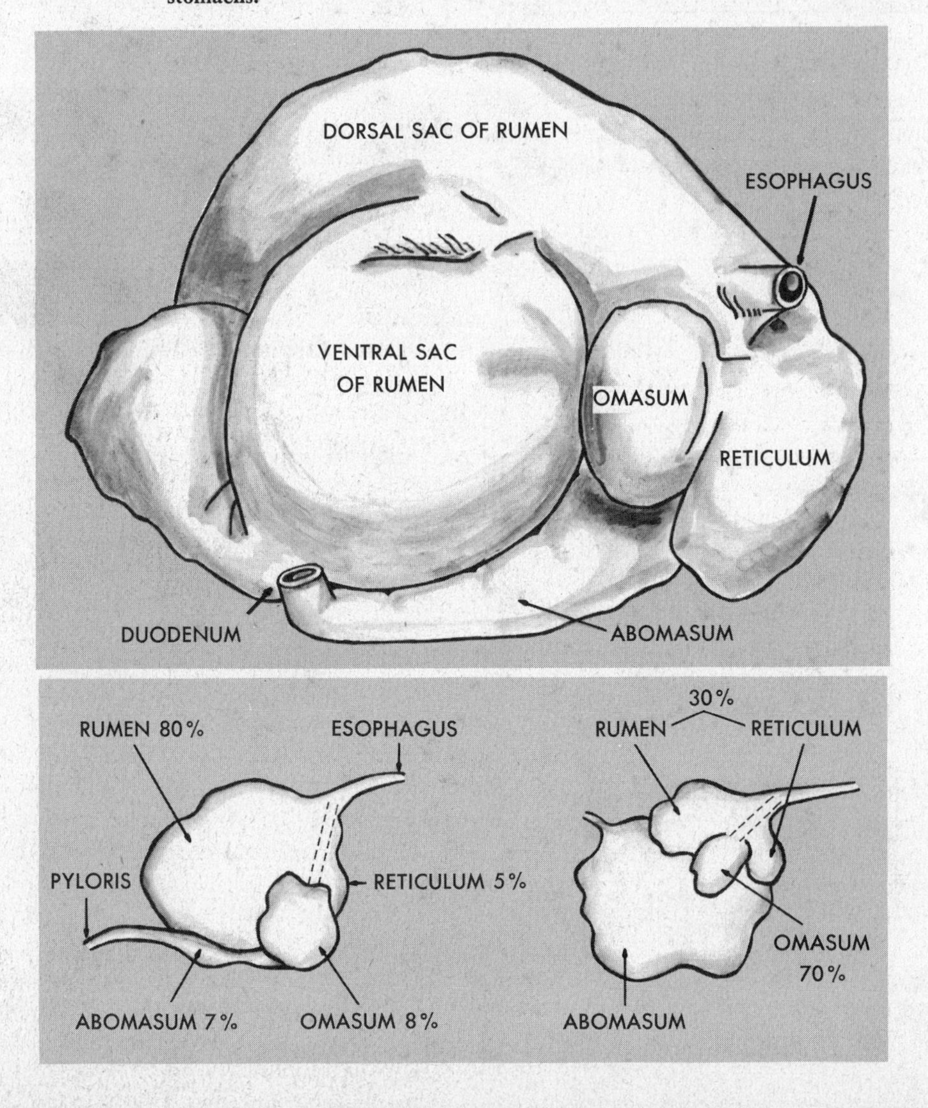

plex vitamins and highly complex protein feeds are seldom necessary in their ration. Proteins are an essential part of the ruminant ration, but they need not be nearly so varied and complete as for swine and poultry. In very young ruminants the rumen is not developed, and during the first few weeks they must receive a diet containing most of the food nutrients.

Reticulum. The reticulum is closely associated with the rumen, and it is here that many foreign bodies, such as wire and nails, are retained. If these objects are not pointed or too sharp, they may be held in the reticulum for long periods without any serious damage. The main functions of the reticulum are to furnish additional storage space and to retain foreign materials that may cause serious damage to the other body organs.

Omasum. The omasum, or third compartment of the stomach, has strong muscular walls. Its function is not too well understood, but seems to be to squeeze the water from the feed before it enters the abomasum or true stomach.

Abomasum. The abomasum is the fourth compartment, or true stomach; its functions are similar to those of the simple stomach in swine or similar animals. Gastric juice, which is necesary in protein digestion, is secreted in the abomasum. When the food leaves the abomasum, it goes into the small intestine, where the digestible portion is absorbed into the bloodstream; the remainder is passed into the large intestine and eliminated as waste.

DIGESTION IN SIMPLE-STOMACHED ANIMALS

In simple-stomached animals food is swallowed directly into the single-compartment stomach, where it is mixed with the digestive juices. Very little bacterial action takes place, and there is no conversion from low-quality to high-quality proteins. Simple-

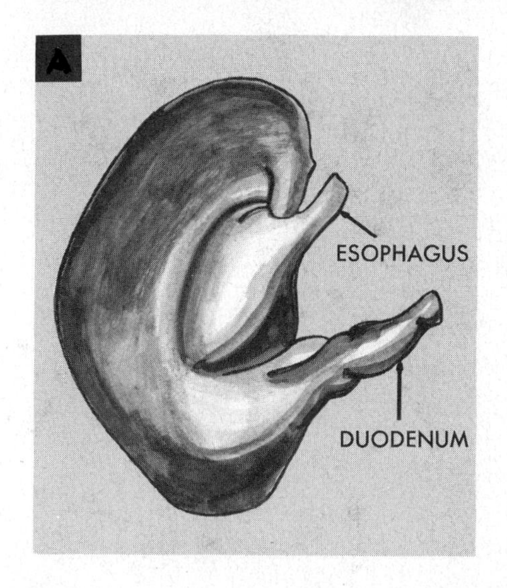

FIGURE 3-3. (A) The stomach of a pig. Note the single compartment. (B) The principal digestive organs of poultry.

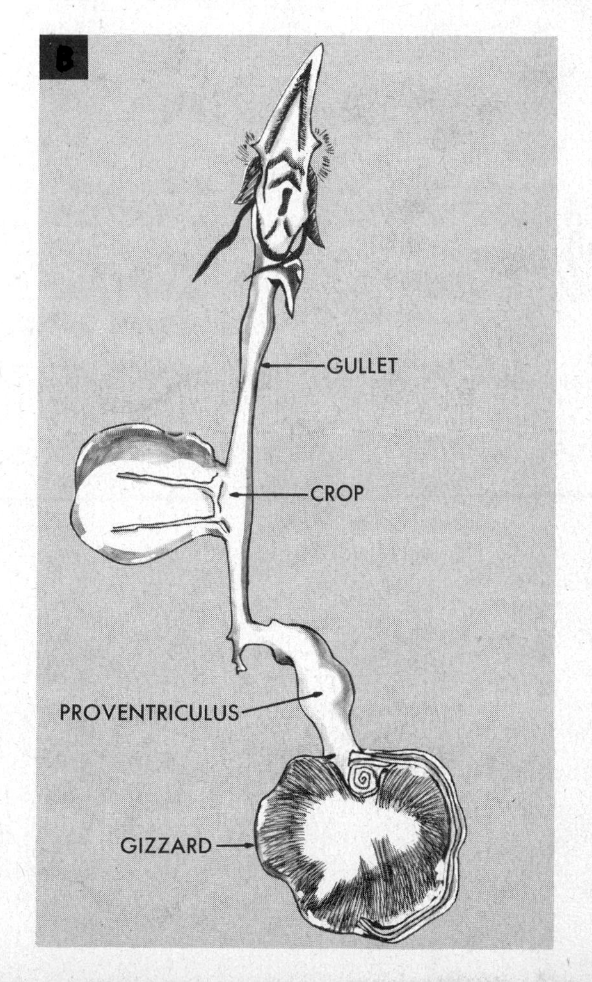

stomached animals have less ability to manufacture vitamins than do ruminants and are unable to digest large quantities of fiber. The digestive process in simple-stomached animals is very similar to the process in the abomasum of ruminants.

We must give much more consideration to the protein, vitamin, and mineral content of rations fed to swine and poultry than to cattle and sheep. It must also be remembered that high-fiber feeds have little value for swine or poultry.

DIGESTION IN POULTRY

Poultry cannot be classed as simple-stomached animals, though their digestive system is somewhat similar. Very little digestion through bacterial action takes place, and poultry digest smaller amounts of fiber than do any other kinds of farm livestock.

The Digestive Organs

The parts of the digestive system in poultry are the *crop*, *proventriculus*, and *gizzard*.

Crop. Food taken in through the mouth moves down the esophagus into the crop. The crop acts as a reserve to hold and moisten feed, but very little digestion takes place.

Proventriculus. The proventriculus is a small organ that receives food from the crop. Here digestive juices are secreted and mixed with the food before it goes to the gizzard.

Gizzard. The chief function of the gizzard is to grind and crush the coarse food before it enters the small intestine, where the digestible portion is absorbed into the bloodstream.

Unlike ruminant feeding, a ration for poultry should include all the essential protein amino acids. All of the vitamins known to be required by animals, except vitamin C, must be included in the ration of poultry.

SUMMARY

The stomachs of ruminants, such as cattle, sheep, and goats, have four compartments. The rumen is, by far, the largest compartment and serves as a storage place for bulky feeds and as a place for bacterial action.

Because of bacterial action during the digestive process, ruminants can digest large quantities of fiber and convert low-quality proteins into essential amino acids. The microorganisms also manufacture most of the B-vitamins required by the animal.

Swine and poultry have a simple digestive system, and neither can convert low-quality to high-quality proteins or manufacture vitamins. They must rely directly upon the feed they consume to obtain all classes of nutrients. They have much less ability to digest fiber than do ruminants.

QUESTIONS

1 Name the four compartments in the ruminant stomach.

2 What is the principal digestive process that takes place in the rumen?

3 Explain why ruminants can convert comparatively low-quality protein feeds and a certain amount of nitrogen compounds into usable amino acids.

4 Why must we supply young ruminants with the B-vitamins?

5 How do ruminants, swine, and poultry rank in their ability to digest fiber?

REFERENCE

Yeates, N. T. M., *Modern Aspects of Animal Production,* Washington, D.C., Butterworth's, 1965.

4 Measuring the Value of Feeds

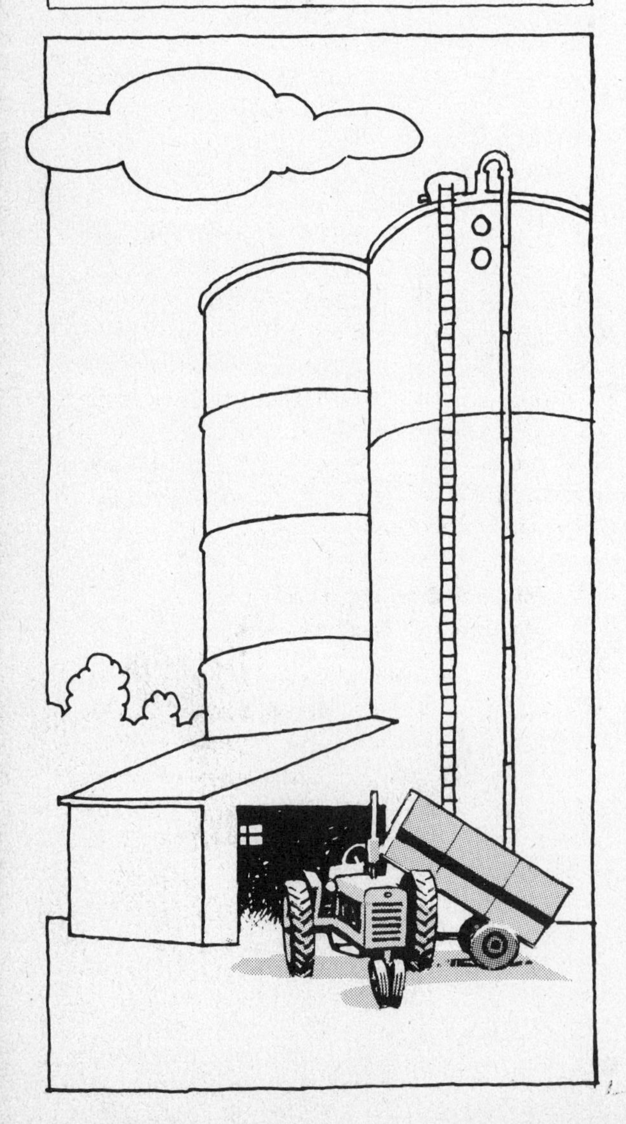

No single method of determining the value of feeds is accurate enough for all purposes. Many factors must be considered if a reasonably correct appraisal of any feed or group of feeds is to be made.

CLASSES OF ANIMALS AFFECT FEED VALUE

The kind of livestock fed is important in determining the value of a feed. Cattle and other ruminants can digest large quantities of fiber, therefore, roughages, particularly fodders, grass hays, and various poor-quality roughages, may have considerable value as feed for cattle but not or very little as feed for swine. Cattle, for instance, can make efficient use of a certain amount of the corncobs or cornstalks, but such feeds would be of little value in swine or poultry feeding.

Cottonseed meal can be given in relatively large amounts to cattle or sheep, often as the only protein concentrate necessary. However, when fed in too large quantities to swine or poultry, cottonseed meal has a toxic effect.

The value of different feeds for the various kinds of livestock will be discussed in the chapters which deal with each animal.

DETERMINING FEED VALUES IN TERMS OF DIGESTIBLE NUTRIENTS

One of the most widely accepted methods of determining feed values is based on the total digestible nutrient content of the feed. Modern feeding standards list the total digestible nutrients and the digestible proteins of the more widely used livestock feeds. A study of the tables in Chapter 2 will tell us the average amount of digestible nutrients in a given amount of any one feed for ruminants. For example, in Table 2-1, corn (No. 2) is shown as having a total digestibility of 80.1 percent. This means that one bushel of shelled corn, which weighs 56 pounds, will provide 44.856 pounds of digestible nutrients. If corn costs $1.50 per bushel, then $1.50 divided by 44.856 will give us the cost of one pound of digestible nutrients supplied by corn.

To determine the cheapest source of digestible nutrients for any kind of animals, the price per bushel, ton, or hundred pounds, whichever is the more common unit of measure, should be divided by the pounds of total digestible nutrients. This will give a comparative cost of various feeds based on feeding value. One must remember the kind of livestock concerned before using this method of evaluating feed. As pointed out in Chapter 1, swine and poultry will digest approximately the same percentage of grains and of grain and animal by-products as cattle and sheep; however, extremely high-fiber feeds, such as corn stalks, straw, or corncobs, would have very little digestibility for swine and poultry and would be considered as having little or no value in their rations. Although the cost per pound of total digestible nutrients is a very good guide to the value of a feed, it is not the whole answer.

Carbohydrates and Fats as Measures of Feed Value

Feeds are generally selected to do a specific job in the animal body. Corn is used as a growing and finishing feed because of its high carbohydrate content, whereas soybean oil meal is fed for its high protein content.

To finish lambs or steers, for example, it is necessary to use one or more of the grains, such as corn, barley, wheat, or oats, along with enough protein to balance the ration. It would be important to know which feed will produce a hundred pounds of gain the most cheaply. In a comparison of finishing feeds, the nitrogen-free extract and carbohydrates are as important as the total digestible nutrients.

Proteins as Measures of Feed Value

Many feeds are purchased primarily for their protein content. They are used to increase the protein percentage in the ration, and the cost should be determined on a digestible protein basis.

In selecting protein feeds, whether the product will give the amino acid balance needed by the kind of livestock to be fed should be considered. Such proteins as cottonseed meal, soybean meal, or linseed meal will produce nearly equal results when fed to ruminants. The cost per pound of digestible protein may be used to determine which one to use. However, in swine and poultry rations, none of these feeds, nor any combination of them, will give the protein balance necessary for best results.

Generally, grains and roughages are not fed primarily for their protein content. Yet, if they are high in this important nutrient, the cost of feeding is generally lower because less protein concentrates are necessary. For example, cattle fed legume hay will require less protein concentrates for the production of beef or milk than will those receiving a low protein roughage. The protein content of roughages and grains is important in determining their feed value.

Mineral Content as a Measure of Feed Value

Livestock feeds are seldom chosen primarily for their mineral content. Minerals are provided by feeding salt, limestone, phosphorus compounds, and trace mineral mixtures. However, the calcium and phosphorus content of good legume forage and the mineral content of animal protein supplements are important considerations when determining their food value.

There are many different grades of limestone, but since calcium is the important mineral it supplies, calcium content is the chief measure of its value.

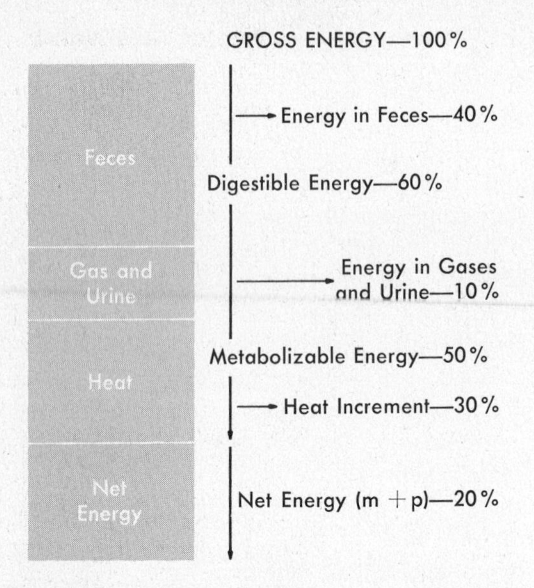

FIGURE 4-1. Energy utilization concepts for ruminant nutrition. (Source: *Iowa Farm Science,* Iowa State University)

Bone meal, when added to the ration in the proper porportion, will give the needed amount of calcium and phosphorus.

Vitamins as Measures of Feed Value

Most farm feeds provide some vitamins, but legume forage is especially valuable for its vitamin content. The high vitamin A content of yellow corn marks the chief difference between yellow and white corn as livestock feed.

Aside from their vitamin content, fish liver oils and other vitamin preparations have very little value.

Net Energy as a Measure of Feed Value

A new measure for evaluating feeds and rations for livestock is to determine the amount of net energy provided by the feed fed to animals. The net energy measure of feed value gives a more accurate value of feeds when compared to the TDN measures. Figure 4-1 outlines the presently accepted energy utilization concepts for ruminants.

Gross energy is the total amount of energy produced. Digestible energy is that energy which disappears on passing feedstuff through an animal's digestive tract. Not all of the digestible energy is available to the animal. A considerable amount is lost in gas production (fermentation) and urine loss. Metabolizable energy is that which remains when gas and urinary losses are taken into account. Net energy is determined by subtracting digestible energy and metabolizable energy from gross energy.

Net energy is that which is available to the animal for production uses (1) to maintain its body, and (2) for production purposes, such as growth and milk. Tables 2-1 through 2-5 in Chapter 2 list the ENE (estimated net energy) values of feeds used in livestock rations.

QUALITY OF FEEDS

Many factors, such as variety, type of soil, weather conditions, and age, affect the quality of a feed and explain the wide range in digestible nutrients which may occur in the same kind of feed.

Corn may vary from 6 to over 9 percent protein, and alfalfa will range from 8 to 19 per cent protein. It is important to consider carefully the quality of the product in judging the value of a feed.

Factors Affecting Quality of Feeds

Varieties of the same crop will differ in food value. The amount of protein in wheat varieties grown under the same conditions will vary from 10 to 16 percent. Oats that produce a plump berry with thin hulls are much higher in digestible nutrients than are those with a large proportion of hulls. State agricultural experiment stations can be relied upon to furnish information regarding varieties of crops for their particular state, together with comparative feeding values.

Soils. Recent experiments and analyses have shown that feeds produced on soils well supplied with plant foods have a higher feeding value than do those produced on poor land.

The mineral content of alfalfa and the protein in corn may vary, depending upon the soil. Land well supplied with nitrogen, from which plants make protein, will produce higher protein feeds than will soils deficient in nitrogen. Alfalfa grown on soils well supplied with lime and phosphorus will be richer in them than will alfalfa grown on deficient soils.

The appearance of a forage or grain may not be the best guide to its food value. A knowledge of the area, or even the farm, where it was grown is important.

Weather. Weather conditions influence feed values. Corn that is high in moisture as a result of cold, wet seasons will have much fewer digestible nutrients per pound than will dry corn. Small grains may have a high percentage of hull to berry and the proportion of fiber will be increased as a result of hot, dry weather, particularly when the kernels are forming.

The same kind of crop can vary in feeding value from year to year. The temperature, the length of the growing season, and the amount and distribution of rainfall all affect the nutrient value of crops.

Balancing a Ration Improves Its Digestibility

Milk, eggs, and meat are produced by animals according to a natural "formula"; the percentage of each nutrient in these products remains about the same regardless of the kind of ration fed. If animals are given a low protein feed, they must eat greater quantities of the feed, and the excess carbohydrates will be eliminated as waste. For example, experiments have shown that hogs fed a ration made up entirely of corn will consume about 13 bushels for every 100 pounds of gain. However, if the corn is balanced with a good protein supplement plus adequate minerals and vitamins, there will be a saving of approximately six bushels of corn per 100 pounds of gain. Balancing a ration will increase the value of all feeds used in the ration.

COMMERCIAL MIXED FEEDS

There are many companies engaged in the business of producing livestock feeds. Therefore, commercially mixed feeds for all classes of livestock are available in every community. When questions arise as to the value of one company's product compared to that of another, we must consider the reliability of the manufacturer and the success that feeders have had with his products. A study of the analysis given on the container or feed tag may be helpful. Most states permit feed companies to sell under either the closed formula or the open formula.

Closed Formula

Under the closed formula, feed companies are generally required to give the minimum percentage of crude protein, crude fat, and nitrogen-free extract, and the maximum percentage of fiber contained in the feed. They must also list the ingredients that were used in making the product, but, except for mineral or ash, they are not required to tell how much of each ingredient has been used. Some states require that the percentage of mineral be listed. For example, a feed may be listed as containing soybean oil meal, tankage, meat scraps, and bone meal. Although this is a true statement of the ingredients used, it does not give the amount of each product in the feed. Many feed companies prefer the closed formula because it prevents duplication of their products by other companies or livestock feeders.

The following is a typical closed formula label that might appear on a feed bag:

Net Weight 100 Lbs.

FAST GROW PIG MEAL

Guaranteed Analysis

Crude protein not less than	26.00%
Crude fat not less than	4.50%
Nitrogen-free extract not less than	43.00%
Crude fiber not more than	8.50%

Ingredients: Alfalfa meal, wheat bran, tankage, soybean oil meal, fish meal, rolled oats, yellow corn, molasses, antibiotic and vitamin supplement, calcium 6.0%, phosphorus 1.3%, salt 5.0%, iodine .0002%.

Open Formula

Feed companies selling open formula feeds must list the minimum percentage of protein, fat, and nitrogen-free extract, the maximum amount of fiber, and the weight of each ingredient contained in the feed. Following is an example of an open formula feed label:

FAST GROW CHICK FEED

Guaranteed Analysis

Protein not less than	21.00%
Fat not less than	3.40%
Fiber not more than	7.40%
Nitrogen-free extract not less than	45.00%

Formula (Pounds per Ton)

470 yellow corn
350 ground oats
150 wheat bran
20 dried brewers' yeasts
350 wheat shorts
40 ground limestone
25 steamed bone meal
150 dehydrated alfalfa meal
320 soybean oil meal
100 meat and one scraps
15 salt
5 trace mineral concentrate
5 vitamin-antibiotic supplement

SUMMARY

Many factors must be considered in evaluating feed. The more important ones are the kind of animal to be fed and the purpose for which the particular feed is being used.

Growing and finishing feeds must be judged primarily for their carbohydrate content and total digestibility. However, their protein, mineral, and vitamin contents are also important because the percentage of these nutrients in the feed determines the amount of other concentrates needed to balance the ration. The quality and percentage of digestible protein in a protein feed is the chief measure of its value. A new method of evaluating a feed is the amount of net energy produced. This is the amount of energy left after digestible and metabolizable energy has been subtracted from the total energy provided by the feed.

Mineral and vitamin concentrates are measured in terms of the essential minerals and vitamins present.

The quality of feeds is affected by weather, soil, method of harvesting, and the different varieties of each crop.

The value of any feed is largely determined by how well the ration is balanced. Feeding more of any nutrient than the animal requires is wasteful. Commercial feeds are sold under open formula or closed formula analysis. The amount of each product in the feed is given on the label when the open formula is used. The value of commercial feeds must be judged largely by the reliability of the company and the success feeders have had with its products.

QUESTIONS

1 How does the kind of livestock to be fed affect the value of the feed?

2 Show by example how you would determine the value of a feed, using total digestible nutrients as a measure.

3 Show how you would determine the value of a protein feed, using digestible protein as a measure.

4 Generally, roughages are not fed primarily for their protein content. Why, then, is the protein content of a roughage important in determining the value?

5 If concentrated amounts of vitamins A, D, E, B_{12}, and other B-vitamins are needed, what sources will furnish them?

6 How do soils and weather affect feed values?

7 Explain why poorly balanced rations waste feed.

8 Explain the chief difference in the two types of labels used for commercial feed fixtures.

REFERENCES

Livestock Feeds and Feeding, Bulletin 384-8, Fort Collins, Colorado, Colorado State College, 1952.

Morrison, Frank B., *Feeds and Feeding,* 23rd edition, Ithaca, New York, The Morrison Publishing Company, 1967.

Recommended Nutrient Allowances for Beef Cattle, Report of the Committee on Animal Nutrition (Revised) 1970, National Research Council, Washington, D.C.

U.S. Department of Agriculture, *Better Feeding of Livestock,* Farmers' Bulletin No. 2052, 1952.

PORK PRODUCTION

5 The Pork Industry

Hog production in the United States is big business. In 1972, 93.7 million head valued at 5.4 billion dollars were produced. This represents about 20 percent of the total world production. China leads the world in hog production.

Hogs are produced on 66 percent of the farms in Iowa and on 47 percent of the farms in Illinois. These two states account for 35.2 percent of the hogs marketed in the nation. About a third of the farm income of some of the Corn Belt states is derived from the sale of hogs. In Iowa, where 68 cents of every farm dollar comes from livestock, nearly 34 cents comes from the sale of hogs.

LEADING STATES IN HOG PRODUCTION

Corn and hog production go hand in hand, so the leading hog-producing states are in the Corn Belt. Iowa, Illinois, Indiana, Missouri, Minnesota, and Nebraska, were the six states producing the highest number of pigs in 1972. These six states accounted for 61.7 percent of the nation's 1972 pig crop. Iowa alone produced more pigs in 1972 than the 44 states lowest in production, and nearly twice as many pigs as Illinois, the state with the second largest production.

Hog producers in the Corn Belt have been concerned by increasing interest in hog production in the South. On December 1, 1972, there were 47.4 million hogs on farms in the twelve North Central States. Only 6.2 million hogs were on farms in the eight South Central States. This, however, was a 7.1 percent increase in the number of hogs in these states since December 1, 1964.

CONSUMER DEMAND FOR PORK AND LARD

The future of pork production in this country will be determined to a large extent by the requirements of the American people for pork and lard, and by their insistence on

FIGURE 5-1. (*Above*) Consumer demand for pork depends on the quality of the product and the price. These market hogs must meet the consumer demands. (*Courtesy* Central Livestock Association, Inc., South St. Paul) FIGURE 5-2. (*Below*) Per capita consumption of beef and veal has increased from 71.4 pounds in 1950 to 117 pounds in 1971. Pork consumption has only increased from 69.2 pounds to 72.5 pounds. (U.S.D.A. Economic Research Service)

a reasonable price for these products. Pork and lard must compete with other animal and vegetable products in palatability, in nutritive value, in availability, in ease of merchandising and storage, and in price.

The average American consumed a total of 182 pounds of red meat and 13.1 pounds of lard in 1953. This was the highest per capita meat consumption on record up to that time. In 1972, the average American consumed approximately 67.4 pounds of pork, 116 pounds of beef, 2.2 pounds of veal, 3.3 pounds of lamb and mutton, 47.9 pounds of broilers and turkey, and 3.8 pounds of lard. The 1972 per capita consumption of red meat and poultry was 24.9 pounds above the previous (1964) all-time high of 211.9 pounds.

The Lard Problem

The substitution of vegetable oils and fats for lard in cooking and the decrease in use of animal fats in soapmaking have lowered the demand for lard. Today consumers

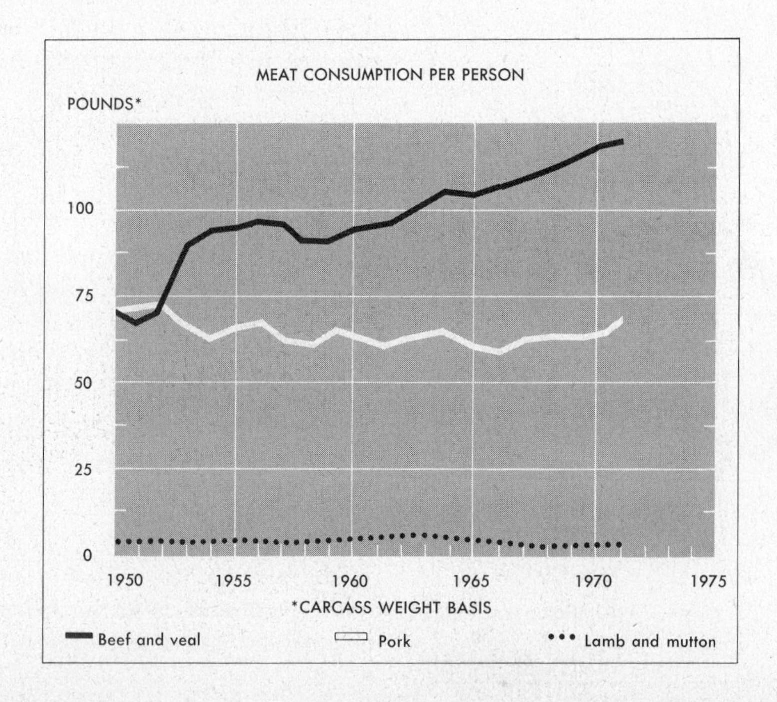

are eating pork, but they are eating the lean cuts—the hams, loins, picnics, and Boston butts. They buy fewer fat cuts and lard. As a result, we are producing more lard than we can sell profitably, even though we export it to other countries. Almost 27 percent of our total lard production was exported in 1964.

Lard exports in 1965 decreased greatly. The United Kingdom purchased 80 percent of our total exports in 1964, but transferred much of their business to European markets in 1965. Less than 15 percent was exported in 1971.

Lard represents about 11.5 percent of the weight of the live hog, but only about 7 cents of each dollar spent for pork products is spent for lard.

The consumption of lard remained fairly constant until 1950. Since that time, the amount of lard consumed has decreased from 12.6 pounds to about 3.8 pounds per person. Lard yield per hog has been lowered through improved breeding, feeding, and marketing practices to about 24 pounds. Though increased amounts of lard are used in manufacturing margarine and shortening products, a major problem of hog breeders and feeders is to find a way to produce a pork carcass with a higher percentage of lean meat and a lower yield of fat cuts and lard.

HOG-CORN PRICE RATIOS

Since corn is the basic feed in producing hogs, and feed costs represent about 65 to 70 percent of the production costs, it is usually possible to determine the extent to which hog production will be profitable by comparing the price of hogs with the price of corn. The term *hog-corn ratio* represents the relationship between the two, based upon the number of bushels of corn that can be bought for the price of 100 pounds of pork.

A break-even hog-corn price ratio is about 13.5 to 1. That is, 100 pounds of live hog should bring the price of 13.5 bushels of corn.

ADVANTAGES OF HOG PRODUCTION

Hog production is well adapted to specialized or diversified types of farming, and the returns come much more quickly than do those from many other enterprises. The investment in swine breeding stock and in equipment is relatively small, and it is possible to get in and out of the business in a comparatively short time.

The feeding of corn and other grains to hogs is a profitable method of marketing these grains. The hog is efficient in producing meat; a pound of pork can be produced on as little as 2½ to 3½ pounds of feed. The young chicken is the only other animal which can produce a pound of meat on this amount of feed.

Hogs can be raised on small or large farms, and in small or large numbers. They make excellent use of pasture, but can be

FIGURE 5-3. A healthy little Hampshire with an inquiring disposition. (*Courtesy* Elanco Products Co.)

produced profitably in confinement. They do not require expensive housing and equipment. The labor requirements in producing hogs are lower than those in dairy and poultry production.

DISADVANTAGES OF HOG PRODUCTION

During recent years a number of diseases have caused heavy losses on some farms and in some areas. Rhinitis, brucellosis, erysipelas, gastroenteritis, anemia, leptospirosis, and swine dysentery losses have been especially severe.

The fact that farmers can get in and out of hog production in a comparatively short time may prove disadvantageous at times. When hog-corn price ratios are wide, 15 to 1 or more, farmers flock in and the increased production may crowd the market.

EFFICIENCY IN PORK PRODUCTION

The profit in producing hogs is determined largely by the efficiency of the grower in production and marketing practices. It has been estimated that the average farmer loses at least 25 percent of the pigs farrowed. However, the cost of maintaining a brood sow from breeding time until the pigs are weaned is about the same regardless of the number of pigs farrowed and weaned.

FIGURE 5-4. Many swine producers got their start at an early age. This Chester White pig is in good hands. (*Courtesy* Chester White Swine Record Association)

The loss of pigs from disease is great, since hogs are subject to many diseases. Some animals die; others are weakened. Considerably more feed is required to feed diseased hogs, additional time is necessary to get them on the market, and they cannot be sold when the market is high. Often the carcass of the animal must be condemned entirely or in part, and sometimes the packer must offer a lower price for the diseased animals. The efficient producer uses practices which prevent disease outbreaks.

Care must also be taken to prevent losses caused by swine parasites. Worms and mange probably cause some loss on most farms.

The efficient producer of hogs must be very careful in the selection of breeding stock and in the breeding of the animals. Growth rate and carcass quality are inherited. Good feeding and disease control cannot entirely offset losses due to poor quality breeding stock or the use of poor breeding methods.

The efficient producer selects feeds carefully and feeds them in proper balance. He makes adequate use of homegrown grains and forages, and uses protein supplements, minerals, antibiotics, and vitamins to supplement the homegrown feeds.

It is possible to produce profitably but at the same time to market inefficiently. Markets are better in some seasons than they are in others. Packers pay higher prices for hogs of high quality than do other buyers. Through careful planning it is possible to have ready for market the kind and weight of hogs that will top the market.

SUMMARY

Hog production is big business in this nation, especially in the Corn Belt States. Iowa, Illinois, Indiana, Missouri, Minnesota, and Nebraska, produced 61.7 percent of the nation's hog crop in 1972. Iowa alone produces more hogs than the 44 low-producing states.

The number of hogs produced annually is affected by the hog-corn price ratio, the feed supply, the supply of hogs at the markets, consumer demand, world economic conditions, supplies of meats which compete with pork, the price being paid, and the swine disease situation.

In 1972 the average person in the nation consumed a total of 236.8 pounds of meat, of which 67.4 pounds was pork. The per capita lard consumption was about 3.8 pounds. Lard represents 11.5 percent of the weight of a live hog, but only 7 cents of each dollar spent for pork products is spent for lard.

A hog-corn price ratio represents the relationship between hogs and corn, based upon the number of bushels of corn that can be bought for the price of 100 pounds of pork. A break-even hog-corn price ratio is 13.5 to 1.

Hog production is a profitable enterprise which requires a comparatively small investment, and the returns come rather quickly. Hogs are efficient converters of feed into food for human consumption.

The market for lard has not kept pace with the market for pork. Pork producers must find a way to produce hogs with more lean meat and less fat.

Profitable hog production is dependent upon how well farmers do the following: (1) use sound methods in selection and breeding, (2) follow efficient feeding and management practices, (3) control diseases and parasites, and (4) use good judgment in marketing operations.

QUESTIONS

1. What percentage of the income on your farm is derived from the sale of hogs?
2. What percentage of the grains produced on your farm are fed to hogs?
3. What percentage of the farm income in your state is obtained from the sale of hogs?
4. About how much meat is consumed in a year by the average American? How much of this amount is pork?
5. What are the leading states in hog production?
6. What do you think can be done to decrease the lard produced on our farms without reducing the production of high-priced pork cuts?
7. What are the advantages and disadvantages of hog production on your farm?
8. What are the essentials of a profitable swine enterprise? Explain.

REFERENCES

Bundy, C. E. and R. V. Diggins, *Swine Production,* 3rd Edition, Prentice-Hall, Inc., Englewood Cliffs, New Jersey, 1970.

Ensminger, M. Eugene, *Swine Science,* 4th Edition, The Interstate Printers & Publishers, Inc., Danville, Illinois, 1970.

U.S. Department of Agriculture, *Livestock and Meat Statistics,* Washington, D.C., 1973.

———, *Livestock and Meat Situation, November, 1971,* Washington, D.C., 1971.

6 Selection of Breeding and Feeding Stock

The pork producer must make a number of decisions before he obtains his breeding stock. He must first decide what type of hogs to raise; then he must determine which breeding program to follow. He usually has at least three plans to consider: (1) he can start with grade or crossbred sows and upgrade them, selling the offspring as market hogs, (2) he can start with purebreds and maintain a purebred herd, selling the offspring to other breeders for breeding purposes or disposing of them as market hogs, or (3) he may start with grade, crossbred, or purebred sows and follow a plan of crossbreeding in producing hogs for the packer market.

After the producer has chosen the type of hog he will raise and the breeding program he will use, he must make a third and more difficult decision: What breed or breeds should he grow? There are numerous breeds of swine available. Which of them is best suited to the breeding program that he plans to follow? The breeding program and the breed or breeds of hogs are both determined in part by the methods he plans to use in marketing his hogs.

A number of factors must be considered in selecting a breed, and the selection of desirable breeding animals within the chosen breed involves additional problems. What body conformation is desired? Should the producer use proven sows and sires or select young, untried breeding stock? In selecting animals, how much importance should he place on rate of gain, economy of gain, prolificacy, and freedom from disease?

Good breeding stock is essential for the most profitable hog production enterprise. It is sometimes possible for a hog producer to overcome, in part, the lack of good breeding stock by efficient feeding, good management, and disease control. For maximum returns, however, it is necessary that he start out with the right kind of breeding animals.

The factors to be considered in selecting breeding and feeding stock are enumerated in this chapter, and suggestions are offered to aid the reader in making wise decisions when choosing animals for the home farm enterprise.

MEAT-TYPE HOGS

The type of hog produced in this country has varied from time to time during the past century. Originally there were two types. The lard type, as the name implies, was a thick-bodied hog, which in market condition carried a large amount of fat. The bacon-type hog was imported from England and the Scandinavian countries, where it was developed to meet consumer demand for "Wiltshire Sides" of bacon.

The competition on the market of vegetable fats with lard has reduced the value of lard, and hog breeders are now trying to develop a hog which will yield a carcass high in lean meat and comparatively low in lard. Regardless of the breed of hog which is being raised, the average farmer is trying to produce a *meat-type* hog.

Meat-type Hog Defined

While swine producers do not completely agree about what is meant by the term "meat-type hog," they generally agree on the following specifications:

1. **When slaughtered, the hog will yield 40 to 45 percent of its live weight in ham, loin, picnic, and butt.**
2. **The animal will weigh from 200 to 210 pounds at five months of age.**

FIGURE 6-1. (*Above*) A champion Poland China boar at the Iowa State Fair about 1920. (*Courtesy* Wallaces' Farmer) FIGURE 6-2. (*Below left*) The present-day meat-type hog. This Duroc was the on-foot Grand Champion Barrow over all breeds at the 1970 National Swine Show. (*Courtesy* United Duroc Swine Registry) FIGURE 6-3. (*Below right*) This Chester White barrow was Grand Champion at the 1971 Iowa State Fair. He had a 5.90-inch loin eye and 48.16 percent of carcass in ham and loin. (*Courtesy* Chester White Swine Record Association)

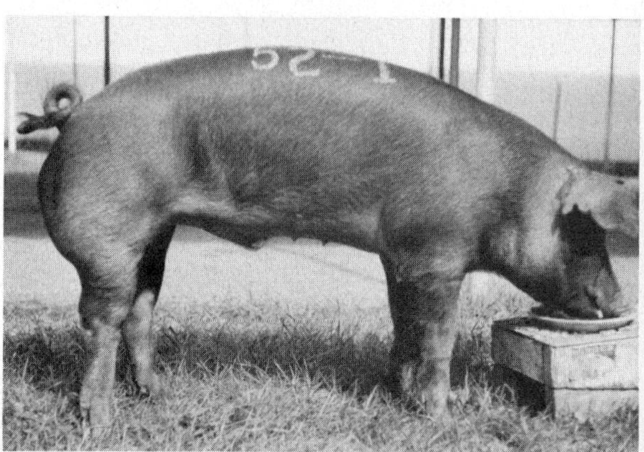

3. The 200- to 210- pound carcass will be 29½ to 31 inches long at five months of age.
4. The backfat thickness will not exceed 1.2 inches.
5. The loin eye will be 4½ to 5 square inches.
6. The ideal carcass should yield about 19 percent ham, 15.2 percent bacon, 17 percent loin, 9.7 percent picnic, and 6 percent Boston butt.
7. The animal will have a feed conversion ability of 1 pound of gain from 3 pounds of feed.
8. The females will produce litters of eight to nine pigs which can be raised to market weight.

Comparison of Meaty and Average Hogs

The two pigs shown in Figures 6-4 and 6-5 were slaughtered and the dressed carcasses frozen in upright position by Iowa State University staff members. Hog A (Figure 6-4) was a meaty gilt weighing 233 pounds. Hog B (Figure 6-5) was an average barrow weighing 225 pounds. The gilt shows much more trimness of jowl, neck, shoulder, and middle. She is better muscled throughout, but especially over the loin, rump, and ham.

Figures 6-6 and 6-7 are the cross sections of the hams of these two pigs. Note the high percentage of muscle and absence of fat in the ham of Hog A. The ham is 81.3 percent lean and 13 percent fat. The ham of Hog B is 47.4 percent muscle and 51 percent fat.

Hog A has 0.77 inches of backfat and 6.5 square inches of loin eye. Hog B has 1.73 inches of backfat and 3.9 square inches of loin eye.

Meat-type Hogs Make Rapid and Economical Gains

Some hog producers, remembering their experiences with the leggy, "hamless wonder" hogs raised in the past, have questioned the rate and economy of gain of meat-type hogs. However, tests have proven that meat-type hogs can be produced as rapidly and as economically as fat-type hogs.

Iowa tests indicated that pigs with less than 38 percent of live weight in lean cuts required 340.5 pounds of feed to produce 100 pounds of gain and weighed 227.5 pounds at five months of age. Pigs with 40 or more percent of lean cuts required only 328 pounds of feed per 100 pounds of gain and weighed 228.5 pounds at five months of age.

FIGURE 6-4. (*Above*) A meaty gilt weighing 233 pounds. FIGURE 6-5. (*Below*) An average barrow weighing 225 pounds. (Figures 6-4 and 6-5 *courtesy* Drs. E. A. Kline and L. L. Christian, Iowa State University)

Packers Pay More for Meat-type Hogs

Most packer-buyers are now bidding from 25 cents to $1.00 per hundredweight above the current market price for uniform loads of meat-type hogs. Some interior markets are buying hogs on the basis of carcass grade and yield. Some farmers have received as much as $1.80 more per hundredweight by selling meat-type hogs according to their carcass value.

BREEDING SYSTEMS

We have already pointed out that farmers select breeding stock to meet the demands of the breeding programs which they are following. Most growers produce hogs for slaughter. It is estimated that less than 4 percent of all growers are raising purebred hogs to be sold for breeding purposes.

The breeding programs followed by producers of purebred hogs are often quite different from those followed by producers of market hogs. These breeders mate purebred boars and sows of the same breed. Some breeders do inbreeding or line breeding. Most breeders of commercial hogs follow upgrading, crossbreeding, or criss-crossing methods, which may involve two or more breeds. A description of each of the various swine-breeding systems may be found in the following paragraphs.

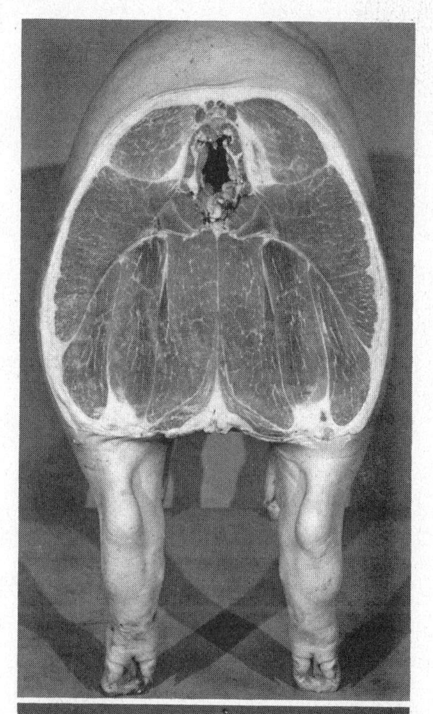

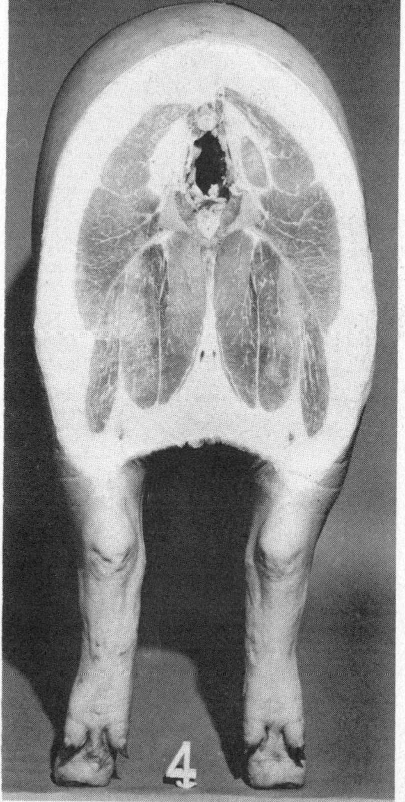

FIGURE 6-6. (*Above*) **Cross section of the hams of the gilt shown in Figure 6-4. Note the muscle. The hams were 81.3 percent lean and 17 percent fat. FIGURE 6-7.** (*Below*) **Cross section of the hams of the barrow shown in Figure 6-5. They contained 51 percent fat and 47.4 percent muscle. (Figures 6-6 and 6-7** *courtesy* **Drs. E. A. Kline and L. L. Christian, Iowa State University)**

FIGURE 6-8. A drove of commercial hogs sired by a Farmers Hybrid boar. (A. M. Wettach photo. *Courtesy* Farmers Hybrid Companies, Inc.)

Upgrading

The system of mating purebred boars with grade sows of the same breed is called *upgrading*. A *purebred* is an animal that is registered or eligible for registration by a breed association. A *grade* animal is one whose sire is purebred but whose dam, or mother, is not eligible for registration. The mating of a purebred Chester White boar with a grade sow herd of the same breed is an example of upgrading.

This method has greatly improved our grade herds in type and productiveness. The development and maintenance of many herds of purebred swine has provided farmers with sources of boars for the production of market hogs.

Purebred Breeding

In the breeding of purebreds, purebred boars are mated to purebred sows of the same breed. The mating of purebred Hampshire boars with purebred Hampshire sows is an example of purebred breeding. Breeders of purebreds produce boars for use by their neighbors who raise market hogs; they also produce purebred animals for use by other breeders of purebreds. Breeders have attempted to improve their herds through careful selection. For many years much of the improvement had to do with hog type, but recently considerable attention has been given to production factors, such as the number of pigs farrowed and weaned, the weight of the litter at weaning time, the rate of gain, economy of gain, and carcass quality.

Crossbreeding

The mating of purebred, hybrid, or inbred boars of one breed with purebred or grade sows of another breed is called *crossbreeding*. From studies made of the hogs received at the packing plants, it was found that from 70 to 80 percent of the hogs possessed the characteristics of more than one breed. A large percentage of commercial hog producers are crossbreeding or are crisscrossing their hogs. An example of this system of breeding is the mating of a Duroc boar and Hampshire sows.

In tests conducted at Iowa State University over a ten-year period, it was found that fewer pigs were born dead among the crossbreds and slightly more of the crossbred pigs lived to weaning age than did the purebred pigs. The crossbred pigs weighed an average of nearly four pounds more at weaning time than did the purebreds. The crossbreds gained more rapidly after weaning and reached the market weight of 225 pounds about ten days earlier than did the purebred. Between 25 and 30 more pounds of feed were required to bring the purebreds up to 225 pounds than were required for the crossbreds.

The crossbred sows were good mothers when bred back to a boar of either of the

TABLE 6-1 ADVANTAGES OF CROSSBREDS FOR COMMERCIAL PORK PRODUCTION

Trait	First Cross (Sows purebred, boars purebred, pigs crossbred)	Multiple Crosses (Sows and pigs crossbred, boars purebred)
	ADVANTAGE OVER PUREBRED	ADVANTAGE OVER PUREBRED
Litter size	None †	12%
Survival	7%	14%
Weight of pigs at 154 days	14%	14%
Litter weight at 154 days	22%	40%
Feed efficiency (other than that due to correlation with gain)	Negligible to slight	Negligible to slight
Meatiness	None	None

† Because sows are purebred.

Courtesy Animal Science Department, Iowa State University, Ames, Iowa.

parent breeds or to a boar of a third breed. The pigs produced compared favorably with the first-cross pigs in rate and economy of gain. Iowa tests are summarized in Table 6-1.

In Ohio tests, pigs produced by crossing purebred Durocs and Poland Chinas weighed 22 pounds more at 180 days than did purebred Duroc pigs. The daily gain was 0.12 pound more, and the litter weight at 180 days was 180.1 pounds heavier. The best gains in the Ohio tests were made by pigs produced by mating the Duroc-Poland-Hampshire gilts with purebred boars of the same breeds.

Criss-crossing

In this system purebred boars of one breed are mated with grade or purebred sows of another breed (as in crossbreeding); then the gilts produced from the first mating are mated with a boar of the same breed as the original sows. The gilts produced from this cross are mated the next year with a boar of the same breed as the boar used in making the first cross. Boars of the parent breeds are then used in alternate years thereafter. A farmer who breeds a group of Spot gilts to a Duroc boar and then alternates the use of boars of these two breeds in the following years is using the criss-cross method.

Rotation Breeding

Swine specialists in several states are now recommending the use of boars of three or four breeds of hogs in rotation. This method is especially recommended when inbred boars are used in producing commercial hogs. The use of four carefully selected breeds maintains the hybrid vigor desired in hog production. The boars used in this system of breeding may be purebreds, inbreds, or crosslines. Crosslines are the progeny resulting from the crossing of two inbred lines of the same breed.

The following is an example of this method: A farmer has a Duroc sow herd which he mates to a Chester White boar.

The gilts from this cross are mated the next year to a Hampshire boar. During the third year he uses a Spot boar, and in the fourth year he uses a Duroc boar. The rotation system is then repeated. A rotation involving Duroc, Poland China, Yorkshire, and Hampshire boars has produced excellent results. A diagram of rotation breeding is shown in Figure 6-9.

Inbreeding

The mating of boars with sows of the same breed which are closely related is called *inbreeding*. The mating of brothers with sisters and of sires with daughters are examples of inbreeding. Purebred and commercial producers have frowned upon inbreeding because usually the offspring are inferior to the parent in growth ability and in prolificacy. After several years of inbreeding, however, it is possible to obtain lines which produce uniform offspring. The crossing of two or more of the inbred lines results in hybrid vigor which is greater than that obtained when two non-inbred purebred lines are crossed.

Line Breeding

Line breeding involves the mating of purebred boars of one line or family with sows of the same line that are not as closely related as those used in inbreeding. The mating of second cousins is an example of line breeding. Many breeders of purebred hogs have used line breeding to intensify the desirable characteristics of certain bloodlines in their herds. An examination of the pedigrees in most herds will produce many examples of line breeding. This method is rarely used in producing market hogs.

Crossline Breeding

When this system is used, inbred boars of one line are mated to inbred sows of another line of the same breed. This method has been used extensively at the U.S.D.A. Agricultural Research Center at Beltsville, Maryland, at the Regional Swine Breeding Laboratory, and at cooperating state agricultural experiment stations.

The crossing of two distinct lines of a breed produces a type of hybrid vigor simi-

FIGURE 6-9. An example of a rotation system of swine breeding. (Drawing by David C. Opheim)

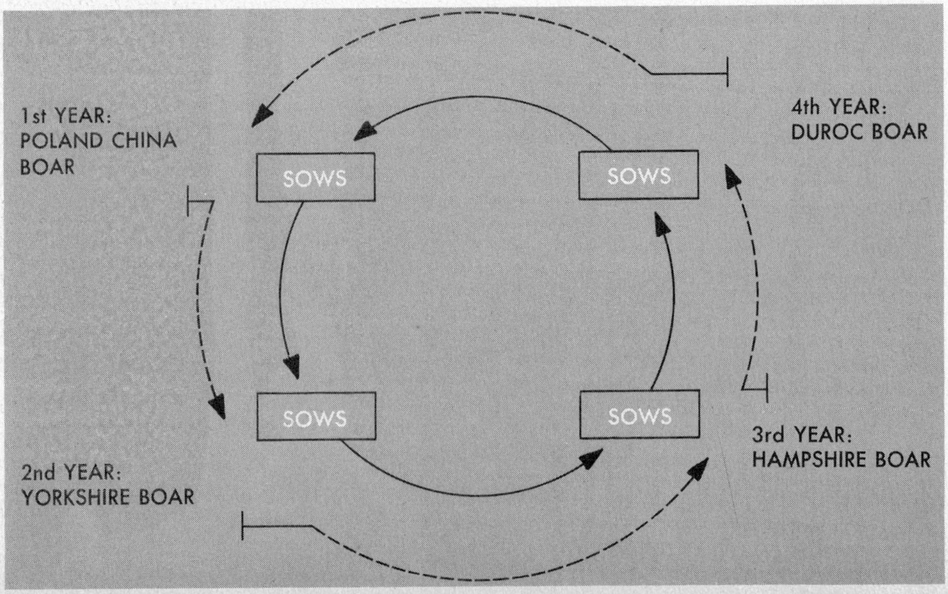

lar to, but in smaller amounts than, that resulting from the crossing of two breeds. Hybrid vigor is increased from about 12 to 15 percent as a result of the crossing of two inbred lines. By crossing four lines, as is done in producing hybrid corn, the hybrid vigor is further increased. By crossing lines of different breeds, it is possible to obtain higher levels of performance than can be obtained by crossing two lines of one breed.

Heritability Estimates

The carcass length, the percentage of ham and fat cuts (based on carcass weight), the backfat thickness, and the loin eye area have 50 to 60 percent (high) levels of heritability. The ham–loin percentage is about 45 percent due to inheritance.

The percent of lean cuts (based on carcass weight), feed efficiency, growth rate (from weaning to market), and five-month weight are medium in heritability, with levels of 25 to 35 percent.

The weaning weight, the number of pigs farrowed, the number weaned, and the birth weight of the pigs have low levels of heritability—5 to 15 percent.

FACTORS IN SELECTING A BREED OF HOGS

Because we have a number of breeds of hogs in this country, it is difficult to determine the breed or breeds which will do best in a breeding program. Each breed has a loyal group of supporters, and there is rivalry among the producers of the various breeds. Each of the breeds has desirable characteristics, yet all of them have some weaknesses. Usually there are as many differences among the individuals within a breed as there are between breeds.

The requirements of a given program will help in selecting the best breed of hogs for its purposes, just as the kind of tractor or seed corn purchased is determined by individual requirements. The goal in hog production is to produce large litters of pigs which can be grown out rapidly and economically, and, when sold, will command the top market price. Pork producers must decide which breed or breeds will fit best in their breeding programs.

The following factors must be given careful consideration in the selection of a breed of hogs: (1) availability of breeding stock, (2) prolificacy, (3) growth ability, (4) temperament, (5) carcass quality, (6) feed conversion efficiency, (7) nicking ability, (8) market demand, (9) disease resistance, (10) feeds available, and (11) personal likes and dislikes of the grower.

CLASSIFICATION OF BREEDS

Up to about 1940, the swine breeds were classified either as *lard type* or as *bacon type*. Most of the hogs produced in this country were of the lard type because there was a ready market for lard. Conditions have changed. Our lard market is partly gone since we have vegetable fats competing with lard on the shortening market. As a result, swine breeders have focused their attention on the production of *meat-type* hogs.

While we can no longer classify hogs as lard type or as bacon type, we find a considerable variation both within and between breeds to the extent that they produce carcasses high in the lean cuts and low in lard and fat cuts. Since the animals within a breed vary in this respect, it appears better to classify the breeds on physical characteristics and recentness of origin.

Shown in Table 6-2 is a classification of most popular old established breeds and of the new breeds according to the predominant color of hair and type of ears.

The American Landrace is listed as an old breed since the Landrace bloodlines obtained originally from Denmark, Norway,

TABLE 6-2 PHYSICAL CHARACTERISTICS OF BREEDS

Breed	Predominant Color of Hair	Type of Ears
OLD ESTABLISHED BREEDS:		
American Landrace	White	Large, slightly drooping
Berkshire	Black with white feet, face, switch	Erect
Chester White	White	Drooping
Duroc	Red	Drooping
Hampshire	Black with white belt	Erect
Poland China	Black with white on face, feet, legs, and switch	Drooping
Spot	Black-and-white spotted	Drooping
Tamworth	Red	Erect
Yorkshire	White	Erect
NEW BREEDS:		
Beltsville No. 1	Black with white spots	Drooping
Beltsville No. 2	Light red	Erect
Lacombe	White	Drooping
Maryland No. 1	Black with white spots	Erect
Minnesota No. 1	Red	Slightly erect
Minnesota No. 2	Black with white spots	Slightly erect
Montana No. 1	Black	Slightly drooping
Palouse	White	Slightly erect to drooping
San Pierre	Black and white	Erect

and Sweden have been recorded by the American Landrace Association, one of the newer swine record associations.

Several other breeds of hogs have been raised in small numbers for many years. The Kentucky Red Berkshire, the Hereford, the Essex, the OIC, and the Mulefoot breeds are still being produced, and record associations are maintained for recording pedigrees.

BREED DIFFERENCES IN CARCASS QUALITY

The market hogs exhibited at the National Barrow Show held annually at Austin, Minnesota, come from many states and represent the bloodlines of the respective breeds. The data obtained from the carcasses of the slaughtered live show winners, and from the entries in the carcass contest give some indication of the ability of the various breeds to produce desirable carcasses.

Table 6-3 is a summary of the data obtained from the carcasses of the first five prizewinners in each of the three weight divisions of each of the separate breed live barrow classes at the 1970 and 1971 National Barrow Shows. A total of 270 barrows, 30 of each of the nine breeds, was included in the study.

Farmland Industries, Inc., of Kansas City, cooperates with swine breeders in conducting testing stations at Lisbon, Eagle

TABLE 6-3 SUMMARY OF CARCASS RESULTS OF LIVE SHOW WINNERS, 1970 AND 1971 NATIONAL BARROW SHOWS

Breed	Length (Inches)	Backfat (Inches)	Ham (%)	Loin Eye (Sq. Inches)	Ham & Loin Index	Age at 220 lbs. (Days)
Crossbred	31.10	1.08	17.96	5.88	125.8	174
Duroc	30.95	1.02	17.89	5.47	133.6	176
Hampshire	31.29	.98	17.56	5.57	131.2	177
Chester White	30.85	1.19	17.89	5.16	130.5	185
Poland China	30.55	1.14	17.50	5.73	132.3	176
Spot	30.89	1.21	17.39	5.08	124.7	182
Yorkshire	31.48	1.17	17.22	5.20	124.2	177
Landrace	31.37	1.25	16.93	4.70	116.3	176
Berkshire	30.97	1.21	16.15	4.96	111.1	185
Average	31.03	1.13	17.39	5.27	126.6	178.5

Grove, and Ida Grove, Iowa and at Clarkson, Nebraska. Presented in Table 6-4 is a summary of all tests completed since their origin.

ORIGIN AND CHARACTERISTICS OF THE OLD ESTABLISHED BREEDS

Five of the nine old established breeds of swine raised in the United States originated in this country. They are the Chester White, Duroc, Hampshire, Poland China, and Spot. The four breeds that are not native to this country are the Berkshire, Landrace, Tamworth, and Yorkshire, imported from Canada and Europe.

American Landrace

Landrace hogs originated in Denmark and were first imported to this country in 1934 for experimental crossbreeding purposes. By government agreement it was not possible at that time to produce and release Danish Landrace stock in this country as purebreds. They were used extensively in crossbreeding, and as a result of this practice many of the new breeds carry Landrace breeding.

In 1972, 6,420 Landrace hogs were recorded by the American Landrace Association, Inc. This organization was incorporated in December, 1950, with headquarters at Noblesville, Indiana. In addition to the importations from Denmark, a number of Landrace hogs have been brought in from Norway and Sweden.

The Landrace has white hair, and the skin is usually white. Small black spots, however, are common. The breed is extremely long, deep-sided, and well-hammered. Usually the animals are flat and sometimes low in the back. The ears are very large and cover much of the face. Some of the Landrace breed have weak pasterns. The breed is prolific, and is efficient in the use of feed. In addition, the carcass of the Landrace hog is meatier than that of many of our American breeds. A total of 95 meat litters was certified in 1971.

Berkshire

This English breed is one of the oldest breeds of swine. For many years the Berkshire was considered the best of the meat breeds because of the excellent carcasses which it produced. The champion barrows and carlots of barrows at national shows often carried Berkshire breeding. During recent years the other breeds have been improved, so that now there is less difference in the quality of the carcasses. The Berkshire still rates well, however, as a producer of a good carcass.

The Berkshire is an extremely long animal and in form and fleshing conforms to the ideal meat-type hog. The breed is black, with white markings usually on the feet, head, and tail; it has long been characterized by a short snout and a wide, dished face. The "pug nose" of the Berkshire of years ago has been well refined through breeding.

Berkshires are slightly smaller than some of the other meat-type breeds at maturity. Mature boars will weigh 900 pounds or more. In 1972 there were 5,529 Berkshire swine recorded in this country. This breed has had general use in crossbreeding programs, with excellent results.

Chester White

The Chester White breed had its origin in Chester and Delaware counties in Pennsylvania. The parent stock used to produce the breed included English Yorkshire, Cheshire, and Lincolnshire bloodlines. The Chester White Swine Record Association was established in 1908.

The Chester White has white hair and skin. Small flecks in the skin are not objectionable, but black or other than white hair is. The Chester White is intermediate in size and mature boars weigh 900 pounds or more.

Barrows of the Chester White breed have made excellent records at state and national shows in both on-foot and carcass tests. This breed has been popular with farmers in some areas. The sows produce and raise large litters which grow out rapidly and make good gains.

There were 20,387 purebred animals registered by the Chester White Swine Record Association in 1972.

TABLE 6-4 AVERAGE PERFORMANCE FARMLAND SWINE TESTING STATIONS

(26 TESTS, 1958–1971)

	BOAR INFORMATION						BARROW INFORMATION				
Breed	Pens	No. Boars	Gain	Eff.	Probe	Index	No. Barrows	HL%	Lgth.	B.F.	L.E.
Berkshire	24	52+	1.92	277	1.02	150	26	37.5	29.7	1.44	4.17
Chester White	71	171+	1.79	274	1.06	141	90	38.0	29.0	1.53	4.12
Duroc	708	2,192	1.99	262	1.06	159	845	37.9	29.2	1.46	4.10
Hampshire	1,204	3,768	1.90	270	.93	159	1,368	40.0	29.8	1.29	4.66
Landrace	106	199+	1.97	267	1.01	156	122	38.4	30.3	1.48	4.12
Poland China	329	730+	1.87	276	1.01	149	401	38.9	29.0	1.40	4.60
Spot	177	465+	1.88	273	1.00	152	212	39.1	29.3	1.41	4.41
Yorkshire	588	1,719	1.93	263	.98	159	691	38.3	30.0	1.49	4.25

Source: Bob Casey, Farmland Foods, Inc.

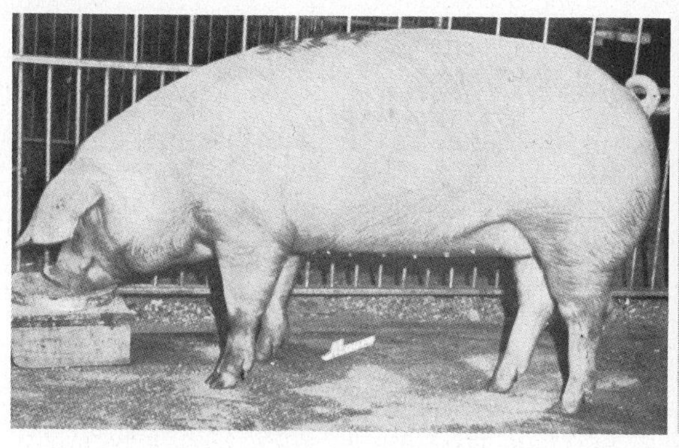

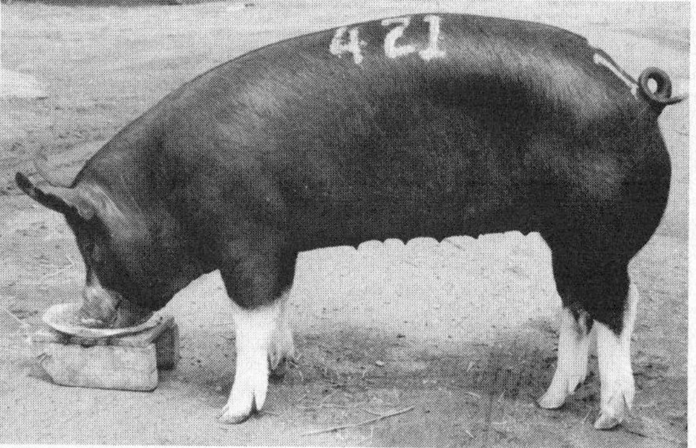

FIGURE 6-10. (*Left*) A Grand Champion Landrace gilt. (Moore photo. *Courtesy* American Landrace Association, Inc.) FIGURE 6-11. (*Right*) A Grand Champion Berkshire gilt. (*Courtesy* American Berkshire Association)

Duroc

The ancestry of this breed is not entirely known, but the Jersey Reds of New Jersey, the red Durocs of New York, and the red Berkshires of Connecticut have contributed to the formation of the breed, which was first called the Duroc-Jersey. Standards were established for the breed in 1885, and although the type has changed several times during the past 80 years, the color standards are still adhered to.

The Duroc is red in color, with the shades varying from a golden to a very dark red. A medium cherry red is preferred. Black flecks may appear in the skin, but large black spots, black hair, and white hair are objectionable. The Duroc is large, with excellent feeding capacity and prolificacy. In type and conformation, the Duroc is similar

FIGURE 6-12. (*Left*) A Champion Chester White boar, seven months old. (*Courtesy* Chester White Swine Record Association) FIGURE 6-13. (*Right*) The Grand Champion gilt at the 1970 National Duroc Congress. (*Courtesy* United Duroc Swine Registry)

FIGURE 6-14. The 1971 All-American Hampshire Senior Spring Gilt. (*Courtesy* Hampshire Swine Registry)

to the Chester White and the Poland China. The sows are good mothers, with good dispositions, and produce large quantities of milk. Farmers have found the breed to be excellent for use in crossbreeding programs.

For the year ending November 1972, there were 66,647 purebred Durocs recorded by the United Duroc Swine Registry. More Durocs were recorded in 1970, 1971, and 1972 than hogs of any other breed. A total of 411 certified litters was recorded in 1970 by the United Duroc Swine Registry of Peoria, Illinois.

Hampshire

The Hampshire breed was developed in Boone County, Kentucky, from hogs imported from England, probably in the early 1800's. The foundation stock, known as the *Thin Rinds* and *Belted Hogs*, had been raised in the New England States.

The breed association was organized in 1893, and although the breed is one of the youngest, it has become very popular. In 1972 Hampshires ranked second among the breeds in the number of animals recorded —56,110.

The Hampshire is a black hog with a white belt encircling the body and including the front legs. The back legs are usually black, and no white should appear above the hock. The head and tail are black, and the ears are erect. No white can appear on the head.

The Hampshire is smaller than some of the other meat-type breeds. It has been bred for refinement, quality, and prominent eyes. The sows of the breed are very prolific and are good mothers. The Hampshire is a good rustler, doing well on pasture and following cattle in the feed lot. A total of 480 litters was certified in 1971.

It is usually shorter-legged than are most other breeds but sound on its feet and legs in most cases. The breed has been used extensively in crossbreeding because of its quality, fleshing, and prolificacy. Nearly one half of market hogs in Illinois have some Hampshire breeding.

Poland China

For many years the Poland China has been considered the largest of the American breeds. Today it is similar in size, type, and conformation to the Duroc, Chester White, and Spot.

The breed was originated between 1800 and 1850 in Warren and Butler counties, Ohio. The White Byfield hog, imported from Russia, and the White Big China hog were used with native hogs in producing the Warren County hog. The use of the Berkshire on Warren County hogs and later the use of boars imported from Ireland produced the Poland China breed.

Since about 1875, the Poland China has been a black hog with six white points—the feet, face, and tip of tail. Mature boars weigh up to 1,000 pounds. The typical Poland China has thick, even flesh and is free from wrinkles and flabbiness. The breed has good length and excellent hams. The head is trim, and the ears are drooping.

Poland Chinas produce excellent carcasses. Barrows of this breed have been winners in both on-foot and carcass contests.

There were 9,079 Poland Chinas recorded in 1972. The sows of this breed are good mothers, but the breed has been best used in crossbreeding. The Poland China crosses excellently with the Duroc, Hampshire, and Chester White.

Spot

Early in the development of the Poland China breed, many of the hogs were spotted, and some breeders preferred this. They were reluctant to adhere to the color standards set up for the Poland China breed, and many continued to grow spotted hogs. Some of these hogs were crossed with the black Poland China and some with Gloucester Old Spots which had been imported from England.

The National Spotted Poland China Record Association was formed in 1914.

The Spot resembles the Poland China in type and conformation. It is a large breed, and the animals are good feeders. To be eligible for registration, the color on the body must be between 20 and 80 percent white, but the desired color is 50 percent black and 50 percent white.

This comparatively new breed has become quite popular. In 1972 the breed association recorded 14,824.

Tamworth

The Tamworth is red in color, with shades varying from light to dark. The head is long and narrow, with a long snout and erect ears. The body is also long and narrow, and the sides are smooth. Usually the Tamworth has a strong back and thin shoulders. The carcass produces bacon of the best quality.

Sows of this breed are prolific and are excellent mothers and foragers. Mature boars weigh up to 700 or 800 pounds. In 1972, 1,415 purebred animals were recorded. Tamworth bloodlines were used in producing the Minnesota No. 1 breed of swine.

FIGURE 6-15. (*Left*) **A Grand Champion mature Poland China sow at the Minnesota State Fair. (*Courtesy* Poland China Record Association) FIGURE 6-16.** (*Right*) **A Champion Spot gilt at the National Spotted Swine Type Conference. (*Courtesy* National Spotted Swine Record, Inc.)**

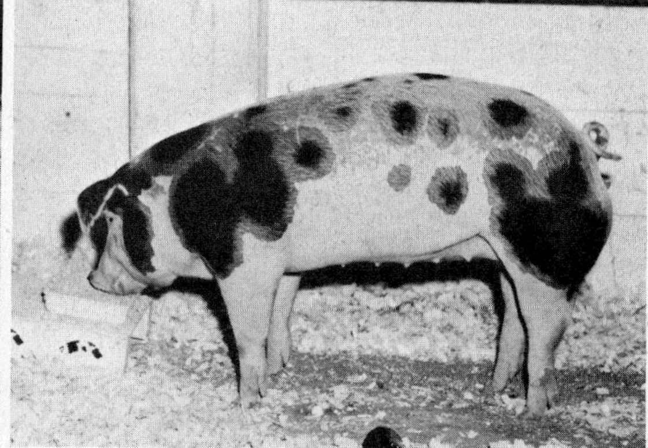

Yorkshire

During the time that our American hogs were classified as *bacon* or *lard* type, the Yorkshire was considered by many as the best bacon type breed. The breed is raised in large numbers in Canada, England, Scotland, and Ireland. It is a native of northern England and was imported to this country early in the nineteenth century.

The Yorkshire is white in color, but occasionally there are black pigment spots in the skin. These spots are objectionable but do not disqualify the animal in the show ring or from recording. The ears are erect. Mature boars weigh from 700 to 1,000 pounds. The Yorkshire is extremely long and deep and has firm flesh.

The American Yorkshire Club, Inc., of Lafayette, Indiana recorded 40,373 animals in 1972. The popularity of the breed has increased greatly since 1952. It has had wide use in crossbreeding. There were 226 litters certified in 1971.

FIGURE 6-17. (*Above*) **A state fair champion mature Tamworth boar. (*Courtesy* Tamworth Swine Association) FIGURE 6-18.** (*Below*) **The Champion Gilt at the 1971 Yorkshire Type Conference. She sold for $7,500. (Moore photo. *Courtesy* American Yorkshire Club, Inc.)**

ORIGIN AND CHARACTERISTICS OF THE NEW BREEDS

A number of new breeds of swine have been developed during the past 25 years in an attempt to produce one that is prolific, will gain rapidly, will make efficient use of feed, will produce a carcass high in lean and low in fat cuts, and has a strong constitution.

Most of the breeds that have been developed to date have had their origin in the U.S.D.A. Agricultural Research Center at Beltsville, Maryland, or at cooperating state agricultural experiment stations. A number of inbred lines of various existing breeds have also been developed.

In the development of new breeds, an attempt has been made to produce animals which possess the desirable characteristics of two or more parent stocks. The Danish Landrace, Yorkshire, and Tamworth have been used in various combinations with other long-established breeds. Following is a description of the most promising results.

New Breeds Developed from Inbred Lines

A summary of the new breeds of swine which have been developed is presented in Table 6-5. Animals of these breeds are recorded by the Inbred Livestock Register Association of Noblesville, Indiana. None of the

TABLE 6-5 NEW BREEDS OF SWINE DEVELOPED FROM INBRED LINES

Breed	Parent Stock	Color	Developer
Beltsville No. 1	74% Landrace 26% Poland China	Black and white spots	U. S. D. A.
Beltsville No. 2	58% Yorkshire 30% Duroc 6% Landrace 6% Hampshire	Light red	U. S. D. A.
CPF	Beltsville No. 1 San Pierre	Black and white	Conner Prairie Farms, Indiana
CPF No. 2	25% Beltsville No. 1 50% Maryland No. 1 25% Yorkshire	Black and white	Conner Prairie Farms, Indiana
Lacombe	55% Landrace 23% Berkshire 22% Chester White	White	Canadian Dep't. of Agriculture
Maryland No. 1	38% Berkshire 62% Landrace	Black and white spotted	U. S. D. A. Maryland Agri. Exp. Sta.
Minnesota No. 1	45% Tamworth 55% Landrace	Red	Minn. Agri. Exp. Sta.
Minnesota No. 2	40% Yorkshire 60% Poland China	Black with white spots	Minn. Agri. Exp. Sta.
Minnesota No. 3	Gloucester Old Spot Welch pig, English Large White, Beltsville No. 2 and others	Red	Minn. Agri. Exp. Sta.
Montana No. 1	45% Hampshire 55% Landrace	Black	U. S. D. A. Mont. Agri. Exp. Sta.
Palouse	Chester White Landrace	White	Wash. Agri. Exp. Sta.
San Pierre	Berkshire Chester White	Black and white	Gerald Johnson, Indiana

Inbred Livestock Register Association

FIGURE 6-19. A litter sired by a Farmers Hybrid boar from a dam also sired by a Farmers Hybrid boar. (*Courtesy* Farmers Hybrid Companies, Inc.)

new breeds is produced in large numbers, yet they have played important parts in the development of crossbreeding programs for commercial hog production.

HYBRID HOGS

A hybrid hog is produced by crossing two or more inbred lines. Some commercial hybrid seed corn is a result of a double-cross process involving four inbred lines. During one growing season line A is crossed with line B ($A \times B$) and line C is crossed with line D ($C \times D$). The next year the progenies of the two crosses are crossed ($AB \times CD$). The corn produced from the latter cross is sold as commercial seed corn.

Hybrid hogs are produced in much the same manner as hybrid corn. Lines of hogs are inbred for several generations and then crosses are made of the inbred lines. The extent to which the hybrid hog is more productive than the parent stock is dependent upon the genetic makeup of the various lines and upon how well they supplement each other when they are brought together.

Hybrid Hog Production

There is no set pattern in the production of hybrid hogs in this country. The hybrid is produced by crossing two or more inbred lines, usually from different breeds. Many farmers and breeders do not understand the true meaning of the term *hybrid*, and as a result some hogs are called hybrids incorrectly.

Hybrid hog production is a complicated procedure and one that involves careful planning and management. Producers of hybrid hogs must understand genetic laws and be able to apply them. Usually there is quite an outlay in breeding stock and equipment. Hybrid hog production has developed beyond the experimental stage. The number of producers of hybrid hogs is small. However, some of them produce large numbers of hogs. The production records of hybrid hogs have demonstrated their usefulness in commercial pork production.

SELECTION OF INDIVIDUAL BOARS AND GILTS

The factors to be considered in selecting boars and sows for the breeding herd are much the same regardless of the breeds and breeding programs involved. It is very important that hog producers carefully select the animals which are to be used in the breeding herd, because it is difficult to grow hogs profitably when inferior breeding stock is used.

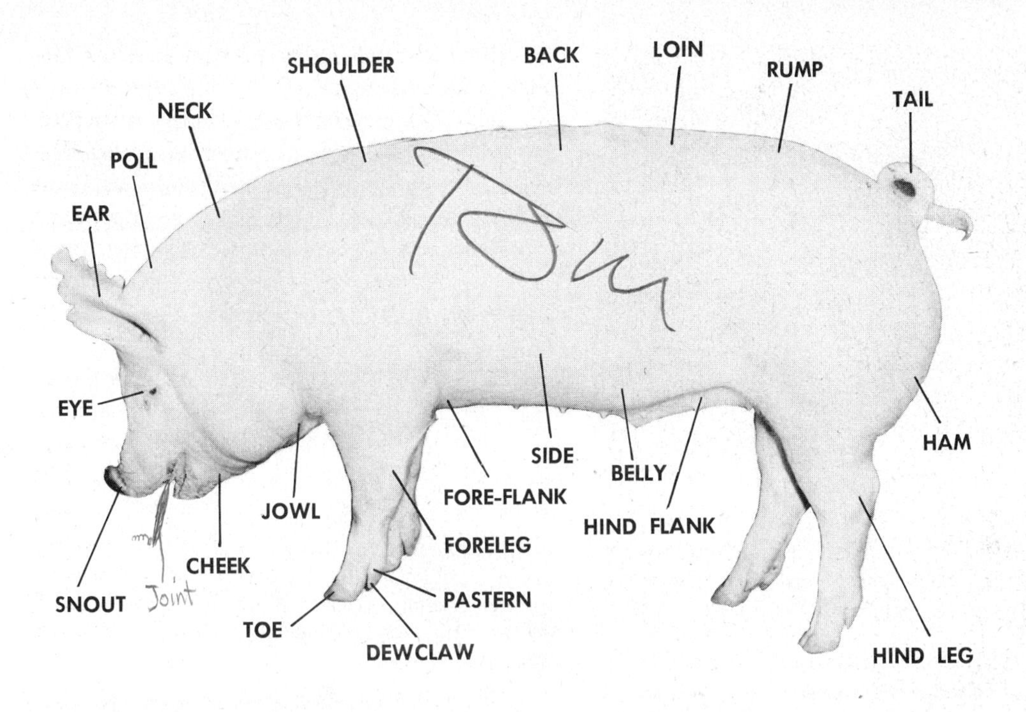

FIGURE 6-20. The parts of a hog. (Original picture *courtesy* American Yorkshire Club, Inc. Artwork by Linda and Jerry Geisler)

Parts of a Hog

Farmers, breeders, and packers use much the same terms in describing hogs. Since these terms will be used repeatedly in the following paragraphs, the reader should become familiar with them. Figure 6-20 shows the various parts of a hog body.

The Ideal Type and Conformation

Breeders of purebreds and commercial pork producers usually have some ideal type of hog in mind in selecting breeding and feeding stock. Usually they do not find animals which possess all the characteristics they are looking for and must select those that are nearest their ideal.

The ideal type changes from time to time with changes in market demands. More attention is usually given to quality of carcass when hogs are plentiful. Prolificacy may be given more emphasis when the price of pork is high than when pork is cheap.

Producers select animals to be mated with other animals. Quite often the animal selected is not the ideal in several respects, but it is sufficiently close to mate well.

Ideal Type of Market Hog. The ultimate goal in all hog production is to produce, efficiently and profitably, a hog that will yield a carcass high in the cuts of pork desired by the consumer. The crossbred barrow shown in Figure 6-21 is considered by many to approach the present concept of the ideal meat-type hog. (For illustrative purposes, a similar animal could have been selected from each breed.) It is the goal of the farmer and hog breeder to select breeding stock which will produce market hogs with the

muscling, conformation, and carcass quality of the illustration.

This barrow has good length and width of body. He carries his depth uniformly from front to rear and he is deep and smooth-sided. He is especially good in the ham (note its plumpness and depth). He gives evidence of being firm-fleshed. There are no signs of flabbiness and he is well muscled, with a trim head and jowl.

This barrow produced an excellent carcass and yielded a good dressing percentage.

FIGURE 6-21. (*Above*) **Grand Champion Barrow at the 1971 National Barrow Show. This crossbred had a 6.65 square inch loin eye, 0.96 inches of backfat, and the ham represented 17.84 percent of the liveweight of the barrow. (*Courtesy* George A. Hormel & Co.) FIGURE 6-22.**(*Below*) **Carcasses of hogs — intermediate, chuffy, and rangy in type. Note the differences in length, meatiness, and backfat thickness. (*Courtesy* Wilson and Company)**

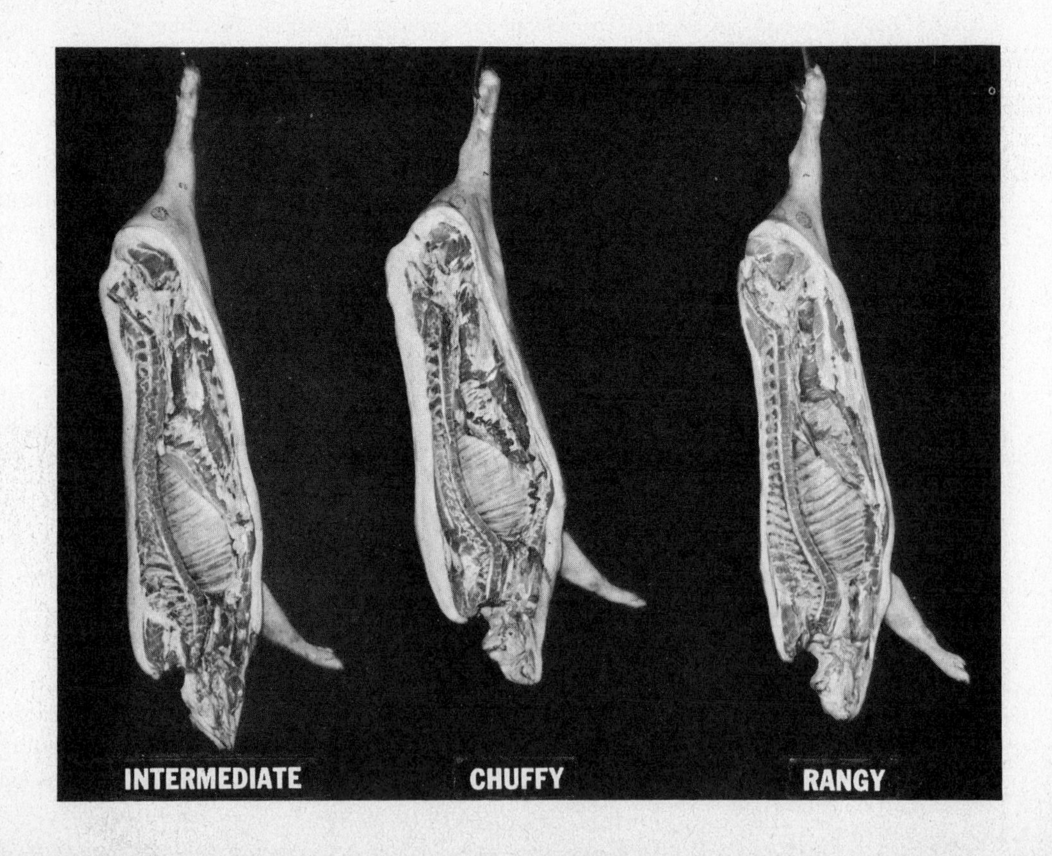

Dressing percentages compare the weight of the carcass after the animal has been slaughtered to the live weight of the hog. Lard-type or exceedingly fat hogs may yield dressing percentages of 75 percent or more. In the meat-type hog a carcass with a high dressing percentage is desired, but it must also be a carcass high in the choice cuts, ham and loin. A barrow should produce a high percentage of lean cuts and a minimum of backfat and lard. Forty-five to 50 percent ham and loin is desirable.

The crossbred barrow shown in Figure 6-21 produced a carcass 31.2 inches long and had a backfat thickness of 0.96 inches. It had a 6.65-inch loin eye and 17.84 percent of the carcass was ham. The ham-loin index was 144.9.

The carcasses of three barrows of different types are shown in Figure 6-22. The carcass on the left is from the intermediate or meat-type hog. It is the type of carcass which we would expect to get from the barrow pictured in Figure 6-21. Note the thickness of the layer of fat over the back and compare it with the backfat on the carcass in the center, which is from a chuffy hog. It has been found that the percentage of lean cuts in a carcass is closely related to the backfat thickness.

The carcass on the right is from a rangy barrow. It has a very thin layer of backfat, but the ham and loin are underdeveloped. A higher percentage of the carcass is in the form of bone and skin. Usually, rangy hogs require more time to be finished for market, and they are marketed at heavier weights.

Today the ideal market hog is the intermediate meat-type hog. Breeders and packers shy away from both the chuffy and the rangy animals. The chuffy hog produces too much backfat and lard; the rangy hog requires too much time to be properly finished and produces a heavy carcass.

Ideal Type of Breeding Animal. The breeder, knowing what type of market hog he wants to produce, can determine the type, conformation, and other qualities desired in the boars and sows to produce this kind of pig. The kinds of feed given and the methods of feeding also influence the growth rate, the economy of gain, and the quality of the carcass produced. Good feeding practices are of little value unless he has the right kind of breeding stock to begin with.

Selection of Females. The mature sow shown in Figure 6-23 possesses many of the characteristics desired in a brood sow. She has a medium-long body and a strong, well arched back. She is deep-sided and has the capacity of the chest and middle that insures good feeding quality and vigor. Note her well-developed, deep, full hams. This sow is good on her feet and legs. She has ample bone, strong but medium-length legs, and short, straight pasterns. A brood sow must have good feet and legs; an inactive or clumsy sow usually is unable to raise a good litter.

This sow has a trim head and jowl. She is feminine and has a prominent eye. Coarse-headed and heavy-jowled sows should not be kept in the herd. Femininity and a rea-

FIGURE 6-23. This Hampshire sow was Grand Champion at the Illinois State Fair and selected as All-American Mature Sow. (*Courtesy* Hampshire Swine Registry)

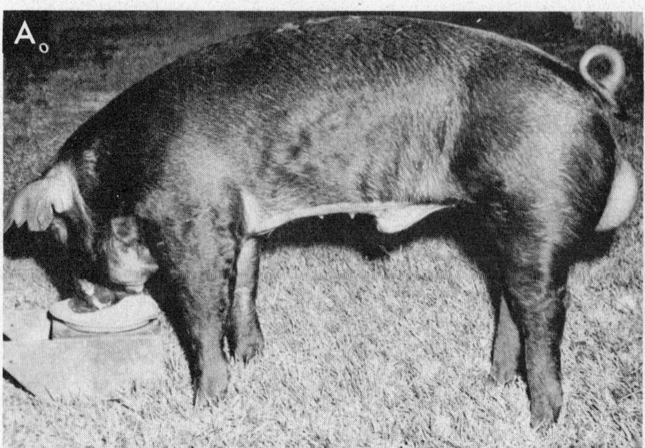

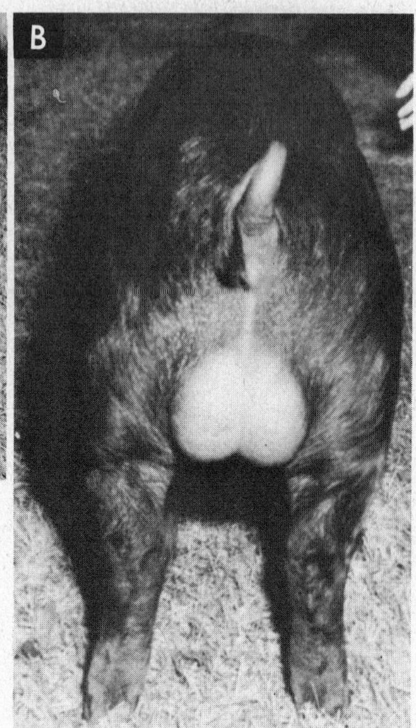

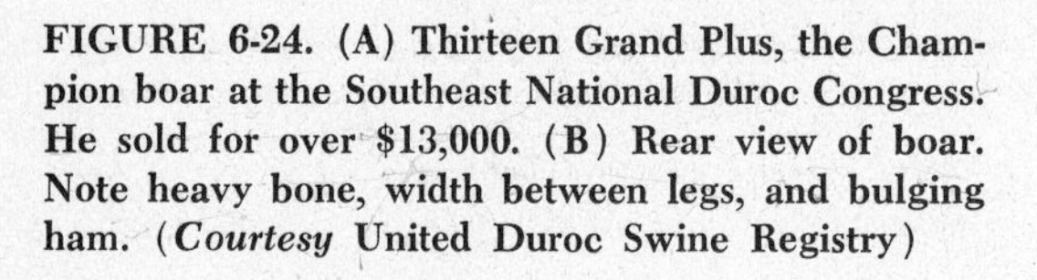
FIGURE 6-24. (A) Thirteen Grand Plus, the Champion boar at the Southeast National Duroc Congress. He sold for over $13,000. (B) Rear view of boar. Note heavy bone, width between legs, and bulging ham. (*Courtesy* United Duroc Swine Registry)

sonable amount of refinement are desirable in brood sows.

This brood sow is smooth in the shoulders, has a wide, well muscled loin, and an excellent hair coat. The refinement of the head and ear, shoulder, and hair coat are important items in selecting female herd material.

The udder of the sow should be well developed, with 12 or 14 sound teats. The teats should be prominent and well spaced. The sow shown in Figure 6-23 has an excellent udder.

Most hog producers select young animals as replacements in their herds, especially in making replacements in the sow herd. Most farmers select gilts that are six to eight months of age.

Selection of Boars. The selection of the herd boar is a major undertaking for most farmers. Genetically, the boar represents half of the herd. It has been through the use of good boars that much progress has been made in swine improvement, since one boar can be mated to a large number of gilts.

Boars should show masculinity and breed-character. They should show cleanness and firmness of jowl, wide open eyes, and the neck blending in well with the shoulders. The shoulders should be smooth and the back strong.

The tail setting should be high and there should be no fat around it. The hams should be deep and full. The lower ham should be smooth and firm. The side should be smooth with a trim middle but good depth at both the fore and rear flank.

Boars should be big for their age, medium long, and rugged. They should have strong bone and stand squarely on all four legs. The pasterns should be short and the legs should be set out on the corners. Shown in Figure 6-24 is an outstanding herd boar prospect.

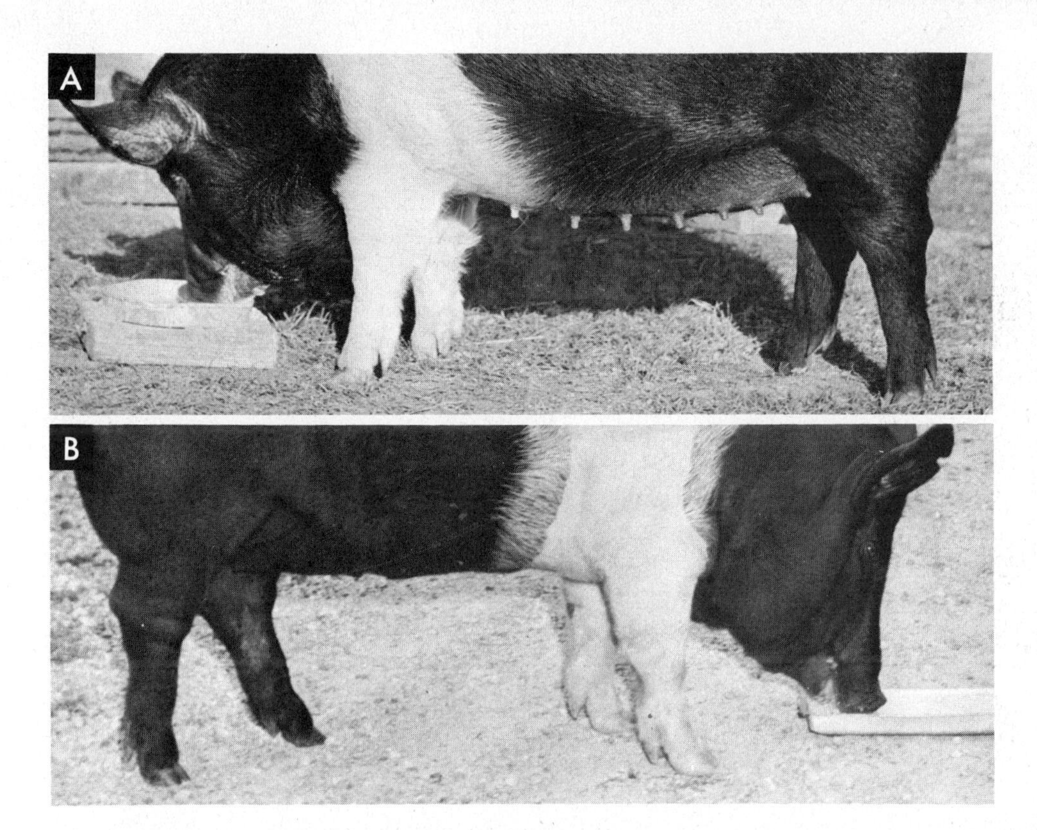

FIGURE 6-25. **(A) A senior gilt with an excellent underline and good feet and legs. (B) A spring gilt with poor underline, feet, and legs. (*Courtesy* American Hampshire Herdsman)**

A boar must have a good constitution. He should have width and depth in the heart area. He should have well developed sex organs. The testicles should be prominent and of equal size.

Underlines. Bidders on boars and gilts in purebred sales usually inspect the teats or rudimentaries on the animal being sold. The teats on boars are called *rudimentaries* or *rudimentary teats.* Sometimes a teat on a sow is not connected with a milk gland and will not produce milk. This is called a *blind teat.* A teat which has an *inverted nipple* is more serious. These teats usually do not come down or produce milk. Boars and sows will pass this characteristic on to their off-spring.

The sow in Figure 6-25(A) has a good underline. The underline of the sow in (B) is poorly developed.

Tests for Carcass Lean and Fat

It is possible to take much of the guess-work out of selecting meat-type breeding stock by using one of four tests which have been developed by swine research specialists.

Backfat Probe. The thickness of backfat on a live animal may be measured by using a sharp knife or scalpel to slit the skin and a metal ruler marked in tenths of inches to measure the backfat. Small rulers are available with sliding clips which can be pushed down through the backfat.

Three probes should be made, each 1½ to 2 inches from the midline of the pig's back. The incisions should be no deeper than ⅜ of an inch and should be crossways to the pig to avoid penetrating the muscle. The incision points are: (1) behind the shoulder, straight above and an inch back of the elbow, (2) in the middle of the back, where the last rib joins the vertebrae, and (3) straight above the stifle in the loin.

The ruler is carefully pushed through the incision into and through the fat until it reaches the back muscle. The clip on the ruler aids in reading the depth of backfat. The probe does not injure the animal and there is little danger of infection.

Purdue Electric Lean-fat Meter. Purdue University researchers have developed an electronic device which will tell the difference in the proportion of lean and fat inside the loin, ham, shoulder, or bacon side of live hogs. This procedure involves no more injury or pain to the animal than the skin-prick of a needle.

FIGURE 6-26. Measuring backfat thickness and loin area by sonoray (high frequency sound). (*Courtesy* Conner Prairie Farms, Noblesville, Indiana)

The needle on the instrument is part of an electrical circuit. When flesh completes the circuit, a high current in lean and low current in fat result. The needle is set for depth, then pushed into the flesh. The meter, which is a part of the instrument, then shows whether the needle is in fat or lean.

Sonoray or Ultrasonics. The use of ultrasonic or high-frequency sound has been proven effective in measuring backfat thickness and loin eye areas in hogs. Sonoray service is available in some states from swine specialists associated with state colleges or from technicians employed by commercial organizations. The equipment and technology necessary will prevent most commercial pork producers from making their own tests.

Sonoray estimates are made from readings on the oscilloscope. The loin eye area is determined by plotting the readings on graph paper. Persons with experience and a thorough knowledge of the anatomy of the area can plot rather accurately the loin eye area. A planimeter is used to measure the estimated square inches of loin eye area.

Ultrasonic estimates made at the University of Missouri have been found to be quite accurate when compared with actual measurements made after the animals were slaughtered.

K-40 Counter. A new tool to help pork producers in the selection of breeding stock has been developed at the Oklahoma State University from research done in England. The tool measures potassium in live animals. Since potassium is located primarily in the muscle part of the body, the instrument shows much promise.

Swine Testing Station Results

Commercial and purebred swine breeders rely heavily on swine testing station results. Stations are in operation in at least 25 states, and the following information is usually available for boars and gilts: (1) daily rate of gain, (2) pounds of feed required per pound of gain, (3) probed inches of backfat, (4) length of littermate barrow carcass, (5) loin eye area of barrow carcass, (6) percentage of ham and loin in barrow carcass, and (7) index score.

Great improvement has come about in swine breeding stock since testing work began. Table 6-6 gives a summary of improvements made in Iowa since tests were started in 1956.

FIGURE 6-27. Checking underlines of pigs entered at the Farmland Swine Testing Station at Eagle Grove, Iowa. (*Courtesy* Bob Casey, Farmland Swine Testing Stations)

Other Factors in Selecting Breeding Stock

Although type and carcass quality are important, there are other factors which should be considered in selecting breeding stock. In selecting both boars and gilts, age, pedigree, performance or production records, health, and disposition should be given consideration.

Age. The use of boars and sows which have proven themselves as producers eliminates the risk encountered in using untried animals. Mature sows usually produce more and larger pigs per litter than do gilts, and are usually better milkers. However, they become large, heavy, and more clumsy than gilts. They require more feed to maintain themselves and need larger farrowing pens. Mature boars usually settle more sows and are more dependable during the breeding season. They do not get the "flu" as easily as do young boars, and are usually in better health during the breeding season.

Most farmers and breeders, however, use young boars and gilts in their breeding operations. The young animals require a smaller investment, need less feed to maintain themselves, and less space in housing. When a young animal dies, there is less loss than there is when a mature animal is lost. The young animal gains in weight during

TABLE 6-6 IOWA TESTING STATION SUMMARY—TWO YEAR AVERAGES, 1956–1970

Two-Year Period	Length (Inches)	Backfat (Inches)	Ham and Loin (Percentage)	Loin Eye (Square inches)	Rate of Gain (Pounds per day)	Feed Efficiency (Pounds per pound of gain)
1956–1957	29.1	1.59	33.4	3.48	1.86	2.94
1962–1963	29.7	1.36	38.7	4.14	1.95	2.71
1969–1970	29.7	1.20	43.9	5.03	2.08	2.50

the production period, and can be sold at a good price after it is no longer useful in the herd.

Pedigree. Every animal has a pedigree, but only the pedigrees of purebred animals can be recorded by a breed association. A pedigree is merely a record of the bloodlines of the ancestors of the animal. The pedigree is very important in the breeding of purebreds, but the breeder of commercial hogs is more concerned with the production record of the sire and dam of the litter than with their pedigrees.

Performance or Production Records. It is possible to judge the performance of a

FIGURE 6-28. (*Above*) **A Hampshire boar at the Iowa Swine Testing Station at Ames that gained 2.18 pounds per day on 2.61 pounds of feed per pound of gain. He probed 0.85 inches of backfat.** (*Courtesy* **Iowa Swine Testing Station) FIGURE 6-29.**(*Below*) **This Yorkshire litter weighed 947 pounds at 56 days. Raised by Eugene Wagner, Newton, Illinois.** (*Courtesy* **Paul Walker)**

mature animal by the size of the litters produced, by the conformation of the pigs, and by the weight of the pigs at 21 or 35 days or at 120 to 150 days of age. Most breeders prefer breeding stock from litters from which eight or more pigs were raised. They want the pigs to average 11 to 15 pounds at 21 days, 20 to 25 pounds at 35 days, or 40 to 50 pounds at 56 days, and 200 to 225 at five months of age. By earmarking the pigs at birth and weighing them at 21 or 35 days, or at 120 to 150 days, it is possible to select breeding animals from the most productive litters.

All the breed associations have production registry programs. The production of the ancestors is recorded on the pedigree in the same way as production records of dairy cows are reported on the pedigrees of cows. The National Association of Swine Records developed a production record plan based on the number of pigs farrowed in the litter and the weight of the litter at 56 days of age. A litter qualified for production registry when there were eight or more pigs farrowed and raised by a mature sow to a 56-day weight of at least 320 pounds and by a first-litter gilt to at least 275 pounds at the same age.

The breed associations now support only one common program, the certified litter program. The Yorkshire and Hampshire breeds support programs encouraging registration of weights at 21 days of age.

The breed associations have a uniform set of standards for a *certified litter.* To qualify, the litter must meet production registry requirements for its breed. In addition, two pigs, each weighing at least 200 pounds at 170 days of age, must yield when slaughtered carcasses 29 inches long, with no more than 1.5 inches of backfat and at least 4½ inches of loin eye at the tenth and eleventh ribs.

A boar that sires five certified litters (CL) from different sows is known as a *certified meat sire* (CMS).

Litter testing of the home herd can help materially in the improvement of the productiveness of the herd. It is good practice to purchase breeding stock from a breeder who has maintained production records.

Health. Breeding animals which are large and heavy for their age usually are in good health, but the best policy is to buy only animals which have been vaccinated for leptospirosis and tested for brucellosis. In some areas it may be desirable to buy animals vaccinated for erysipelas. The herd from which the animals come should be inspected, and breeding and feeding stock should be purchased only from disease-free herds.

Careful hog producers inspect the herds and the farms on which they are raised before buying breeding stock. Many breeders will not buy breeding or feeding animals which have been marketed through public sales barns or stockyards. Buying only disease-free stock and keeping it away from other hogs for two or three weeks is a sound policy.

Disposition. The disposition of a sow may materially affect the number of pigs she will save at farrowing time. A sow that is nervous and easily disturbed is more likely to lie or step on pigs than is a sow with a quiet disposition. Quiet animals and those that like to be scratched usually have good dispositions and are good gainers. A good brood sow should permit you to enter the pen at any time, even during farrowing.

The disposition of a boar is equally important because the boar will be the sire of the next year's sow herd. A boar should be friendly in disposition, active, and a good rustler. Inactive boars are usually slow breeders.

FEEDER PIGS

The number of farmers who buy feeder pigs has increased in recent years. Cattle feeders have always purchased large numbers of feeder pigs to follow cattle in fattening lots because it is easier for them to buy the pigs as they are needed than it is to maintain a breeding herd. But we now have a new group of buyers. Many farmers who have been unsuccessful in raising pigs because of disease losses or who do not have facilities to farrow litters are now buying pigs in fairly large numbers.

It has been estimated that any farmer who cannot raise an average of 6½ pigs per litter on his farm will find it to his advantage to buy pigs rather than to try to produce them himself. The average number of pigs weaned per litter in this country is rarely above 6½ to 7 pigs.

Feeder pig production has become a profitable business, especially in areas where feed is not available to grow out the pigs for market. It is estimated that approximately 564,000 feeder pigs are sold annually in Wisconsin.

Several of the Corn Belt states import 500,000 to 1,000,000 feeder pigs each year in addition to the feeder pigs that are produced within the state for sale to other swine producers. It is estimated that from 20 to 30 percent of the hogs marketed were farrowed on farms other than those on which they were finished.

The National Feeder Pig Producers Association members handle about 1½ million pigs a year. Its membership includes the Wisconsin Feeder Pig Marketing Cooperative, the Missouri Farmers Livestock Association, Tennessee Livestock Association, and other similar marketing groups.

Selection of Feeder Pigs

Regardless of the source, extreme care must be taken in the selection and purchase of feeder pigs. The reputation of the breeder, the breeding stock used in producing the pigs, the type and conformation, the health of the animals, and the purchase price must be carefully considered.

Feeder Pig Grades. The U.S.D.A. established, effective April 1, 1969, revised standards for six grades of feeder pigs: U. S. No. 1, U. S. No. 2, U. S. No. 3, U. S. No. 4, U. S. Utility, and U. S. Cull. The first four grades should produce carcasses of like grades. The pigs graded "Utility" should produce carcasses that would grade U. S. 2, 3, or 4. Unthrifty pigs are graded "Cull." It is advantageous to both buyer and seller that the pigs be graded.

Price of Feeder Pigs. How much is a feeder pig worth? There is no one answer to this question. The price varies with the needs of the prospective buyer, the availability of the pigs, the price of farm grains, the market outlook for pork products, and the crop outlook. In 1971, 40-pound feeder pigs which had been vaccinated, wormed, and castrated were selling for $18 per head delivered to the farm. The top price on No. 1 market hogs at that time was $24 per hundredweight. Efficient producers could then raise their own feeder pigs for less money. The inefficient producers, and those without breeding stock and facilities, purchased the pigs at $18 per head and felt that they had made a good buy.

SUMMARY

Good breeding stock is essential for the most profitable hog enterprise. Lard-type hogs are no longer profitable; meat-type hogs must be grown.

A meat-type hog is one which will yield 45 to 50 percent of its carcass in ham and loin. It will weigh 200 pounds in five months and have a carcass which is 29½ to 31 inches long, has 4½ to 5 inches of loin eye, and has a maximum of 1.2 inches of backfat. It will produce a pound of pork on 3 pounds of feed. The litters will average from 9 to 10 pigs raised.

Tests indicate that meat-type hogs made as rapid and as economical gains as the fat or chuffy type of hog.

There is no one best breed of hogs. The differences among the animals within a breed may be greater than those between breeds. The following factors should be considered in selecting a breed: availability of good breeding stock, prolificacy, growth rate, temperament, carcass quality, nicking ability, market demand, disease resistance, feeds available, and personal likes and dislikes.

Of the nine old established breeds, the Hampshire and Duroc are most popular among farmers.

The Berkshire, Hampshire, Tamworth, and Yorkshire have erect ears. The Chester White, Duroc, Hereford, Landrace, Poland China, and Spot hogs have drooping ears.

The Beltsville No. 1, Beltsville No. 2, Maryland No. 1, Minnesota No. 1, Minnesota No. 2, Montana No. 1, and San Pierre are new breeds developed in this country by crossing two or more of our existing breeds or by crossing Danish Landrace hogs with one or more of our old established breeds.

The American Landrace breed is an outgrowth of importations of Landrace hogs from Denmark, Norway, and Sweden. It has been used extensively in the production of new breeds.

We need to develop longer hogs with trimmer jowls and heads. Boars and gilts should be selected from large uniform litters. Breeding animals should be large for their age, have good length and uniform width and depth, and be firm-fleshed. There should be no signs of wasteness or wrinkling. The feet and legs should be straight and set out on the corners. Sound underlines with 12 or more nipples are preferred. Good eyesight and a good disposition are musts.

FIGURE 6-30. A group of thrifty feeder pigs sold at a Maryland market. (*Courtesy* National Hog Farmer)

QUESTIONS

1. What breeds of hogs are most common in your community?
2. Why do you raise the breed of hogs which you have on your farm?
3. What factors should be considered in selecting a breed?
4. What use can be made of the new breeds of hogs in your community?
5. What kind of a production testing program should you use on your farm to improve the productiveness of your herd?
6. Describe the type and conformation of the kind of barrow which will top the market.
7. What are the advantages of crossbreeding over upgrading and breeding of purebreds?
8. Describe the factors that you would consider and what you would look for in selecting open gilts.
9. What additional factors would you consider in selecting bred sows?
10. Outline the qualities you would seek in selecting a young boar for use in your home herd.
11. What is the difference between a blind and an inverted nipple?
12. In selecting breeding and feeding stock, how can you be certain that the animals are from a disease-free herd?
13. What are the official grades of feeder pigs?
14. What is meant by the term *certified meat sire?*

REFERENCES

Briggs, Hilton M., *Modern Breeds of Livestock,* 3rd Edition, The Macmillan Company, New York, New York, 1969.

Bundy, C. E. and R. V. Diggins, *Swine Production,* 3rd Edition, Prentice-Hall, Inc., Englewood Cliffs, New Jersey, 1970.

Ensminger, M. Eugene, *Swine Science,* 4th Edition, The Interstate Printers & Publishers, Inc., Danville, Illinois, 1970.

Hunsley, Roger E., *Livestock Judging and Evaluation,* AS-388, Purdue University Studies, Lafayette, Indiana, 1971.

Krider, J. L. and W. E. Carroll, *Swine Production,* 4th Edition, McGraw-Hill Book Company, New York, New York, 1971.

Russell, H. G. and G. R. Carlisle, *Evaluating Swine Carcass Information,* AS-626, University of Illinois Press, Urbana, Illinois, 1965.

7 Feeding and Management of the Breeding Herd

The profit or loss from a swine-breeding herd is often determined by the number of pigs weaned and marketed per sow. An average of six or seven pigs marketed per sow is usually necessary to come out even. For profitable hog production the number of pigs farrowed, weaned, and marketed must be increased.

It has been estimated that out of 100 pigs farrowed only 65 will live to be weaned, and only 55 will be marketed. Tests conducted at Iowa State University indicated that when 11 pigs were weaned per litter, the average cost of each eight-week-old pig was about $10.45. When nine pigs were weaned the cost was $12.01 per pig. Pigs from litters of seven weaned cost $14.52, and pigs from litters of five cost $17.92 at weaning time. These figures indicate that unless seven or more pigs per sow can be saved, buying feeder pigs should be considered. Otherwise, little profit can be anticipated from the breeding enterprise.

The feeding and management of the herd at breeding time and during the gestation period greatly influence the number of pigs farrowed and weaned. Good production methods will result in the following:

1. **More pigs being farrowed per sow.**
2. **Larger and healthier pigs at birth.**
3. **Fewer dead pigs, runts, and abnormal pigs per litter.**
4. **Better production of milk by the sows.**
5. **More and heavier pigs weaned per litter.**

FEEDING AND CARE DURING THE BREEDING SEASON

A farmer may suppose that the small litters his sows produce are due to the lack of prolificacy in his breeding stock. Usually he is wrong in making this assumption. Small litters are very often a direct result of carelessness in the feeding and management of the herd at breeding time.

Age to Breed

Gilts should be bred to farrow when they are 11 to 13 months of age if they have been well grown out. The maturity of the gilt is more important than its age. Most gilts which have done well reach puberty and come in heat when they are five to six months of age. It is not a good policy to breed gilts during their first heat periods. Larger litters usually result from gilts bred during their third or fourth heat periods.

Gilts should be well grown out and weigh from 200 to 250 pounds at breeding time. Boars should be a little heavier if they are to receive heavy service. Some breeders prefer to use mature animals, fall boars, or early spring boars in their breeding programs. The older boars are more dependable, especially in settling mature sows.

Heat Period

The heat period, which usually lasts two or three days, is the time during which the sow will accept the boar. It is better to breed the sow during the second or third day of the heat period than during the first day. Sows that are in heat are usually restless and frequently mount other sows. The vulva is usually enlarged and inflamed. Sows that

FIGURE 7-1. (*Above*) **This Hampshire litter is evidence that the dam has received an adequate and balanced ration.** (*Courtesy* **Hampshire Swine Registry**) **FIGURE 7-2.** (*Below*) **The boar and gilt should be well grown out before breeding, especially when lot-breeding is practiced.** (*Courtesy* **Kent Feeds**)

are not bred will usually come in heat at intervals of about three weeks. Breeding gilts twice, 12 to 24 hours apart, will usually increase conception rate and litter size by 10 percent.

Gestation Period

The period of time between the breeding of the sow and the farrowing of the litter is known as the *gestation* or *pregnancy period.* The length of the period varies somewhat but is usually from about 112 to 115 days. Breeders quite often figure 114 days as the gestation period. Older sows usually have longer gestation periods than gilts.

Time to Breed

Farmers plan their breeding operations to have the pigs farrow when the temperature, housing, pasture, and labor conditions are favorable. They also plan farrowings so that the pigs can be grown out and ready for a desirable market.

In the ten Corn Belt States, which produce about 75 percent of the nation's pig crop, 33.8 percent of the 1970 pig crop was farrowed during the months of March, April, and May. Only 17.7 percent were farrowed during December, January, and February. Almost equal numbers were farrowed during the other two quarters of the year—23.8 percent during June, July, and August, and 24.5 percent during September, October, and November.

Farmers who produce two litters of pigs per sow, or who produce litters in both the spring and the fall, may breed their sows to farrow in August and September as well as in February and March. Table 7-1 shows the dates of breeding swine and the corresponding farrowing dates, based upon a 114-day gestation period.

Flushing

The condition of the sow or gilt at breeding time affects (1) the regularity of the heat period, (2) the number of eggs, or reproductive cells, produced, and (3) the conception, or the settling, of the sow at the first service. *Flushing* is the feeding of the sows to insure their good health and a gain of 1 to 1½ pounds per day from about three weeks before breeding until after they are bred. Gilts or sows which are recovering from "flu" at the time of breeding sel-

TABLE 7-1 SWINE BREEDING AND FARROWING DATES

Date Bred	Date Due to Farrow	Date Bred	Date Due to Farrow
January 1	April 24	July 1	October 22
January 15	May 8	July 15	November 5
February 1	May 25	August 1	November 22
February 15	June 8	August 15	December 6
March 1	June 22	September 1	December 23
March 15	July 6	September 15	January 6
April 1	July 23	October 1	January 22
April 15	August 6	October 15	February 5
May 1	August 22	November 1	February 22
May 15	September 5	November 15	March 8
June 1	September 22	December 1	March 24
June 15	October 6	December 15	April 7

dom produce good litters. It is better to flush these sows and breed them for a later litter.

The ration to be fed during the flushing period will vary with the age and condition of the sow and with the feeds and pastures available. It is important that the ration be well balanced (14 to 16 percent protein) and fed in the proper amounts. Plenty of proteins, minerals, vitamins, and forages is desired. Gilts usually should receive 2 to 2½ pounds of feed daily per 100 pounds of live weight, whereas mature sows may need less than 1½ pounds daily for every 100 pounds of live weight. Usually a ration consisting of ½ to ¾ of a pound of a balanced protein and mineral supplement, 1 pound of oats, and enough corn or other grains fed each day to produce daily gains of about 1½ pounds will meet the needs of a sow during the flushing period. Sows which are not on pasture should be self-fed alfalfa hay or should receive from 1 to 2 pounds of alfalfa meal daily. However, sows and gilts should not become overfat. Iowa State University tests show that gilts fed 5 pounds of feed per day before and during breeding farrowed more live pigs than gilts fed 7 pounds per day. It is important that gilts and sows be somewhat thin but gaining in weight at breeding time.

A recent breeding experiment at the Ohio Agricultural Experiment Station showed that after being fed alfalfa before breeding, sows produced more eggs during the critical heat period. Alfalfa and other legume pasture or high-quality alfalfa hay should be fed to both the boar and the sow before breeding. University of Missouri tests show that the feeding of ½ gram of antibiotic per sow daily for 10 to 14 days, beginning 3 to 5 days before breeding, increases the litter size by 1.74 pigs.

FIGURE 7-3. Boars should be kept away from the breeding herd. Pictured are a house and shade for a boar on an Illinois farm. (*Courtesy* American Hampshire Herdsman)

Feeding and Management of the Boar

Normally, if the boar is in good, thrifty condition and is well managed, he will not affect unfavorably the size of the litters which he sires. If his vitality is too low or if he is used very heavily, the sperm, or male reproductive cells, may be so weak that fewer of the eggs produced by the female will be fertilized, and small litters may result.

On some farms the boar is neglected. He is purchased at breeding time and placed in a pen in the hog house. He may have been used previously by other farmers. He is in poor physical condition. Quite often he is not fed enough protein, mineral supplement, and vitamins. Such a boar will be low in vitality and a poor breeder.

The boar should receive about the same kind of ration fed to the gilts during the flushing period. A 16 percent protein ration is recommended. He should not be fattened but kept in good, thrifty condition. The following ration is suggested:

TABLE 7-2 ILLINOIS BOAR RATION 16 †

Feed	Units
Ground shelled corn	1,550 lbs.
Soybean meal (50%)	400 lbs.
Steamed bone meal *or* dicalcium phosphate	20 lbs.
Ground limestone	20 lbs.
Trace mineral salt with zinc	10 lbs.
Vitamins:	
Riboflavin	1 gm.
Pantothenic acid	5 gms.
Niacin	15 gms.
Choline	100 gms.
Vitamin A	3 million units
Vitamin D	0.3 million units
Vitamin B_{12}	16 mg.
Antibiotics	20 gms.

† To be hand fed.

Source: University of Illinois

The boar should be placed on a lot from ¼ to ½ acre in size, which has been seeded to alfalfa, red clover, ladino clover, or some other forage crop. A movable house should be provided at one end, and the boar should be fed and watered at the other end of the lot. Exercise is important in caring for a boar.

If properly managed, yearling and other mature boars can be mated to 50 or 60 gilts during a breeding season. An eight-month-old boar should not usually be expected to service more than 20 or 30 sows unless he is very carefully managed.

Pen- Versus Lot-Breeding

It is usually better to bring the boar to the sow or the sow to the boar at the time of service rather than to permit the boar to run with the sows. One good service is normally sufficient, and a record can be kept of the breeding date. If gilts are mated during their first heat period, or if boar power is available when bred later, they should be bred twice at 12- to 24-hour intervals. A young boar turned in with a large number of sows may be punished and become shy. This is especially true if mature sows are involved.

A mature boar should not be mated to more than three sows during one day, and a young boar should not serve more than two sows per day. Some farmers who practice pen-breeding bring a sow to the boar in the morning and another sow during the evening. Breeders of purebreds often space services six or eight hours apart and breed as many as four sows to an outstanding sire in one day. This plan cannot be followed throughout the entire breeding season.

Ranty and Inactive Boars

Some boars are very active during the breeding season and may be difficult to manage. They pace back and forth along fences or even go through the fence. Some

FIGURE 7-4. Pasture, silage, and hay can make up the greater share of the ration for bred sows and gilts during the first months of gestation. (*Courtesy* American Yorkshire Club, Inc.)

breeders place a barrow or a bred gilt in the pen with the boar with excellent results. A better practice is to keep the boar in a lot at some distance from the sows.

Other boars are inactive—quite often because of improper feeding and management—and are slow breeders. It is a good idea to try out a boar a month or so before the breeding season begins, to make certain the boar will serve and settle a sow. If he does not serve or settle the sow, the situation can be remedied before time to breed the main herd.

A boar may be inactive because he has become too fat or has not had sufficient exercise. Many so-called nonbreeders are boars of this type. Be certain your boar is in good breeding condition. It is a good policy to turn two boars together in case one is inactive. The boar should be familiar with his surroundings, because quite often a boar will not service a sow in new quarters. Some boars will not serve a sow when other hogs or human beings are around. Pen-breeding is preferable from this standpoint.

Some boars breed only at night or early in the morning. It is a good idea to try out the boar and bring gilts to him at the time of day when he is most active. It is not a good idea to start a young boar by mating him to mature sows. It is much better to mate him to gilts of his own age and size.

A breeding crate may be helpful in getting a boar to serve a sow. This is especially true in mating mature boars with young gilts. A temporary crate made of straw bales and panel gates or a commercially made crate may be used.

A number of products are sold as remedies for the inactivity of boars, but in most cases they are not effective. A veterinarian should be called in if it is necessary to use drugs or hormones.

Artificial Insemination

Rapid progress is being made in the use of artificial insemination (AI) in swine breeding. Several large firms are now providing AI service. Their boar studs consist largely of certified meat sires, champion show-ring boars, or boars from certified litters.

In the past, four problems have hampered the use of AI in swine breeding: (1) inability to get sows to come in heat at the same time, (2) inability to freeze boar semen, (3) poor heat detection in sows by producers, and (4) the limited number of services per boar.

Ovulation Control. A chemical developed in England and known as Aimax or ICI 33,828 showed promise of partially solving the first problem. The chemical is fed to sows and gilts in their ration for 20 days. It blocks the normal heat cycle. When the chemical is removed from the feed, the females come in heat and can be bred in 5 to 7 days. The use of this drug would synchronize swine breeding so that all sows and gilts might be bred at one time and would then farrow at one time. Aimax was used experimentally in this country with success. It was found, however, that deformed pigs would result when the drug was fed to pregnant sows. It is not being marketed now.

Other methods of estrous control have been successful. Injecting 1,000 I. U. of pregnant mare serum (PMS) in sows at weaning time and following 96 hours later with an injection of 500 I. U. per sow of human chorionic gonadotropin (HCG) usually causes ovulation 30 to 40 hours after the HCG injection. Best conception and litter size result when injected sows are mated or inseminated 28 hours after the HCG injection.

Confined gilts will usually show estrus after an injection of 750 I. U. of PMS followed in 96 hours with an injection of 500 I. U. of HCG. A change in environment after the PMS injection is recommended. The gilts should be inseminated 28 hours after the HCG injection.

Timing of Semen Use. Although frozen semen losses its potency, it has been found that, by the use of proper diluents and methods, semen can be used two or three days after it has been collected. However, most semen is collected and used the same day. Many breeders inseminate each female a second time, 12 to 18 hours later, to insure settling and large litters.

Recognizing Estrus. Most breeders have discovered that the third problem is not serious. They can detect when females are in heat either by observing a boar placed in an adjacent pen or by putting pressure on the backs of the animals. If the sows remain standing, they are in heat.

Service Potential of Boars. A boar at time of service may ejaculate 15 to 50 billion sperm. From 2 to 4 billion sperm are used in inseminating each female. Semen may be collected from each boar at least twice a week, and each ejaculation is sufficient to breed 10 to 25 sows.

The actual insemination process is simple. Many vocational agriculture instructors have trained swine breeders to do the job, and increased use of AI in swine breeding is anticipated.

Breeding Records

A record of the breeding dates for the sows in the herd is very helpful in checking on the ability of the boar to settle sows and in selecting sows to be kept in the herd. It is also helpful in selling bred sows. The buyer will need to know when to expect the sows to farrow.

The breeding record is helpful at farrowing time in deciding when to pen up the sows. It is a good practice to mark the gilts, as they are bred, by the use of ear tags or by clipping hair on the hip or back, and to record the mark and the date of breeding.

MULTIPLE FARROWING

Packers and marketing specialists have encouraged farmers to follow a multiple farrowing system of breeding in order to make economical use of the capital invested in breeding stock and equipment and to distribute the marketing of hogs throughout the year. At one time, a large percentage of the hogs produced in this country were farrowed in the spring (from January through April) and sold during the following fall and winter. The supply during the marketing period exceeded the packer demand, and low prices resulted. The tendency during the past several years has been to increase the number of fall litters and to decrease the number of spring litters.

Farrowing Programs

Two Litters Per Sow Multiple-farrowing Program. In order to prevent seasonal runs on the market and the resulting low prices, many producers are now splitting the sow herds into three, four, or five groups and having each group farrow twice each year. This trend has caused a leveling of hog production during the year. For example, in Indiana and Ohio there are now about as many pigs farrowed from June through November as from December through May, and in Iowa the number of spring pigs farrowed is now only 26 percent larger than the number of fall pigs.

FIGURE 7-5. (*Above*) **Bred sows should receive limited grain while on pasture. (*Courtesy* American Berkshire Association) FIGURE 7-6.** (*Below*) **Excess weight is not desirable during gestation and at farrowing time. (*Courtesy* Kent Feeds)**

Three Litters Per Sow Per Year. The use of prestarted pig rations has made it possible to wean pigs when they are 10 to 21 days of age. As a result, it is possible to rebreed sows and raise almost three litters from each sow each year.

FEEDING AND CARE DURING GESTATION

The vigor of the pigs and the number produced in the litter at farrowing time are determined by the number of eggs fertilized by the boar at the time of breeding and by the methods used in feeding and managing the sows during pregnancy. Not all eggs fertilized at breeding time produce live, healthy pigs at farrowing time. Some of the fertilized eggs produce embryo pigs which die early in the pregnancy period and are absorbed. Others die later, and are farrowed as dead pigs. Those that live but are weak are sometimes called squealers.

The ration fed the sow has much to do with the type of litter which she will farrow. The methods used in feeding, exercising, and housing the sow also influence the number of healthy pigs which she will farrow. The kind of litter that we should strive to produce is shown in Figure 8-4. Note the number and the health of these pigs. Tests have shown that a pig that weighs 3½ pounds at birth has a six times better chance to reach weaning than a 1½-pound runt.

Rations for Bred Sows and Gilts

Bred sows use feed to maintain their bodies and also to produce litters of pigs. Young gilts also need feed to grow and become mature sows. If the rations are inadequate, the sows will be ineffective in maintaining themselves and in producing strong litters. Weak pigs may result. Young sows may remain small or stunted in growth and be poor milkers. The feeding of a balanced ration in adequate amounts during pregnancy is a must in profitable pork production.

The age and condition of the sow determine the amount and kind of ration which she should receive. Thin sows require more feed than do sows in good flesh. Large sows need more feed than do small sows and young sows must have more feed per hundred pounds of live weight than most mature sows. The stage of pregnancy also must be considered. An average requirement is 4 pounds per day.

Nutritional Allowances. Pregnant sows and gilts should receive rations containing 14 to 16 percent protein, 0.7 to 0.9 percent calcium, 0.6 to 0.7 percent phosphorus, 1,500 to 2,000 units of vitamin A, and 200 to 300 units of vitamin D per pound. The following milligrams per pound of the B vitamins and antibiotics should also be fed: riboflavin, 1.5; niacin, 8.0; pantothenic acid, 6.0, and antibiotics, 5.0. In addition, the ration should contain about 0.5 percent salt and 5 to 7.5 micrograms of vitamin B_{12} per pound.

Gain Desired. A gain of 75 to 125 pounds in gilts during pregnancy will allow for the growth of the gilt and her litter. Mature sows should gain 75 to 100 pounds during gestation. These figures can be reduced somewhat if the litter is to be weaned at one to three weeks of age.

Amount of Feed Required. Until 1953, it was assumed that a thin mature sow would require from 1 to 1¼ pounds of feed daily and a young gilt 1½ to 2 pounds daily for every hundred pounds of live weight. Recent tests, however, indicate that we can reduce the amount of concentrates by feeding an abundance of legume hay, corn or grass silage, or by providing adequate pasture.

Hand-Feeding Versus Self-Feeding. Swine producers must control the rate of gain of bred sows and gilts. They can be fed by hand the right amounts of feed daily or self-fed a bulky ration. Either method may be used successfully. The self-feeding method requires less labor, and hand-feeding usually takes less feed.

If the sows gain too rapidly, the amount of corn or grain in the ration can be reduced and ground alfalfa can be added.

Bred sows and gilts should receive about ½ to one pound of supplement each day.

Rations. The ration best suited to any farm must be based upon the kinds of feeds and forages available and the prices of these feeds in the community. The key to feed efficiency is the use of high-quality, cheap, farm-grown feeds, properly supplemented with proteins, minerals, vitamins, and antibiotics.

Following are rations recommended by swine specialists. Grain sorghum may be substituted for corn or oats in the rations, but it should be ground. Rye may be fed in small quantities, but it cuts down the palatability of the ration.

Pasture

Sows that have access to alfalfa, ladino, rape, or other good pasture need little or no grain during the first half of the pregnancy. They will need only about half as much grain during the latter part of the pregnancy period as they would have needed had they been in dry lot. While on pasture sows should receive from ⅓ to ½ pound of protein supplement per head daily and have free access to minerals.

Silage

Corn or grass silage can be fed to bred sows. Corn silage should be supplemented with 1½ pounds of a good protein supplement per sow per day. Grass silage should be supplemented with one pound of protein supplement and two pounds of corn daily. Sows will consume from 10 to 14 pounds of silage daily.

TABLE 7-3 PREGESTATION, BREEDING, AND GESTATION RATIONS
(For boars, sows, or gilts being fed 4 lbs. per day)[1,2]

Percent protein	Ingredient	1	2	3	4
8.9	Ground yellow corn[3]	1,510	1,410	1,535	1,490
44.0	Solv. soybean meal[4]	400	400	300	250
17.0	Dehydrated alfalfa meal	. .	100	. .	100
50.0	Meat and bone meal	. .	. .	100	100
	Calcium carbonate (38% Ca)	15	15	10	5
	Dicalcium phosphate (26% Ca, 18.5% P)	45	45	25	25
	Iodized salt	12.5	12.5	12.5	12.5
	Trace mineral premix	2.5	2.5	2.5	2.5
	Vitamin premix	15	15	15	15
	Feed additives[5]	. .	. .	. .	. .
	Total	2,000	2,000	2,000	2,000

Calculated analysis

		1	2	3	4
Protein	%	15.5	15.92	15.93	15.48
Calcium	%	0.93	0.99	0.97	0.93
Phosphorus	%	0.73	0.72	0.72	0.71
Lysine	%	0.81	0.84	0.79	0.75
Methionine	%	0.26	0.26	0.27	0.26
Cystine	%	0.26	0.26	0.24	0.23
Tryptophan	%	0.18	0.20	0.16	0.16
Metabolizable energy	Cal./lb.	1,294	1,252	1,290	1,256

Feeding Directions

[1] These rations can be used for gilts on pasture during gestation since they require 3 to 4 lbs. of feed daily. These rations can also be used for interval-fed sows or gilts if the average daily intake is approximately 4 lbs.

[2] To simplify the feeding program these rations can also be used for lactation.

[3] Ground oats can replace corn up to 20 percent of the total ration. If more than 20 percent oats is used in the ration, the level of feeding should be increased because of the low energy content of oats. Ground milo, wheat or barley can replace the corn.

[4] If 48.5 percent soybean meal is used instead of 44, use 25 to 50 lbs. less soybean meal and 25 to 50 lbs. more corn. If whole cooked beans are used instead of 44 percent soybean meal, use 100 to 150 lbs. more beans and 100 to 150 lbs. less corn.

[5] Feed additives are not generally recommended during gestation or for gilts during the developer period after selection unless specific disease problems exist. High levels of feed additives (100 to 300 gm/ton) may be beneficial 2 to 3 weeks prior to breeding and 2 to 3 weeks prior to farrowing.

Source: Iowa State University

TABLE 7-4 GESTATION-LACTATION RATIONS

INGREDIENT	RATIONS	
	Gestation-Lactation (Hand-fed) lbs.	Gestation (Self-fed) lbs.
Ground milo or corn [1]	1148	854
44% Soybean meal	384	180
17% Dehydrated alfalfa meal	100	176
Alfalfa hay (good quality)	..	176
Wheat bran	100	150
Ground oats	..	400
Ground dried beet pulp	200	..
Dicalcium phosphate	32	32
Ground limestone	4	..
Salt (iodized) [2]	10	10
Trace minerals [2]	2	2
Vitamin premix [3]	20	20
	2000	2000

[1] If constipation is a problem before and immediately after farrowing the additional substitution of 5 to 10% wheat bran or dried beet pulp for corn or milo is recommended.

[2] The trace mineral and/or iodized salt should supply 90 grams of zinc; 0.15–0.20 grams of iodine; 90 grams of iron and 10 pounds of salt per ton of feed.

[3] The vitamin premix should supply the following amounts per ton of complete feed for gestation-lactation and self-fed gestation rations; Vitamin A 3,000,000 I.U.; Vitamin D 400,000 I.U.; riboflavin 2.0; niacin 12.0 grams; calcium pantothenate 6.0 grams; choline chloride 200 grams; Vitamin B_{12} 20.0 milligrams.

Source: University of Nebraska

Water

Plenty of water should be fed to sows during pregnancy, either in the form of slop feeds, in troughs twice daily, or provided in automatic waterers. Water is especially important during the summer when the temperature is high. Fresh water should be available to the sows at all times.

Exercise

During the early stages of gestation it is desirable to have the sows forage in the cornstalk fields or in the pasture. Brood sows need exercise, and if they do not rustle out into the fields themselves, they should be fed at a distance of from 10 to 15 rods from their housing quarters. It is not a good policy to feed brood sows in the hog house or just outside their housing quarters.

Shelter

During the summer months pregnant sows need only a wooded area, open sheds, or shades to protect them from the sun and rain. Portable houses with open doors or straw sheds are satisfactory during the winter months in many areas. In the northern Corn

Belt warmer houses may be necessary. Several sows may sleep in one house, but they should not be crowded. Bred gilts should have 11 to 14 square feet of shelter per head in cold weather and 15 to 22 square feet of shade per head in warm weather. Mature sows need 16 to 20 square feet of shelter per head in cold weather and 20 to 30 square feet of shade per head in warm weather.

Sows should not be kept in the same lots or buildings with other types of livestock. Sows heavy with pig may be injured by cattle or horses, or may injure themselves by crowding or by going up steep inclines, under creeps, or in low doors. Sows should not be in the cattle feed lots.

SUMMARY

It is usually more profitable to raise two litters from each sow each year. By early weaning it is possible to obtain five litters from a sow in two years.

Gilts should be bred to farrow when they are 11 to 13 months of age. It is usually best to breed gilts during the second day of the third or fourth heat period. The gestation period is 114 days.

Boars and sows should be in gaining condition but should not be fat at breeding time. Rations high in protein, minerals, and vitamins are important during the flushing or prebreeding period. Legume pasture or high-quality legume hay should be provided.

Pen-breeding is preferred and breeding records should be kept. Not more than one or two sows should be mated to a young boar in one day. Mature boars may be mated to three or four sows in one day if matings are properly spaced.

The effectiveness of AI will be greatly increased if PMS and HCG or Aimex are used to promote and synchronize the heat periods of sows and gilts.

Gilts should gain from 75 to 125 pounds, and sows 75 to 100 pounds during pregnancy if they are to suckle litters for six to eight weeks.

With limited forage sows need from 1 to 1¼ pounds of feed per head and gilts 1½ to 2 pounds daily. Sows and gilts on good pasture need little or no feed during the first part of pregnancy. Sows and gilts should not be permitted to get fat. Sows and gilts may be hand-fed grain and supplement in amounts necessary to keep them in good condition. Alfalfa hay may be self-fed. Pregnant sows and gilts may be self-fed a bulky ration. It should contain 25 to 35 percent alfalfa. Gilts may be fed 8 to 12 pounds and sows 10 to 15 pounds of corn or grass-legume silage. Alfalfa and ladino clover pastures will provide for most of the nutritional needs of bred sows and much of the needs of bred gilts.

It is good practice to separate bred sows and gilts from other hogs and livestock.

Sows and gilts should be fed at some distance from sleeping quarters, or be allowed to glean a cornstalk field or good pasture.

From 11 to 14 square feet of dry, draft-free housing should be provided for each bred gilt and 16 to 20 square feet for each bred sow during winter. From 15 to 30 square feet of shade should be provided during the summer.

QUESTIONS

1 How often will gilts come in heat?

2 In the heat period, when is the best time to breed gilts?

3 Which is better on your farm, a one-litter-per-year or multiple-litter breeding program? Why?

4 When should sows and gilts be bred in order to farrow litters in February and March?

5 What care and management should be given the boar during the breeding season?

6 What use can you make of artificial insemination in swine breeding on your farm?

7 What is meant by *heat synchronization?*

8 Plan a good ration for flushing a group of 225-pound gilts.

9 What nutritional allowances are recommended for bred sows and gilts during gestation?

10 How much should sows and gilts gain during pregnancy?

11 Describe the best methods of making effective use of silage in feeding pregnant sows and gilts.

12 How much alfalfa hay should be included in a self-fed ration for pregnant sows?

13 What are the grain and supplement needs of sows and gilts on alfalfa and ladino pasture during gestation? Explain.

14 Describe the methods that you would use in managing the sow herd on your farm to provide adequate rations, exercise, housing, and protection from injury.

REFERENCES

Carlisle, G. R. and H. G. Russell, *Your 1969 Hog Business Ration Suggestions,* AS-377, University of Illinois Press, Urbana, Illinois, 1969.

de Baca, Robert, R. E. Rust, and L. N. Hazel, *Crossbreeding for the Commercial Pork Producer,* Pamphlet 286, Iowa State University Press, Ames, Iowa, 1961.

Hodson, Harold, Jr., *et al., Life Cycle Swine Nutrition,* Pm-489, Iowa State University Press, Ames, Iowa, 1970.

Krider J. L. and W. E. Carroll, *Swine Production,* 4th Edition, McGraw-Hill Book Company, New York, New York, 1971.

Morrison, Frank B., *Feeds and Feeding,* 23rd Edition, The Morrison Publishing Company, Ithaca, New York, 1968.

National Academy of Sciences-National Research Council, *Nutrient Requirements of Swine,* Washington, D.C., 1968.

Omvedt, I. T., *et al., Improving Swine Through Breeding,* Extension Bulletin 306, University of Minnesota Press, Minneapolis, Minnesota, 1965.

8 Feeding and Management of Growing-Finishing Pigs

Critical periods in hog production occur at farrowing time, during the suckling period, at weaning time, and during the period from weaning until the pigs have been marketed. The successful producer plans carefully the practices which will be followed during each of these periods. If he fails to care for his hogs properly during any one of the periods, serious losses may take place. He may have fewer pigs per sow at weaning or at market time, the pigs may be unthrifty and require a long period of time and a large quantity of feed to get them ready for market, or the quality of the hogs produced may be so poor that they cannot be sold as No. 1 hogs.

CARE AND MANAGEMENT AT FARROWING TIME

Time of Farrowing

Some farmers have been successful in producing only one litter of pigs per sow per year by having the sows farrow in late May or June in timber areas or on legume pasture with temporary shelters or houses. Sows need less attention during warm weather than during the winter or early spring. These farmers give their sows little or no attention at farrowing time, and plan on saving fewer pigs per litter. Therefore, they keep a few more sows than they would if they gave their sows more attention.

Whether this practice is economically sound depends upon the price of hogs, the capital and facilities available, the labor supply, and the feed situation. Many farmers now plan on two or more litters per year, which usually necessitates having sows farrow during the cold months of January, February, and March, and again during the hot months of July, August, and September. With the latter system, special care of the sows and litters at farrowing time is necessary.

Housing

The portable hog houses or the pens in the permanent, central-type house should be cleaned and ready for use several days before the sows are due to farrow. The sow should be penned up a few days before farrowing so that she will be familiar with the surroundings and less likely to be nervous. The size of the pen needed will vary with the age and size of the sow, and whether or not guardrails, pig brooders, or farrowing stalls are used. Pens 7 or 8 feet wide and from 8 to 10 feet long are preferred. In central-type houses the short side of the pen should be next to the wall or alley.

The house should be clean, dry, relatively warm, and well ventilated. The floors of movable houses should not be drafty; floor cracks should be repaired. It is sometimes a good policy to bank the sides and ends of movable houses with baled straw or with earth.

Some farmers raise the front side of the movable hog house about 6 to 8 inches so that the sow will lie in the pen with her back toward the door. This arrangement prevents the sow from lying on many pigs. Some breeders have tilted floors in permanent hog houses. A floor slope of one inch to the foot is recommended.

Most movable houses have openings at the ends for ventilation. Windows for light are not necessary in movable houses and are not satisfactory as means of providing ventilation.

A large central-type house for farrowing purposes is not necessary on some farms. Movable houses may be pulled up to the farmstead, where electricity is available for use at farrowing time, and then moved to the pasture to house the pigs during the summer. The investment is much less when movable houses are used, and this practice makes desirable housing available during the summer months. Some producers use both permanent and movable houses.

FIGURE 8-1. (*Above*) **A 24′ x 124′ farrowing stall and finishing house on the Bob Hamilton farm in Iowa.** (*Courtesy* **B. G. Thraikill,** *Des Moines Register and Tribune*) **FIGURE 8-2.** (*Below*) **Portable houses used for farrowing and finishing on rotated pastures. (J. C. Allen photo)**

Slat Floors. A pig likes quarters where he is warm, dry, and draft-free. It is important that the facilities be convenient, easy to clean, as labor-saving as possible, and economical. Swine producers who farrow large numbers of litters continuously during the year can well afford to provide a modernized central farrowing house.

Following are ideas for consideration in planning new farrowing quarters:

1. **A complete floor of concrete slats 3 inches wide on top with ⅜-inch slots between. Slats 8 feet long should run parallel to the sow. No bedding is necessary.**
2. **Feed and water in the farrowing stalls, so that there is no need to turn sows out each day.**
3. **A manure pit about 2 feet deep below the slats, pumped out twice each year.**
4. **Long, narrow farrowing pens, 4½ to 5 feet by 12 to 14 feet. A pig creep at one end, sow feeder and waterer at the other.**
5. **Three-inch corrosion-resistant steel slats front and rear. A 3-foot solid section in the middle for pigs to sleep on. Slots of ⅜ inch.**
6. **Slat floors in pull-together movable houses. In winter use plywood sheeting where pigs lie. Bank the ends of the house.**

FIGURE 8-3. Farrowing quarters should be scrubbed with hot water containing soap and a germicide or lye. (*Courtesy* John W. Simpson, Missouri)

Sanitation

Losses caused by swine diseases and worms can be avoided in part by properly cleaning the farrowing crate or pen before it is used. All dust, dirt, and litter should be removed, and the floors and walls from 1½ to 2 feet from the floor should be scrubbed with boiling lye water. One pound of lye should be added to every 20 gallons of water. After the pen has been cleaned, it should be sprayed with a good disinfectant. Commercial disinfectants or a mixture of one pint of cresol solution to about four gallons of water may be used. The lye water loosens the dirt and kills the worm eggs, while the disinfectant kills disease germs. Equipment used in the pen should also be cleaned and sprayed.

The sides, underline, feet, and legs of the sow should be brushed and washed with soap and warm water before she is placed in the farrowing pen. Any dirt left on the sow may contain worm eggs which, if eaten by the pigs, could produce worm infection and runty pigs. In moving the sows or pigs from one lot to another, it is best to haul them so that worm eggs or disease germs will not be picked up enroute.

Bedding

Coarse-ground corncobs, wood shavings, fine straw, or sawdust may be used for bedding. Too much bedding or coarse hay and straw bedding may result in the loss of pigs. A deep depression in a heavily bedded pen will cause the pigs to lie too close to the sow and will hinder them in getting away from the sow when she moves. A bushel basket or two of bedding in a pen is much better than a 4-inch layer covering the entire pen. The bedding should be kept clean, dry, and well distributed. Removing the sow from the pen each morning and evening for brief exercise helps to keep the pen clean and dry.

Guardrails

A number of devices have been developed to keep sows from lying on their pigs. A simple device is the guardrail which is placed on the three rear sides of the pen about 8 to 10 inches from the wall and 8 to 10 inches from the floor. These may be made of metal pipe, native poles, or 2 × 4's. In any case, they should be installed several days before the sow is due to farrow so that she will be used to them. They permit the pigs to get back of the sow without being crushed between the sow and the wall.

FIGURE 8-4. (*Above*) Comfortable, clean, and roomy quarters are essential. Farrowing stalls are preferred over pens with guardrails. (*Courtesy* Hampshire Swine Registry) FIGURE 8-5. (*Below*) These stalls provide space for pigs to be on either side of the sow. Stalls often save one more pig per litter. (*Courtesy* Kent Feeds)

FIGURE 8-6. Little pigs do not have a chance to live under these conditions. Note the unprotected drop cord and the absence of guardrails.

Heat Lamps or Brooders

The temperature in the farrowing house should range from about 50 to 60 degrees. When the temperature drops below 50 degrees, the little pigs chill and may take cold. The use of heat lamps or pig brooders is recommended for the first week or so after farrowing if weather conditions are unfavorable. The heat lamp or brooder not only provides heat but also attracts the pigs away from the sow so that they will be less likely to be stepped or lain upon.

The infrared heat lamp usually consists of a drop cord with a mechanical support, a socket, a heat bulb, and a protective reflector with a screen over the end of the reflector. The bulb should be a 250-watt lamp. A hard Pyrex heat-resisting bulb is preferred. In using heat lamps, care should be taken to avoid fire. The cord should be heavy and rubber-covered and have a lock-type connector. The socket should be porcelain and keyless, and the lamp should be supported by a chain, wire, or bracket. The electric cord should never be used to support the heating unit.

The unit should be placed about 30 inches above the top surface of the bedding, and care should be taken that water cannot drip down on the heat lamp. To prevent the sow from getting into contact with the lamp, a barrier made of 2 × 4's should be built in the corner of the pen with only enough room for the pigs to get under the lamp for heat. The barrier should be pen height, or 36 inches.

The hover-type brooder has had general use for several years and is an effective means of providing heat and protection. The brooder is usually built in the corner of the pen so that only the little pigs can crawl underneath.

The top of the shelter should be 12 inches from the litter. It should be solid, except for a 12- to 14-inch hole in the middle, where the reflector and bulb are placed. Usually a 75- or 100-watt bulb is adequate. A bulb larger than 150 watts should never be used. One-fourth-inch hail screen should be placed over the opening in the top of the shelter to prevent injury to the bulb and the pigs. A barrier should be constructed to prevent the sow from coming in contact with the bulb assembly.

In wiring a building for pig brooders, provide a permanent and separate circuit with a maximum of 1,500 watts on each circuit protected by a 15-ampere fuse. The wire should be No. 12 size or heavier. The extension cords to the brooders should not be longer than 6 feet, and each pen should have a separate outlet. The pig brooder, like the guardrails, should be installed several days before time for the sows to farrow.

Farrowing Stalls

The farrowing stalls in Figure 8-7 have both guardrails and heat lamps. This is the most effective means of providing heat and protection for the pigs. It is estimated that farmers can save at least one more pig per litter by using such stalls.

When farrowing stalls are used, the sow is confined in the stall, which is 24 inches wide, except when she may be let out for feed, water, and exercise. The little pigs can run around the sow and under heat lamps on either side. Each heat lamp may be used by two litters at the same time, since a divider is provided to keep the litters separated.

The stall may be 6 feet long, but a length of 8 feet is preferred. A 2 × 4 across the back of the stall may be used to regulate length. The partition panels should be 10 to 12 inches above the floor, depending on the size of the sow. The same precautions should be taken in the use of heat lamps as discussed previously in this chapter.

Sows may be left in the farrowing stall for only a few days or for several weeks. Moving the sows to regular pens after two or three days makes it possible to use the same stall for several litters. The sow should be removed from the stall every morning and evening for exercise, feed, and water.

Feeding and Management

The sow should be fed a bulky, laxative ration in moderate amounts just previous to farrowing. She should receive all the water she can drink, but cold water should be avoided. It is usually best to reduce the ration just before farrowing and to give her

FIGURE 8-7. Interior of a farrowing house with stalls, each equipped with a feeder and waterer. The sows face the alley. (*Courtesy* Brower Better Built Equipment)

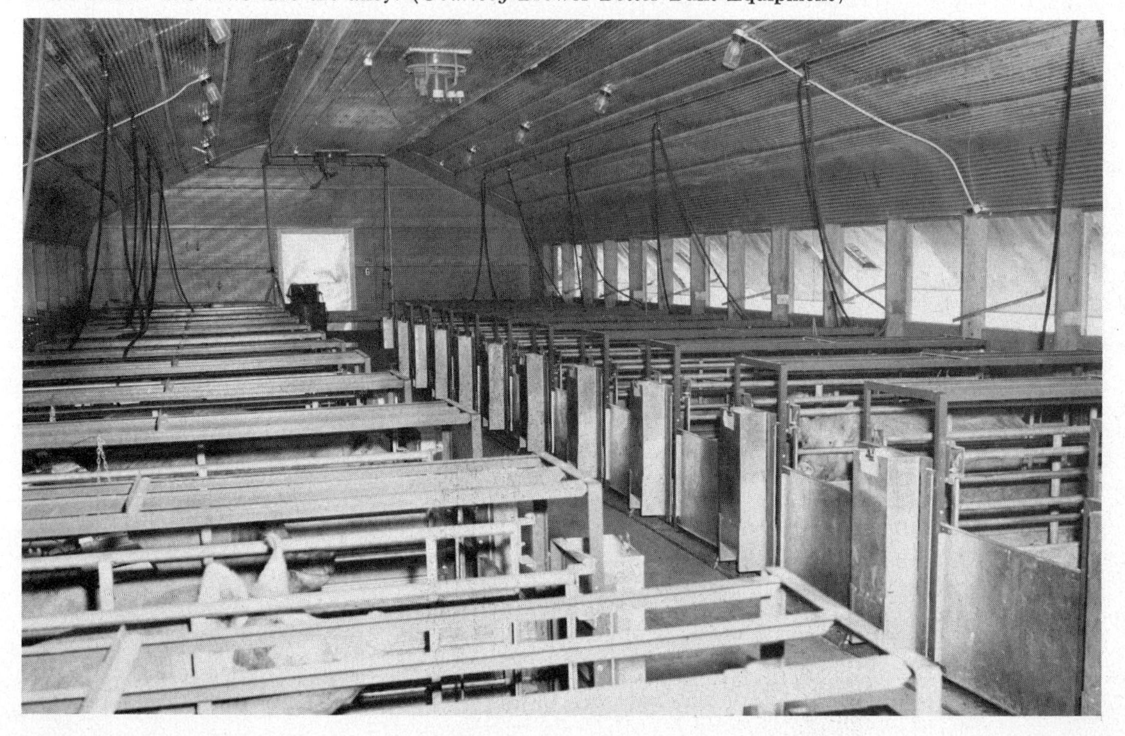

no feed for 12 hours after farrowing, unless she is nervous and appears hungry.

Rations at Farrowing Time. The following rations are recommended for sows at farrowing time:

Ration 1 (University of Illinois):

Ground shelled corn	300 lbs.
Ground oats	300
Wheat bran	300
Dry lot sow supplement	100
	1,000 lbs.

Ration 2 (Purdue University):

Ground oats	600 lbs.
Wheat bran	150
Alfalfa meal	150
Protein supplement	100
	1,000 lbs.

The sow may be fed the same ration after farrowing as before. It is usually best to feed about a half ration the first day and to increase the ration gradually until she is on full feed. Heavy milking sows should be hand-fed for the first week after farrowing.

FIGURE 8-8. Clipping the points of the upper and lower incisor teeth of a pig. (*Courtesy* U.S.D.A. Bureau of Animal Industry)

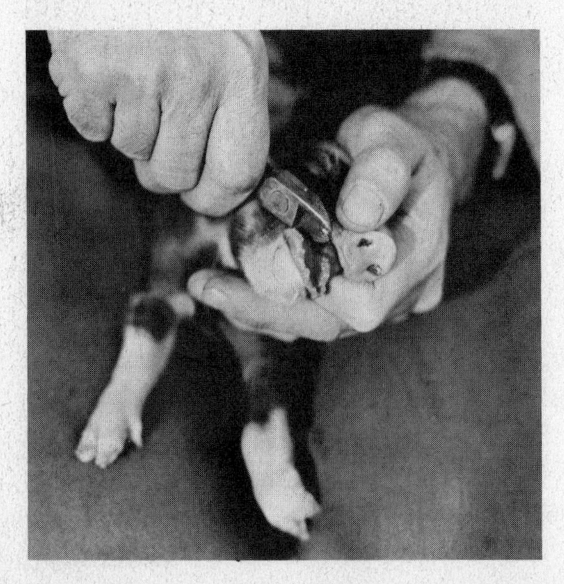

Illinois Ration 16 (Table 7-2) may be fed during the farrowing period in reduced amounts, but this ration may be too high in protein for heavy milking sows.

Assistance at Farrowing Time. Although most sows do not need assistance at farrowing time, it is a good policy to be on hand. Sows become nervous as they approach farrowing time and may pace the pen and scrape up bedding materials. Most sows farrow within 24 hours after milk develops in the nipples, but some sows have milk in the nipples for two days or longer before farrowing.

If the pen is cold at farrowing time you may need to dry the pigs and place them under the brooder or heat lamp as they are farrowed. If the pen is not equipped with a brooder or heat lamp, you can keep the pigs warm by putting them in a box or basket and taking them to the house or by placing heated bricks or a jug of warm water in the basket. The pigs should be returned to the sow as soon as possible to suckle. Pigs normally suckle every two or three hours. If pigs are removed from the sow, they should be returned every two or three hours for nourishment. Some pigs may not show interest in nursing, even when they are a day or two old. These pigs may respond to a hand-fed solution of half sirup and half water or to an injection of 5 to 10 cubic centimeters of a 25-percent glucose solution.

A good hogman will have worked with his sows so that they will not be nervous when he is present at farrowing time. If they are nervous, it is best to leave them alone or to work very quietly.

Generally speaking, when sows have difficulty farrowing, it is best to have a veterinarian render assistance. Many instruments are on the market for use in helping sows at farrowing time, but they must be used with caution, or serious injury to the sow may result.

A man with a small hand may straighten out a pig which a sow has been unable to deliver; however, under no conditions should one attempt to do this without using a rubber glove coated with vaseline. Brucellosis in hogs is readily transferable to man, and undulant fever may result.

Navel Cord. Sometimes the navel cord is long, and it is impossible for the pig to move about freely. The cord should be cut and the navel daubed with a tincture of iodine solution. This solution should be used on all pigs regardless of the length of the cord.

Needle Teeth. The clipping of needle teeth is a controversial issue. Some hogmen do not clip the teeth because they believe the mouth and gums may be injured, allowing infectious organisms to enter the body. The practice, if done properly, will not cause disease infections. It may even prevent them because, when fighting, pigs will not be able to inflict wounds about the face that would permit necrotic and rhinitic organisms to enter the body.

There are four of these tusklike teeth on each jaw. Clipping should be done with a pair of sharpened side cutters or a cuticle clipper in such a way that there is a clean, smooth break with no injury to the gums and no jagged edges. Many hog breeders have developed skill in clipping needle teeth and make it a regular practice.

Ear-notching. The notching of the pigs' ears at farrowing time is a universal practice among breeders of purebred hogs and among many commercial producers. It is the most practical method of identifying the pigs of a litter so that the productiveness of a sow can be determined and considered in the selection of breeding stock. All sow-testing programs begin with the ear-notching of the pigs.

In the fall, farmers often select the largest gilts to keep as brood sow replacements. The gilts may have been large because there were only three or four pigs raised in the litter. Unless the litters are marked, there is no way of knowing the kind of litter the gilts came from. The pigs should be marked as they are farrowed, with all the pigs of one litter getting the same ear notch.

Most producers number the litters in the order in which they are farrowed. The first litter farrowed is No. 1, the second litter farrowed is No. 2, and so forth. The notches may be made with a notching instrument or a scissors. It is also a good practice to weigh the pigs as they are being ear-notched.

Purebred and commercial swine breeders and swine research specialists may wish

FIGURE 8-9. Ear-notching a pig shortly after birth is recommended. (Hufnagle photo. Reprinted from *Successful Farming*)

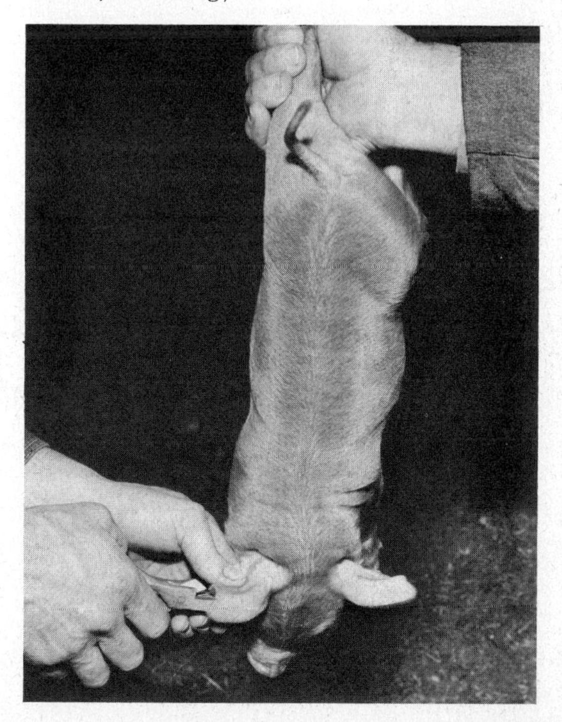

to mark each pig so that it can be identified by litter number and by number within the litter. This is accomplished by a dual system of ear-notching. One system is used to identify the litter while the other is used to identify the various pigs of the litter. Figure 8-10 illustrates the system adopted by the 1964 National Pork Industry Conference. This system must be used in recording Chester White, Hampshire, and Poland China swine. For registration purposes, Yorkshires must be ear-notched according to the Standard system shown in Figure 8-10 or the Yorkshire system shown in Figure 8-12.

MANAGEMENT DURING THE SUCKLING PERIOD

The weight of the pigs at weaning time is a good indication of the inherited growth ability of the pigs and of the ability of the sow to produce milk. It has been estimated that 25 to 30 percent of pigs that are farrowed fail to reach a weaning age of eight weeks. Although many of the pigs die because they are lain upon, injured, diseased, or chilled, we know that many of them also die because they do not receive sufficient milk and other feeds. The ration given the sow, the ration provided for the pigs, the pasture available, and the management given the sow and litter determine whether or not a high percentage of the pigs in the litter can be saved and have a heavy weight at weaning time.

Feeding the Sow

A sow, when fed a good ration, may produce 6 to 8 pounds of milk per day. When the size of the litter is small, 5 or 6 pounds

FIGURE 8-10. (*Left*) **Standard ear-notching system. Notches in the right ear indicate litter number; notches in left ear indicate pig number. The lower drawing shows notches for pig No. 7 of litter No. 11. FIGURE 8-11.** (*Right*) **Alternate ear-notching system. Notches in right ear indicate litter number; notches in left ear indicate pig number. Pig No. 8 of litter No. 14 is illustrated in the lower drawing. (Original drawings by Linda and Jerry Geisler)**

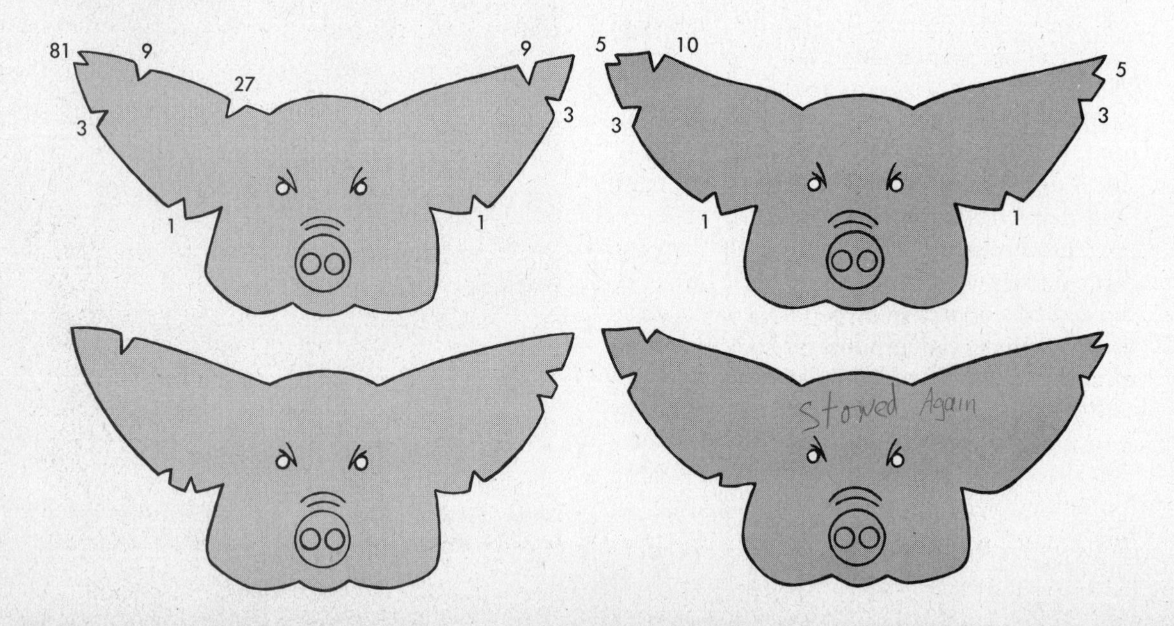

of milk may be sufficient. However, sows which are nursing from 10 to 12 pigs must produce the maximum of milk to meet the needs of the pigs satisfactorily. As we increase the size of the litters produced by our sows, we must also improve the rations fed to both the sows and their litters. A sow's milk contains about 81 percent water, nearly 6 percent fat, slightly more than 6 percent protein, about 6 percent lactose sugar, and about 1 percent ash or mineral. The sow must receive feeds containing these nutrients in sufficient amounts to produce the milk required by the litter.

The bulky, laxative ration recommended for the sow at farrowing time is usually continued for several days after farrowing. The amount is increased daily. Sometime during the second week after farrowing the sow should be on full feed. By that time the pigs require large amounts of milk, and usually changes must be made in the ration. Most sows by this time can, and should, be self-fed a ration containing about 15 percent protein.

The condition of the sows, the number of pigs being nursed, and the type of pasture available influence the amount and kind of ration which should be fed. The following rations are recommended for sows during the suckling, or lactation, period.

Iodized Casein. Feeding iodized casein to nursing sows during the first week after farrowing increases milk production. Iodized casein contains thyroxine, a natural hormone produced by animals. It should be fed at the rate of 100 milligrams per pound of total ration from the 110th day of gestation through the first weeks of lactation. An injection of oxytocin at farrowing time will achieve the same result.

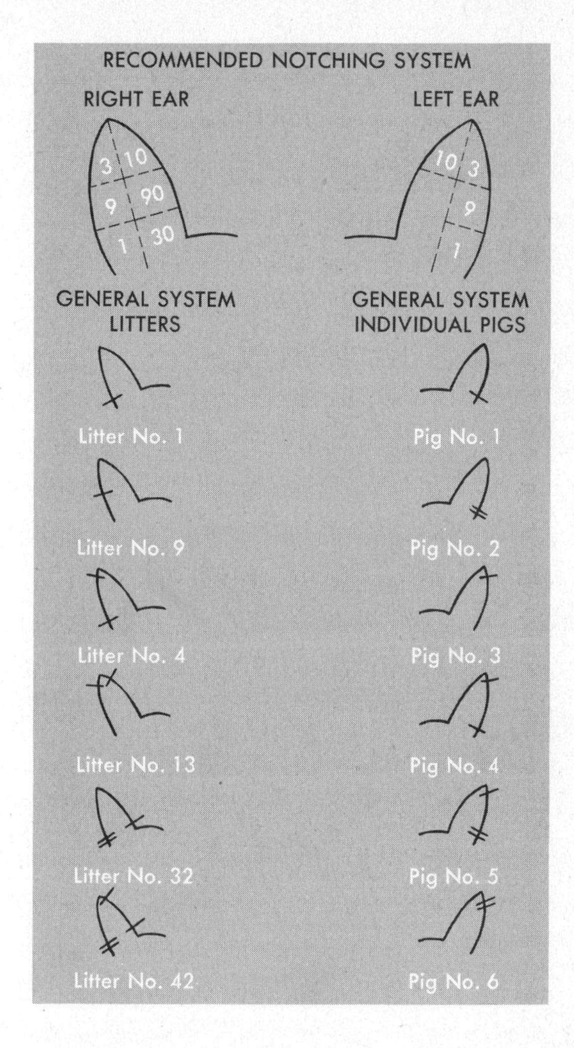

FIGURE 8-12. Ear-notching system recommended by the American Yorkshire Club, Inc. (*Courtesy American Yorkshire Club, Inc.*)

Creep-feeding Young Pigs

Pigs will begin to nibble at feeds when they are a few days old and can consume a considerable amount of feed by the time they are two or three weeks old. The milk production of the sow usually declines after the third week, and it is important that the little pigs be fed a palatable ration in ample amounts. This is especially true when sows are suckling large litters or when the sows are in poor condition.

The little pigs need a more palatable ration, higher in protein and in antibiotics

TABLE 8-1 LACTATION RATIONS[1]

Percent protein	Ingredient		1	2	3	4
8.9	Ground yellow corn[2]		1,507	1,407	1,407	1,387
44	Solv. soybean meal[3]		400	400	400	250
8	Dried beet pulp		..	100	..	..
14.5	Wheat bran		..	..	100	100
17	Dehydrated alfalfa meal		..	..	..	100
50	Meat and bone meal		..	..	..	100
	Calcium carbonate (38% Ca)		15	15	15	5
	Dicalcium phosphate (26% Ca, 18.5% P)		45	45	45	25
	Iodized salt		10	10	10	10
	Trace mineral premix		3	3	3	3
	Vitamin premix		20	20	20	20
	Feed additives[4]		..	..	..	..
	Total		2,000	2,000	2,000	2,000
Calculated analysis						
	Protein	%	15.51	15.46	15.79	15.75
	Calcium	%	0.93	0.96	0.93	0.93
	Phosphorus	%	0.72	0.72	0.77	0.76
	Lysine	%	0.81	0.83	0.83	0.77
	Methionine	%	0.26	0.25	0.26	0.26
	Cystine	%	0.25	0.25	0.26	0.24
	Tryptophan	%	0.18	0.18	0.19	0.17
	Metabolizable energy	Cal./lb.	1,292	1,274	1,266	1,229

Feeding Directions

[1] These rations may be full fed from a few days after farrowing until the baby pigs are weaned.

[2] Ground oats can replace corn up to 15 percent of the total ration. Ground milo, wheat or barley can replace the corn.

[3] If 48.5 percent soybean meal is used instead of 44 percent soybean meal, use 50 lbs. less soybean meal and 50 lbs. more corn. If whole cooked beans are used in place of 44 percent soybean meal, use 100 to 150 lbs. more soybeans and 100 to 150 less corn.

[4] A high level of feed additive (100 to 300 gm/ton) may be beneficial 2 to 3 weeks before farrowing until 7 to 10 days after farrowing, but will be of little benefit thereafter.

Source: Iowa State University

and lower in fiber content, than their mothers. As a result, the feeding of milk replacer or starter rations is recommended. These rations should be fed in a trough or feeder in a pen separate from that of the mother. The gates may be adjusted so that the pigs can get in but the sows cannot.

The use of milk replacer and starter rations is very important if the pigs are to be weaned from the sows at an early age.

Milk Replacer and Starter Rations. There have been a number of important developments in the creep feeding of pigs. A common practice among hog producers at one time was to feed young suckling pigs rations of rolled or hulled oats, cracked wheat, or cracked corn.

Much research was done in developing synthetic milk rations in the hope that pigs could be weaned from the sows when only

TABLE 8-2 LACTATION RATION

Ingredients	Pounds	Composition	
Milo-western yellow	1607.8	Protein %	14.0
Soy meal (50%)	208.6	Calcium %	0.60
Dehydrated alfalfa meal (17%)	100.0	Phosphorus %	0.50
Dry skim milk	—	Histidine %	0.36
Tankage	50.0	Isoleucine %	0.70
Whey	—	Lysine %	0.64
Molasses (wet)	—	Phenylalanine %	0.64
Dicalcium phosphate	12.4	Threonine %	0.46
Calcium carbonate	7.3	Tryptophan %	0.16
T.M. Salt	10.0	Vitamin A. I.U./lb.	4000.0
Vit. min-antibiotic (1) supp.	4.9	Niacin (mg./lb.)	23.8
		Pantothenic acid (mg./lb.)	7.7
Total	2000.0	Riboflavin (mg./lb.)	2.2
		Vitamin B_{12} (mcg./lb.)	5.0

Source: Oklahoma State University

a few days old. While excellent feeds were developed, they required more exacting procedures than the average farmer was willing to use or had the time for, and the cost of the feeding program was high.

Swine nutritionists have concentrated their efforts on building dry pig starters that are more palatable and better fortified with vitamins, minerals, and antibiotics than creep rations previously developed. A number of excellent rations are now available.

Baby-Pig Feeding Program. Shown in Table 8-3 is the Iowa State University life-cycle swine-feeding program, which utilizes three rations in starting and growing out pigs.

Prestarter Period. Shown in Table 8-4 are formulas of milk replacer rations. No more than 10 to 15 pigs should be penned together and the pigs should be of the same size. Approximately 3 to 4 square feet of floor space per pig should be provided if the floors are partly slatted, and floor drafts should be avoided. Solid pen walls should be used. Heat lamps or brooders should be provided and the bedding should be kept clean and dry. Young pigs will need plenty of fresh water. Automatic fountains are preferred. They should be clean and sanitary at all times.

FIGURE 8-13. A handmade farrowing stall with special feeder for the pigs. (*Courtesy* Eli Lilly & Company)

Starting Period. As the pigs reach 10 to 12 pounds in weight, the ration is shifted from the milk replacer to a highly fortified starter feed which has been pelleted. The formulas of starter rations for weaned and unweaned pigs are shown in Table 8-5.

TABLE 8-3 LIFE CYCLE SWINE FEEDING PROGRAM

Stage of cycle	Length of feeding program	Season	Complete ration % Protein	Complete ration Lbs./day	Lbs. corn or grain/day	Lbs. supplement/day
Boars	From time purchased at 5-6 months of age	Summer	13-16	4-6	3-5	0.8-1.0
		Winter	12-14	5-7	4-6	
		Increase intake 1-2 lbs. during heavy breeding season.				
Gilts						
Pregestation	From time selected at 5-6 months until breeding at 7-9 months of age	Summer	13-16	4-6	3-5	0.8-1.0
		Winter	12-14	5-7	4-6	
Flushing and breeding	For 3 weeks prior to breeding (do not continue after breeding)		12-14	6-9	5-8	1.0-1.2
		Increase corn intake approximately 2 lbs./day.				
Gestation		Summer	13-16	4-5	3-4	0.8-1.0
		Winter	13-14	5-6	4-5	
		Increase intake 1-2 lbs. last 3-5 weeks of gestation if gilts too thin.				
Lactation	Wean at 3-5 weeks after farrowing		14-16	10-14	Full feed complete ration.	
Sows						
Breeding and gestation	Breed back at first heat period after weaning (flushing is not beneficial with sows)	Summer	15-18	3-4	2-3	0.8-1.0
		Winter	13-16	4-5	3-4	
		Increase intake 1-2 lbs. last 3-5 weeks of gestation if gilts too thin.				
Lactation	Wean 3-5 weeks after farrowing		14-16	12-16	Full feed complete ration.	
Pigs						
Milk replacer	Use only if weaned before 3 weeks and feed until pig weighs 12 lbs.		20-24		Full feed complete ration.	
Starter	Use as a creep feed and continue after weaning until pigs are 8 weeks of age or 40 lbs.		18-20		Full feed complete ration.	
Grower-finisher	From 8 weeks of age or 40 lbs. until market weight		13-16	Full feed (may limit feed after 125 lbs.)	With free choice, corn consumption varies depending on weight of pig. Supplement intake should be approx. 0.75 lbs. regardless of weight.	

Source: Iowa State University

TABLE 8-4 MILK REPLACERS[1]
(For baby pigs before 3 weeks of age)

Percent protein	Ingredient		1	2	3	4
8.9	Ground yellow corn		507	652	551	530
48.5	Solv. soybean meal		450	500	600	500
15	Rolled oat groats		..	..	..	200
33	Dried skim milk		400	400	200	200
12	Dried whey		400	200	400	400
31	Fish solubles		50	50	50	..
	Sugar		100	100	100	100
	Stabilized animal fat		50	50	50	20
	Calcium carbonate (38% Ca)		10	10	10	10
	Dicalcium phosphate (26% Ca, 18.5% P)		5	10	10	10
	Salt		5	5	5	5
	Trace mineral premix		3	3	3	3
	Vitamin premix		20	20	20	20
	DL-methionine		..	..	1	2
	Feed additives[2] (gm/ton)		100-300	100-300	100-300	100-300
	Total		2,000	2,000	2,000	2,000
Calculated analysis						
	Protein	%	22.95	23.61	23.48	21.69
	Calcium	%	0.73	0.72	0.69	0.69
	Phosphorus	%	0.61	0.62	0.61	0.61
	Lysine	%	1.57	1.58	1.56	1.40
	Methionine	%	0.42	0.43	0.44	0.45
	Cystine	%	0.34	0.35	0.36	0.34
	Tryptophan	%	0.29	0.30	0.30	0.28
	Metabolizable energy	Cal./lb.	1,439	1,429	1,418	1,405

Feeding Directions

[1] The milk replacer ration is normally fed in only limited amounts. It should be used for pigs weaned prior to 3 weeks of age until they reach approximately 12 lbs. Then they can be switched to a starter ration. It is a good ration to feed orphan pigs when the sow dies, extreme disease outbreak (TGE) occurs, or the sow is agalactic.

[2] The feed additive may be part of the vitamin premix, or if a separate premix, it should replace an equal amount of corn.

Source: Iowa State University

These rations are to be used either as creep rations or the only ration for pigs weaned early (3 weeks or older). They should be fed until the pigs weigh 40 pounds.

Pigs like pellets better than meals or crumbled feeds. They do not readily eat dusty or finely ground feeds. It has been found that pigs will consume nearly twice as much feed in pellet form as in the form of meal.

While being fed the starter ration, the pigs should not be crowded. At least 4 inches of feeder space should be provided for every two pigs. An adequate supply of fresh water should be available at all times. Partial slat floors are recommended.

TABLE 8-5 PIG STARTER RATIONS[1]

Percent protein	Ingredient	1	2	3	4	5
8.9	Ground yellow corn	966	1,191	1,112	1,062	862
48.5	Solv. soybean meal	450	450	500	..	..
44	Solv. soybean meal	..	..	..	550	..
37	Soybeans, whole cooked	..	..	..	..	750
33	Dried skim milk	50	50	..	..	..
12	Dried whey	300	200	300	300	300
31	Fish solubles	50	..	..	..	..
	Sugar	100	25	25	25	25
	Stabilized animal fat	20	20	..	..	..
	Calcium carbonate (38% Ca)	10	10	10	10	10
	Dicalcium phosphate (26% Ca, 18.5% P)	25	25	25	25	25
	Iodized salt	5	5	5	5	5
	Trace mineral premix	3	3	3	3	3
	Vitamin premix	20	20	20	20	20
	DL-methionine[2]	1	1	..	..	..
	Feed additives[3] (gm/ton)	100-300	100-300	100-300	100-300	100-300
	Total	2,000	2,000	2,000	2,000	2,000

Calculated analysis

		1	2	3	4	5
Protein	%	18.66	18.28	18.87	18.62	19.51
Calcium	%	0.73	0.69	0.71	0.72	0.75
Phosphorus	%	0.64	0.62	0.64	0.63	0.66
Lysine	%	1.12	1.06	1.12	1.11	1.16
Methionine	%	0.36	0.35	0.30	0.30	0.30
Cystine	%	0.30	0.30	0.31	0.31	0.31
Tryptophan	%	0.22	0.22	0.23	0.23	0.26
Metabolizable energy	Cal./lb.	1,359	1,375	1,352	1,309	1,422

Feeding Directions

[1] The pig starter ration should be used as a creep ration before weaning and fed after weaning until the pigs reach approximately 40 lbs. Then the pigs can be switched to a grower-finisher ration.

[2] Rations 3, 4, and 5 are borderline in methionine content. The addition of 1 lb. of DL-methionine to these rations may improve performance and feed efficiency slightly, although pigs will perform very satisfactorily on these rations without the DL-methionine addition.

[2] The feed additive may be part of the vitamin premix, or if a separate premix, it should replace an equal amount of corn.

Source: Iowa State University

FIGURE 8-14. (*Left*) **Pigs like pelleted feeds.** (*Courtesy* **Eli Lilly & Company**)
FIGURE 8-15. (*Right*) **Sod is provided to prevent anemia.** (*Courtesy* **Rath Packing Company**)

PRE-WEANING MANAGEMENT

Anemia Control

Many small pigs die each year from anemia. Anemic pigs usually show a loss of appetite, become weak, and sometimes have a swollen condition around the head and shoulder. They have trouble breathing and may "thump," making a dull sound with each breath.

Anemia is a blood condition which is caused by a lack of copper and iron. The condition can be prevented by giving the pigs exercise and an opportunity to pick up copper and iron from the soil. If the pigs are confined and cannot get exercise, sod can be brought to them. It can be taken from a disease-free field or from the roadside.

Another method of controlling anemia is to use drugs obtained from the drugstore or veterinarian. Iron and copper pills, which can first be given each pig when it is about five days old and weekly thereafter, are available. Pigs may be given an injection of a solution of copperas (ferrous sulfate), which can be purchased at any drugstore. A recommended procedure is to supply 100 to 150 milligrams of iron per pig by injecting an iron-dextran compound in the fleshy top part of the neck when the pigs are two or three days old. Injection in the neck prevents lameness, running out of injected material, and staining in the ham. A second injection after the third week may be necessary for pigs weaned late.

Castration

Pigs should be castrated when they are two to three weeks old. They recover from the operation very quickly at this age. Under no conditions should the pigs be castrated, wormed, vaccinated, and weaned at the same time. These operations should be spaced at two-week intervals.

Unless heat is provided by brooders or heat lamps, pigs should be castrated when the weather is warm. The hog house and pen should be clean, dry, and well bedded.

Two persons are usually needed in castrating—one to hold the pig while the other operates. The pig may be held by a front and a hind leg on opposite sides with its back on the floor, or by the hind legs with its head and shoulders between the assistant's knees. The scrotum should be washed with a mild antiseptic solution or with soap and water. The sharp knife used in castrating and the operator's hands or rubber gloves should be carefully disinfected.

The incision may be made over each testicle, or in-between the testicles, parallel to the middle line of the body. The incision should pass through the skin near the top of the testicle and through the testicle covering. The testicle is pulled slowly through the incision, and the attachments are separated with little bleeding. Both testicles are removed in this manner.

Care should be taken to make the incision fairly long and low, so that it will drain well. As much of the cord as possible should be removed. Usually there is no need to apply disinfectant or healing oil after the operation.

When mature boars are castrated, it is usually necessary to throw them and tie their legs. The operation is similar to that on pigs, but there is more danger of bleeding.

Sometimes pigs are ridglings or "originals." They have only one testicle in the scrotum. It is often best to turn these pigs, as well as any that are ruptured, over to a veterinarian.

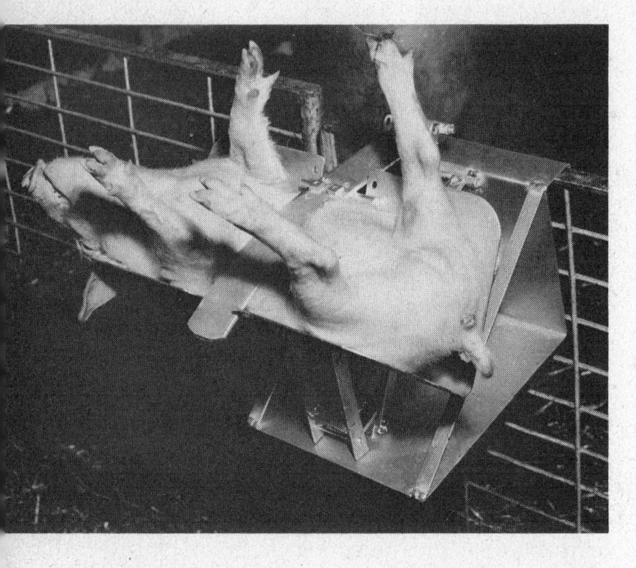

FIGURE 8-16. (*Above*) **The "Pig Grip" is a handy piece of equipment for castrating and vaccinating pigs.** (*Courtesy* **Caswell Manufacturing Company)**
FIGURE 8-17. (*Below*) **A healthy Yorkshire litter. Greatest benefits in feeding antibiotics are derived when they are fed to unthrifty pigs.** (*Courtesy* **American Yorkshire Club, Inc.)**

Vaccination

In some hog-producing areas it is advisable to vaccinate pigs for leptospirosis and erysipelas. These diseases are discussed in Chapter 9. Usually only healthy pigs should be vaccinated and the two vaccinations should be done separately about two weeks apart. Some veterinarians begin the vaccinations when the pigs are six or seven weeks of age; others wait until the pigs are eight to ten weeks of age.

Antibiotics

Antibiotics increase the growth rate of growing-finishing pigs by 5 to 10 percent and increase feed efficiency from 3 to 5 percent. They aid in the control of nonspecific scours, in coping with stress situations of weaning, and in confinement feeding.

The practical level of feeding antibiotics in pig starters seems to be 20 to 40 grams of antibiotics per ton of total ration, or 10 to 20 milligrams per pound. (A pound equals 453.6 grams, and a gram equals 1,000 milligrams.) Early experimental work indicated that Aureomycin, Terramycin, and penicillin were equally effective in increasing the rate of gain and in protecting pigs from disease.

Recent research has shown that mixtures of penicillin and streptomycin, ASP-250, CSP-250, and new antibiotics like tylosin may be more effective than older antibiotics used singly. Forty grams of mixture per ton is recommended in 16-percent grower rations, and 10 grams of antibiotics is recommended for finishing rations.

Aureo S. D. 250 and CSP-250 are trade names for three-way combinations of chlortetracycline, sulfathiazole, and penicillin. These combinations are especially effective in feeding pigs up to 30 or 40 pounds in weight. They are more expensive than single antibiotics. Tylosin has shown up exceedingly well when fed both alone and in combination.

Pasture

Some hog producers move the sows and litters to clean legume pasture as soon as possible after the pigs are a week or two old. In cold weather it may be necessary to postpone moving the litter. Pigs can be raised efficiently while confined, but care must be taken to provide sanitary quarters and a ration which contains the minerals, vitamins, and proteins that are provided by pasture crops.

Care should also be taken in moving pigs to pasture so that no infections are picked up enroute. It is better to haul the sow and her litter than to drive them. On some farms concrete runways have been provided for this purpose. They can be cleaned easily and permit the pigs to travel back and forth from the farrowing house to the pasture without picking up infections in old lots.

FIGURE 8-18. These Hampshire sows and litters are making good use of pasture. (*Courtesy* Hampshire Swine Registry)

Weaning

With the use of fortified pig starter rations it is possible to wean pigs before the usual age of 8 weeks. Following are weaning pointers:

1. **Pigs less than 10 days of age should not be weaned from the sow unless adequate equipment, heat, labor, and highly fortified rations are available.**
2. **The tendency is to wean pigs when they are 3 to 5 weeks old.**
3. **It is a good idea to wean pigs by weight rather than by age. Wean the larger pigs from the litter first.**
4. **Avoid making any changes in the ration at weaning time.**
5. **When the entire litter is weaned at one time, the sow should be moved from the pigs, rather than the pigs from the sow.**
6. **Make certain that feed and water are fresh and plentiful.**
7. **Comfortable quarters are essential. Warm, dry, well ventilated quarters are necessary during winter months. Shade and cool, well ventilated quarters are important in hot weather.**
8. **It is best to separate pigs by weights and keep not more than about 20 pigs of the same size in a pen at weaning time.**

FEEDING AND MANAGEMENT FROM WEANING TO MARKET

Approximately one third of the total cost of producing a 220-pound market hog is involved in raising the pig to weaning age. It takes the remaining two thirds of the total cost to grow out the pig from weaning time until it is ready for market. From the production standpoint, the period from weaning to market is an important one, since more than twice as much capital is involved in this period as is involved in producing weaning pigs. Serious losses can occur during this period as a result of a disease outbreak or poor feeding and management.

The efficient hog raiser plans his production program carefully. He provides adequate rations and feeds them in the proper amounts. He uses the methods of feeding best suited to his farm. He provides pasture of good quality in the desired amount, if he uses pasture. He is efficient in housing his hogs and in controlling diseases and parasites. The profit from the enterprise can be judged by the cost of production and by the selling price. Usually he can influence the cost of production more easily than he can the selling price.

Worming

Most farms are contaminated with roundworms, which are injurious to swine. This parasite is discussed fully in Chapter 9, but it is mentioned here because most pigs must be wormed when they are placed on pasture or moved to finishing pens.

Roundworms may be controlled by the use of rotated pastures, but it is never certain that the mature animals in the herd are free from the parasites. The only safe procedure is to treat the pigs for worms even when the clean-ground system of rotated pastures is used.

Materials for Worming. Many drugs and other materials for worming purposes are sold commercially. Among the best wormers are dichlorvos (Atgard), piperazine compounds, hygromycin B, and levamisole HCI (Tramisol). The wormers are safe and convenient, and control more kinds of worms than any other wormers on the market. The instructions of the manufacturer should be followed.

Pasture Versus Confinement Feeding

Each farmer must evaluate his facilities and resources to decide whether he should grow his pigs on pasture or in confinement. Some farmers prefer to use pastures in their production programs. Following are the advantages of the two systems.

Advantages of Pasture

1. **Pork can be produced on clean hog pastures with 15 to 30 percent less of concentrates than is required in confinement feeding.**
2. **The protein-supplement feed bill can be cut by one-third to one-half when good legume pasture is used.**
3. **There is less chance for losses from diseases and parasites when pigs are on "clean" legume pasture.**
4. **The equipment necessary to feed and manage hogs under sanitary conditions is less when they are on pasture than when they are confined.**
5. **There is no bedding problem during the summer months.**
6. **Hogs on pasture will spread their own manure. Less labor is involved.**
7. **Most tenants must rely upon use of pastures in growing out hogs. Few landowners will provide desirable facilities for confinement-feeding programs.**
8. **It takes less know-how to grow pigs on pasture than in confinement.**
9. **Pasture-fed animals are usually better breeders.**

Advantages of Confinement Feeding

1. **It may not be economical to use land valued at $500 an acre or more as hog pasture.**
2. **More rapid gains can be produced when pigs are in confinement, if adequate rations and good management are provided.**
3. **Confinement facilities can be used during more months of the year than can pasture facilities.**
4. **There is usually less labor required in hauling and handling feed when pigs are confined.**
6. **Fence problems will be minor, compared to pasture feeding.**
7. **The confinement method is best for growing out a large number of hogs on a small farm.**

Kinds of Pasture Crops

A number of pasture crops have been tested and their value for swine has been determined. Shown in Table 8-6 is a summary of tests conducted by the Missouri Agricultural Experiment Station.

According to the Missouri tests, alfalfa proved superior to all other crops in pounds of pork produced per acre and in the value of the pork produced. Red clover and rape ranked second and third.

TABLE 8-6 VALUE OF VARIOUS PASTURE CROPS FOR SWINE †

Crop	Number of days Pastured	Number of Hogs per Acre	Pounds of Pork per Acre of Forage	Value per Acre with Pork @ $20 per Cwt.
Alfalfa	163	10	592	$118.40
Red clover	130	12	449	89.80
Rape (Dwarf Essex)	82	23	395	79.00
Sorghums	87	15	275	55.00
Bluegrass	136	12	274	54.80
Soybeans	25	17	175	35.00
Cowpeas	32	13	149	29.80

† The rations of the pigs were limited to ½ to ⅔ of a full feed.

Source: University of Missouri

FIGURE 8-19. (*Above*) **More than 55 percent of the nation's pig crop is raised on pasture. Each sow on the farm pictured had an individual house.** (*Courtesy* **American Yorkshire Club, Inc.**) **FIGURE 8-20.** (*Below*) **These purebred Poland China pigs are grown out on excellent rotated pasture.** (*Courtesy* **Poland China Record Association**)

Birdsfoot trefoil ranks with alfalfa in palatability, in feed value, and in ability to stand drouth. It does well seeded with bluegrass as a permanent pasture crop. Often two to three years are required to get a stand, but it will last for many years. The usual rate of seeding is from 4 to 5 pounds of inoculated seed per acre.

The nonleguminous crops—oats, rape, bromegrass, Sudan grass, and rye—should be used only as emergency crops when the legumes winterkill or when new seedlings do not survive dry weather. Rye provides a very early pasture in the spring but dries up in the early summer. Rape does not provide an early pasture but does make an excellent late pasture.

Ladino clover, a comparatively new clover which resembles white Dutch clover, is rated in several states as the best pasture for swine. In Purdue tests, it was found superior to alfalfa as a pasture for full-fed hogs, as indicated by a more rapid gain and a 40-percent saving in protein supplement. The feed cost of 100 pounds of gain was $9.35 on the ladino pasture and $10.42 on the alfalfa.

Pasture Management

The management given the pastures and the hogs which graze them may greatly influence the effectiveness of the pastures in hog production programs. Many good pastures are not managed properly.

Rotated Pastures. In order to prevent disease and parasite outbreaks, it is desirable to provide a new pasture area each year. Farmers who make new seedings each year have a new pasture area for each crop of pigs.

Farmers who use permanent or semipermanent pastures accomplish the same result by dividing the pasture area into three or four parts, and then using the divided areas in rotation.

Pigs per Acre. The number of pigs or hogs on an acre depends on the kind of pasture, fertility of the soil, and weather conditions, and on whether a full or limited feeding program is followed. An acre of legume or legume-grass pasture should provide adequate pasture for 20 to 30 growing-fattening pigs on full feed. Frequent clipping of pastures during the summer months is recommended if the pigs do not keep the growth down.

Location of Feeders and Waterers. Considerable labor may be saved by locating the feeders and waterers near the fence so that they may be filled by power equipment without driving into the field. Large feeders which are protected from the weather are recommended.

Tests conducted at the South Dakota Station indicate that the feeders and waterers should be located close together. When they were located more than 300 feet apart the average daily gain decreased.

Ringing Pigs. It is sometimes necessary to nose-ring pigs to keep them from rooting the sod in the pasture. The feeding of balanced rations containing animal proteins and minerals tends to reduce rooting. Care should be taken in ringing pigs to avoid injury to the bone structure of the nose.

Feeding Pigs from Weaning to Market

Most rapid gains are usually brought about by full-feeding pigs, both those in confinement and those with access to good legume pasture. When feeds are scarce or high in price, it is sometimes desirable to feed a limited ration and make more effective use of the pasture. Each hog raiser must decide which of the two systems will be more profitable on his own farm.

Methods of Feeding. Pigs may be hand fed or self-fed. Self-feeding saves labor and usually produces more rapid and economical gains than does hand feeding.

FIGURE 8-21. (*Above*) **Many pigs in the South are grown out on pasture along wooded areas. Bulk delivery trucks auger feed directly into the feeders.** (*Courtesy* **Allied Mills, Inc.**) **FIGURE 8-22.** (*Below*) **A walk-in type of feeder.** (**Plan No. 10.** *Courtesy* **Doane Agricultural Service, Inc.**)

Self-feeding promotes sanitation and provides large feed storage space. Self-feeders, however, are expensive and are sometimes neglected by irresponsible caretakers. Sometimes the use of the self-feeders decreases the effect of good pasture because the appetites of the pigs are satisfied while they are at the feeders. Some farmers prefer to hand-feed their pigs. They can watch them more closely and can give types of feeds which cannot be self-fed, such as ear corn.

It is possible to get pigs to consume more feed by mixing it with water, providing it is fed as a wet mash, rather than as a thin slop. Paste or wet mash feeding encourages the pigs to consume more water. Mechanical systems for paste feeding are available.

If pigs are to be hand-fed, give only the amount of feed that they will clean up in an hour or two. This is especially important if paste or wet mash feeding is practiced. Stale feed in the trough is not only wasteful but also unappetizing, and it decreases the amount of feed the pigs will consume.

Self-feeders. A good self-feeder should make available to the pigs a supply of clean, dry feed at all times. It should be solidly constructed and should protect the feed from wind or rain. It should be large enough to store sufficient amounts of feed to last the pigs for several days. It should have an adjustable throat or opening and some type of agitator, so that various types of feeds will feed down in the desired amounts. The trough should be constructed in such a manner that little feed will be wasted. Doors or lids over the troughs are desirable. If the ration is fed as a complete mixed ration, only one feeder will be needed. If feeds such as shelled corn, ground oats, and supplement are fed separately, separate feeders or separate compartments in a large feeder will be required.

The use of large feeders that will handle 250 bushels of shelled corn and 1,500 pounds of supplement will cut down the cost of feeding equipment and will save labor.

There should be 1 foot of feeder space for every three to four hogs. If supplement is fed separately, 15 to 25 percent of the total feeder space should be reserved for supplement. Heavier hogs require less supplement and less feeder space.

Automated systems and equipment are available and recommended, but users of automated equipment should have sufficient volume of production continuously during the year to justify the investment.

Rations. Farm grains should make up the main part of the ration. Corn, when available, produces the most rapid and economical gains. Barley, wheat, oats, and grain sorghums may be fed, but should be ground. These feeds produce slower gains but may produce better carcasses.

A common practice in the Corn Belt is to feed rations made up largely of corn supplemented with tankage and soybean oil meal, or with a mixed protein feed. From 10 to 30 percent of the ration may be made up of grains other than corn.

In areas where corn is not obtainable, it is possible to build good rations by using ground barley or ground grain sorghums.

Since the grains and pasture cannot provide all the proteins, vitamins, and minerals necessary, we must supply sufficient feeds to meet the nutritional needs of the pigs. Protein and mineral supplements fortified with vitamins and antibiotics must be added.

Proteins. It has been pointed out that proteins are made up of amino acids and that 10 of these amino acids are essential for animal health. Corn, which is the basic grain fed to hogs in much of the Corn Belt, is low in two of these essential amino acids, lysine and tryptophan. Therefore, it is necessary to supplement corn with protein feeds rich in these two amino acids. Most proteins of plant origin are deficient in lysine. The exception is soybean oil meal, which is a good source of lysine. Animal proteins, such as tankage, meat scraps, and fish meal, are also good sources. A good protein supplement, then, should contain soybean oil meal or a meat-origin protein, or both.

Plant proteins, in general, are good sources of tryptophan, whereas meat and bone scraps, if fed along with corn, will not provide an adequate amount of this amino acid for maximum growth. Most mixed protein and mineral supplements contain several protein feeds of plant origin which provide an adequate supply of tryptophan.

Value of Protein Supplements. Numerous tests have been conducted to determine the value of protein supplements in feeding pigs. In general, protein feeds are more important for pigs weighing 45 to 85 pounds than for those weighing 100 pounds and over. Light pigs fed corn alone will gain less than ⅓ pound per day. The same pigs fed a protein supplement will gain more than a pound a day. Heavier pigs fed corn alone will require about 200 pounds more feed to produce 100 pounds of gain than will pigs fed corn and a protein supplement. Tests show that 100 pounds of tankage may save from 500 to 600 pounds of corn.

Protein feeds are more important in feeding pigs in confinement than in feeding those on pasture. The use of good legume pastures can greatly reduce the amount of protein concentrates required in the ration.

Until the discovery of vitamin B_{12}, it was necessary to provide rations containing 18 to 20 percent protein for weanling (50-pound) pigs. By adding B_{12}, we can reduce this to 15 to 16 percent, providing the ration is well balanced. The amount of protein necessary for heavier hogs can be reduced proportionally by the addition of this vitamin. Pigs weighing 50 to 125 pounds require 13- to 14-percent protein rations; heavier hogs, weighing 125 to 200 pounds, require 12 to 13 percent if the ràtions are balanced and vitamin B_{12} has been added.

Growing-finishing pigs may be fed the complete mixed rations given in Table 8-11 or may be fed a ration of grain, protein supplement, minerals, and other feed additives given separately. Large producers tend to feed complete mixed rations.

Table 8-7 gives examples of recommended protein and mineral supplements for pigs on pasture and in confinement. Formulas for complete ground and mixed rations using 35- and 40-percent protein supplements are presented in Table 8-8.

Antibiotics, arsenicals, and minerals are discussed later in this chapter. It is suggested that these materials be mixed with the protein supplement or in a complete mixed ration, and be self-fed. A hog will balance his own ration if given an opportunity to do so. For pigs weighing 75 to 150 pounds, 83

FIGURE 8-23. These pigs are fed a paste or gruel ration fortified with antibiotics. Note the partial slatted floor. (*Courtesy* Big Dutchman Co.)

TABLE 8-7 40-PERCENT SUPPLEMENTS [1, 2]

Percent protein	Ingredients	1	2	3	4
48.5	Solv. soybean meal	1,628	.. (lbs.)	1,238	1,458
44	Solv. soybean meal	..	1,223	..	..
17	Dehydrated alfalfa meal	..	..	100	..
50	Meat and bone meal [3]	..	550	400	..
61	Fish meal, menhaden	..	..	..	200
	Calcium carbonate (38% Ca)	90	45	55	80
	Dicalcium phosphate (26% Ca, 18.5% P)	160	60	85	140
	Iodized salt	50	50	50	50
	Trace mineral premix	12	12	12	12
	Vitamin premix	60	60	60	60
	Feed additives [4]	..	..	..	..
	Total	2,000	2,000	2,000	2,000

Calculated analysis

		1	2	3	4
Protein	%	39.48	40.66	40.87	41.46
Calcium	%	3.95	4.02	3.96	3.98
Phosphorus	%	2.01	2.05	2.02	2.05
Salt added	%	2.50	2.50	2.50	2.50
Lysine	%	2.67	2.55	2.60	2.86
Methionine	%	0.55	0.56	0.56	0.66
Cystine	%	0.59	0.48	0.52	0.58
Tryptophan	%	0.55	0.46	0.49	0.56
Metabolizable energy	Cal./lb.	1,123	968	1,047	1,122

Feeding Directions

[1] These supplements can be used to make growing-finishing, gestation or lactation rations.

[2] Supplements 2 and 3 may be self-fed free choice with shelled corn.

[3] The meat and bone meal was considered to have 8.10 percent calcium and 4.10 percent phosphorus. If meat and bone meal with a higher concentration of calcium and phosphorus is used, the amount of dicalcium phosphate should be reduced accordingly.

[4] The level of feed additives will depend on the type of ration in which the supplement is going to be used, but should be 4 to 6 times higher than desired in the complete ration.

Source: Iowa State University

pounds of corn and 17 pounds of 40-percent protein supplement will make the recommended 14-percent protein ration. For pigs weighing more than 150 pounds, 87 pounds of corn and 13 pounds of 40-percent supplement will make a 13-percent ration. The amounts of protein supplement and corn needed to make up rations of different levels of protein are shown in Table 8-9.

Antibiotics. The feeding of antibiotics to pigs from weaning time to market weight will normally result in an increase in gain of 5 to 20 percent, will save 20 pounds of feed in producing the first 100 pounds of gain, will reduce the problem of scours in young pigs, will result in more uniform pigs in the litter, and will have the added advantage of almost eliminating the runt-pig problem.

TABLE 8-8 COMPLETE RATIONS

	Percent protein in complete ration							
	13		14		15		16	
Ground yellow corn	1,725	1,675	1,650	1,600	1,600	1,525	1,550	1,450
40% supplement	275	..	350	..	400	..	450	..
35% supplement	..	325	..	400	..	475	..	550
	2,000	2,000	2,000	2,000	2,000	2,000	2,000	2,000
	Calculated analysis for complete rations (using an average analysis for the 40% or 35% supplements)							
Protein %	13.2	13.1	14.3	14.1	15.1	15.1	15.9	16.1
Calcium %	0.56	0.58	0.71	0.71	0.81	0.84	0.91	0.97
Phosphorus %	0.49	0.49	0.56	0.55	0.60	0.61	0.64	0.66
	Suggested uses							
Growing finishing	x	x	x	x	x	x	(Ca High)	
	(after 125 lbs.)							

Source: Iowa State University

Experimental work at the Iowa and Illinois Experiment Stations indicates that Aureomycin, Terramycin, and penicillin antibiotics have yielded good results, but that mixtures of penicillin and streptomycin, or ASP250, and new antibiotics like tylosin and oxytetracycline-oleandomycin, tylosin-streptomycin combinations produce greater increases in gains over control groups. Antibiotics give the greatest benefit when fed to pigs from birth to 125 pounds. Feeding antibiotics to heavier pigs is not usually recommended, unless the pigs are unthrifty. The antibiotic increases the rate of gain, but has little effect on feed efficiency after pigs weigh 75 to 125 pounds. Ten to 25 grams per ton of complete ration is the usual recommendation, when needed.

While the feeding of antibiotics will increase the rate of gain in healthy pigs by 5 to 20 percent, it will increase the gain in unhealthy pigs by as much as 100 percent. The higher the disease level, or the more infection present, the greater the value of feeding antibiotics.

The feeding of antibiotics to healthy pigs is good insurance against scours and intestinal diseases, and some increases in gains will result. Antibiotics fed to young pigs on high-disease-level farms will produce maximum results. Tests also show that antibiotics can be effective only when fed with a balanced ration.

Antibiotics may be purchased in the form of vitamin-antibiotic premixes in 10-, 25-, and 50-pound packages. It is possible to add them to the other feeds included in the mixed ration. The directions of the manufacturer or distributor should be followed closely, and care should be taken to mix the materials uniformly with the other feeds.

A popular and easy way to obtain feeds containing antibiotics is to buy a ready-mixed feed. Most feed companies have fortified their rations with both antibiotics and vitamins.

Vitamins. Pigs require thiamine, riboflavin, niacin, pantothenic acid, pyridoxine, biotin, folic acid, vitamin B_{12}, and choline for normal health. These are water-soluble vitamins.

In addition to the water-soluble vitamins, pigs require vitamins A, D, and E,

TABLE 8-9 AMOUNT OF SUPPLEMENT OF DIFFERENT PROTEIN PERCENTAGES NEEDED WITH CORN OR MILO TO FORMULATE RATIONS OF DIFFERENT LEVELS OF PROTEIN (Grain figured at 8.5% protein)

Percent Protein in supplement		Percent Protein in Total Ration 10	11	12	13
30	Grain	1,860 lbs.	1,767 lbs.	1,674 lbs.	1,581 lbs.
	Suppl.	140	233	326	419
	Lbs. grain per lb. suppl.	13.3	7.6	5.1	3.8
31	Grain	1,867	1,778	1,689	1,600
	Suppl.	133	222	311	400
	Lbs. grain per lb. suppl.	14.0	8.0	5.4	4.0
32	Grain	1,872	1,787	1,702	1,617
	Suppl.	128	213	298	383
	Lbs. grain per lb. suppl.	14.6	8.4	5.7	4.2
33	Grain	1,877	1,796	1,714	1,633
	Suppl.	123	204	286	367
	Lbs. grain per lb. suppl.	15.3	8.8	6.0	4.4
34	Grain	1,882	1,804	1,725	1,647
	Suppl.	118	196	275	353
	Lbs. grain per lb. suppl.	15.9	9.2	6.3	4.7
35	Grain	1,887	1,811	1,736	1,660
	Suppl.	113	189	264	340
	Lbs. grain per lb. suppl.	16.7	9.6	6.6	4.9
36	Grain	1,891	1,818	1,745	1,673
	Suppl.	109	182	255	327
	Lbs. grain per lb. suppl.	17.3	10.0	6.8	5.1
37	Grain	1,895	1,825	1,754	1,684
	Suppl.	105	175	246	316
	Lbs. grain per lb. suppl.	18.0	10.4	7.1	5.3
38	Grain	1,898	1,831	1,763	1,695
	Suppl.	102	169	237	305
	Lbs. grain per lb. suppl.	18.6	10.8	7.4	5.6

14	15	16	17	18	19	20
1,488 lbs.	1,395 lbs.	1,302 lbs.	1,209 lbs.	1,116 lbs.	1,023 lbs.	1,070 lbs.
512	605	698	791	884	977	930
2.9	2.3	1.9	1.5	1.3	1.0	1.2
1,511	1,422	1,333	1,244	1,156	1,067	1,022
489	578	667	756	844	933	978
3.1	2.5	2.0	1.6	1.4	1.1	1.0
1,532	1,447	1,362	1,277	1,191	1,106	1,021
468	553	638	723	809	894	979
3.3	2.6	2.1	1.8	1.5	1.2	1.0
1,551	1,469	1,388	1,306	1,224	1,143	1,061
449	531	612	694	776	857	939
3.5	2.8	2.3	1.9	1.6	1.3	1.1
1,569	1,490	1,412	1,333	1,255	1,176	1,098
431	510	588	667	745	824	902
3.6	2.9	2.4	2.0	1.7	1.4	1.2
1,585	1,509	1,434	1,358	1,283	1,208	1,132
415	491	566	642	717	792	868
3.8	3.1	2.5	2.1	1.8	1.5	1.3
1,600	1,527	1,455	1,382	1,309	1,236	1,164
400	473	545	618	691	764	836
4.0	3.2	2.7	2.2	1.9	1.6	1.4
1,614	1,544	1,474	1,404	1,333	1,263	1,193
386	456	526	596	667	737	807
4.2	3.4	2.8	2.4	2.0	1.7	1.5
1,627	1,559	1,492	1,424	1,356	1,288	1,220
373	441	508	576	644	712	780
4.4	3.5	2.9	2.5	2.1	1.8	1.6

TABLE 8-9 (Continued) *(Grain figured at 8.5% protein)*

Percent Protein in supplement		Percent Protein in Total Ration 10	11	12	13
39	Grain	1,902 lbs.	1,836 lbs.	1,770 lbs.	1,705 lbs.
	Suppl.	98	164	230	295
	Lbs. grain per lb. suppl.	19.4	11.2	7.7	5.8
40	Grain	1,905	1,841	1,778	1,714
	Suppl.	95	159	222	286
	Lbs. grain per lb. suppl.	20.1	11.6	8.0	6.0
41	Grain	1,908	1,846	1,675	1,723
	Suppl.	92	154	215	277
	Lbs. grain per lb. suppl.	20.7	12.0	8.3	6.2
42	Grain	1,910	1,851	1,791	1,731
	Suppl.	90	149	209	269
	Lbs. grain per lb. suppl.	21.2	12.4	8.6	6.4
43	Grain	1,913	1,855	1,797	1,739
	Suppl.	87	145	203	261
	Lbs. grain per lb. suppl.	22.0	12.8	8.9	6.7
44	Grain	1,916	1,859	1,803	1,746
	Suppl.	84	141	197	254
	Lbs. grain per lb. suppl.	22.8	13.2	9.2	6.9
45	Grain	1,918	1,863	1,808	1,753
	Suppl.	82	137	192	247
	Lbs. grain per lb. suppl.	23.4	13.6	9.4	7.1

which are fat-soluble. Yellow corn and green hays and pastures supply adequate amounts of vitamin A. Sunlight provides vitamin D. Cereal grains and green hays or pastures provide vitamin E.

Some of the water-soluble vitamins are available to the pigs in farm grains and forages. Thiamine or vitamin B_1, pyridoxine, biotin, and folic acid are available in cereal grains or in forage crops, and need not be provided from other sources. The other water-soluble vitamins must be provided from sources other than the normal farm feeds.

Riboflavin should be provided at the rate of 1 milligram per pound of ration for pigs weighing 50 to 200 pounds. Niacin, or nicotinic acid, is required the rate of 7.5 milligrams for pigs weighing 50 to 200 pounds.

14	15	16	17	18	19	20
1,639 lbs.	1,574 lbs.	1,508 lbs.	1,443 lbs.	1,377 lbs.	1,311 lbs.	1,246 lbs.
361	426	492	557	623	689	754
4.5	3.7	3.1	2.6	2.2	1.9	1.7
1,651	1,587	1,524	1,460	1,397	1,333	1,270
349	413	476	540	603	667	730
4.7	3.8	3.2	2.7	2.3	2.0	1.7
1,662	1,600	1,538	1,477	1,415	1,354	1,292
338	400	462	523	585	646	708
4.9	4.0	3.3	2.8	2.4	2.1	1.8
1,672	1,612	1,552	1,493	1,433	1,373	1,313
328	388	448	507	567	627	687
5.1	4.2	3.5	2.9	2.5	2.2	1.9
1,681	1,623	1,565	1,507	1,449	1,391	1,333
319	377	435	493	551	609	667
5.3	4.3	3.6	3.1	2.6	2.3	2.0
1,690	1,634	1,578	1,521	1,465	1,408	1,352
310	366	422	479	535	592	648
5.5	4.5	3.7	3.2	2.7	2.4	2.1
1,699	1,644	1,589	1,534	1,479	1,425	1,370
301	356	411	466	521	575	630
5.6	4.6	3.9	3.3	2.8	2.5	2.2

Source: Iowa State University

Pantothenic acid is required by growing pigs at the rate of 4 milligrams per pound of ration. Much of the choline may be supplied by corn, which contains 200 milligrams per pound, and oats, wheat, and barley, which contain about 450 milligrams per pound. Meat scraps, fish meal, and soybean meal are good sources of this vitamin.

Vitamin B_{12}, a byproduct in the manufacture of antibiotics, was discovered to be an important part of the animal protein factor (APF). About 5 micrograms (1,000 micrograms equals 1 milligram) of vitamin B_{12} are needed per pound of dry ration to promote the most rapid gains among young growing pigs.

The best means of providing the vitamins that cannot be supplied by farm grains, protein supplements, and forages is to give them in the form of a vitamin-antibiotic

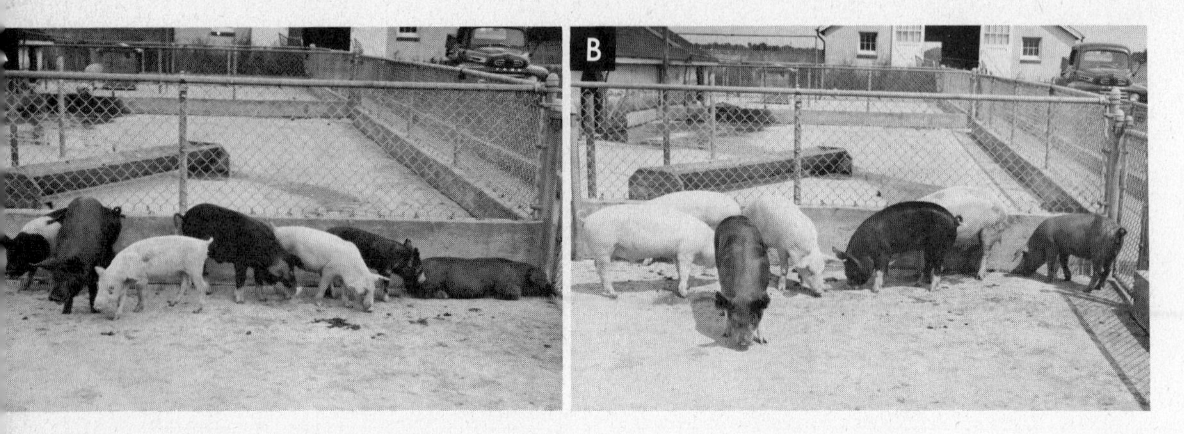

FIGURE 8-24. **(*A*) These pigs received no vitamin B_{12} or antibiotics in ration. (*B*) Pigs from the same herd that received both vitamin B_{12} and antibiotics in their ration. (*Courtesy* U.S.D.A. Bureau of Animal Industry)**

premix, which can be mixed with the protein supplement or with the complete ration, or to buy a ready-mixed supplement that contains the needed vitamins and antibiotics.

The information on the tag or carton should be checked carefully to determine the content of the mix and the cost.

Minerals. Pigs that are on good legume pasture and receive skim milk or bone meal usually need little additional mineral feeds. Pigs that are on poor pasture and are fed plant proteins need a considerable amount of minerals. Because cereal grains are very low in mineral content, it is a good idea to self-feed a mineral mixture or a protein supplement to which minerals have been added. Calcium, phosphorus, and salt, the most important mineral materials in feeding hogs, are easily obtained. Ground limestone is an excellent source of calcium, and steamed bone meal supplies both calcium and phosphorus. A low-fluorine calcium phosphate may also be used, but an excess of 0.4 percent fluorine is toxic to the pigs.

Swine rations should contain about ½ pound of salt per 100 pounds. In tests conducted at Purdue University, a penny's worth of salt saved 287 pounds of feed. Salt may be fed free choice in a feeder, it may be mixed with the mineral, or it may be mixed with a complete ration. Usually it is mixed with the mineral, which is fed free choice.

Small quantities of so-called trace minerals should also be supplied in the mineral mixture. Iodine, iron, copper, zinc, manganese, and cobalt are considered most important, and they are most easily provided by purchasing trace mineral premixes, ready-mixed mineral feeds, or ready-mixed protein and mineral supplements.

Mineral mixtures which have been recommended for self-feeding free choice or for use in mixed rations are shown in Table 8-10.

Arsenicals. Arsenicals act like antibiotics in stimulating gains and increasing feed efficiency in pigs, especially when pigs are fed on dry lot or under high-disease-level conditions. In Iowa tests arsanilic acid added to the rations of pigs on dry lot resulted in a 7-percent increase in daily gain and a 7-percent saving in feed. Pigs on dry lot fed 3-Nitro gained 16 percent more quickly on 3 percent less feed than a control group

TABLE 8-10 COMPLETE VITAMIN-MINERAL MIXES FOR CORN-SOYBEAN MEAL RATIONS[1, 2]

Ingredients	1	2
Calcium carbonate (38% Ca)	500	400
Dicalcium phosphate (26% Ca, 18.5% P)	1,000	. .
Defluorinated treble phosphate (32% Ca, 18% P)	. .	1,100
Iodized salt[3]	400	400
Trace mineral premix[4]	75	75
Vitamins[5]	25	25
120 million IU vitamin A		
16 million IU vitamin D		
80 grams riboflavin		
320 grams pantothenic acid		
600 grams niacin		
400 milligrams B_{12}		
Total	2,000	2,000
Calculated analysis (to be adjusted for guaranteed analysis)		
Elemental phosphorus %	22.50	25.20
Elemental calcium %	9.25	9.90
Salt %	20.00	20.00

[1] For growing-finishing pigs, 50 lbs. of the vitamin-mineral mix can be mixed with 250 to 450 lbs. of 44 percent soybean meal and 1,500 to 1,700 lbs. of ground corn to make 1 ton of a complete, well-balanced ration. For a gestation ration to be fed at the level of 5 lbs. per day, 50 to 60 lbs. of the mix should be mixed with 300 lbs. of 44 percent soybean meal and 1,640 to 1,650 lbs. of corn. Also, 80 lbs. of the mix can be mixed with 400 lbs. of 44 percent soybean meal and 1,520 lbs. of ground corn for a gestation ration to be fed at the level of 4 lbs. per day or for a lactation ration.

[2] **Do not** add or feed free choice additional vitamins or minerals with any of the suggested balanced swine ration formulas in Tables 7-3 through 8-6, since they contain sufficient vitamins and minerals. These mixes can be fed free choice to sows on pasture or in other instances where free-choice vitamins and minerals are needed.

[3] If plain salt is used, iodine should be added to give a level of 0.001 percent in the vitamin-mineral mix.

[4] The trace mineral premix added should contain approximately 7.0 percent iron, 0.45 percent copper, 5.5 percent manganese and 8.0 percent zinc. If individual sources of the trace minerals are used, they should be added at levels to give the following concentrations in the vitamin-mineral mix: iron, 0.3 percent; copper, 0.02 percent; manganese, 0.2 percent; zinc, 0.3 percent.

[5] High-potency vitamin sources should be added rather than a vitamin premix, since the large amount of carrier used in most premixes will take up critical volume in the vitamin-mineral mix.

Source: Iowa State University

of pigs. For growing pigs on pasture, 3-Nitro both increased average daily gains and reduced feed per pound of gain by 5 percent. Arsanilic acid, when fed to pigs on pasture, reduced feed consumption 4 percent but failed to increase the rate of gain. Similar results have been obtained in Minnesota and Texas tests.

Arsenicals are valuable in controlling scours. The University of Illinois recommends the addition of 90 grams of arsanilic acid to a ton of complete ration, or 450 grams to a ton of supplement. When 3-Nitro is used, 22 grams per ton of complete ration or 100 grams per ton of supplement is recommended.

Arsenicals should not be fed for a period of seven to 10 days before the hogs are to be marketed, thus permitting the arsenic to disappear from the liver and muscle tissue of the animal. The directions of the manufacturer should be followed closely in feeding arsenicals.

Copper Compounds. There is some evidence that copper may have an effect similar to that of antibiotics in the ration of growing pigs. Copper is toxic and, when fed in excess, accumulates in tissue. It appears to have value as a treatment for intestinal disorders that do not respond to antibiotics or arsenicals. The maximum amount to add to a complete ration is 125 parts per million. This amount may be obtained by adding 250 grams (9 ounces) of cupric carbonate, 160 grams (6 ounces) of cupric oxide, or 500 grams (18 ounces) of cupricate sulfate to a ton of complete ration.

In adding the copper to the supplement, 1,100 grams (40 ounces) of cupric carbonate, or 700 grams (25 ounces) of cupric oxide, or 2,200 grams (80 ounces) of cupric sulfate in each ton of supplement will provide the desired amount.

Copper compounds should not be used in the feed when manure is disposed of in a lagoon; the copper in the manure will kill the bacteria in the lagoon which decompose the organic waste.

Complete Ration for Growing and Fattening Hogs. There are two common systems of feeding pigs from weaning to market: (1) self-feeding corn or other grains and a balanced supplement, free choice, and (2) self-feeding a complete mixed ration to attain market weight.

The first system is easy to follow and requires less work than other systems, but its efficiency varies with the palatability of the grain, supplement, and pasture. Undereating or overeating of supplement is common. As a result the cost and rate of gain may vary.

It is possible to formulate a complete ration which will provide for the food nutrient needs of the pigs by mixing ground corn or other grains with a balanced supplement in proper proportions. According to several swine specialists, this is the recommended system of feeding pigs, whether on pasture or in confinement.

Grower-Finisher Ration. The grower-finisher ration may be fed to weaned or unweaned pigs. Weaning is a normal process if the pigs are provided with adequate amounts of the creep feed rations.

Shown in Table 8-11 are the formulas of grower-finisher rations. These rations, like the starting rations, may be purchased from reputable feed manufacturers as mixed feed or can be mixed at the local elevator.

It is best to sort pigs according to size and to feed not more than 20 pigs in one lot if they are confined. Each pig should have 8 square feet of floor space if it is confined. Fresh water and adequate housing are essential.

Grain sorghum, barley, wheat or wheat mids, or molasses may be cheaper feeds than corn at times and may be substituted for it. Other protein feeds may also be substituted for soybean oil meal when price permits.

It is sometimes more economical to feed *least-cost* than *least-time* rations. The added income obtained by getting hogs marketed while prices are high may justify added costs. At other times it may be more profitable to use least-cost rations even though more time is required to get the pigs up to market weight.

Least-cost rations may be formulated by determining the most economical combination of corn, or other grain, and soybean oil meal to produce 100 pounds of gain. Up to a certain point, a pound of protein will substitute for more than a pound of corn. The first pounds of protein substituted for corn have the most value. A pound of pro-

TABLE 8-11 GROWER-FINISHER RATIONS (for pigs from 40 to 240 pounds)[1]

Percent protein	Ingredients	1	2	3	4	5
8.9	Ground yellow corn[2, 3]	1,688-1,488	1,713-1,538	1,538-1,338	1,680-1,480	1,723-1,523
44	Solv. soybean meal[4]	250-450	..	..	175-375	175-375
48.5	Solv. soybean meal, dehulled[4]	..	225-400	..	..	..
37	Soybeans, whole cooked[5]	..	..	400-600	..	..
17	Dehydrated alfalfa meal	..	..	..	50	..
50	Meat and bone meal	..	..	..	50	..
61	Fish meal, menhaden	..	..	..	..	50
	Calcium carbonate (38% Ca)	15	15	15	8	10
	Dicalcium carbonate (26% Ca, 18.5% P)	25	25	25	15	20
	Iodized salt	10	10	10	10	10
	Trace mineral premix	2	2	2	2	2
	Vitamin premix	10	10	10	10	10
	Feed additives[6] (gm/ton)	0-100	0-100	0-100	0-100	0-100
	Total	2,000	2,000	2,000	2,000	2,000
Calculated analysis						
	Protein, %	13.01-16.52	13.08-16.54	14.25-17.05	13.00-16.51	13.04-16.55
	Calcium, %	0.65- 0.67	0.64- 0.66	0.67- 0.69	0.61- 0.64	0.60- 0.63
	Phosphorus, %	0.52- 0.55	0.52- 0.55	0.54- 0.57	0.51- 0.54	0.52- 0.56
	Lysine, %	0.61- 0.88	0.61- 0.87	0.70- 0.91	0.58- 0.85	0.62- 0.89
	Methionine, %	0.23- 0.28	0.23- 0.27	0.24- 0.27	0.23- 0.27	0.25- 0.30
	Cystine, %	0.22- 0.27	0.22- 0.27	0.23- 0.27	0.21- 0.26	0.21- 0.26
	Tryptophan, %	0.14- 0.19	0.14- 0.19	0.16- 0.21	0.13- 0.18	0.13- 0.19
	Metabolizable energy, Cal./lb.	1,326-1,310	1,346-1,345	1,383-1,401	1,308-1,292	1,334-1,317

Feeding Directions

1 Start with the higher level of soybean meal (lower level of corn) with 40-lb. pigs, and decrease the soybean meal (increase the corn) in 50- to 100-lb. increments such that you reach the lower level at approximately 125 lbs. If you prefer, one level of protein can be fed from 40 lbs. to 240 lbs. with essentially equal results as with the varying levels. In this case use a level of soybean meal and corn which is approximately the midpoint of the listed range (for example, in ration No. 1 you might use 1,588 lbs. of corn and 350 lbs. of soybean meal). If barrows and gilts are separated, use the higher end of the range for soybean meal for the gilts and the lower end for the barrows.

2 Ground milo, wheat or barley can replace the ground corn. Ground oats can replace corn up to 20 percent of the total ration

3 If the ration is to be pelleted, 25 to 50 lbs. of molasses or binder can replace 25 to 50 lbs. of corn.

4 Three pounds of L-lysine and 1 lb. of DL-methionine can be substituted for 100 lbs. of soybean meal.

5 Since the high fat content of whole cooked soybeans increases the energy content of the ration, the protein level in a ration utilizing whole cooked soybeans should be approximately 1 percent higher than a similar ration with soybean meal in order to maintain the same energy-to-protein ratio.

6 The feed additive may be part of the vitamin premix, or if it is a separate premix, it should replace an equal amount of corn

Source: Iowa State University

FIGURE 8-25. A portable pig wallow. (*Courtesy Hampshire Swine Registry*)

tein substitutes for less corn as the pig grows heavier. When the price of protein is high compared to the price of corn, feed costs can be reduced by lowering the amount of protein and increasing the amount of corn in the ration. When the price of protein declines relative to the price of corn, it becomes economical to increase the protein level of the ration.

Water. If pigs are to make fast and economical gains, they must have plenty of fresh water available at all times. A pig needs ½ gallon to 1½ gallons of water daily for every 100 pounds of weight. Eighty pigs weighing 150 pounds on pasture in July may consume as much as 80 to 150 gallons of water in one day. Young pigs and sows nursing litters have high water requirements, and more water is needed during the hot months of the summer than during the winter.

Although water may be supplied in the form of slop feeds, it is preferable that the pigs have access to automatic waterers. Hand-feeding of water in clean troughs is satisfactory if done often enough, and if the pigs can be kept out of the troughs. Slats may be nailed across the trough to keep the pigs out. The labor necessary to provide a sufficient amount of water by hand is usually not available.

Shade. The use of rotated pastures during the summer months makes it necessary to provide some form of shade to protect the pigs from the hot sun. Pigs pastured in wooded areas or with access to the farm buildings usually need no other shade unless the pasture area is at some distance from the buildings.

The shades should be from 4 to 5 feet above the ground to permit circulation of air and should be located on high ground. A 10 by 12 foot shade will accommodate 10 to 20 pigs, depending upon their size. The sunshades may be temporary structures or they may be pieces of equipment which can be moved from field to field.

Tests conducted at Davis, California, showed that 100-pound pigs did best when the temperature was around 70 degrees. Some hog producers have found it profitable to install cooling devices in central-type hog houses used during the summer months.

Hog Wallows. Sanitary wallows may be used during the summer months to aid in keeping the pigs cool and healthy. Portable wallows of the type shown in Figure 8-25 may be constructed.

Feeding and Management of Pigs in Confinement

An increasing number of pigs are being produced in confinement. It is estimated that 20 percent of the hogs marketed are grown out in confinement. The development of complete and fortified rations has made it possible to provide growing pigs with the nutrients, vitamins, and minerals commonly supplied by pasture. Vaccination and the feeding of arsenicals and antibiotics have aided materially in the control of disease.

FIGURE 8-26. A confinement feeding layout using open-front housing, a mechanical waterer and a mechanical feeder. (*Courtesy* Edward Lodwick, Cincinnati Gas and Electric Co.)

Confinement production has made rapid strides and some producers have invested heavily in housing and equipment. Others, especially in the South, are growing pigs in confinement with very little investment in housing and equipment.

Summer and Fall Pigs. Most summer and fall pigs are well beyond the weaning stage by the time the pastures dry up in the fall. The pigs will weigh from 75 pounds upward, and can be fed the rations recommended for confinement feeding in Table 8-11. Confinement rations contain more supplement than do pasture rations. The supplement recommended for confinement feeding contains more animal protein, more alfalfa, more vitamin supplements, and fewer minerals than that recommended for pasture feeding.

The use of self-feeders and automatic waterers is recommended in confinement feeding as in pasture feeding. The big problems in confinement feeding, aside from those of nutrition, are in providing adequate and sanitary quarters. It is much more difficult to care for a large number of pigs in a hog house, on a feeding floor, or in a small lot than it is to manage pigs on pasture, unless slatted floors and manure removal equipment are available. Bedding, ventilation, and sanitation problems are often troublesome.

Housing. Buildings open on only one side for ventilation are satisfactory in the South. Most finishing houses in the northern states are insulated buildings with built-in ventilation and sanitation control. The use of lagoons and slatted floors has almost eliminated manure handling. The entire floor may be slatted or only the back quarter or one third of the pen area. Pens 6 feet by 16 feet will accommodate 20 pigs of 50 to 75 pounds, 16 pigs of 75 to 125 pounds, or 12 pigs weighing more than 125 pounds. When tail biting is a problem, it is suggested that pens be made smaller and that fewer pigs be kept in each pen.

Fans should be provided for ventilation. It is important that both intake and exhaust fans be used to maintain a uniform temperature of 68 to 70 degrees.

No heating system is necessary if the building is insulated and properly ventilated.

Feeding. With the entire floor slatted, self-feeders and automatic waterers are used. One linear foot of feeder space should be available for every 4 pigs of 50 to 75 pounds, every 3 pigs of 75 to 125 pounds, and every 3 pigs weighing more than 125 pounds.

An automatic waterer should be available for every 20 or 25 pigs.

Floor feeding of complete pelleted rations is recommended when sufficient labor and a concrete floor are available. Floor feeding reduces waste of feed, encourages the pigs to keep the pen area clean, and reduces the investment in feeding equipment.

FIGURE 8-27. A confinement feeding facility with mechanized feed-handling equipment. (*Courtesy* Lester's Inc.)

Feeding Floors. A paved feeding floor is the answer to many sanitation problems in confinement feeding. Concrete feeding floors can be cleaned quite easily, and the pigs do not have to use their energy in wading through mud during the wet fall and spring seasons. Considerable labor and feed are saved in feeding and in managing hogs on feeding floors. More of the fertilizer value of the manure is saved, and more sanitary conditions can be maintained. A good feeding floor is almost a "must" in confinement feeding, especially during the wet weather.

Most hog producers like feeding floors of strips of concrete at least ten feet wide. A 150- to 200-pound hog requires 10 to 15 square feet of area. A feeding floor 10 by 60 feet will provide space for 40 to 60 hogs weighing 150 to 200 pounds, or for about 80 younger pigs.

A strip of concrete the width of the portable house and 10 to 12 feet deep is usually considered adequate for a sow and litter.

Hogging Down Corn

The practice of hogging down corn is less popular today than it was a few years ago. With the multiple-litter system of breeding, the spring pigs are farrowed in February and March and are marketed before the new crop of corn is ready for hogging down. The fall pig crop may be large enough to glean a field following the corn picker, but the pigs are usually too small to break down the stalks in the hogging-down process.

In hogging down corn, best results are obtained by using thin, active pigs weighing 90 to 150 pounds. Heavier hogs waste more corn and do not make economical gains.

An early maturing variety of corn is best for hogging down, and the pigs should be turned in after the corn is well dented. A protein-mineral supplement and water should be available at all times.

Garbage Feeding

Most states have passed ordinances prohibiting the feeding of raw garbage to hogs. The cooking of garbage kills disease organisms, but it is costly and the cooked garbage is less palatable to the pigs.

A ton of raw garbage formerly produced 50 to 80 pounds of pork. With the changes that have come about in kitchen and restaurant management, a ton of garbage today may produce only 20 to 30 pounds of pork. Unless considerable amounts of grains are fed with the garbage, a poor carcass will be produced.

Garbage-fed hogs were responsible for the rapid spread of vesicular exanthema during the past few years, and cases of trichinosis have occurred in some communities where raw garbage was fed to hogs. In feeding garbage, extreme care should be taken to maintain sanitary quarters. Concrete feeding floors are recommended. The yards should be thoroughly cleaned and disinfected regularly. Thorough cooking of pork products is necesary to kill any disease organisms that may be present in the meat.

There has been concern that African fever might be brought in from Cuba and spread through garbage-feeding. In August, 1971, 6,430 premises were feeding garbage to 596,147 hogs. Raw garbage was being fed on 97 premises to 4,207 head. No garbage was being fed in Illinois, Wisconsin, North Dakota, and South Dakota. One third of the garbage fed in the country is fed in Texas, Florida, and Georgia. Several states have made it illegal to feed garbage in any form. The feeding of uncooked garbage is illegal in all states.

SUMMARY

From 25 to 30 percent of the pigs farrowed never reach weaning age, and 80 to 90 percent of death losses occur shortly after farrowing time.

Sows should be penned up two or three days before farrowing time. The farrowing pen or stall should be thoroughly cleaned and disinfected. Farrowing stalls are recommended over farrowing pens and guardrails. Heat lamps should be provided on each side of the sow. Farrowing quarters should be bedded lightly with dust-free dry material.

A bulky, laxative ration in moderate amounts should be fed just prior to farrowing. Fresh unchilled water should be available. It is good practice to be on hand at farrowing time. The navel cord should be clipped and treated with tincture of iodine. Pigs should be earmarked at birth to facilitate record keeping.

Needle teeth may be clipped in case pigs are inclined to fight.

FIGURE 8-28. A 16-nozzle water sprinkling system for cooling hogs on a hot day. (*Courtesy* Edward Lodwick, Cincinnati Gas and Electric Co.)

Sows should be fed lightly after farrowing but brought to full feed in a week to 10 days.

Pigs should be fed a 20- to 24-percent protein, highly fortified milk replacer feed until they weigh 12 pounds. They should be fed an 18- to 20-percent protein starter feed until they weigh 40 pounds.

Pigs like and do better on pellet feeds than on meals. Creep feeds should contain 5 to 15 percent sugar and be highly fortified with vitamins, minerals, and antibiotics.

Anemia may be controlled by providing clean sod, by using iron and copper pills, or by injecting fluid. Boars should be castrated by the time they are two or three weeks old.

Vaccination for leptospirosis and erysipelas should be done during the first eight weeks. Vaccination, castration, and weaning should not be done at the same time.

Weaned pigs should be wormed before they are moved to clean ground. Pigs can be raised profitably in confinement, if adequate rations and sanitation are provided.

Farm grains must be supplemented in feeding growing-fattening hogs. Protein, mineral, vitamin, and antibiotic feeds are essential. Corn in the Corn Belt and grain sorghum in the Southwest are usually the best grains. A mixed protein is considered better than a single protein feed. Pigs weighing 50 to 125 pounds require rations containing 15 to 16 percent protein. Heavier hogs need rations with 13 to 14 percent protein.

FIGURE 8-29. The open-front houses and lagoon of the Farmland Swine Testing Station at Lisbon, Iowa. (*Courtesy* Farmland Swine Testing Stations)

Pigs weighing 100 to 200 pounds will show an increase of about 5 percent in gains when antibiotics are fed. Pigs doing poorly will show greater increases.

A mixture of steamed bone meal (38 pounds), limestone (38 pounds), salt (20 pounds), and trace mineral premix (4 pounds) will meet the mineral needs of growing-finishing hogs.

Pastures are used by many hog producers. They save 15 to 30 percent of concentrates used by pigs in confinement. Rotated pastures aid in controlling diseases and parasites.

Confinement feeding may be more economical on very high-priced land and when large numbers of hogs are produced. Rations for pigs fed in confinement must be more highly fortified than those for pasture feeding.

Ladino clover and alfalfa are the two best legume pastures. Ladino is superior to alfalfa in some areas.

It pays to self-feed a complete ration to growing-finishing pigs. A *least-cost* ration should be fed when a stable market is anticipated. When a lower market is anticipated, it may pay to feed a *least-time* ration in order to market the hogs before the break in price.

Shades and wallows should be used during hot weather. Pigs do better when the temperature is 60 to 70 degrees. Fresh water should be available at all times; each pig will consume ½ gallon to 1½ gallons per day per 100 pounds of weight.

Slatted or partial slatted floors are recommended in feeding pigs in confinement, and lagoons are recommended for manure disposal.

QUESTIONS

1. Outline the plan which you should use on your farm to care for brood sows at farrowing time.
2. Make a diagram of an economical and serviceable farrowing stall.
3. How much attention should be given pigs at farrowing time?
4. Outline a system which you can use on your farm in earmarking the pigs.
5. Contrast milk replacer, starter, and growing rations.
6. Which antibotics should be fed to young pigs and in what amounts?
7. At what age should pigs be castrated? vaccinated for erysipelas? for leptospirosis?
8. How can anemia be controlled in young pigs?
9. Outline a program of feeding and management of the sows and litters on your farm through the weaning period.
10. Which is the better system of feeding hogs on your home farm, in confinement or on pasture? Why?
11. What should be the protein content of rations for 25- to 50-pound pigs? 50- to 125-pound pigs? 125- to 200-pound pigs?
12. Outline a good protein mixture for use in balancing the rations for the growing-finishing hogs on your farm.
13. What is the place of antibiotics in feeding growing-finished hogs? Explain.
14. Which of the vitamins needed by growing-finishing hogs must be provided in the form of vitamin premixes?
15. Outline a simple mineral mixture which will meet the mineral requirements of growing finishing hogs.
16. Outline a program that will provide adequate hog pasture for the swine enterprise on your home farm.
17. Outline a complete balanced ration that you could use as a "least-time" ration on your farm.
18. List essentials of a good confinement growing-finishing program.
19. What is the place of copper compounds and arsenicals in swine production?

REFERENCES

Bradley, C. M., *et al.*, *Missouri Feeder Pig Manual*, University of Missouri, Columbia, Missouri, 1970.

Bundy, C. E. and R. V. Diggins. *Swine Production*, 3rd Edition, Prentice-Hall, Inc., Englewood Cliffs, New Jersey, 1970.

Carlisle, G. M. and H. G. Russell, *Your Hog Business Ration Suggestions*, Circular 1023, University of Illinois, Urbana, Illinois, 1971.

Coopersmith, R. L., *Producing and Marketing Feeder Pigs,* C-349, Kansas State University, Manhattan, Kansas, 1965.

Feeder Pig Production Guide, Extension Folder 223, University of Minnesota, St. Paul, Minnesota, 1965.

Hodson, Harold, Jr., *et al., Life Cycle Swine Nutrition,* Pm-489, Iowa State University, Ames, Iowa, 1970.

National Academy of Sciences-National Research Council, *Nutrient Requirements of Swine,* Washington, D. C., 1968.

Sewell, H. B., *Profitable Pork Production,* Circular 734, University of Missouri, Columbia, Missouri, 1961.

9 Disease and Parasite Control

Swine diseases and parasites are responsible for losses amounting to millions of dollars each year. It is estimated that one third of the pigs farrowed die before they are weaned, and another third are stunted or are unprofitable because of diseases or parasitic conditions. Only about one third of the pigs farrowed are grown out as healthy pigs. On many farms the profit or loss from the enterprise is determined largely by the extent to which disease and parasite losses have been controlled.

PREVENTION IS BETTER THAN CURE

Most diseases, ailments, and parasitic conditions of hogs may be prevented, and preventive measures are much more effective than are cures. Treating diseased pigs is expensive because of medicine and veterinarian costs and because feeds are wasted when fed to unthrifty pigs. A stunted or runty pig requires a long feeding period and a large amount of feed to get him ready for market. It is cheaper to prevent the unhealthy condition than it is to remedy it.

Farmers who raise large numbers of hogs year after year on the same lot usually have disease and parasite losses. The control of diseases and parasites is largely a matter of sanitation. The use of disease- and parasite-free breeding stock, rotated legume pastures, clean and disinfected houses, and good balanced rations fortified with vitamins and antibiotics can do much to reduce losses. Some diseases and parasites, however, must be controlled by vaccination and medication.

There is much movement of hogs within and across state lines. These animals may be carriers, and present laws make it very difficult to trace the source of infection. In some states, hogs moving from farm to farm must be identified, usually by ear tag or tattoo, but no identification is required if they pass through a market back to another farm. Many hogs are found to be diseased by

FIGURE 9-1. Health problems are sometimes associated with confinement hog production. Note the slatted floor area, the automatic waterer, and the self-feeder. With clean, dry bedding material and even temperature, these pigs should do well. (*Courtesy* Big Dutchman Co.)

packers, who are unable to identify the herd from which the animal came unless the animals were sold on a yield and grade basis. A move is being made to require all hogs marketed or moved from the farm to be identified by ear tag or tattoo bearing the code number of the producer.

SWINE DISEASES

There are a large number of infectious diseases of swine and their prevalence varies from community to community and from year to year. Erysipelas, coliform enteritis, vibrionic dysentery, bordetella rhinitis, brucellosis, flu, anemia, leptospirosis, baby-pig disease, MMA, PPLO, and gastroenteritis produce most of our disease problems.

Hog Cholera

Losses in the United States due to hog cholera in the past have varied from $10 million to $60 million anually. The National Hog Cholera Eradication Program launched in 1959 has resulted in a decrease of reported cholera outbreaks in the nation from about 5,000 in 1959 to 822 in 1968, and only 207 cases in 1972. Forty-one states have been declared cholera free and the remaining 9 states are in Phase IV of the eradication program. There was only one case of cholera reported in the nation in February of 1973.

Cause. The disease is caused by a virus so small that it cannot be seen under a microscope. It is present in the blood and body tissues of infected animal, and it is spread through the urine and feces, as well as through nose and mouth secretions.

Young hogs are more susceptible to the disease than are older hogs. The disease is usually introduced by new animals brought to the herd, or by virus carried to the farm on the shoes of attendants. It can also be spread by trucks, by birds, and by streams.

Symptoms. The incubation period is usually from three to seven days. Infected animals first show fever and loss of appetite. Later the eyes become filled with a sticky discharge, and the hogs prefer dark quarters. They lose weight, and the underside of the neck and abdomen may show dark red or purple coloration. Infected animals cough and have difficulty breathing.

Usually the first cases that appear are acute, and the animals die in three to seven days. Chronic cases last longer. Some hogs do not die but make only partial recoveries from the disease.

Treatment and Prevention. All infected animals are being destroyed. The government shares with the state government in making payment for animals destroyed. All hog cholera virus and vaccine have been eliminated. December 1972 was the date set by experts as the time when they expected the nation to be cholera free.

Erysipelas

Hogs may contract this disease in the acute form and die quickly, or they may have the chronic form, which causes general stiffness and swelling of the joints in the legs. It is sometimes called the *diamond-skin disease.*

Cause. The disease is produced by a microorganism or bacterium similar to those which cause arthritis in some animals. Diseased animals, and others which carry the disease, pass off the organism in urine and feces, and other hogs pick it up by eating contaminated feed or drinking contaminated water.

Symptoms. Hogs with acute form have a high body temperature and may have a redness of skin as in cholera. They may die within 24 hours or in up to four days. Those with the less acute form run high temperatures for a couple of days. When the temperature drops, red diamond-shaped patches, which disappear in time, show on the skin. Mortality from this form is low.

Those that have the chronic form are stiff in the joints and lame. They become sluggish and are easily fatigued. Often they rest on their haunches or on their breastbones.

Treatment and Prevention. In treating erysipelas an injection of anti-erysipelas serum in the early stages of the disease is helpful. Exceptionally good results have been obtained when penicillin was injected with the serum in both the acute and chronic forms. Since the disease is difficult to identify, it is best to rely on the local veterinarian for diagnosis and treatment.

Sick animals should be segregated from the herd immediately, and the healthy animals moved to clean ground. As a preventative, the entire herd may be vaccinated with the serum and penicillin.

FIGURE 9-2. An animal infected with swine erysipelas. (*Courtesy* U.S.D.A. Bureau of Animal Industry)

Erysipelas Bacterin. Killed vaccine bacterins have been developed which may be used in controlling erysipelas without danger of spreading the disease or infecting human beings. The new vaccine has replaced the old double-treatment method, which permitted transfer of the disease from vaccinated to unvaccinated animals.

Both oral and injection types of vaccines are available. The former requires less labor, but the latter insures the same treatment for every pig. Pigs should be vaccinated when 8 to 10 weeks of age. Breeding stock should be revaccinated every 6 to 8 months—sows before breeding time.

Coliform Enteritis, or Necro

This disease is often associated with other diseases and with wormy pigs. It is the cause of large losses of pigs each year, as well as the cause of a large number of runt pigs. It produces an inflammation of the intestines and is very infectious, easily spread, and chronic.

Cause. The disease is caused by a microorganism, usually *E. coli bacteria.*

Symptoms. Coliform enteritis usually begins with a rise in temperature, a decrease in appetite, and diarrhea. The temperature drops after a few days, and the pigs begin to eat again. They remain unthrifty, however, and fail to gain normally. They are usually weak and thin and have rough hair coats. Mortality of pigs infected is high during the first week and 20 to 40 percent in those infected after three weeks of age.

Treatment and Prevention. Prevention consists of practicing strict sanitation, feeding adequate rations to both sow and litter, and feeding antibiotics and sulfa drugs to the sows before farrowing. Infected pigs should be fed an adequate milk replacer or starter ration and drugs recommended by the veterinarian. Nitrofurans, chloramphenicol, sulfas, and the antibiotics Aureomycin, streptomycin, neomycin-polymixin, and tylosin combinations are recommended.

Sanitation is important in controlling necrotic infections. Rotated pastures, sanitary quarters, and the isolation of any sick pigs will help prevent losses.

FIGURE 9-3. A typical example of a pig with vibrionic dysentery. (*Courtesy* Abbott Laboratories)

Vibrionic Dysentery

This disease is sometimes called *bloody scours* or *swine dysentery*, and may be confused with coliform enteritis. It is almost always associated with unsanitary conditions.

Cause. The cause is unknown, but vibrionic dysentery is recognized as a specific disease. Healthy pigs fed feeds contaminated with materials from the intestines of diseased animals become infected. Veterinarians have associated the disease with a bacterium called *Vibrio*. They now believe that spirochete bacteria may be the cause.

Symptoms. This disease is very acute and infectious, and the main symptom is a bloody diarrhea or black feces. The animals may or may not go off feed. They have some fever, but it is not high. Some pigs die within a few days; others linger for several weeks or longer. It is more serious in young pigs than in older ones. The disease affects the caecum and the colon, which become inflamed and bloody.

Treatment and Prevention. The feeding of 3-Nitro, or other arsenicals, and tylosin has proven helpful in controlling vibrionic dysentery. Recent Kentucky tests indicate that nf-180 furazolidone is also effective in stopping vibrionic dysentery in swine. A veterinarian should be called and directions followed closely when commercial treatments are used.

Isolating sick animals and moving healthy pigs to clean quarters are recommended, as in controlling coliform enteritis.

Virus Pig Pneumonia (VPP)

Virus pig pneumonia is considered one of the most serious hog diseases. While deaths are few, slow gain and poor feed-conversion losses are high. The incubation period is 10 to 16 days. The disease is chronic and may last for months. It is spread from sows to pigs by contact.

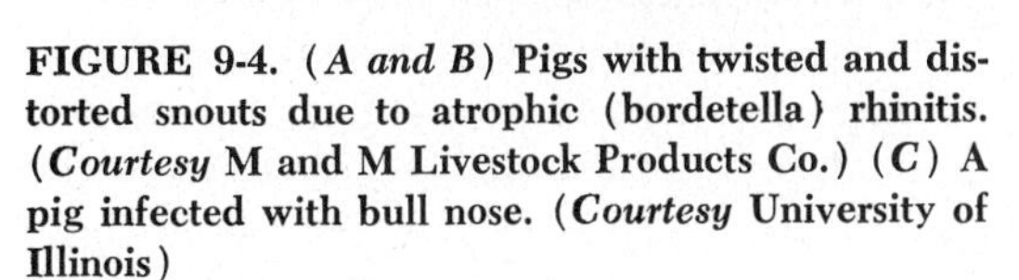

FIGURE 9-4. **(*A and B*) Pigs with twisted and distorted snouts due to atrophic (bordetella) rhinitis. (*Courtesy* M and M Livestock Products Co.) (*C*) A pig infected with bull nose. (*Courtesy* University of Illinois)**

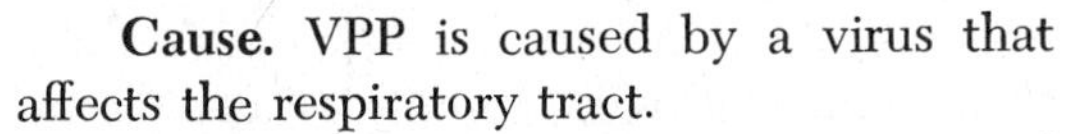

Cause. VPP is caused by a virus that affects the respiratory tract.

Symptoms. Pigs may stop growing, and they will sneeze and cough.

Treatment and Prevention. There is no cure, but tylosin and other antibiotics help keep down secondary infections. The best prevention is to clean up the premises and restock with Specific Pathogen Free (SPF) breeding stock. SPF stock is obtained originally by taking pigs from the sow by hysterectomy and raising them in isolation to avoid reinfection. SPF stock is considered free of virus pig pneumonia and bordetella rhinitis, and is becoming available in large numbers in various breeds.

Bordetella Rhinitis (Atrophic)

This disease, though prevalent somewhat longer, has caused most losses since 1950. It is now widespread among herds of both purebred and commercial hogs. It is quite often confused with *bull nose.* Rhinitis is very infectious, whereas bull nose is not.

Cause. The disease is thought to be caused by *Bordetella bronchoseptica* bacteria which get into wounds or scratches in the mouth or nose of the pig. It spreads from one pig to another through contaminated feed or water or by body contact.

Symptoms. The infection seems to start when the pigs are a few days or a few weeks old. They first show signs of sneezing, which becomes more pronounced as they grow older. At 4 to 10 weeks of age the snout begins to wrinkle and may bulge or thicken. Sometimes the snout becomes twisted or distorted.

The hair coat is usually rough, and infected pigs are poor doers. Quite often there is a discharge from the nose which may be clear and thick or puslike and bloody. Some pigs develop a short, deformed snout similar to that of the early Berkshires.

Treatment and Prevention. There is no known cure for the disease, but several practices are recommended as control measures.

1. **The nasal cavities of adult swine should be swabbed and the smear tested for bordetella.**
2. **Gilts and boars from litters in which there are pigs with rhinitis should not be saved for breeding purposes.**
3. **The dams of infected pigs should be sold as soon as possible.**
4. **Bred gilts or sows showing outward signs of rhinitis should be sold in advance of farrowing.**
5. **Early-weaned pigs fed highly fortified prestarter and pig-starter rations will be less likely to pick up atrophic rhinitis infection.**
6. **Tests indicate that rations high in vitamins and antibiotics with aid materially in rhinitis control. Aureo SP250 containing both antibiotics and sulfa drugs is effective.**
7. **Tests also indicate that an imbalance of calcium and phosphorus in the ration may encourage rhinitis infection. Increased quantities of calcium may be desirable.**
8. **Boars, gilts, and bred sows purchased and brought onto the farm should come from rhinitis-free or SPF herds and be kept in isolation.**

Brucellosis

This disease has been known as *contagious abortion of swine* and as *Bang's disease.* Surveys have indicated that 1 to 3 percent of the hogs in the United States are infected, and each year many persons become infected with brucellosis. In man the disease is known also as *undulant fever* and as *Malta fever.*

Losses caused by brucellosis in swine herds are due to the abortion of pigs, pigs born weak, sterility or infertility in sows and gilts, and sterility in the boar when the reproduction tract becomes infected.

Cause. The disease is caused by an infectious organism, *Brucella suis,* and is spread from animal to animal through contact and by contaminated feed, water, and afterbirth.

Symptoms. There are no symptoms which can be easily recognized with certainty. Many animals carry the disease but appear to be normal in every respect. The symptoms which *may* appear are premature abortion of litters, sterility, and inflammation of the joints and the uterus or testicles. Small litters may also be a symptom, but they often result from other causes.

The only sure way to determine whether or not your herd is infected is to have the entire herd blood-tested. A herd of swine may be considered free from infection when, on two tests made 30 to 60 days apart, no animal reacts, no previous evidence of infection (such as abortion or weak pigs) exists, and no new stock has been added to the herd during the past three months.

Treatment and Prevention. There is no treatment. The herd should be tested and all reactors marketed. Retests should be made every 30 to 90 days until the herd is free from the disease.

Breeding and feeding stock should be selected from brucellosis-free herds and isolated for at least four weeks after purchase.

PPLO

Mycoplasmosis (PPLO infection) is a disease of pigs 3 to 10 weeks of age. It can infect older hogs weighing 125 to 175 pounds.

Causes. The disease is caused by mycoplasma organisms. PPLO is an abbreviation of the term *pleuropneumonia-like organism.* The organism lives inside the nose in the tissues that line the cavities. It is believed that the disease spreads by nose-to-nose contact.

Symptoms. Infected pigs show signs of pneumonia or arthritis. Diagnosis is difficult, since other diseases can cause the same symptoms.

Treatment and Prevention. Depopulation of the infected herd and restocking with noncarrier animals is the only method of eliminating the disease. At present there is no test to determine carriers.

The use of tylosin mainly as a preventative is recommended. Forty to 60 grams per ton of the feed grade tylosin is recommended for pigs weighing about 70 pounds. In acute cases, higher levels of tylosin may be added to the feed. Injections of the same drug may be helpful. Good management with low stress conditions serve as excellent preventative measures. On a veterinarian's recommendation, tetracycline may be given in feed or water, or by injection.

Flu

In a survey made in one state, 16 percent of the farmers interviewed said that their hogs had had flu during the preceding year. Very few animals die with the flu, but it can take the profit out of the hog business.

Cause. Flu in hogs, much like flu in human beings, is caused by a virus and is passed readily from one animal to another. It is most likely to show up when the resistance of hogs has been lowered by fatigue, exposure to temperature changes, changes in feeds, or movement to new quarters. Hogs confined in a poorly ventilated building may become overheated and may chill when exposed to the air. Drafty and extremely cold sleeping quarters can cause just as much trouble as buildings that are too hot.

Symptoms. Infected pigs usually go off feed and become listless and inactive. They appear to be distressed and have difficulty breathing. There may be a discharge from the eyes and a cough. At later stages, the cough is deep and loud. Infected hogs have fever for a few days. The sickness usually lasts less than a week if permitted to run its course, but treatment may shorten this time.

Treatment and Prevention. Veterinarians can inject porcine bacterin No. 2, which may aid in the control of flu, but there is no effective vaccine or serum for the disease. Hogs should be provided with warm, clean, well ventilated quarters and plenty of fresh water. Healthy pigs fed good rations are less likely to get the flu.

Anemia

This disease was discussed in the preceding chapter.

Transmissible Gastroenteritis

Outbreaks of this disease have caused mortality as high as 100 percent among pigs in some localities. Infected older animals suffer a short-lived diarrhea, which is followed by loss in weight, but they recover in a few days.

Cause. Research indicates that the disease is caused by a virus which can spread from pig to pig by direct and indirect con-

FIGURE 9-5. The stomach and intestinal tract of a pig infected with transmissible gastroenteritis. (*Courtesy* University of Illinois)

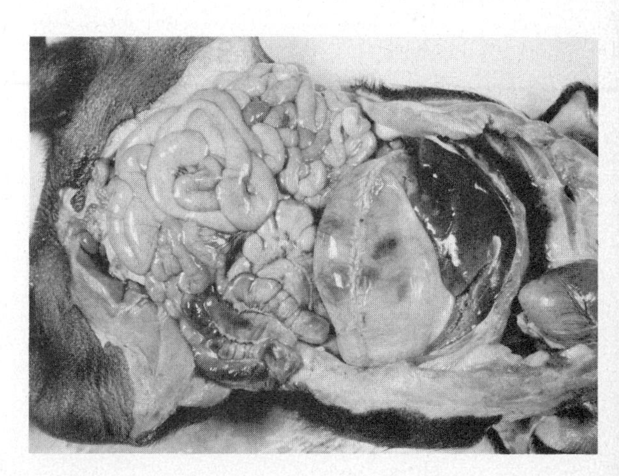

tact. The incubation period is very short—18 to 24 hours.

Symptoms. The infected pigs vomit and have diarrhea. The feces appear to be partially curdled milk. The pigs become thin and die within three or four days. Postmortem examinations of the pigs show inflammation of the stomach and intestines.

Treatment and Prevention. While no treatments for infected animals have proven completely successful, some experimentation indicates that early weaning of the pigs and the use of prestarters, very highly fortified with antibiotics and arsenicals, may be the best treatment for transmissible gastroenteritis (TGE).

A vaccine for the prevention of TGE has been developed. It is available from veterinarians and is about 70 percent effective. Sows and gilts should be vaccinated twice, 60 and 30 days before farrowing.

One means of prevention is to scatter the farrowing places over a wide area. Sows should be moved to new quarters away from the other hogs. Any bred sows brought onto the farm should be isolated, and sanitation, as recommended for the control of other diseases, should be practiced.

Sows which contract TGE during gestation and have recovered two to three weeks before farrowing will farrow litters that usually are immune or resistant to the disease. These sows should be kept for future breeding purposes. Frozen TGE capsules have been used in Illinois to produce immunity in sows.

Multiple farrowing will aid in the control of TGE. In most cases, the virus causing the disease will disappear during the interval between farrowings, if this is two months or more. Sunlight and temperatures of 80 to 90 degrees will kill the virus.

The disease spreads rapidly. Visitors should not be permitted on or near the hog lots.

Leptospirosis

This disease is a very important one to all livestock farmers because it affects cattle, swine, sheep, horses, and man. In Illinois tests it was found that nearly 29 percent of the swine herds had infected animals.

Cause. The disease is caused by a species of bacteria known as *Leptospira*. The organism may be found in the kidney or urinary tract.

Symptoms. Diagnosis of the disease is difficult because the symptoms resemble those of cholera and erysipelas. The disease causes fever, loss of appetite, loss in weight, jaundice, anemia, abortion, and reduced milk flow. Abortions, as well as high percentages of dead pigs, are common. Pigs that are farrowed and live may show signs of anemia and scouring. Many may die during the first two weeks.

Treatment and Prevention. A veterinarian should be called in as soon as symptoms are apparent. Blood and tissue tests should be made. If abortion has taken place, tests for Bang's disease should also be made.

The test for leptospirosis is not accurate. Some infected animals do not react to the test and will be carriers. Terramycin fed at a rate of 500 grams per ton of feed for a 14-day period is effective in eliminating the carrier stage of leptospirosis in swine and controls the acute infection.

Vaccination is the best method of fighting leptospirosis. Breeding animals should be vaccinated with *L. pomona bacterin* two or three weeks prior to breeding, and pigs to be kept for breeding purposes should be vaccinated at weaning time.

MMA

The MMA syndrome (metritis-mastitis-agalactia) has become a major swine disease. The disease takes a heavy toll of baby pigs.

The infection is especially serious in herds that are kept in confinement.

Cause. The agent that causes MMA has not been determined. The syndrome represents a number of disease symptoms and there may be several causes. MMA is really a group of diseases. It may be transmitted by boars at breeding time or may be due to contamination of the birth canal with manure at farrowing.

Symptoms. Metritis is an inflammation of the milk-producing glands in the udder. Agalactia has to do with poor milk flow.

MMA usually appears a day or two after farrowing. Sows lie on their bellies and the pigs cannot nurse. Constipation may be present. The udder may be swollen, hard, and hot to the touch. There is little milk. A yellow or white discharge from the vulva of the sow is present.

The pigs appear hungry and weak from lack of milk. Many will die unless treatment of both sow and pigs is started early.

Treatment and Prevention. Treatment consists of practices to overcome constipation, stop infection, and start milk flow. Injections are recommended as follows:

1. 3 to 5 cc postpituitary hormone.
2. 10 cc of a combination antibiotic.
3. 4 to 6 cc of cortisone product (5 cc B_{12} complex).

The injections should be repeated in 12 hours and daily until satisfactory results are obtained. A veterinarian is necessary in the control of MMA. An abundance of water and liquid feeding is recommended. Weak pigs may be hand-fed a sugar solution until the sow is milking.

Jowl Abscess

Federal meat inspectors condemn parts of over 4 million swine carcasses each year because of abscesses in the neck and other areas of the carcass. It is estimated that because of such abscesses $12 million worth of pork is lost. In addition there is some loss in feed efficiency on infected farms.

Cause. Jowl abscesses are caused by streptococcus bacteria which invade the tonsils, where they grow, multiply, and spread. The infection spreads through feed, water, and pig-to-pig contact.

Symptoms. Localized abscesses appear in the throat, neck, and head. They commonly appear at one side of the jowl. Suckling pigs appear to be more resistant to the infection than weaned pigs.

Treatment and Prevention. Breeding animals showing signs of abscesses should be culled. The breeding herd should be put on a ration containing 50 grams of antibiotics per ton of complete ration for a week before farrowing and during the suckling period. The pigs should be given a starter feed containing 100 grams of antibiotics per ton and continue on this antibiotic level through weaning.

At 10 to 11 weeks of age the pigs should be vaccinated with Jowl-Vac. The oral vaccine should be given to pigs with no feed or water for two hours before or after treatment. Strict sanitation should be practiced.

SWINE PARASITES

A parasite is something that lives on, and gets its food from, some other plant or animal. Hogs have a number of parasites. Some live on or under the skin and are called *external parasites.* Others live within the organs of the body and are called *internal parasites.* The latter group of organisms is most injurious to swine.

Roundworms

Large intestinal roundworms, or ascarids, cause hog raisers heavy losses every year. These losses are due to stunting, development of a potbelly, general weakness, and sometimes death of the pigs.

In tests conducted by the U.S.D.A., it was found that worm-free pigs fed a well fortified ration gained an average of 161 pounds in 169 days. Littermate worm-infested pigs fed the same ration gained an average of only 119 pounds.

Life History. The roundworm is a large, thick, yellow or pink worm, about the size of a lead pencil. The adult normally lives in the small intestine. The female produces thousands of eggs daily. These are eliminated from the body in droppings. Pigs become infested with roundworms by swallowing the eggs with feed or water. The eggs are abundant in old hog lots and pastures. The young worms hatch out in the pig's intestines. They penetrate the wall of the intestine and travel in the bloodstream to the liver, and on to the lungs. From the lungs, they move up to the mouth where they are swallowed. They return to the intestine, where they mature in about two to two-and-a-half months.

Damage. The roundworms in the intestine consume nutrients that the hog needs, and also cause digestive disturbances. Worms are more injurious to pigs on poor rations than to pigs on good rations. During the time that the young worms are in the lungs, the pig has difficulty breathing, and, if exposed to dust or changes in weather, is subject to pneumonia. The liver is also damaged by the worms, and it can function only in a limited capacity. The worms weaken the pigs, making them susceptible to the many swine diseases.

Treatment and Prevention. Dichlorvos, sold as Atgard V, Hygromycin, piperazine compounds, Thiabendazole, and levamisole

FIGURE 9-6. Life cycle of a large swine roundworm. (*Courtesy* U.S.D.A. Bureau of Animal Industry)

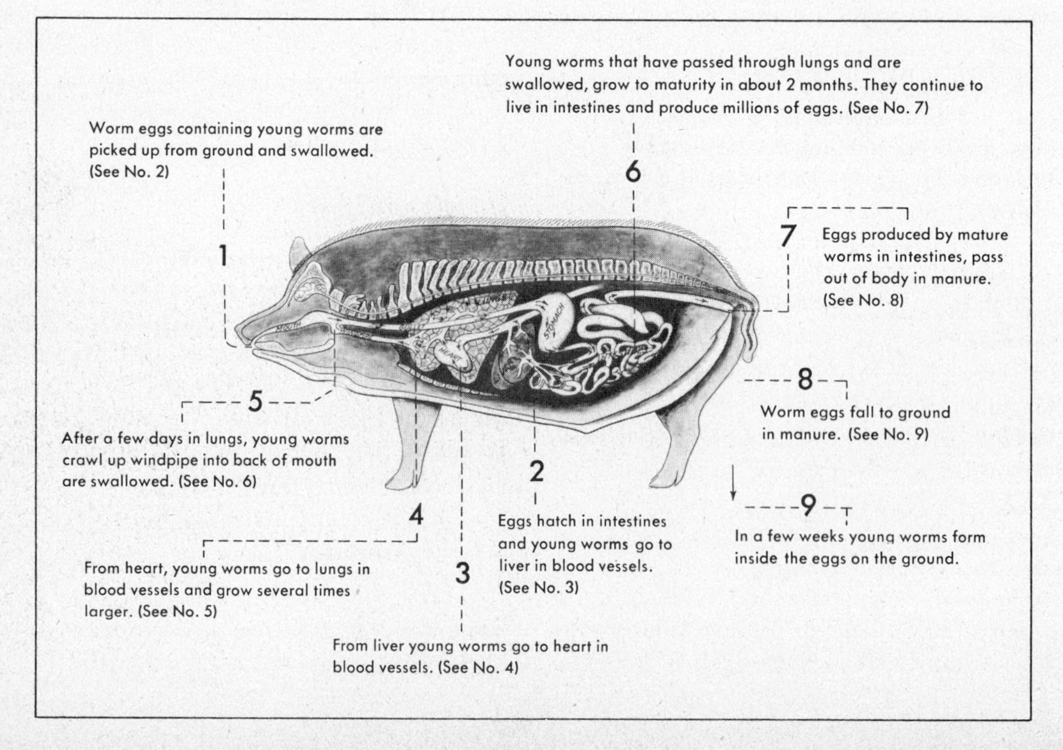

HCl are the best wormers now in use. Dichlorvos is an organic phosphate. It kills both immature and mature roundworms, nodular worms, and whipworms. Hygromycin is an antibiotic and should be fed continuously in the ration from the time the pigs start eating until they are marketed. It controls all three major hog worms—roundworms, whipworms, and nodular worms.

Piperazine compounds should be fed at the rate of ⅕ ounce of anhydrous piperazine per 100 pounds of animal weight. The pigs should be wormed at weaning time and at 50-day intervals until they are marketed. Both Hygromycin and piperazine compounds may be used in worming pregnant sows.

The preventive measures for controlling worms have been described in Chapter 8.

Lungworms

These worms are long, slender, whitish in color, ½ inch to 2 inches long, and threadlike in diameter. They are found in the windpipe, more often in the two branches of the lower windpipe, and in the lungs.

Life History. The female lungworm produces large numbers of eggs, which are coughed up, swallowed, and eliminated with the droppings. The lungworm eggs are swallowed by angleworms. They hatch and develop in the angleworms, and the pigs become infested when they eat the angleworms while rooting. In the intestines of the pig, the lungworm penetrates the intestinal wall and follows the bloodstream to the heart and back to the lungs.

Damage. Lungworms weaken the pigs and cause them to cough, to breathe with difficulty, and to grow more slowly. Severe infestations may cause death. Infected animals are susceptible to other diseases and to other parasites.

Treatment and Prevention. Levamisole HCl (Tramisol) was approved in 1971 for use on swine in the control of lungworms. It also controls round and nodular worms. It is used in the drinking water as a one-dose treatment.

Lungworms can be controlled in part by the use of rotated pastures and by sanitary practices during farrowing time and until the pigs weigh 75 pounds.

Mange

Mange is a highly contagious skin disease and, although very few animals die as a result of mange, it is the most serious of the external parasites.

Cause. Mange is caused by a very small mite that spends its entire life on the hogs. It feeds on the tissues of the skin and blood and burrows into the skin, causing a dry, rough, scaly hide.

Damage. Hog mange causes intense itching, which motivates the hog to bite and rub itself. The infection usually starts around the eyes and ears and along the underline, where the skin is tender. It may

FIGURE 9-7. The result of using a medicated wormer on a badly infested gilt. (*Courtesy* Moorman Manufacturing Company)

FIGURE 9-8. Pig being sprayed with malathion for mange and lice control. (Gordon photo. *Courtesy* Wallaces Farmer)

spread until the entire body is covered with a red rash or a heavy, scaly hide.

Treatment and Prevention. The mite can live in infested quarters for several weeks. Control measures must include a thorough cleaning of the shelters and houses. All manure should be removed. All surfaces should be disinfected with chemicals or with boiling water.

The chemicals malathion and Toxaphene are excellent for use in controlling mange. Both should be used with water to make a 0.5 percent solution. About half a pound of commercial detergent should be added to the solution. The pigs should be sprayed with this solution at the rate of about 2 to 4 quarts per animal. Sows and pigs should be sprayed in warm weather. Brood sows should be sprayed several weeks before farrowing, so that the young pigs will not become infested. When nursing sows are sprayed, they should be thoroughly dry before the pigs are allowed to nurse.

Lice

Hog lice, by their bloodsucking habits, cause some loss to swine producers, and they may be responsible for the spread of infection.

The louse is about 1/4 inch long and grayish-brown in color. During the winter months it may be found in the ears, in folds of skin around the neck, and around the tail.

The female lays several eggs a day during the winter. These eggs are attached to the hair and hatch in two to three weeks. They mature in another two weeks.

Treatment and Prevention. The malathion and toxaphene treatments for mange will also kill hog lice. The hogs should be treated as often as necessary.

SUMMARY

Swine diseases and parasites can take much of the profit out of the hog business. Erysipelas, brucellosis, coliform enteritis, vibrionic dysentery, bordetella rhinitis, gastroenteritis, leptospirosis, flu, anemia, and roundworms cause the heaviest losses. Gastroenteritis, leptospirosis, MMA, and PPLO are the newest diseases and are proving to be real problems to swine producers.

Erysipelas, TGE, and leptospirosis can be best controlled by vaccination. Vaccination of the gilts and sows for erysipelas before breeding and the erysipelas bacterin vaccination of pigs after weaning are recommended.

The feeding of arsenicals, antibiotics, highly fortified pig milk replacers and starters, and milk products is the best means of controlling enteritis. Sanitation and rotated pastures are essential in controlling diseases of the digestive tract.

Rhinitis and brucellosis may be prevented by the isolation of diseased animals and by careful selection of breeding and feeding stock. All swine-breeding stock should be tested for brucellosis and leptospirosis, and only animals which have had negative results on two tests, given at least 30 days apart, should be kept.

Vibrionic dysentery (black scours) in pigs can be controlled by feeding 3-Nitro or other arsenicals.

Flu in hogs can best be controlled by the feeding of well-balanced rations and by proper housing. There is no treatment for gastroenteritis, but it can be controlled by vaccination, in part by isolating new animals brought to the farm, and by dividing the herd and pasturing it in areas apart from other hogs.

When pigs are aborted or die, the breeding animals should be tested for leptospirosis, as well as for brucellosis. Infected animals should be sold. It is possible to vaccinate for leptospirosis.

Dichlorvos, Hygromycin, piperazine compounds, Thiabendazole, and levamisole HC1 should be used in treating pigs for roundworms. Rotated pastures and sanitary practices are effective in preventing both roundworm and lungworm losses. Levamisole HC1 is approved for lungworm control.

Mange and hog lice can be controlled by dipping or spraying with Malathion or Toxaphene.

QUESTIONS

1. What would you do if erysipelas was discovered in your home herd?
2. How does bordetella rhinitis differ from bull nose?
3. How does vibrionic dysentery differ from coliform enteritis?
4. How can transmissible gastroenteritis losses be controlled on your farm?
5. What is the extent of losses in your community due to leptospirosis and how may they be reduced?
6. How would you identify and control MMA and PPLO on your farm?
7. Outline a plan for controlling roundworms on your farm.
8. What methods are best in controlling mange?
9. Outline a disease- and parasite-control program for your farm.
10. Is there a place for SPF hogs on your farm?

REFERENCES

Dunne, Howard W., *Diseases of Swine*, 3rd Edition, Iowa State University Press, Ames, Iowa, 1970.

Dykstra, R. R., *Animal Sanitation and Disease Control*, The Interstate Printers & Publishers, Inc., Danville, Illinois, 1961.

Heeney, Marvin W., *Swine Production*, Colorado State University, Fort Collins, Colorado, 1971.

Jones, J. R. and A. J. Clawson, *Raising Hogs in North Carolina*, North Carolina State University, Raleigh, North Carolina, 1969.

Schwartz, Benjamin, *Internal Parasites of Swine*, Farmers Bulletin No. 1787, U. S. Department of Agriculture, Washington, D. C., 1952.

Seiden, Rudolph, *Livestock Health Encyclopedia*, 3rd Edition, Springer Publishing Co., Inc., New York, New York 1968.

10 Marketing Hogs

The efficient hog raiser tries to plan his breeding and feeding operations so that he will have hogs of the weight and conformation desired by the packer and consumer ready for market when the price is most favorable. With feed and pasture available, and no serious disease problems, it is possible to have hogs ready for market in 4½ to 5½ months. By planning breeding operations and allowing a 4½- to 5½-month growing-out period, it is possible to plan the time the hogs will be grown out and ready for market.

The season when the price will be best cannot always be accurately predicted. Hogs of certain weights sell better during some months than during others. The number of hogs going to market varies seasonally, and when hogs are plentiful the price goes down. When few hogs are being marketed, the price is higher. Farmers who follow the markets from year to year are usually able to anticipate the best time to market their hogs.

Most farmers find it necessary to make changes in their breeding and feeding methods in order to have their pigs at marketable weights at the time when the price is best. At times it is necessary to decide whether the increased selling price of the hogs will offset the costs of the changes in production. Each farmer must analyze his swine enterprise and decide on the production and marketing program best suited to his farm. Top production demands efficiency.

Swine producers, through breeding and feeding practices, can control the type, quality, and finish of the hogs they produce. A check of the prices paid by packers for various grades and weights of hogs will indicate the kind of hog that will bring the most money. The efficient hog raiser produces the kind of hog desired by the packer and by the consumer. He plans ahead. He has the desired type of hog at the preferred weight when the packer and producer need it for slaughter.

CHANGES IN THE HOG MARKET

It was pointed out in Chapter 5 that the consumer today is buying less lard and fat pork than he purchased a few years ago. The housewives who buy meat are guiding farmers in the production of quality pork by demanding leaner cuts. They bypass pork chops with an inch of fat for lean chops, or they purchase poultry, veal, lamb, or beef instead.

With 117.8 million head of cattle on our farms on January 1, 1973, we can expect beef to be plentiful for the next few years. This will mean cheaper beef and more competition for pork.

The fast-growing broiler industry is another competitor. In 1940 about 816 million pounds of live birds was produced. More than 1.5 billion pounds was produced in 1950 and 11.4 billion pounds was produced in 1972. A total of 129.0 million turkeys was raised in 1972.

Pork products are in a competitive market. The future of the enterprise is in part dependent on the quality and quantity of pork products produced and the methods used in merchandising.

The Lard Problem

At one time, wholesale lard sold for more than live hog prices. The situation has changed. Wholesale lard during the past few years has been selling for considerably less than the price of live hogs. The prices of lean cuts of pork have risen, and the prices of fats have declined.

In 1972 the wholesale selling price of pork loins was more than three times the wholesale selling price of lard. Table 10-1 shows the wholesale prices of various pork products in Chicago from 1965 to 1972. In April 1972, the average cost of hogs in Chicago was $22.42 per hundredweight. Lard at that time sold wholesale for $15.70 per hundredweight. The average per capita consumption in 1972 was 67.4 pounds of pork and only 3.8 pounds of lard.

Hog raisers and packers have a joint responsibility to make available to consumers a better-quality pork product. Hogs with more lean meat and less lard must be produced, and they must be marketed at a resonable weight. There are two ways of doing this: (1) by producing a better-muscled meat-type hog, and (2) by marketing hogs at lighter weights.

TABLE 10-1 AVERAGE WHOLESALE PRICES OF PORK CUTS AND LARD, CHICAGO 1965–1972 (Expressed in Cents per Pound)

Year	Loins	Whole Ham	Sliced Bacon	Lard
1965	51.8	56.2	50.0	17.6
1966	56.2	61.0	49.7	17.1
1967	49.7	49.8	44.7	13.6
1968	49.7	48.1	42.9	13.2
1969	55.3	54.2	47.6	15.4
1970	53.5	51.7	48.4	16.8
1971	49.8	44.6	46.1	15.7
1972	62.9	51.5	46.7	15.8

Source: U.S.D.A.

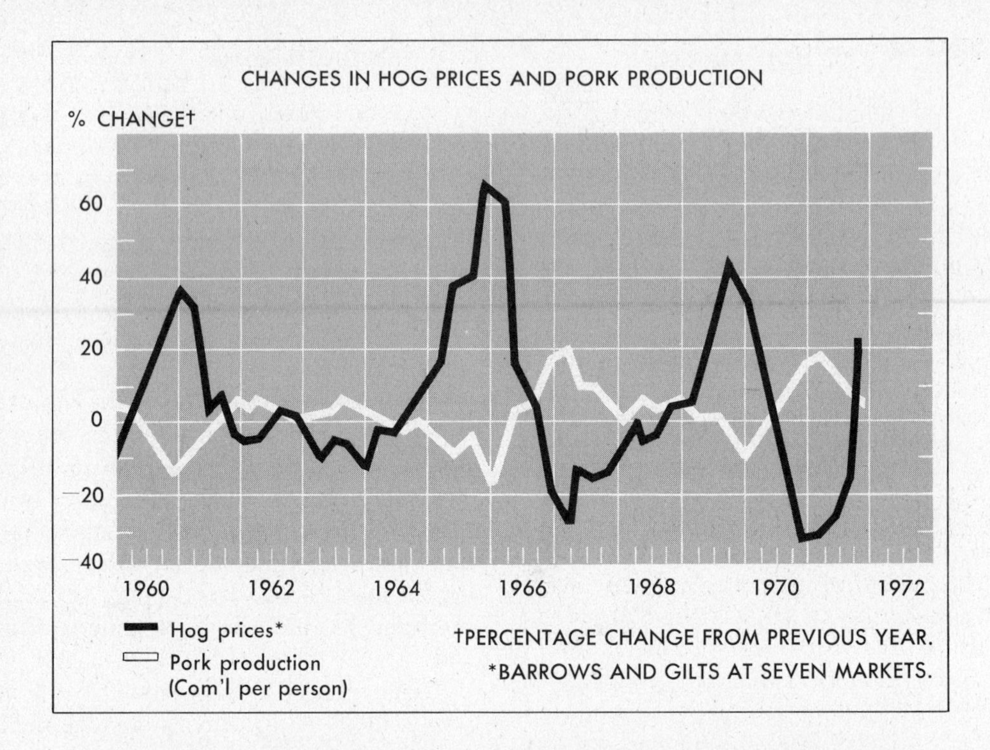

FIGURE 10-1. (*Above*) **Hog prices and pork production trends, 1960–1971. Note small changes in production greatly influence the price. FIGURE 10-2.** (*Below*) **The spring pig crop is larger than the fall crop, but the difference has been decreasing.** (**Figures 10-1 and 10-2** *courtesy* **U.S.D.A. Economic Research Service**)

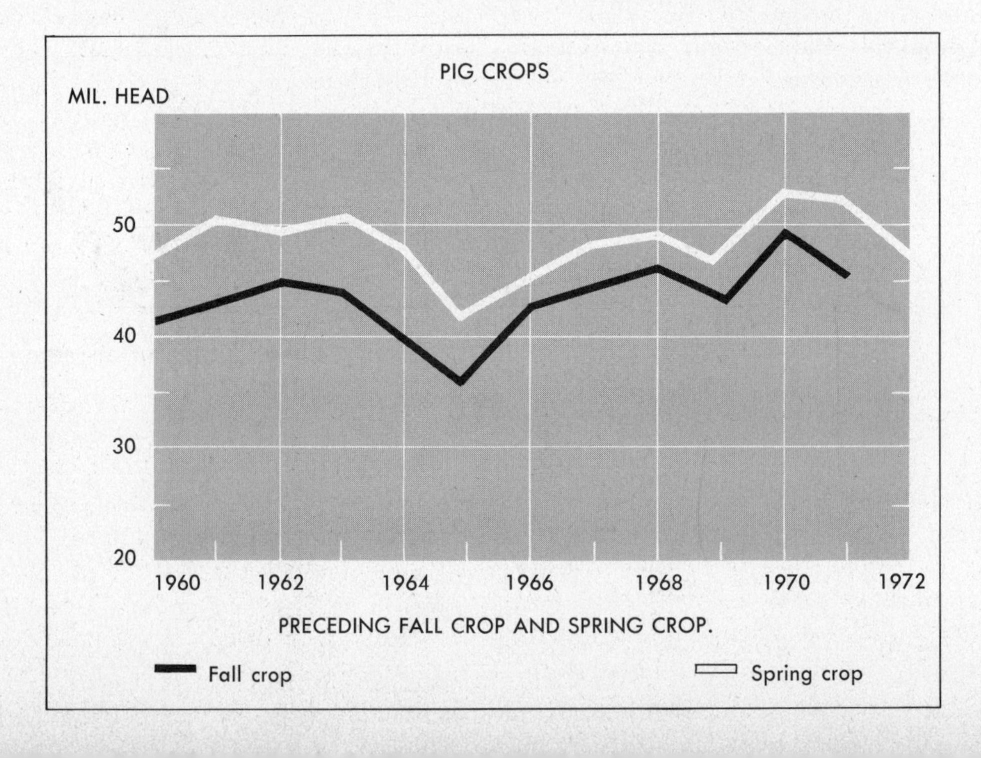

THE MEAT-TYPE HOG

The meat-type hog is being developed to meet the changes in demand for pork products. It is not necessarily a new breed; we have meat-type hogs in all our standard breeds.

Many farmers believe that it costs more to produce a heavily-muscled meat-type hog than it does to produce a hog carrying more lard. This is not true. In tests conducted at Iowa State University, hogs with less than 36.9 percent lean cuts and 26.4 percent fat cuts weighed 229 pounds at five months of age and used 340 pounds of feed to gain 100 pounds. Pigs which had more than 41 percent lean cuts and 22.9 percent fat cuts weighed 227 pounds at five months and used only 333 pounds of feed to gain 100 pounds.

TIME TO SELL

Two pig crops are normally produced each year. One is farrowed in the late winter and early spring and is ready for market in the summer and early fall. The other is farrowed in the summer and fall and sold in the winter and early spring. The bulk of the year's hog production is marketed during these two periods. A large share of the summer- and fall-farrowed pigs is marketed in February, March, and April, and the majority of the spring pigs is marketed in September, October, and November. The supply of hogs during these two marketing periods is such that packers can buy hogs at lower prices than they can during other seasons.

Seasonal Price Variations

The variation in monthly prices of market barrows and gilts at eight major markets combined during the 1965-to-1970 period is shown in Figure 10-3. Since the markets varied widely in 1969 and 1970, the average prices paid during the 1965-to-1968 period show clearly the seasonal price influence. During the four-year period, highest prices were paid during July ($23.40), August ($22.92), June ($22.88), and September ($21.40). Lowest prices were paid during April ($19.20), March ($19.75), November ($19.90), and October ($20.30).

The spread in prices between the high and low months was $3.56 in 1968, $7.16 in 1969, $12.56 in 1970, and $4.20 during the four-year period 1965 to 1968.

Since more pigs are produced in the winter and spring than in the summer and fall, there is less variation in the prices received for summer and fall pigs. During the 1965-to-1968 period, the best time to market summer and fall pigs was during the months of December and February. Pigs farrowed in late July, August, and September can easily be fed out in time to get them on the high market. The spread in prices paid for hogs between February and April was about $1.80 per hundred pounds.

WEIGHT TO SELL

The consumer wants smaller and leaner cuts of pork and less lard, and the consumer must be satisfied if pork is to meet the competition of other meats. Consequently, farmers must grow meatier hogs and market them at proper weights; hog raisers now produce more litters per year than in the past, to feed out home-grown grains and to realize the same gross income.

Heavy Hogs Are Discounted

Roughly speaking, there is nearly twice as much lard in the carcass of a 300-pound hog as there is in the carcass of a 200-pound pig. A 200-pound live hog will produce 26 to 28 pounds of lard and backfat. A 300-pound hog will yield about 50 pounds.

Following are the average prices paid by interior Iowa and Southern Minnesota packers during 1970 for U.S. No. 1, 2, and 3 barrows and gilts of various weights, according to U.S.D.A. data:

200–220 pounds	$22.59
220–240 pounds	$22.12
240–270 pounds	$21.05
Sows	$18.29

The average weight of barrows and gilts at seven major markets in 1972 was 237 pounds. The average in August was 230 pounds. Heaviest hogs were marketed in May, weighing 242 pounds, and in June, 241 pounds. Tests in Missouri and Wisconsin indicate that it may be profitable to finish high-quality meaty hogs at weights of 240 to 250 pounds, especially when wide hog-corn price ratios exist (16 to 1 or wider).

Light Hogs Make Cheaper Gains

Swine producers are interested in economical use of feed, and it is known that young animals make cheaper gains. Tests conducted at the University of Minnesota indicated that lighter hogs used less feed and made cheaper gains than did heavier hogs. The data were as follows:

Weight of pigs (Pounds)	*Pounds Feed for 100 Pounds Gain*
Birth–100	304
100–200	359
200–300	415
300–400	470
400–500	510

These data applied to 1971 prices show the cost of feed in producing the 200-pound hog to be $9.67 per hundredweight. The cost in producing the 300-pound hog would be $11.05 per hundredweight.

MARKET CLASSES AND GRADES OF HOGS

Market hogs are classified in terms of sex, use, weight, and value. Classes are pro-

FIGURE 10-3. Average prices received for barrows and gilts throughout the year. (*Courtesy* U.S.D.A.)

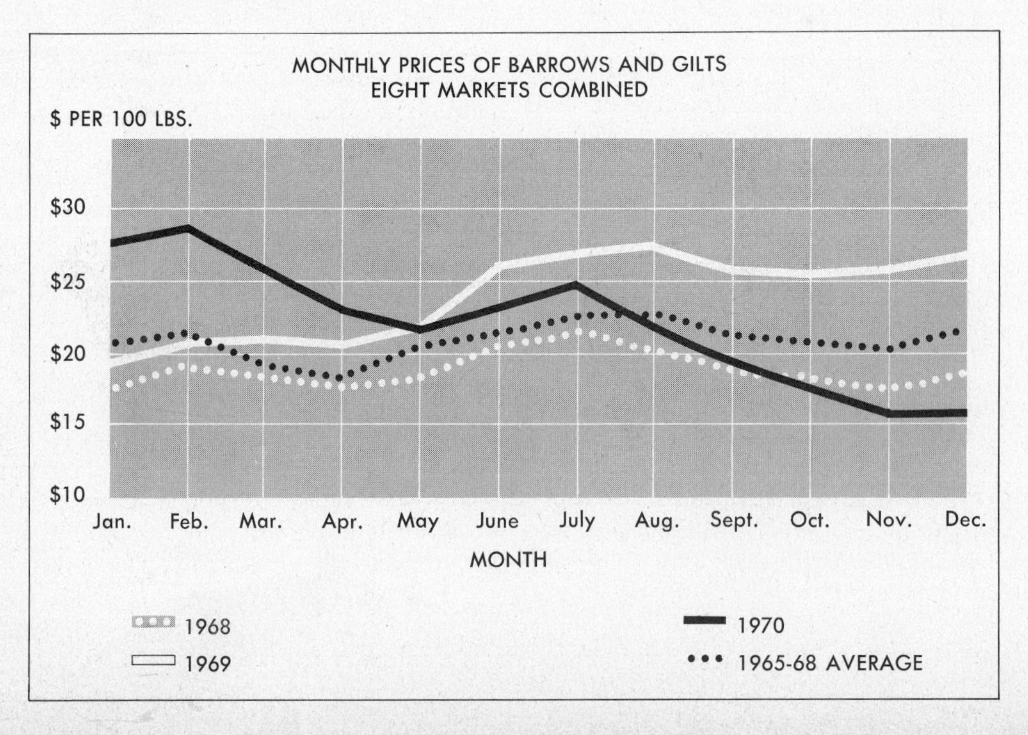

vided for barrows and gilts, sows, stags, and boars. Animals are classified according to use, as slaughter hogs, slaughter pigs, stockers, and feeders. The weights vary with the classes according to sex and use.

Most hogs are sold on the basis of grade standards established by the U.S.D.A. in 1952 and amended in 1955 and 1968. The five official grades, U.S. No. 1, U.S. No. 2, U.S. No. 3, U.S. No. 4, and U.S. Utility,

TABLE 10-2 MARKET CLASSES AND GRADES OF HOGS AND PIGS

Use	Sex	Weights (Pounds)	Grades
Hogs:			
Slaughter hogs	Barrows and gilts	Under 180 180–240 240–300 300 and over	U. S. No. 1 U. S. No. 2 U. S. No. 3 Utility
	Sows	300 and over 270–300 300–330 330–360 360–400 400–450 450–500 500–600 600 and over	U. S. No. 1 U. S. No. 2 U. S. No. 3 Utility
	Stags	All weights	Ungraded
	Boars	All weights	Ungraded
	Unclassified	All weights	Ungraded
Feeder and stocker hogs	Barrows and gilts	120–140 140–160 160–180	Choice Good Medium Common
Pigs:			
Slaughter pigs	All classes	Under 30 30–60 68–80 80–100	Ungraded Ungraded Good Medium Cull
	Barrows and gilts	100–200	Choice Good Medium Cull
Feeder pigs	Barrows and gilts	Under 80 80–100 100–120	U. S. No. 1 U. S. No. 2 U. S. No. 3 U. S. No. 4 Utility Cull

apply to barrows, gilts, and sows sold as slaughter animals. Three sets of standards have been developed, one for barrows and gilts, one for sows, and one for feeder pigs.

A systematic grading procedure makes it possible for hogs to be marketed according to the value of their carcasses or according to the value of the animals as stockers or feeders. Market grades of hogs serve the same purposes as grades of corn or grades of butter. Prices are quoted for the various grades, and it is possible for both the buyer and the seller to make comparisons before making market transactions.

Shown in Table 10-2 are the market classes and grades of hogs and pigs. Variations in weight and grade classifications among markets are common.

U.S.D.A. Hog Grades

The five grades established by the U.S.D.A. apply to hogs on foot and on the hook. The degree of finish, the quantity and quality of the lean meat, and the percentage of fat determine the grade. The four top grades are numbered from 1 to 4, according to the percentage of lean meat in the carcass. The other grade, "Utility," is so called because of underfinish and poor quality of lean-meat cuts. The use of these grades is voluntary.

The thickness of backfat in relation to the length, weight, and dressing percentage of the carcass serves as the basis for determining the grade. A summary of the U.S. hog carcass grades is shown in Table 10-3.

The official specifications for the U.S. standards for grades follow.

U.S. No. 1. Hogs in this grade will produce carcasses with superior lean quality and belly thickness, and a high percentage of lean cuts. The width through the hams is nearly equal to the width through the shoulders and both are wider than the back. The sides are long and smooth. The rear flank is slightly full and has less depth than the foreflank.

The chilled carcass should produce about 53 percent in the four lean cuts. Backfat thickness may vary from 1.3 inches to 1.6 inches as carcass length increases from 27 to 36 inches or as carcass weight increases from 120 to 255 pounds. Superior muscling may compensate for slight overdevelopment of fatness.

U.S. No. 2. Carcasses from hogs of this grade have an acceptable quality of lean, and a slightly high yield of lean cuts. From 50 to 52.9 percent of the chilled carcass will be in the four lean cuts. The average backfat thickness will increase from 1.6 to 1.9 inches with an increase of carcass length from 27 to 36 inches, or increased weight of carcass from 120 to 255 pounds.

TABLE 10-3 EXPECTED YIELDS OF THE FOUR LEAN CUTS BASED ON CHILLED CARCASS WEIGHT, BY GRADE

Grade	Yield
U.S. No. 1	53 percent and over
U.S. No. 2	50 to 52.9 percent
U.S. No. 3	47 to 49.9 percent
U.S. No. 4	Less than 47 percent

Source: U.S.D.A.

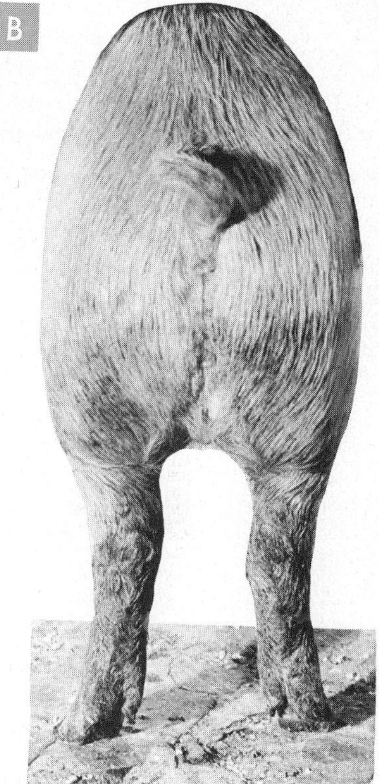

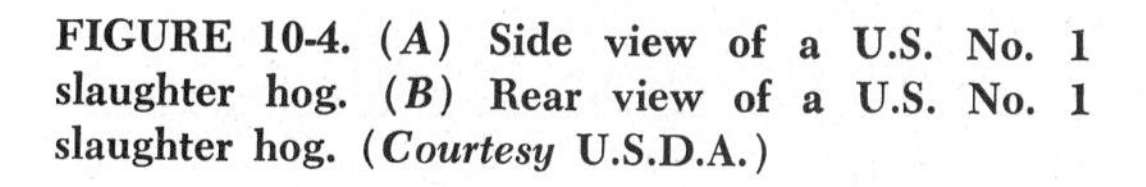

FIGURE 10-4. (*A*) Side view of a U.S. No. 1 slaughter hog. (*B*) Rear view of a U.S. No. 1 slaughter hog. (*Courtesy* U.S.D.A.)

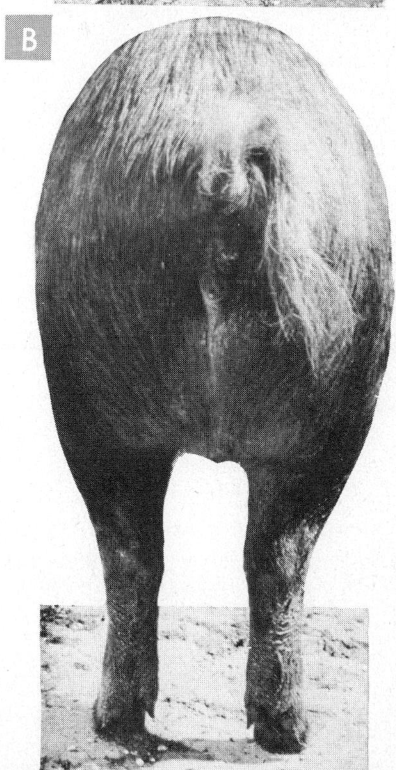

FIGURE 10-5. (*A*) Side view of a U.S. No. 2 slaughter hog. (*B*) Rear view of a U.S. No. 2 slaughter hog. (*Courtesy* U.S.D.A.)

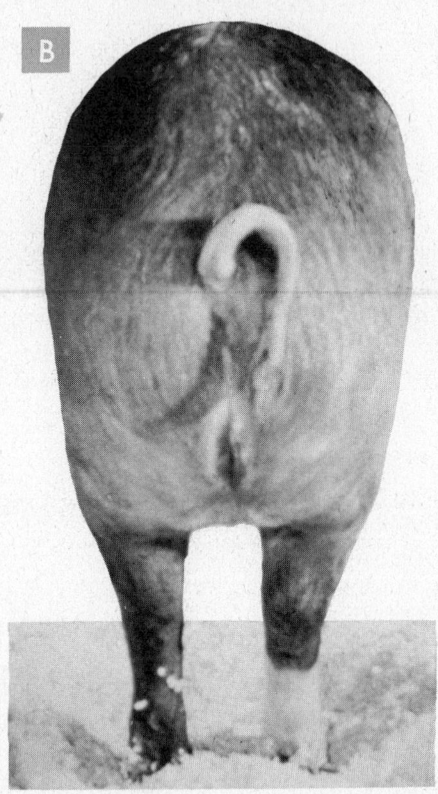

FIGURE 10-6. (*A*) Side view of a U.S. No. 3 slaughter hog. (*B*) Rear view of a U.S. No. 3 slaughter hog. (*Courtesy* U.S.D.A.)

U.S. No. 3. These hogs have an acceptable quality of lean and a slightly low yield of lean cuts—47 to 49.9 percent. The maximum average backfat thickness increases from 1.9 to 2.2 inches with increases in carcass length from 27 to 36 inches, or increases in weight from 120 to 255 pounds. A slight excess of fat may be compensated by superior muscling.

U.S. No. 4. Barrows and gilts of this grade have carcasses acceptable in quality of lean, but with a lower expected yield of lean cuts—less than 47 percent. The average backfat thickness increases from 1.9 to about 2.5 inches as carcass length increases from 27 to 36 inches, or as carcass weight is increased from 120 to 255 pounds.

U.S. Utility. This grade includes all hogs that have characteristics indicating less development of lean quality than the minimum requirements for the U.S. No. 4 grade. Hogs will usually have a thin covering of fat. The sides may be wrinkled and the flanks shallow and thin.

U.S. grades for feeder pigs were established in 1966 and revised in 1969. The six grades are similar to those for slaughter animals and are based on the slaughter potential and thriftiness of the animals.

SELLING HOGS ACCORDING TO CARCASS WEIGHT AND GRADE

The government grades represent an attempt to satisfy hog producers who have felt that packers were not paying for quality hogs. Market quotations have usually been based on the weight of the hogs, rather than on quality. It is difficult to grade hogs on foot, and most packers have preferred to buy them by the pound in droves or loads. There was little incentive for hog breeders to improve the carcass quality of their hogs when they all sold at the same price.

Advantages of Selling Hogs by Carcass Grade

Selling hogs on the basis of carcass grade and yield encourages farmers to

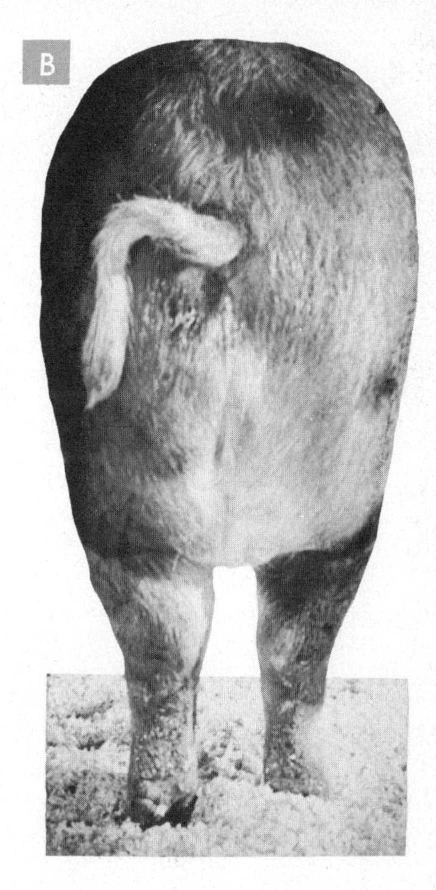

FIGURE 10-7. (*A*) Side view of a U.S. No. 4 slaughter hog. (*B*) Rear view of a U.S. No. 4 slaughter hog. (*Courtesy* U.S.D.A.)

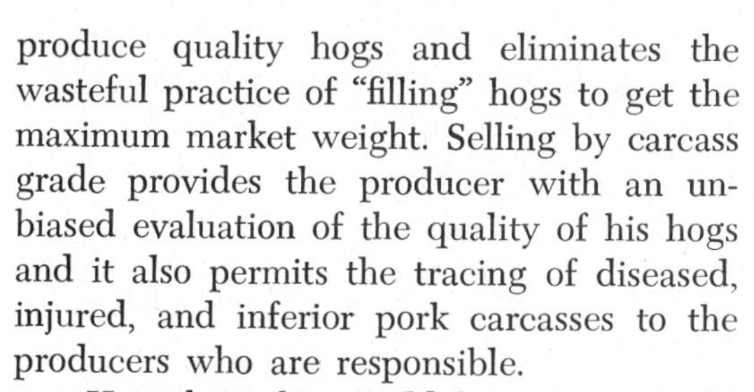

produce quality hogs and eliminates the wasteful practice of "filling" hogs to get the maximum market weight. Selling by carcass grade provides the producer with an unbiased evaluation of the quality of his hogs and it also permits the tracing of diseased, injured, and inferior pork carcasses to the producers who are responsible.

Hogs have been sold by carcass grade in Denmark, Sweden, Great Britain, and Canada, and the method has proved efficient and practical. Some packing plants in the Corn Belt have purchased hogs on this basis for several years. In general, breeders with high-quality hogs profit by selling their hogs on a carcass-grade basis, whereas hog producers with below-average hogs profit by selling them in the traditional manner.

Disadvantages of Selling Hogs by Carcass Grade

One of the chief reasons for delay in selling hogs by carcass grade has been that the packer-buyers had neither the facilities nor the personnel to do the job. Buying by carcass grade is less flexible than buying by weight. There is less opportunity for both buyer and seller to bargain. Quite often the grading is done in the absence of the seller, who may later question the results. Many farmers are not sufficiently informed in regard to carcass quality to question the packer's judgment. Another objection to the system is that the seller must wait until the hogs have been slaughtered and processed before he can be paid.

Future of Merit Buying of Hogs

Most interior packers in this nation are now buying hogs on a live-grade or grade-and-yield basis. As the carcass value increases $1.00 to $1.50 per hundredweight, the packer may pay a premium for each increase in grade. The spread between the price paid for U.S. No. 1 and U.S. No. 3

hogs has sometimes been too narrow to encourage producers to improve their breeding and feeding methods. Merit and grade-and-yield buying have increased rapidly as packers have become equipped and have obtained personnel qualified to do the job.

Less than 4 percent of the 1968 total marketings of hogs was sold on a grade-and-yield basis. Nearly 90 percent of those sold on this basis was sold in the Corn Belt. Nearly 16 percent of those sold in Missouri, 10 percent of those in Kansas, and 6 percent of those sold in Iowa was sold on a grade-and-yield basis. Many producers received an average of $1.50 or more per hundredweight about the price offered at the traditional market.

FIGURE 10-8. **(*A*) Hams, (*B*) loins, and (*C*) sides of bacon from U.S. No. 1, U.S. No. 2, and U.S. No. 3 slaughter hog carcasses. (*Courtesy* Rath Packing Company)**

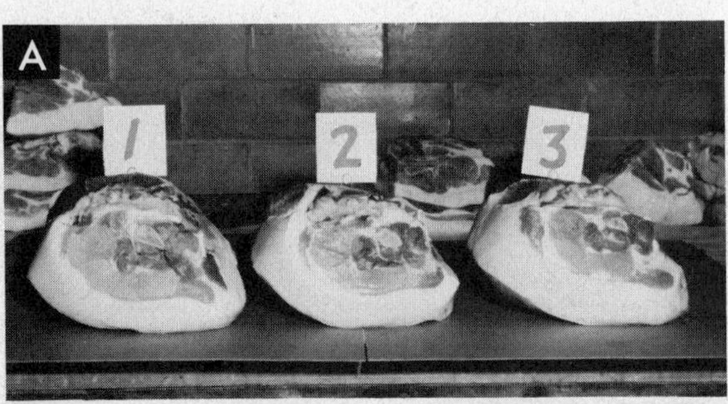

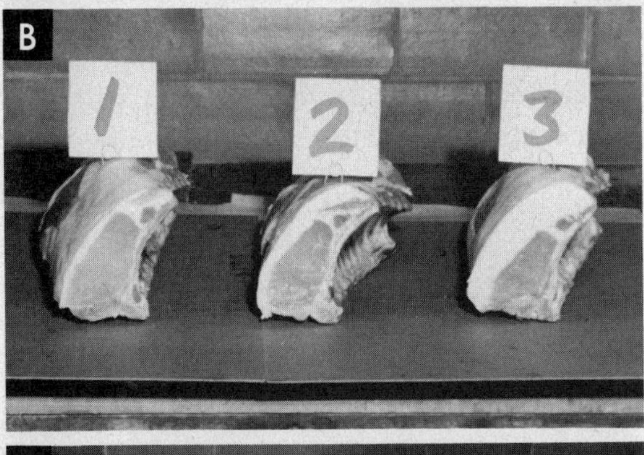

CHOOSING A MARKET

If it is assumed that during early March 87 fall-farrowed pigs are ready for market as U.S. No. 1 slaughter hogs, where should they be sold? Several times each year almost every hog raiser is confronted by this or a similar problem regarding the best hog market. The answer depends upon the locality.

Markets Available

In every community there are local hog buyers, dealers, and livestock auctions. In many communities hogs can be consigned and sold through a cooperative shipping association. Some meat packers have concentration yards or buying plants in the community. Hogs can be trucked to interior packing companies located 20 to 50 miles away, or they can be trucked to public stockyards at a central market, such as Joliet or Kansas City. They can be shipped direct to the packer or sold through a commission firm. What is best?

Factors in Choosing a Market

A number of factors determine the selection of a market. Distance, transportation problems, shrinkage, methods of grading, handling and selling charges, dependability, and price quotations are perhaps the most important.

To market effectively, one must know the price at each available market of the grade to be sold, and a choice will have to be made from among a half dozen or more markets. The U.S.D.A. Market News Service, the various cooperating radio stations, and the newspapers supply market information. At times, a telephone call may result in several dollars more in profits.

In addition to the price at the various markets for the kind of hogs to be sold, the cost of transporting them and the shrinkage to be expected must be known. Consignment to a terminal market requires a commission or handling charge. This may amount to 20 to 30 cents per hundredweight. The transportation cost will vary with the number of hogs to be transported and with the distance to be covered. The shrinkage and loss due to injury or death will vary with the method of transportation and the distance. These losses may be as little as 10 cents or as large as 60 cents per hundredweight.

Sixty-six percent of the hogs sold in 1968 was sold direct to packing plants or their buying stations or through country points. Terminal markets accounted for less than 20 percent and auction markets about 14 percent of the sales. The ten major packers bought nearly 75 percent of their purchases direct.

CARE IN MARKETING HOGS

A study completed by Livestock Conservation, Inc., indicated that nearly 10 percent of the hogs that arrived at packing plants were bruised. The average loss in dollars per animal bruised was $1.10. The swine industry's loss due to bruises in one year was estimated at $15.5 million.

Location of Bruises

Sixty-one percent of the bruises was on the hams, the source of one of the highest priced cuts. More than one-fourth of the bruises were on the back and loin, the source of rib and loin chops. Figure 10-10 shows bruised hams with the bruised sections removed. Hams and loins that have been trimmed because of the removal of bruised sections cannot be sold at the prices received for unbruised cuts.

FIGURE 10-9. Slaughter hogs being marketed at a central market. (*Courtesy* Central Livestock Association, Inc., St. Paul)

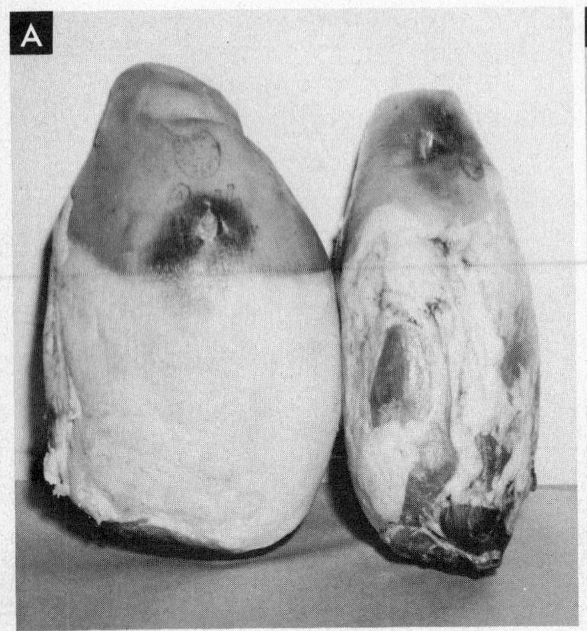

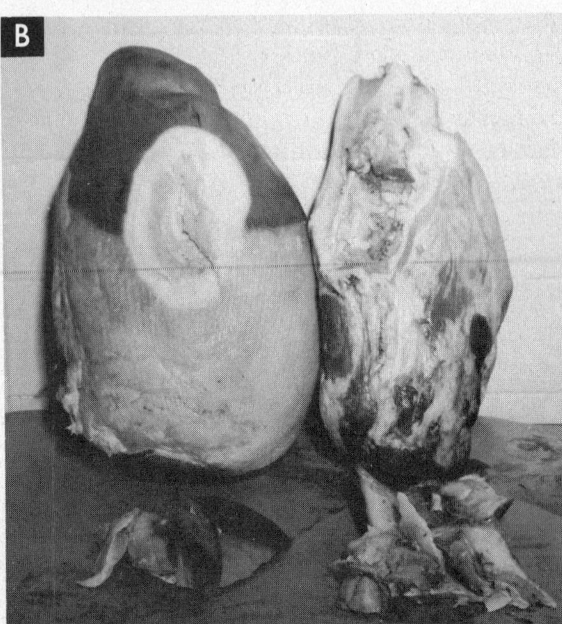

FIGURE 10-10. (*A*) Hams showing bruises and (*B*) the removal of the bruised sections. (*Courtesy* Rath Packing Company)

Cripples and Dead Animals

Surveys indicate that about 13 hogs in every 1,000 are dead on arrival at the market. It is estimated that, besides this loss, another 13 hogs in every 1,000 marketed arrive crippled.

Prevention of Losses

Care in preparing animals for shipment, in loading, and in handling hogs in shipment will greatly reduce losses due to bruises, crippling, and death. Following are suggestions for the proper handling of hogs at market time.

1. Do not feed hogs heavily before shipping.
2. Allow 3½ square feet of floor space in the truck for each 225-pound hog.
3. Clean the truck or car before loading.
4. Use sand for bedding in hot weather and straw—or sand and straw—during cold weather.
5. Separate the heavy hogs from the light hogs in the truck or car.
6. Wet down the hogs and bedding during hot weather.
7. Remove all boltheads, nailheads, and other obstructions from the loading chute and truck.
8. Have adequate loading equipment.
9. Handle the hogs quietly and with care.
10. Use canvas slappers in loading.
11. Do not put too few or too many hogs in a truck.
12. Separate hogs from other types of livestock being transported in the same truck.
13. In hot weather, move the hogs at night.
14. Close the side openings and put a cover over the truck during cold weather.
15. Drive carefully and avoid sudden stops.

WITHDRAWAL TIME FOR FEED ADDITIVES, ANTIBIOTICS, AND PESTICIDES

Hog producers are responsible for any chemical residues found in the carcasses of their hogs when slaughtered. The 1970 Federal law provides for a prison sentence and/or a $10,000 fine for offenders. It is important that care be taken in drug use. The following are musts in the proper use of drugs in swine production:

1. **Follow the label instructions.**
2. **Follow the withdrawal requirements.**
3. **Feed only the recommended dosages.**
4. **Do not mix antibiotics and drugs unless this is specifically recommended by the manufacturer.**
5. **Make certain that all feed bins and feeders are empty of medicated feed during the withdrawal period.**

Table 10-4 gives a list of feed additives, antibiotics, and pesticides and the withdrawal time required for each.

USING THE FUTURES MARKET IN HOG MARKETING

The futures markets on pork bellies and live hogs offer producers, packers, and others a means of taking much of the price risk out of the business, and also a means of establishing prices in advance. Many believe that, as pork production becomes more specialized with fewer but larger operations, increased participation in the futures market will take place.

Factors Affecting Participation in Futures Markets

Three factors will determine the extent to which hog producers will use futures markets:

1. **The magnitude or scope of the hog enterprise and its relationship to the total farm business.**
2. **The willingness and ability of the producer to bear the risk of falling prices.**
3. **The margin, or per unit profit, expected and necessary for successful operation of the enterprise.**

Advantages of Using the Futures Market

The futures market can protect the swine grower against declines in the cash market for slaughter hogs. It can provide the feeder-pig producers with an indication of the profitability of selling pigs as feeder pigs or as slaughter hogs.

Disadvantages of Using the Futures Market

Only a small percentage of the swine producers in this country are sufficiently knowledgeable about the futures market to use it well. Serious mistakes could be made. There is a cost in using the market. Some small growers do not have the necessary volume of production to use the futures market effectively.

Hedging

Hedging is the primary reason for futures markets. It is the sale or purchase of a futures contract against a presently held commodity or against price fluctuations. A farmer who will have 600 hogs to sell in January could hedge by selling these hogs on a futures market when the sows are bred, when the pigs are farrowed, when the pigs are weaned, or when feeder pigs are purchased. He would sell January futures and buy back a like contract when he sells the hogs.

If the market goes down, he can buy back for less than the contract price. If the market goes up, he has his guaranteed prices

TABLE 10-4 WITHDRAWAL TIME FOR SWINE FEED ADDITIVES, ANTIBIOTICS, AND PESTICIDES

Feed Additive	Withdrawal Time
Arsanilic Acid	5 days
Aureo SP-250 (chlortetracycline-sulfamethazine-penicillin)	7 days
Chlortetracycline	None
Bacitracin (all forms)	None
Hygromix (Hygromycin B)	48 hours
Furazolidone	None
Neomycin	None
Neo-Terramycin (neomycin-oxytetracycline)	None
Nitrofurazone	None
Oleandomycin	None
Penicillin (Procaine)	None
Penicillin and streptomycin	None
Piperazine	None
Roxarsone (3-Nitro)	5 days
Sodium Arsanilate	5 days
Terramycin (oxytetracycline)	None
Tylan (tylosin)	None
Tylan + Sulfa (tylosin + sulfamethazine)	5 days

Injectable Antibiotic	Withdrawal Time
Dihydrostreptomycin "combiotic"	30 days
Oxytetracycline (Terramycin)	10 days
Tylosin	4 days
Penicillin "combiotic"	5 days
Erythromycin	48 hours
Lincomycin	48 hours
Sulfonamides (oral also)	10 days *

Pesticide	Withdrawal Time
Toxaphene	28 days
Lindane (spray)	30 days
Lindane (dip)	60 days
Coumaphos (Co-Ral)	14 days
Carbaryl (Sevin)	7 days
Malathion	None

* Unless otherwise directed

Source: Iowa State University

but loses an opportunity to make the profit above the contract price.

Contract Provisions

The new live-hog futures contract is an agreement to deliver a unit of 30,000 pounds of U.S.D.A. No. 1 or 2 barrows and gilts averaging 200 to 230 pounds. Delivery is made at Chicago, with delivery at other specified markets allowable at 50 cents per hundredweight discount. Trading is done through the Chicago Mercantile Exchange. The contract allows up to 90 hogs of U.S. No. 3 and 8 head of U.S. No. 4 grade hogs in the 30,000-pound unit.

Live-hog contracts are traded every month in sequence. The commission charge is $30 per unit of 30,000 pounds. The initial margin demanded is $400 per unit with maintenance of $200, except during delivery month, when the two fees are $500 and $300 respectively.

Pork-belly contracts are for 30,000 pounds. The commission is $36, with the initial margin demanded being $700 and maintenance $400, except during month of delivery, when the margin is $1,000 and the maintenance deposit is $700. These contracts are also traded every month in sequence.

SUMMARY

Your breeding and feeding operations should be planned carefully in order to have U.S. No. 1 hogs ready when the market is best. A leaner, better-muscled market hog should be produced and should be sold when it weighs from 200 to 240 pounds. Because the market for lard has declined, less should be produced. A 300-pound hog produces nearly twice as much lard as does a 200-pound hog.

A meat-type hog can be produced more cheaply than can a lard-type hog. Light hogs produce 100 pounds of gain on less feed than do heavy hogs. Heavy hogs sell for less on the market than do 210-to-220-pound hogs.

To get the best price, spring pigs should be sold in July, August, and September and fall pigs in January, February, and June. The difference between high and low markets may be as much as $4 to $6 per hundredweight.

The U.S.D.A. grades are No. 1, No. 2, No. 3, No. 4, and Utility. Hogs should be sold as they reach the No. 1 grade. The U.S.D.A. grades are based largely upon leanness of carcass, backfat percentage, length of carcass, finish, and weight. Selling hogs by grade will encourage farmers to improve the quality of hogs they produce.

Packers who buy hogs according to grade and yield usually pay higher prices for choice hogs and lower prices for lardy hogs than do buyers who do not grade the hogs. A market should be chosen carefully. The net return after transportation, shrinkage, and other costs have been deducted is the determining factor.

About two-thirds of the hogs marketed are sold direct to packers. Less than 20 percent is sold through terminal markets, and about 14 percent pass through the livestock auctions.

Hog producers must follow the directions of suppliers of feed additives, antibiotics, and pesticides to make certain that drug withdrawal requirements are met. Producers are held responsible for residues found in the dressed carcasses.

Swine growers may hedge their enterprise earnings by the sale or purchase of a futures contract for a presently held commodity or for commodity needs, to protect them against price fluctuations.

Nearly 10 percent of the hogs arriving at packing plants is bruised. About 13 hogs in every thousand arrive at the plant crippled and 13 arrive dead. Care in trucking and shipping will prevent losses, injuries, and bruises. The truck should be carefully bedded down and should not be underloaded or overloaded. Each 200-pound hog should have 3½ square feet of floor space. Heavy hogs should be separated from light hogs, and hogs should be separated from other livestock in the truck.

QUESTIONS

1. Why does the packer pay more per hundredweight for a 225-pound hog than for a 280-pound hog?
2. What percentage of the carcass of a hog should be hams and loins?
3. What is the difference between the present U.S.D.A. grades and the old grading system?
4. What qualifies a hog to be graded U.S. No. 1?
5. Which will make the most economical gains, a 100-pound pig, a 200-pound hog, or a 300-pound hog?
6. Does it cost more to produce a meat-type than a lard-type hog?
7. In which months should spring pigs be sold to bring in the most money? When should fall pigs be sold?
8. How much variation occurs in the price of light hogs during the year?
9. How will you select the market through which you will sell your hogs? What factors will you consider?
10. What precautions do you need to take in transporting market hogs?
11. Would you prefer to sell hogs by carcass grade or by weight? Why?
12. What precautions must a producer take to make certain that there are no residues of drugs in the carcasses of hogs that he sells?
13. What use can you make of the futures market in the marketing of hogs on your farm?
14. How can you improve the income from your home hog enterprise through better marketing practices?

REFERENCES

Bundy C. E. and R. V. Diggins, *Swine Production*, 3rd Edition, Prentice-Hall, Inc., Englewood Cliffs, New Jersey, 1970.

Fowler, Stewart H., *The Marketing of Livestock and Meat*, The Interstate Printers & Publishers, Inc., Danville, Illinois, 1961.

Snowden, O. L. and Alvin Donahoo, *Profitable Farm Marketing*, Prentice-Hall, Inc., Englewood Cliffs, New Jersey, 1960.

U. S. Department of Agriculture, *How Do Your Hogs Grade?*, Marketing Bulletin No. 16, Washington, D. C., 1961.

———, *Livestock and Meat Statistics,* Washington, D. C., 1973.

———, *Livestock Slaughter,* MtAn 1-2-1, Washington, D. C., 1970.

———, *U.S.D.A. Grades for Slaughter Swine and Feeder Pigs,* Marketing Bulletin No. 51, Washington, D. C., 1970.

———, *U.S.D.A. Grades for Pork Carcasses,* Marketing Bulletin No. 49, Washinging, D. C., 1970.

BEEF PRODUCTION

11 The Beef Industry

"We're having steak for dinner." These words seldom have to be repeated to bring an American family to the dinner table. Beef is one of the more important meats in the American diet. It is high in nutritional value, and many people consider it first in flavor. We consumed an average of 114.5 pounds of beef per person during 1971.

With our ever-expanding population, there will be an even greater demand for beef in the future. Beef producers must meet this challenge. They must produce at a profit more beef per acre of farm land, and they must furnish beef to the consumer at a price he can afford to pay.

Opportunities in beef production are numerous. Almost any young person with a love for livestock, an eagerness to learn, and a willingness to work can find a life of satisfaction in some phase of the beef production industry.

CLASSES OF BEEF CATTLE PRODUCERS

The three main divisions of the beef cattle production industry are: (1) the production of feeders, (2) the production of finished or slaughter cattle, and (3) the purebred beef cattle industry. It is also possible under certain conditions to develop a program combining two or more of the above phases of the beef industry.

Production of Feeders

The production of feeders is carried on primarily by producers located on lands that are not suitable for heavy crop production other than grass.

These farmers and ranchers maintain herds of cows to produce calves. The calves are usually dropped in the spring and run with the cows on the pasture or the range during the grass season. In the fall the calves are weaned from their mothers and either sold as feeder calves or carried through the winter on hay, pastured the next summer, and sold as yearlings to be finished for slaughter.

It is not necessary to have grain for feed in feeder cattle production. The cow herd may be successfully maintained on pasture in summer and on hay or pasture during the winter; therefore, such a program is well adapted to land areas not suitable for rotation crops.

Cattle can grow and produce on rations containing mostly roughages. This makes it possible to develop a profitable business on what may otherwise have been waste land.

Many young farmers with capital too limited to purchase high-priced crop land

FIGURE 11-1. (*Above*) Skills and knowledge required for success in the beef business, combined with a quality herd of cattle, offers a satisfaction that cannot be entirely evaluated in terms of dollars and cents. (*Courtesy* American Hereford Association) **FIGURE 11-2.** (*Below*) The Corn Belt offers a wide variation in the different classes of beef production. This quality purebred herd is owned by Leveldale Farms, Calamus, Iowa. (*Courtesy* American Angus Association)

may find opportunities in the purchase of low-priced grassland which can be used in the production of feeder cattle. Successful feeder cattle production depends largely upon having a good supply of cheap forage crops available.

Production of Slaughter Cattle

Farmers who make a business of buying feeder cattle and feeding them for market are called "cattle feeders." Since grain is usually essential in finishing high-quality beef, most cattle feeders are located in a feed grain area.

Profits from feeding cattle come from two sources: (1) the selling price over purchase price, which is known as margin, and (2) the value of the increased weight over the cost of the feed. If a feeder buys cattle weighing 700 pounds at $28 per hundredweight and sells them after finishing for $30 per hundredweight, he has made a margin of $2 on each 100 pounds of original weight, or a total margin of $14 per head. If the cost of the feed and other expenses of the feeding operation amounts to $20 per 100 pounds of gain, then the feeder has made a profit of $10 per 100 pounds of gain. If he increases the weight of these cattle to 1,100 pounds, representing a gain of 400 pounds per animal, his profit on gain is $40. His marginal profit is $14, giving a total profit of $54 per head.

The principal reasons cattle feeders buy and finish cattle are: (1) to receive a higher price for their grain when sold as beef, and (2) to increase soil fertility by spreading the manure on the land. If the manure is properly handled, a large part of the plant food removed by the crops may be returned to the soil. Farms that have had a cattle-feeding program over a period of years are generally high in fertility.

The Purebred Beef Cattle Industry

The breeder of purebreds produces high-quality bulls and cows to improve the breed and to provide bulls for use by the

FIGURE 11-3. A large feed lot operation. Note the grain and roughage storage facilities in the background. (*Courtesy* Hawkeye Steel Products, Inc.)

producers of feeder cattle. He maintains a pedigreed herd of breeding stock. He sells to feeder cattle producers and to other breeders. Quality purebred animals usually bring premium prices. Commercial cattlemen turn to the breeder for bulls, and occasionally for cows or heifers for use in improving their herds. The breeders of purebreds can rightfully claim most of the credit for the improvement of the beef cattle breeds.

Probably more skill, knowledge, and patience are required for success in the purebred business than for any other phase of beef production. The breeder of purebreds must know the type of cattle that are in demand. He must keep in mind the kind that will make the most money for the producer and the feeder, and the type that will produce a carcass that can be cut to suit the consumer.

More capital is needed per animal for purebred production than for any other phase of the beef cattle business. Foundation stock is usually expensive, and equipment needs per animal are greater than they are for feeder or slaughter cattle production. Improvement is slow, and it may be many years before a herd of quality animals is developed. Before the breeder can reap substantial returns from his effort he must have a large number of quality animals to sell. Although it is not essential, experience in growing and finishing commercial cattle may be helpful before attempting to develop a purebred herd.

Large amounts of grain are not essential in purebred production. Bulls and females that are being fitted for shows need grain in order to produce the finish necessary to bring out the type and quality of the animal. Animals carrying a medium degree of finish are usually more attractive to prospective buyers.

Combining Two or More Enterprises

Under certain circumstances, two or more beef enterprises may be combined profitably. An example might be the program on a half-section farm with 160 acres

FIGURE 11-4. Purebred Angus bulls bred and raised by Angus Valley Farms, Tulsa, Oklahoma. These bulls will be sold for breeding purposes to improve both purebred and grade herds. (*Courtesy* American Angus Association)

of land capable of producing good grass, but too hilly for rotation cropping; the other quarter-section may be level land, in rotation crops. A herd of stock cows may be kept to consume the pasture and part of the hay, but since there probably will not be enough corn to feed out the calf crop, some of the calves may be sold as feeders. The remainder may be placed in the feed lot and finished for slaughter.

SUMMARY

There are three major divisions of the beef cattle industry:

1. **The production of feeders.**
2. **The production of slaughter cattle.**
3. **The purebred beef cattle industry.**

Lands which are not suitable for rotation crops but which will produce grass are best adapted to feeder cattle production. Feeders are generally sold to commercial feedlots or grain-producing farmers who finish them for slaughter. The farmer who feeds cattle depends upon the difference in the price of beef over feed and operating costs, or the difference (margin) in the buying and selling price for his profit.

The breeder is responsible for furnishing improved stock for commercial cattlemen and for other breeders. Breeders of purebreds generally need a higher capital outlay per animal than do ranchers or cattle feeders. Breeders require considerable knowledge of type and mating if they are to be successful. Income from purebred herds may be high because quality animals usually bring premium prices. Farms producing a combination of grain and grass in large quantities may be suited to two or more beef enterprises.

QUESTIONS

1 Which beef cattle enterprise is best adapted to your farm? Why?

2 What type of land or farm is best suited for developing each of the three types of beef cattle enterprises? Why?

3 Give an example of how two beef enterprises could be combined successfully on one farm.

4 Why is more knowledge and skill required for success in breeding beef cattle than in the other beef production programs?

5 Which program requires the most capital per animal? Why?

6 What determines the profit in each type of the beef production enterprises?

REFERENCES

Diggins, Ronald V., and C. E. Bundy, *Beef Production*, Englewood Cliffs, New Jersey, Prentice-Hall, Inc., 1971.

Ensminger, M. E., *Beef Cattle Science*, Danville, Illinois, The Interstate Printers and Publishers, 1968.

Fowler, S. H., *Beef Production in the South*, Danville, Illinois, The Interstate Printers and Publishers, 1968.

Snapp, Roscoe R., and A. L. Neuman, *Beef Cattle*, Sixth Edition, New York, John Wiley and Sons, Inc., 1969.

12 Selection of Breeding and Feeding Stock

Upon entering the beef cattle business, the breed of cattle to be produced must be decided. The term *breeder* generally applies to the producer of improved purebred animals for breeding purposes. Most of the early developmental work on beef cattle breeds was started in the eighteenth century by British breeders. During this period, British cattle breeds were imported into the United States and until about 1900 were the only important breeds used in the improvement of cattle in this country. Since 1900, cattle from India, France, Switzerland, Germany, Italy and Australia have been imported to develop breeds that could be adapted to certain sections of the United States. Many of these breeds have found favor among cattlemen who feel that they have a definite place in American agriculture.

SELECTING THE BREED

Before a start is made on a cattle breeding program of either purebred, grade or crossbred animals careful consideration should be given the following: (1) personal likes, (2) availability of breeding stock, (3) outlet for surplus animals, and (4) environmental conditions under which the animals will be raised.

Personal Likes

Most cattle breeders develop a liking for certain breeds of livestock. Unless other conditions make it inadvisable to select a breed of personal choice, that factor should be given due consideration before making a start in breeding cattle.

Availability of Breeding Stock

The cattle breeder must be continually on the lookout for animals he can bring into his herd for the purpose of improving his own cattle. Other breeders of the same breed are his main source of supply. If there are few in his area, he will have to travel long distances to obtain replacement animals and they will cost more.

Outlet for Surplus Animals

The breeder of purebreds depends upon the other purebred and commercial or non-purebred breeders for the sale of bulls and surplus females. It is important that demand for the breed be considered before selecting one.

Grade or crossbred herd producers generally find cattle feeders their best market outlet. Although cattle feeders are less concerned about breed than are cattle breeders, certain areas tend to show partiality for certain breeds. If the producer of commercial cattle has a good market in a certain locality, he will be wise to give consideration to the breed that is most popular among those feeders.

Environmental Conditions

Weather, grazing, disease, and insect conditions are important factors in selecting a breed of cattle. Brahman cattle, for example, are able to make good use of poor forage, are not particularly bothered by flies, ticks, or mosquitoes, and are resistant to Texas fever, all of which makes them and their crosses especially valuable in certain Southern areas. The Hereford is noted for hardiness and foraging ability but is subject to certain diseases and cannot be raised where they are prevalent. The Angus, Charolais, Limousin, Shorthorn, and Simmental also have special merits under certain circumstances. Cattle best adapted to the conditions under which they will be raised should be considered by the prospective producer.

BREEDS OF BEEF CATTLE DEVELOPED IN EUROPE

Aberdeen-Angus

The Aberdeen-Angus is one of the most popular breeds of beef cattle in the United States. They originated in northern Scotland in a cool damp climate. The first importations of these Scottish cattle to play an important part in the United States was made in 1873 by George Grant, a native of Scotland then living in Victoria, Kansas. Many more importations were made during the latter part of the nineteenth century and have continued up to the present time. However, as the breed became well established in America, importations gradually declined. Today most of the great sires of the breed in the United States were produced in this country.

The Aberdeen-Angus cattle are black; white is not permitted except on the underline behind the navel and there only to a moderate extent. The breed is polled (without horns), which has contributed to Angus popularity among many breeders. The polled characteristic in Angus cattle is so well established that when Angus are crossbred with horned breeds most of the first cross offspring are polled. Angus cattle have shown more resistance to certain eye diseases, particularly cancer eye and pinkeye, than have some of the other breeds. Calves from Angus cows are usually smaller at birth than are calves from other breeds, but the weights are equal at weaning to most other breeds. The smaller calves at birth cut down calving difficulties, and there are fewer cows and calves lost at this time.

The body form of Angus cattle is smooth, trim, medium in size, adequate in length of side, and well muscled.

Red Angus

Occasionally a red animal crops out from a herd of black Angus even though both parents are black. This is because many black Angus carry a red gene. (See Chapter 42.) Red Angus have existed for years but it is only recently that a group of breeders has been organized for the purpose of breeding Red Angus in the United States. The Red

Angus breeders have adopted high standards for registration. Each animal must be inspected by a committee and approved before registration.

The general characteristics are similar to those of the black Aberdeen-Angus (since they have the same ancestry), except for color. The Red Angus has a deep red color that is very attractive.

Devon, or Ruby Reds

Devon cattle were developed in England, and are one of the oldest of the English breeds. Records show that several im-

FIGURE 12-1. (*Above*) Note the muscle development of this young individual bred by the Kaydell Angus ranch, Watsonville, California. The goal of producing high-quality, tender, juicy red meat is within reach when breeding stock such as this is used. (*Courtesy* American Angus Association) FIGURE 12-2. (*Below*) A satisfied feeling comes to individuals in the beef industry when pictures like this one, taken at the Meier Angus Farm, Jackson, Missouri, exhibit modern beef quality. (*Courtesy* American Angus Association)

portations of Devon cattle were made into this country during its early history. The Devon is red in color. The skin is yellow and the head supports medium-sized, upward-curving horns that are creamy white with black tips.

Cattle placed in the finishing lot have produced high quality beef. Devon cows are good milkers, producing milk testing about 4.5 percent butterfat.

Galloway

Galloway cattle are native to Scotland. This breed is considered one of the oldest of the British breeds. They were probably introduced into the United States about 1860. For a period following their introduction they became popular through the North Central states. However, probably because of the slower development of the Galloway, their popularity has steadily declined.

Galloways are good rustlers and extremely hardy, able to stand cold weather conditions. They are the smallest of the beef breeds, black in color with long, curly hair. The breed is polled, has short legs, and is blocky and compact in type.

Hereford

Hereford cattle are native to England. They originated in the county of Hereford, which lies in the fertile valley between the Severn River and the eastern boundary of Wales.

The first breeding herd of Herefords to play an important part in establishing the

FIGURE 12-3. (*Right*) The meat type. This Red Angus bull was named a certified Meat Sire by Performance Registry International. To qualify, ten or more of his progeny had to meet PRI's standards for weight, age, carcass quality, and carcass cutability. (*Courtesy* Red Angus Association of America) FIGURE 12-4. (*Below left*) This Galloway bull has proven to be one of the very top herd sires of the breed. (*Courtesy* American Galloway Breeders Association) FIGURE 12-5. (*Below right*) An excellent Hereford bull. Note the desirable length of body and degree of bone. (*Courtesy* American Hereford Association)

FIGURE 12-6. (*Left*) **Production records and applied management help in producing breeding stock with quality traits.** (*Courtesy* **American Hereford Association**)
FIGURE 12-7. (*Right*) **A Polled Hereford female that would improve most herds.** (*Courtesy* **American Polled Hereford Association**)

breed in the United States was that of William H. Sotham and Erastus Corning of Albany, New York, in 1840. The Hereford breed grew rapidly in the United States, and today there are more Herefords than any other breed in this country. The breed is popular from coast to coast and constitutes, by far, the largest percentage of the cattle found on the Western range.

Hereford cattle are easily distinguished by their red-colored bodies and white faces. The accepted color is a rich red with white face, and they are often referred to as "white-faced cattle." The white is also found on the flank, underline, breast, crest, tail switch, and below the hock and knees on both fore and hind legs.

In form Hereford cattle are muscular, moderate to long in length of side, adequate in length of leg, large in size, trim, and smooth. They are well developed in the regions of valuable cuts—the back, loin, and hind quarters or round.

The Hereford breed is well known for its vigor and foraging ability.

Polled Herefords

In 1900, Warren Gammon of Iowa wrote to nearly every breeder of Herefords in the United States asking if they had any cattle which did not develop horns. He succeeded in securing 13 head of purebred Herefords that were polled. From this small beginning the polled Hereford breed was established.

The breed has become very popular among breeders who desire the Hereford form but dislike the horns. Polled Hereford originating from registered Hereford stock may be registered in both breed associations.

In form and characteristics, the polled Herefords closely resemble their ancestors, the Herefords. The distinguishing difference is the absence of horns.

Shorthorns

The Shorthorns originated in northeastern England in an area which includes the counties of Durham, Northumberland, and York.

FIGURE 12-8. (*Left*) **An ideal Shorthorn bull. Note the desired length of body and trimness throughout.** (*Courtesy* **American Shorthorn Association)** **FIGURE 12-9.** (*Right*) **Big white bulls like this one are examples of the ideal type of Charolais cattle.** (*Courtesy* **American-International Charolais Association)**

The breed, introduced in 1783 by Miller and Gough of Virginia, was the first to be established in America. These cattle gained rapidly in popularity, and are found throughout the United States today. They represent one of the three most popular breeds.

In form, Shorthorns are large, rectangular, and well muscled. They range in color from red to white and all combinations of these colors, such as spotted or roan. Shorthorns are well liked by many commercial cattlemen for crossing on other breeds for the production of feeder cattle.

Polled Shorthorn

The Polled Shorthorns were developed by a cross and from naturally polled Shorthorns found in the breed. Most present-day Polled Shorthorns are descendants of purebred Shorthorn cattle, and are eligible for registration in the American Shorthorn Breeders' Association herd books.

In form and color the Polled Shorthorns are similar to the Shorthorn except for the polled characteristic.

Charolais

The Charolais originated in France and is one of the most important breeds of French cattle. Only a small number of Charolais cattle have been imported into the United States, and most of them have gone to Texas, Louisiana, and Florida, where they have been used for crossing purposes, especially with Brahmans. Charolais cattle are now found in all areas which produce beef cattle. They have been used to a large extent in crossbreeding programs.

Charolais are light cream colored and are one of the largest of all beef breeds. They are big, long-bodied, heavily muscled animals but lack the smoothness of the British breeds.

Scotch Highland

The Scotch Highland breed of beef cattle was developed in the Hebrides, a group of islands near the west coast of Scotland.

The breed has not been popular in the United States, but a few have been imported from time to time. They have found favor among some ranchers in the northern plains for crossing on other breeds to produce animals more capable of withstanding the long, hard winters.

Scotch Highland cattle are small but exceedingly hardy. They have a long, coarse outer hair coat and a soft, thick undercoat which gives them natural body protection against severe weather conditions. Acceptable colors are black, brindle, red, light red, yellow, dun, and silver.

Limousin

The Limousin breed originated in France in an area similar to our Ozark region. Genetically the breed is unique in that it seems to have originated and developed in relative isolation with very little outside genetic influence. The first Limousin bull was imported to Canada in 1967 and the first Limousin-cross calves arrived in North America in the summer of 1969.

The breed is distinguished by its rich, red-gold color over the back, shading to light buckskin or straw color under the belly and around the legs and muzzle.

In size they are slightly smaller than the French Charolais but average heavier and larger than British breeds. In conformation they are long-bodied, heavy-muscled, trim-middled, and relatively light-boned, therefore claiming remarkable carcass yield and cutability. Calving has not been a problem since birth size is smaller than some of the other imported breeds. The purpose of the breed in the United States will be for beef crossbreeding.

Simmental

The name Simmental is derived from the name of the valley of origin in Switzerland, Simmen Valley. The Simmental, one of continental Europe's most widely distri-

FIGURE 12-10. (*Left*) **A typical Highland female, showing depth of heart, good bone, and straight top.** (*Courtesy* **American Scotch Highland Breeders Association)**
FIGURE 12-11. (*Right*) **An excellent example of the modern Limousin breed.** (*Courtesy* **North American Limousin Foundation)**

buted breeds of cattle, has made a dramatic entrance into the beef production picture of North America.

In color they are light red or cream with faces much like the Herefords. They usually have some white spots or a white band over the shoulders. The breed is known for its large size—mature cows weigh from 1,400 to 1,800 pounds and bulls from 2,400 to 2,800 pounds. Because of their size, muscling, docile dispositions, and milk production they are popular for beef crossbreeding.

Maine-Anjou

The newest of the so-called "exotic" breeds, the Maine-Anjou, a product of France, is the result of continued crossings between the Shorthorn and the Manceau. French breeders began shortly before 1830 to combine the Manceau's hardiness, vigor and finishing traits with the Shorthorn's early maturity.

The coat color is red, red with white spots, or roan. Maine-Anjou cattle are considered larger than any other French breed and are long-bodied as well as heavily muscled. These traits are useful in beef crossbreeding programs.

Chianina

The Chianina originated in the Chiana Valley of Italy and is thought to have been used as a sacrificial beast when Rome was in its glory. The main purpose of the breed is to produce meat, but they also function as draft animals in some parts of the country.

Chianina cattle are white with a black switch on the tail. Their skin pigment is uniformly black. They are quite possibly the largest cattle in the world, with many bulls weighing over 4,000 pounds. They are docile in disposition and adapt well to many pasture and climatic conditions. They are noted for rapid growth and well-marbled, fine-textured meat. In the United States the breed should find its way into many crossbreeding programs.

BEEF CATTLE DEVELOPED IN INDIA

Brahman

Several breeds of cattle exist in India. Most of them have been named after the Indian province in which they have been developed. In Europe and South America, they are known collectively as Zebu, and in the United States they are called Brahman. They are the oldest existing breed of domestic cattle.

The first cattle of this breed to play a part in the development of Brahman cattle in this country were two bulls given to Richard Barrow of Louisiana in 1854.

In recent years, considerable interest has been shown in the development of Brahman cattle in the South. A number of Brahman crossbred feeder cattle have reached

FIGURE 12-12. Simmental cattle are increasing in popularity, especially in crossbreeding programs. (*Courtesy* American Simmental Association, Inc.)

FIGURE 12-13. A Brahman cow. American Brahmans are good mothers and heavy milkers, and raise big, thrifty calves. (*Courtesy* American Brahman Breeders Association)

FIGURE 12-14. Progeny from this bull have been consistent with excellent scale, fleshing, eveness, thickness of hindquarters, and well-balanced frames. (*Courtesy* Murray Grey Breed Association)

Midwest feed lots and have given good results as feeder cattle.

Brahman cattle are characterized by a large hump over the shoulders and loose skin in the area of the dewlap. They have drooping ears, and instead of the "moo" of other cattle they produce a sound resembling a grunt. The most prevalent color is some shade of gray, although red is also acceptable. Another breed trait is longevity. Productive cows 15 and 20 years old are not uncommon.

In form the Brahman cattle are more upstanding, less compact, and lacking in the smoothness of other breeds.

Brahman cattle are resistant to Texas fever, can stand heat well, and are bothered little by flies, ticks, and mosquitoes. They are able to produce beef when grazing on poor-quality forage on which many other breeds would fail. These characteristics have made Brahman and crossbred strains, developed by using Brahman on other breeds, very popular in areas of the Southern part of the United States. The crossbred calves have produced gains and carcass quality equal to those of any of the other breeds. Like the Angus, they show resistance to cancer eye and pinkeye.

BREEDS OF BEEF CATTLE DEVELOPED IN AUSTRALIA

Murray Grey

The Murray Grey is native to Australia and gets its name from its gray color and place of origin, the Upper Murray River region. The animals are polled and come originally from Shorthorn and Angus cross.

Murray Grey cattle are solid dark to silver gray in color. In size the breed is larger than our British breeds. They are known for their rapid gains, superior carcasses, easy calving, and docile dispositions.

Direct imports of cattle from Australia are not permitted because of health regulations. The first Murray Grey semen was imported to the United States in the summer of 1969.

BREEDS OF BEEF CATTLE DEVELOPED IN THE UNITED STATES

Several breeds of beef cattle have been developed in the United States. Some of these breeds were developed by using Brahman crosses on European breeds. The objective was to combine the Brahman ability to graze poor-quality forage and their resistance to insects and heat with the smoother, heavier-muscled qualities of the European breeds.

The need for beef cattle that could withstand the hot, humid climate, the pests, and diseases prevalent in many sections of the South was the primary factor in creating an interest among Southern farmers, ranchers, and experiment stations toward the development of new breeds. These breeds have played an important part in changing much of Southern agriculture from a one-crop system to a cropping and livestock program. The results have been improved soil fertility, conversion of the forage of untillable land into beef, and increased income to the farmers and ranchers of the South.

Santa Gertrudis

This breed was developed on the Santa Gertrudis division of the King Ranch in southwest Texas. It resulted from crossing Brahman beef-type bulls on beef-type Shorthorns.

The Santa Gertrudis is approximately three eighths Brahman and five eighths Shorthorn. They are large beef animals; mature cows attain weights of 1,600 pounds and mature bulls 2,000 pounds on pasture. They are solid cherry-red in color and horned. The ears are somewhat pendulant. They are smoother and more compact than the Brahman, but retain the loose hide and underline skin folds characteristic of their Brahman ancestry.

The breed is especially adapted to subtropical climates and semiarid grazing conditions. They are noted for their ability to make large gains on grass and to rustle on areas of sparse forage, and for their tolerance to heat and insects.

FIGURE 12-15. A Santa Gertrudis bull. (*Courtesy* Santa Gertrudis Breeders International)

Other breeds developed from Brahman and European crosses are the Brangus (Brahman × Angus), Beefmaster (Brahman × Hereford × Shorthorn), Charbray (Charolais × Brahman), and Braford (Brahman × Hereford).

Crossbred Cattle

For commercial cattle breeders, crossbreeding has brought out some definite advantages over purebreeding or straightbreeding. Experimental work of the U.S.D.A. and state experiment stations has shown that a rotation system of crossing three breeds results in somewhat heavier calves at weaning time and faster feed-lot gains. Crossbred heifers reach puberty at an earlier age, have a higher conception rate, and have a greater calving percentage during the first thirty days of calving season.

There are also disadvantages to be considered when crossing breeds of cattle. Crossbreeds usually are mixed in color and lack the uniformity in type of the straightbred cattle. Uniformity in color, type, and size adds to the appearance of a group, and uniform cattle generally sell better, especially if sold as feeders.

SELECTION OF FOUNDATION BREEDING STOCK

After having decided upon the breed, individual animals for foundation stock must be selected. Good and inferior animals exist in all breeds. Regardless of the breed, certain general characteristics that contribute to beef production should be understood and used as a basis of selection. The breeder of purebreds will be concerned with individual breed characteristics which animals must have if they are to be eligible for registration. This information can be secured from the breed associations.

The producer of grade animals is usually less concerned about breed disqualifications, but should carefully consider those factors that contribute to economical beef production.

The building of a good herd of breeding cattle is a long-time proposition. It should be remembered that the cost of feeding and managing an inferior herd is equal to that of a good herd. While the initial cost of superior foundation animals may be high, the long-time cost, in relation to income from the herd, will be less than that for inferior foundation stock. It is better to buy a few good animals than a large number of poor ones, if the goal is the establishment of a high-quality herd.

In selecting foundation stock, the breeder should consider two sets of factors: (1) those he can see in sizing up the individual animal, and (2) those in which

FIGURE 12-16. Lack of uniform color, evident in this picture, has been a drawback in selling crossbreeds to some feeders. Rate of gain is usually equal or superior to straightbred cattle. Most people cannot tell when viewing the carcass what color the hide was. (*Courtesy* Elanco Products Company)

he must rely upon production records for his information.

Determining Desirable Body Conformation

One can judge desirable conformation by closely inspecting the animal in question. If possible, a similar inspection should be made of the sire, dam, sisters, brothers, and other closely related animals. By inspecting close relatives of the animal being considered for foundation stock, one can determine, to some extent, whether the line breeds true to type. It is not uncommon to find an attractive animal from a strain in which few good ones exist. Such an animal may be disappointing in the quality of his or her offspring. Since animals may inherit poor qualities from ancestors several generations back, it is important to observe as many representatives of the line as possible to determine to what extent undesirable qualities are cropping out.

Points to Consider

Fast-growing animals are important to economical production. A foundation animal should be at least average in size for its age and breed. The animal should move freely and be heavily muscled. In type it should have moderate length of body and legs with uniform depth from front to rear. While it adds little to carcass value, a tail head that blends smoothly into the rump improves appearance. Good body depth and a well sprung rib increase feed capacity. A straight top and underline is considered desirable. The leg should be straight, the bone clear cut, dense, and of moderate size.

The hind quarters should be wide, plump, heavily muscled, and well developed in proportion to the rest of the animal. The thighs should be thick, carrying flesh down to the hocks. The twist (distance between hind legs to the top of rump) should be deep, not cut up between the hind legs. The animal should be well muscled over the back, loin, and rump. On cows or heifers of calving age the udder should not be fleshy. It should extend well forward and well up behind, with teats squarely placed, well apart, and of good size.

The desired head is broad, short, slightly dished, and clean cut. The eyes are full and expressive with good width between them. Distance from the eyes to nostrils should be of moderate length. The muzzle should be wide and flaring and the nostrils open. The shoulders should be smooth and broad on top. The brisket must be wide, moderately deep, and free from flabby flesh and wrinkled skin, and not too prominent. The forelegs should be wide apart, allowing for good width on the chest floor. Bulls should possess pronounced masculinity with a well-developed crest. Females should show refinement and should give indication of being good producers.

As a final step in breeding stock selection step close to the animal and, keeping the hand flat, feel down over the back, loin, and ribs. This procedure will tell the amount and uniformity of fleshing over the region of valuable cuts. Breeding cattle should not carry too much finish. However, the covering should be uniform and free from patches and lumps. Lean meat, as indicated by heavy muscling, is more important than excessive fat. The hide should be pliable and of medium thickness and the hair should be fine and soft.

Passing Final Judgment on Breeding Stock

Remember there is no perfect animal. However, after carefully considering all

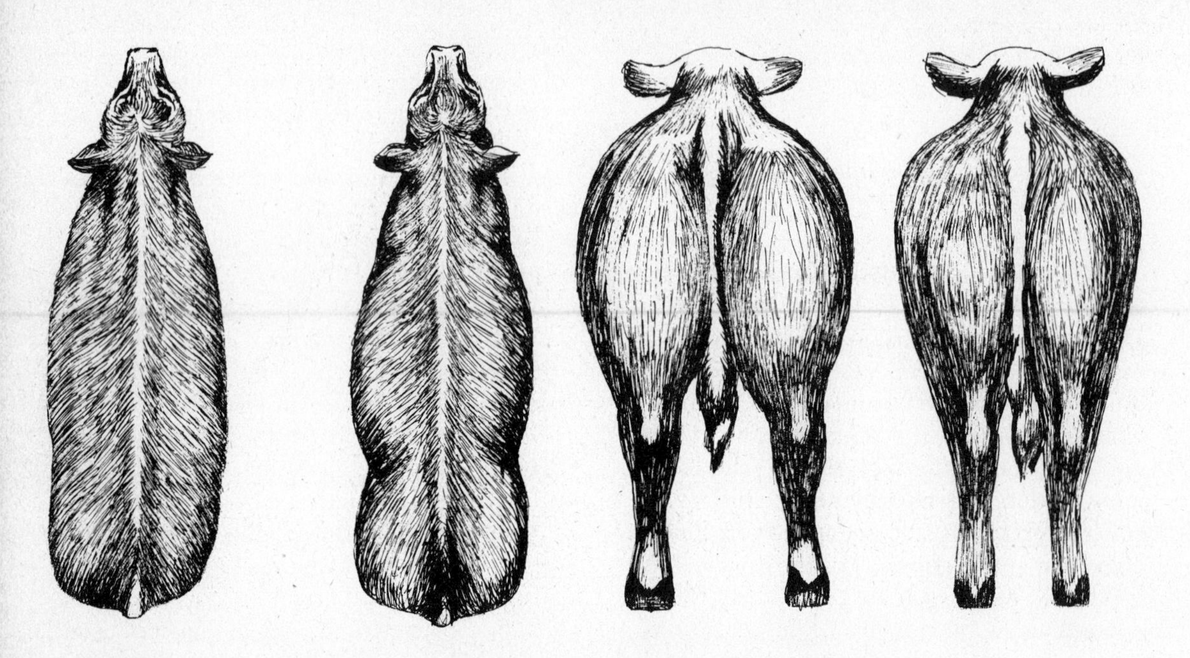

FIGURE 12-17. (*Left*) **The back, loin, and rump should be uniform in width and should show muscle development. FIGURE 12-18.** (*Right*) **A thick, heavily muscled quarter is desirable. A shallow, narrow, flat quarter is undesirable.**

points listed here, if the animal has scored well compared to the average of the breed, then we may be sure the general type is satisfactory.

It is important to try to correct weaknesses that exist in the herd. For example, in buying a bull to breed with cows that are rough in the tail head, a bull should be selected that will help to correct this fault. Many times in buying additions to the breeding herd other characteristics may have to be compromised on in order to correct a herd fault.

Most cattlemen will agree that type, development in the regions of the most valuable cuts, size for age, health, and vigor should receive the most emphasis when buying breeding stock for commercial herds. Breeders of purebreds will have to give special consideration to characteristics that may disqualify an animal for registration. The other factors mentioned should be considered but are secondary to those that have most to do with economical beef production.

Selection Based Upon Production Records

Research shows there is little association between body conformation of an animal and its ability to gain weight or to produce a desirable carcass. While conformation is important, it should not be the sole means by which a cattleman selects his breeding stock. Conformation characteristics and gaining ability are inherited. However, an animal that has inherited outstanding body conformation may not necessarily have inherited good gaining ability. The value of the animal as a sire of cattle exhibiting good gaining ability must be largely determined by the growth and production records of the

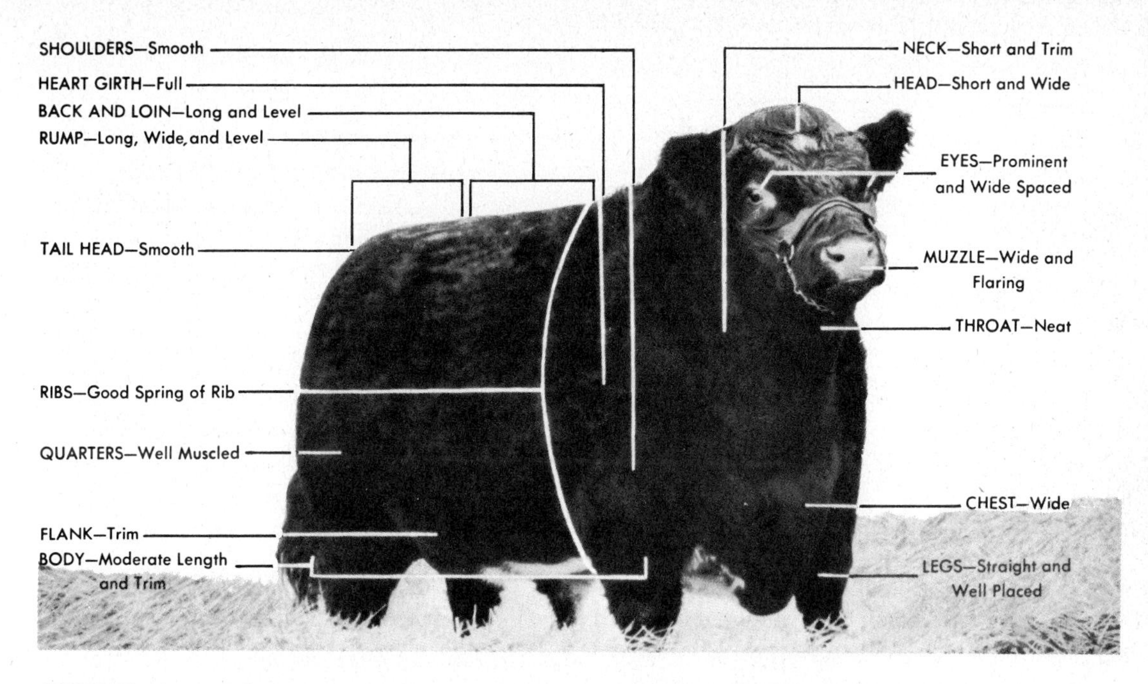

FIGURE 12-19. Before one can become a good judge of cattle, he must know the various parts that make up the conformation of the animal. (Abernathy Photo. *Courtesy* American Shorthorn Association)

progeny. The outward body conformation of the live animal does not always indicate the amount of muscling or the proportion of lean meat to fat.

Progeny and production records will reveal the following information: (1) inherited growth ability, (2) milking qualities of the cows, (3) fertility record of the herd, (4) percentage of lean to fat, and (5) quality of carcass.

The term *progeny testing* means the testing of the offspring from certain breeding animals. Since the bull contributes half the characteristics inherited by the calf crop, most progeny testing has been done on herd sires. The purpose is to determine the ability of a bull to produce fast-growing, efficient calves that will produce desirable carcasses.

The birth weight of calves has been found to be a fairly accuate method of determining their growth rate and also the ability of a sire to transmit this desirable quality to his offspring. By selecting at random at least ten calves sired by the same bull from different cows, and recording their rate and efficiency of gain and carcass information, a fairly accurate record of the transmitting ability of a sire can be determined. A combination of both birth weight and growth records of the progeny of a bull will give sufficient evidence to prove or disprove his value as a sire of fast-gaining calves. When sufficient evidence has been obtained of the ability of a bull to sire fast-growing calves with desirable carcasses, he is known as a "proven sire." When the services of a proven sire cannot be secured, young animals sired by a proven bull are next best.

The weaning weight of calves is a good indication of the milking ability of the dams. If feeding and environmental conditions are uniform, the heaviest calves at weaning time indicate the growth ability and good milking qualities of the mother. The fertility record of the herd and individuals in the herd is important in selecting breeding stock.

DWARFISM

Dwarfs are usually compact and stocky, but very much under-sized. They generally develop a large stomach and heavy shoulders. Later they may become swaybacked and develop crooked legs. Dwarfs represent an almost 100 percent economic loss.

Cause of Dwarfs

Dwarfism is inherited. The parents may be perfectly normal but, when mated, produce a dwarf. The problem is to determine which animals in the herd are dwarf carriers. When a bull, normal in appearance, but a dwarf carrier, is mated to a similar cow, on the average one calf in four will be a dwarf. Of the remaining three calves, two will be carriers of the dwarf factor, but will appear normal, and one will be free of the dwarf factor. As an example of how great the loss would be, suppose that in a herd of 16 cows eight were dwarf carriers. If the eight dwarf carrying cows were mated to a dwarf carrying bull, two dwarf calves per year could be expected. Of the remaining six calves, four would carry the factor and, if kept in the herd, might produce more dwarf calves. In addition, the non-dwarf carrying cows if bred to the same bull would produce an average of four calves carrying the dwarf factor, although they would be normal in appearance. To produce dwarf calves both parents must be carriers of the factor.

Prevention of Dwarfism

To purge the herd of dwarfs is not simple. If a dwarf results from any mating, both parents carry the factor, and they should be eliminated. This does not tell us how many more cows may be carriers. The ratio of dwarf calves from parents that are cariers is one in four. This is only a ratio based on large numbers, and it is conceivable that a carrier cow mated to a carrier bull for ten years may never produce a dwarf.

After having disposed of the carrier bull, one is confronted with the problem of securing a replacement that is not a carrier. If the prospective bull has been in service and has not sired any dwarfs, that is some indication he may be free of the factor. If replacement animals are selected from a herd where the record has shown no dwarfism, they probably do not carry the factor.

At present, scientists are trying to perfect methods whereby dwarf carriers may be detected before going into service and contaminating cattle herds. Prospects of new discoveries in detection methods such as X-rays and other devices are promising, but it is too early to come to any definite conclusions.

SELECTION OF FEEDER CATTLE

Feeder cattle are those that are unfinished or that do not carry enough condition to make the slaughter grade of which they are capable. Such cattle are usually purchased from the range by cattle feeders and put into the feed lot to be finished. The feeder depends upon the margin and value of the gain over feed costs and other expenses for his profits.

Feeder cattle vary considerably in age, weight, body conformation, ability to make rapid gains, and amount of condition they carry. These large variations in feeder cattle create a number of problems for the purchaser of cattle for finishing. There will be price variations depending upon weight, quality, and sex, but the prices will not vary in the same proportion each year.

Certain weights and qualities of cattle may be better adapted to the amount and kinds of feed available. Also the future market outlook must be considered. Cattle feeders' problems are many, and only those with good judgment are likely to succeed financially over a long period.

CLASSES AND GRADES OF FEEDER CATTLE

Feeder cattle are classified and graded according to age, sex, weight, and conformation.

Age

Cattle are classified according to age as calves, yearlings, two-year-olds, cows, bulls, and stags. All cattle are designated as calves until they are one year old. Between the ages of one year and two years they are classified as yearlings. Animals over a year but less than 18 months old are sometimes referred to as short-yearlings, whereas the term long-yearlings is applied to animals over 18 months, but under two years of age. Two-year-olds are cattle between the ages of two and three years. Cattle three or more years of age seldom enter into consideration in the feeder trade, except for cows, bulls, and stags.

Sex

Cattle are classified as steers, heifers, heiferettes, cows, bulls, and stags.

Steer. A steer is a male animal that was castrated at an early age and before he reached sexual maturity.

Heifer. A heifer is a female animal that has not developed the mature form of the cow and has not had a calf. Usually females under three years of age are classed as heifers. The term heiferettes refers to young cows, usually those that have not had more than one calf.

Cow. A cow is a mature female that has had one or more calves. A barren female (one that fails to conceive) that has reached maturity is also classed as a cow.

Bull. A bull is an uncastrated male of any age.

Stag. A stag is a male animal castrated after he has developed the physical characteristics of a mature bull.

Weight

Steers and heifers are classified as heavy, medium, and light. There is no weight classification for cows, bulls, and stags. The weight class a steer or heifer falls in is determined by the age of the animal. For example, to be classified as heavy a steer calf would need to weigh from 450 to 500 pounds. A heavy yearling feeder steer would weigh from 700 to 800 pounds, whereas a heavy two-year-old feeder steer would weigh around 1,100 pounds. The amount of condition is important in determining the weight classification of feeder cattle. Some cattle have a frame large enough to classify as heavies, but owing to a short feed supply may be thin and not classify heavier than medium or, in extreme cases, as light weights. Heavy calves could be heavier than light yearlings. This is especially true if the yearlings are thin. Cattle have attained most of their growth by the time they are two and a half years old. Weight beyond 1,000 pounds is largely due to finish. Yearlings on good pasture could be as heavy as two-year-olds that have less abundant grazing conditions.

Grade

Feeder cattle grades apply to steers, heifers, and cows. They are prime, choice, good, standard, commercial, utility, and inferior.

Bulls and stags make up such a small percentage of the market that they will not be considered here.

Prime Cattle. The term prime is applied to only a small percentage of feeder steers

FEEDER STEERS
U.S. GRADES

PRIME

CHOICE

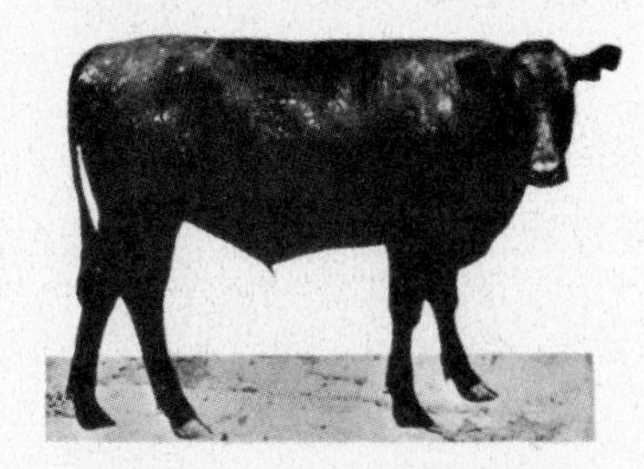

GOOD

STANDARD

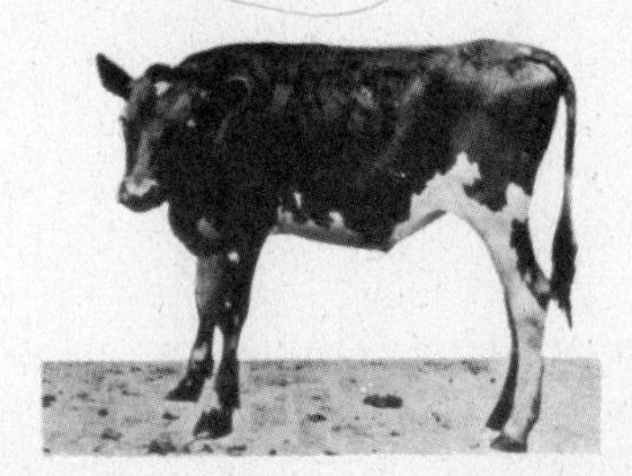

UTILITY

and heifers. Those grading prime show exceptional smoothness and body conformation. Such cattle, especially the steers, are in demand by showmen who finish them out for exhibition at the market cattle shows. They usually sell for such a high price that the investment would not be profitable for commercial feeders.

Choice Cattle. Choice is the highest practical grade of feeder cattle. They are superior in conformation, natural finish, and quality. They are cattle with long bodies, moderate depth, well-turned tops, long rumps, and trim briskets and flanks. Well-developed in the quarters, they show evidence of having only the high-grade or purebred ancestry of strictly beef cattle breeds. One who buys choice young cattle can expect them to finish into high choice or prime slaughter grades and bring the higher price paid for the top grade of slaughter cattle.

Good Cattle. Cattle grading good show less muscling and are more upstanding and not as smooth as the higher grades. They lack development of the more valuable cuts such as the back, loin, and hind quarters. Feeder cattle grading good can be expected to finish into good to choice slaughter cattle.

Standard Cattle. Standard grade cattle are upstanding, and uneven in the top and underlines. Their hip bones are prominent and they are somewhat narrow over the back and light in the hind quarters. They may show some evidence of dairy breeding. Standard steers may not be expected to grade higher than good as slaughter cattle.

Commercial Cattle. Commercial feeder cattle are restricted to cattle that will be too mature for good or standard grade when they reach slaughter potential. They are usually cows which are thin muscled, angular, and rough.

Utility Cattle. Utility cattle are unthrifty, small for their age, thinly muscled, narrow, and shallow bodied. The legs are long and often crooked.

Inferior Cattle. Cattle inferior to utility are graded inferior.

Cows, Bulls, and Stags. Cows, bulls, and stags do not make up a large percentage of the feeder cattle trade. However, when cows become old or nonproductive, they go to market. Some cattle feeders make a practice of buying cows. They feed them to improve the slaughter grade, thereby increasing the value per pound. The profit on cows usually comes from margin because cows are not very efficient users of feed.

Bulls usually come to the market as singles, and are sold by breeders and commercial producers when they can no longer use them for breeding. A very small percentage of the bulls are castrated and sold as stags.

SELECTING THE CLASS AND GRADE OF FEEDER CATTLE

The kind of cattle to buy depends upon so many factors that only a few can be considered here. However, the more important considerations are: (1) kind and amount of feed available, (2) price of feed, (3) difference in prices between feeder and slaughter cattle, (4) future market outlook, (5) difference in prices between grades of feeder cattle, and (6) the length of time the cattle are to be fed.

Feed

Some grain and high-quality forage is essential for the production of the higher grades of slaughter cattle. One who plans to

FIGURE 12-20. U. S. Grades of feeder cattle are prime, choice, good, standard, commercial and utility. Commercial and inferior grades are not shown here. (*Courtesy* U. S. Department of Agriculture)

TABLE 12-1 CLASSES AND GRADES OF FEEDER CATTLE

Sex	Age	Weight	Grade
Steers	Calves	Heavy Medium Light	Prime, choice, good, standard, utility, inferior
	Yearlings	Heavy Medium Light	Prime, choice, good, standard, utility, inferior
	2-year-olds	Heavy Medium Light	Prime, choice, good, standard, utility, inferior
Heifers	Calves	Heavy Medium Light	Prime, choice, good, standard, utility, inferior
	Yearlings	Heavy Medium Light	Prime, choice, good, standard, utility, inferior
	2-year-olds	Heavy Medium Light	Prime, choice, good, standard, utility, inferior
Cows	All ages	Any weight	Choice, good, commercial, utility, inferior

market choice slaughter cattle should select feeders of at least good grade. Choice cattle weighing from 800 to 1,100 pounds generally draw the highest prices in the slaughter cattle market. Experimental work at Iowa State University indicates that cattle grading good or even standard as feeders may produce choice carcasses when fed a high concentrate ration, but that the high grades of feeder cattle tend to put on more fat than desirable when too many concentrates are fed. Feeding for grade will be discussed later.

The cattle chosen may be either steers or heifers. If heifers are purchased, the price should be less than that of steers in the same grade. Heifers finish somewhat faster on the average than steers, but will not equal steers in dressing percentage. The market price of finished heifers is usually less than that of steers in the corresponding grade.

It takes both quality cattle and quality feed to produce top beef. Extremely low-grade feeders, regardless of how well fed, will not make high-grade slaughter cattle. If an abundance of low-quality roughage, such as corn fodder, sorgo fodder, and hay that is coarse or stemmy, is available, it may be used profitably by feeding it to cattle. Cattle may be bought in the fall and used to clean up a corn field behind the picker or run out on the pastures as long as there is sufficient feed. Medium-to-heavy yearlings, weighing from 600 to 750 pounds, and commercial-to-standard grades would be well suited for this purpose. After the rough feed in the fields has been eaten, the cattle may be put in the feed lot and full-fed roughage with enough protein to meet their needs. Grain should be limited to 3 to 5 pounds per day until the last 30 to 60 days of the feeding period. A full feed of grain may be used the last month or so. These cattle will market as standard-to-good slaughter steers, with some grading choice, and will not require more than 15 to 25 bushels of grain.

It is important that lower-grade cattle be bought at a price considerably under the price of choice grades. Such cattle will require more feed to make slaughter animals, and the cattle feeder will receive less for them on a hundredweight basis. They do offer an opportunity to convert low-quality feed into a marketable product. High-grade cattle fed on low-quality feed will generally not produce the highest grade of beef. The kind of feed available should determine to a large extent the grade of cattle to buy.

Feed Prices

When prices of feed are high in proportion to finished beef and when feeder prices are low, the cattle feeder generally expects to make his profit on the margin rather than on the gain in weight. Under these conditions heavy cattle weighing from 800 to 900 pounds may be the best. These cattle will be ready for market after a gain of from 200 to 300 pounds. The purpose of the gain is to finish the cattle to meet the requirements of their slaughter grade and to bring the feeder an increased price per hundred over the cost of the feeders. If he makes a marginal profit of $2 to $3 per hundredweight, he has a profit of $16 to $27 per head if he can break even on the gain. If the reverse is true—that is, low-priced feed and high-priced feeders—then the chance for profit is greater on the gain than on the margin. When feed is cheap, light yearlings or calves weighing from 300 to 500 pounds will cost less. With low-priced feed the feeder may put from 600 to 800 pounds of gain per head on the cattle. His profit will result from the gain rather than from the margin.

Spread in Feeder and Slaughter Cattle Prices

If the price of feeders is low compared to the price of slaughter cattle, again the margin is important. Heavy cattle will usually bring in the most profit.

Future Market Outlook

Price prospects for slaughter cattle are important in determining which weights and grades of feeders to buy. If the immediate future outlook is good, but the long-time outlook is very uncertain, then heavy cattle, which will meet the grade most in demand for slaughter cattle, are more certain to make a profit. Such cattle may be short-fed from 90 to 120 days and sold before the expected price break.

Length of Feeding Period

The amount of feed available usually determines the length of time cattle are fed. Cattle feeders with feed they want to market as beef generally desire light animals of a grade that will correspond to the quality of feed available.

These cattle are often calves bought in the fall, wintered to gain a pound a day, which is only a good growth gain, and turned on pasture the next spring. Some grain may be fed to the cattle while on grass. The second fall they are put into the feed lot and finished according to their grade.

Compensatory Growth

The economic value of compensatory growth in feeding cattle has become important to the feeder. If the genetic gaining ability of the animal has been held back because of environmental or nutritional factors, and then it is placed in a more desirable environment or fed a nutritionally balanced ration, the fast gain made by the animal is called compensatory growth. Cattle in poor flesh could give the cattle feeder such a growth, but don't confuse this with a run-down, diseased condition. Experience in buying feeder cattle is a basic factor in the success of the beef cattle feeding enterprise.

SUMMARY

A cattle breeder is one who produces animals either for breeding purposes or for the feed lot. However, the term breeder generally refers to one who tries to improve the breed and produces purebred animals primarily for breeding purposes.

British breeders are credited with most of the early developmental work on beef cattle breeds.

When selecting a breed, the prospective breeder must consider personal likes, availability of breeding stock, outlet for surplus animals, and environmental conditions.

The most important breeds of cattle in the United States are three British breeds: Aberdeen-Angus, Hereford, and Shorthorn. Other European breeds found in the United States include the Galloway and the Devon, both British breeds, the Charolais, Limousin, and Maine-Anjou are French breeds, the Simmental, a Swiss breed, and the Chianina, a breed from Italy.

Brahman cattle are native to India. The breed differs considerably from European cattle, being more upstanding and less compact, and having drooping ears. They are resistant to many insects and diseases affecting European breeds and can graze and gain on forage too scant for the survival of many other breeds.

The Brangus, Charbray, Beefmaster, Santa Gertrudis, and Braford are all breeds developed in the United States, most of them in the South. They were developed by blending Brahman blood with Angus, Charolais, Hereford, and Shorthorn cattle. These breeds are credited by their supporters with having exceptional vigor, size, fast growing ability, and a tolerance to extremes in climatic conditions, insects, and diseases. They are good rustlers and can forage successfully over areas of scant vegetation.

Crossbred cattle have shown some advantages in gaining ability over straight-bred cattle.

Foundation stock should be selected on the basis of conformation, progeny, and production records.

Dwarfism is an inherited abnormality that has caused considerable economic loss in many herds. Dwarf calves come from mating animals which carry the dwarf factor.

Feeder cattle are unfinished cattle that are generally either sold to cattle feeders or put into the feed lot by the producers and finished for market.

Feeder cattle vary considerably as to age, weight, and grade. Most feeder cattle are steers or heifers. The weight classes are light, medium, and heavy. The age classes are calves, yearlings, and two-year-olds. Steers and heifers are graded prime, choice, good, standard, commercial, utility, and inferior. Feeder cows, bulls, and stags have a top grade of choice.

In selecting feeder cattle one should give careful consideration to the available feed, feed prices, difference in feeder and slaughter cattle prices, future market outlook, and the length of the feeding period.

QUESTIONS

1 Define the term *cattle breeder.*

2 List the important decisions a prospective cattle breeder must make.

3 Describe the European breeds of beef cattle found in the United States.

4 What are the three most popular breeds of cattle in the United States?

5 How does the Brahman differ from the European breeds of beef cattle?
6 List the breeds of beef cattle developed in the United States.
7 How were the breeds listed in Question 6 developed?
8 Tell how you would obtain breeding stock.
9 Describe what you would look for in breeding stock.
10 What information will production records reveal?
11 Discuss the cause and prevention of dwarfism.
12 What are feeder cattle?
13 What are the age classifications of feeder cattle?
14 What are the weight classifications of feeder cattle?
15 List the grades of heifers, steers, cows, bulls, and stags.
16 Describe the choice feeder steer.
17 What are the chief factors that determine the grades of feeder cattle?
18 Give an example of when heavy cattle would be a good buy. Light cattle.
19 Under what conditions would you recommend the purchase of choice feeders? the plainer grades?
20 What are the important factors one should consider in determining the weight and grade of cattle to buy?
21 Define the term *compensatory growth.*

REFERENCES

Baker, Marvel L., Leslie E. Johnson, and Russell L. Davis, *Beef Cattle Breeding Research at Fort Robinson,* Miscellaneous Publication 1, Agricultural Experiment Station, Lincoln, Nebraska, University of Nebraska, 1952.

Beef Cattle Feeding and Breeding Investigations, Kansas Agricultural Experiment Station, Reports 37, 38, 40, Kansas State College, Manhattan, Kansas.

Feeding and Breeding Test with Beef Cattle, Reprinted from Miscellaneous Publication No. MP-34, Feeders' Day Report, Oklahoma Agricultural Experiment Station, Stillwater, Oklahoma, Oklahoma A & M, 1954.

Gregory, P. W., W. C. Rollins, and F. D. Carroll, *Heterozygous Expression of the Dwarf Gene in Beef Cattle,* Reprint from the Southwestern Veterinarian, Volume 5, No. 4, Summer, 1952, pp. 345–349.

Holzman, Henry P., *Beef Performance Testing,* Circular 551 (revised), Brookings, South Dakota, South Dakota State College, 1965.

Iowa Beef Cattle Improvement Program, Agricultural Extension Service, A. H. 810, 1960, Iowa State University, Ames, Iowa.

Roubreck, C. B., N. W. Hilston, and S. S. Wheeler, *Progeny Studies with Hereford and Shorthorn Cattle,* Wyoming Agricultural Experiment Station, Bulletin 307, Laramie, Wyoming, University of Wyoming, 1951.

U. S. Department of Agriculture, *Livestock Breeding Research at the U. S. Range Livestock Experiment Station,* Agriculture Information Bulletin No. 18, Washington, D. C.

United States Department of Agriculture, *Official United States Standards for Grades of Feeder Cattle,* S.R.A.— *C & MS 183,* 1965.

13 Feeding and Management of the Breeding Herd

The breeding herd must be properly fed and managed if a good calf crop is to be obtained. The size of the calf crop, the vigor and size the calves attain by market time, and the feeding efficiency of the herd largely determine the profit realized. Recent studies and research in cattle nutrition have shown that cows can utilize an amazingly large amount of low-quality roughage, when properly balanced with minerals, vitamins, and proteins, and still produce a strong, healthy calf crop.

FEEDING THE HERD ON PASTURE

When an ample amount of good pasture is available, summer feeder problems of the breeding herd are easily solved. Young growing plants are generally high in most food nutrients and represent a fairly well balanced ration for two-year-old heifers and cows.

Studies reveal that cattle will spend not much more than eight hours a day grazing. Whenever a pasture fails to provide, in eight hours of grazing, sufficient forage to satisfy the animal's requirements, additional feed should be provided.

Supplementing the Grass

Whenever the pasture is insufficient to provide adequate nutrients to meet the needs of the cattle, it is necessary to supplement the pasture with additional feed for best results. Ten pounds of good legume hay, 30 pounds of legume silage, or 20 to 25 pounds of sorghum or corn silage will replace about half of the pasture requirements for mature cattle. If a lower quality forage is used, the amount will need to be increased. If available, silage is an excellent supplement when the grasses are dry or mature. Dry forages are generally considered superior to silage when the pasture grasses are green and succulent.

Many farmers use extra forage to supplement the pasture early in the season to prevent the cattle from cropping the grass too short. When it is known that there is insufficient pasture to be the sole forage for cattle during the season, the amount may be extended by feeding additional forage from the start of the pasture season.

Grain may be used to replace part of the pasture requirements for breeding cattle. Three to five pounds of corn, barley, wheat, or sorghum grain will replace about half the pasture requirements for mature breeding cattle. If oats are used, from five to seven pounds will be required.

Salt should be within easy reach of the cattle at all times. As an insurance against mineral starvation and since mineral mixtures are relatively cheap, a good mineral mixture should be kept before the cattle at all times. Following are some good mineral mixtures recommended for various sections of the United States.

Mineral mixtures for areas where only additional salt, calcium, and phosphorus needs to be provided:

1. **200 pounds steamed bone meal**
   100 pounds common salt
2. **50 pounds steamed bone meal**
   25 pounds ground limestone
   25 pounds common salt

FIGURE 13-1. (*Right*) The quality and quantity of pasture has a direct influence on the number of acres required to furnish feed for an animal. (*Courtesy* American Brahman Association) FIGURE 13-2. (*Below*) Pasture production can be supplemented with dry forages during dry weather. Hay or silage can be used with excellent results. (*Courtesy* American Brahman Breeders Association)

Mineral mixtures for areas where one or more trace minerals, in addition to salt, calcium, and phosphorus, may be deficient:

1. 100 pounds common salt
25.5 pounds ground limestone
50.0 pounds steamed bone meal
25.0 pounds red oxide of iron
2.5 pounds pulverized copper sulfate
1.0 ounce cobalt sulfate
2. 23 pounds iodized trace mineral salt
25 pounds ground limestone
50 pounds steamed bone meal
2 pounds iron (ferric oxide)
3. 50 pounds bone meal
25 pounds ground limestone
20 pounds iodized salt
5 pounds trace mineral premix*

*** Trace minerals can be purchased as a trace mixture and added to the other ingredients.**

It is seldom necessary to supplement good pasture with any vitamin supplement. Vitamins A and D are the only ones ever likely to be deficient in the rations of cattle a month or more of age. Green plants, containing an abundance of vitamin A and sunlight, will provide the vitamin D requirements. Cattle can store vitamin A in abundant quantities and can draw upon this reserve for several months. Only when cattle have been grazed for several months on poor-quality dry pastures is there a danger of a vitamin A deficiency.

Cattle suffering from a vitamin A deficiency show an inability to see in dim or subdued light, a staggering gait, and excessive running from the eyes and nose. If a vitamin A deficiency is suspected, they may be fed one pound per day of fresh, dehydrated alfalfa meal or two pounds of high-grade alfalfa hay. When these feeds are not available a vitamin supplement, prepared commercially, containing 3,000 international units of vitamin A per gram, may be fed at the rate of ¼ pound per animal per day. If this supplement is fed over a period of 10 to 15 days, enough vitamin A will be provided to correct the condition. After that, one or two feedings per week should prevent a recurrence of the trouble.

When an ample amount of green pasture forage is available, mature breeding cattle seldom need additional protein. If the pasture is short or if poor-quality and low-protein forage is used to supplement the pasture, from ¼ pound to one pound per head of protein concentrates such as cottonseed cake or meal, soybean, linseed, or peanut oil meal will be needed to balance the ration.

The practice of using salt mixed with the protein supplement, as a means of controlling the intake of protein when the mixture is self-fed to cattle, is gaining in popularity among ranchers in some sections of the range country. The amount and quality of the pasture determines the needed level of protein intake. As the salt is increased or decreased in proportion to the protein meal, consumption of the mixture increases or decreases. Constant adjustment of the salt-protein ratio is necessary to provide for the correct intake of protein. The salt-protein mixture permits the self-feeding of protein, which reduces farm labor and provides for a more even consumption of the supplement and more uniform grazing of the range than does hand feeding. It is necessary to have plenty of water available when a salt-protein mixture is fed.

Commercially prepared protein blocks have been successful in providing protein for cattle on pasture. The blocks are hard and require some extra effort to eat, and this helps limit consumption to the amount required.

In tests carried out by the Oklahoma Agricultural Experiment Station, daily consumption of protein meal by 700-pound cattle was held to 2 pounds per day by mixing ⅞ of a pound of salt with each 2 pounds of meal. California tests showed that consumption could be controlled by shifting the salt content from 10 to 30 percent in a ration made up of equal parts of cottonseed meal and barley.

Water

Water is the cheapest and most essential element in livestock nutrition. The need for plenty of fresh, clean water within easy reach of the cattle cannot be overestimated. Cattle make faster gains and utilize feed more efficiently when plenty of water is available. Range cattle tend to feed more in the vicinity of the watering place. When watering facilities are too far apart the range is grazed unevenly. Some areas will be grazed so closely that the grass will be destroyed and will be replaced by undesirable weedy plants.

Pastures Compared

There is a wide variety of pasture plants. Weather, soil, and moisture conditions determine the pasture crops best adapted to each locality. Pasture crops are divided into two general classes: legumes and grasses.

Legumes. Legumes have the ability to utilize atmospheric nitrogen. This element is necessary in the building of proteins, and legumes are usually higher in protein than are the grasses. Legumes are heavy users of phosphorus and calcium and, when grown on

FIGURE 13-3. (*Above*) Grasses and legumes combine to make excellent pasture for economical beef production. Combinations of adapted varieties can be successfully grown in your area. (*Courtesy* American Hereford Association) FIGURE 13-4. (*Below*) A natural water supply like this one on the Lazy J C Ranch, Cottonwood Falls, Kansas, provides an essential nutrient in livestock nutrition. (*Courtesy* American Angus Association)

land well supplied with these elements, are rich in these essential minerals. Legumes are usually higher in total food value than are grasses. In general, legumes are very palatable to cattle. However, when legumes are used alone as a pasture crop, there is danger of bloat. The loss from bloat has been so great that straight legume pastures cannot safely be recommended unless an antifoaming agent is fed daily. A chemical compound called poloxalene has shown remarkable success in preventing bloat when fed at the rate of 10 grams per head daily. It is also available in the form of blocks which may be placed in the pasture. The chemical is combined with molasses and other ingredients and is very palatable, but due to the hardness of the blocks, there is no danger of the cattle consuming too much. Even though pure legumes may be pastured safely if poloxalene is also consumed, a combination of legumes and grasses provides more forage production per acre.

Wherever legumes are adapted to the locality, they should make up part of the pasture mixture. Experiments conducted at several stations have shown that when a legume and grass pasture is used, it is superior to grass alone. Some of the advantages of legumes used in conjunction with grasses are: (1) faster gains, (2) longer grazing season, and (3) greater carrying capacity of the pastures.

Some of the more common pasture legumes are: alfalfa, alyce clover, alsike clover, trefoil, crimson clover, kudzu, ladino clover, lespedeza, red clover, sweet clover, vetch, and white Dutch clover. Most areas of the United States have one or more legumes adapted to the particular region. Legumes do not grow well under arid or semiarid conditions. In the drier sections they may be grown on irrigated land as hay or pasture crops.

Grasses. As with legumes, there are many types of pasture grasses. Various areas have a number of grasses adapted to their climatic conditions. In selecting grasses for permanent pasture seeding, those that will produce the most forage over an extended period and that are palatable to livestock are recommended. Some of the better grasses to be used in mixtures for temporary and semi-permanent pastures in the midwest are: bluegrass, brome grass, tall fescue, orchard grass, redtop, perennial rye grass, sudan grass, and timothy. All of these grasses are perennials except sudan grass, which is an annual.

There are many combinations of legume and grass seed mixtures used for seeding pastures. The length of time a pasture is expected to last is important in selecting a mixture.

There are such wide varieties of soil and climatic conditions in the United States that no attempt will be made to give recommended pasture seed mixtures for all areas. This information may be secured from the local vocational agriculture instructor, the county extension director, or the agricultural college. Some recommended mixtures for the Midwestern states that have wide adoption in other sections are listed in Table 13-1.

For economical beef production, plans should be made to pasture the cattle as much of the year as possible. Studies at Purdue University showed that 55.4 percent of the feed cost for an entire year occurred during a 127-day wintering period. The cost of wintering rations was 2½ times as much as the cost of pasture per day.

Nutrient Value of Pasture Varies

The nutrient value of pasture crops does not remain the same during the season. As plants become more mature, the fiber content increases and the protein and vitamin contents decrease. The same is true when the growth of plants is retarded because of dry weather. Many pastures that have provided the entire ration successfully during one part

TABLE 13-1 GRASS AND LEGUME MIXTURES RECOMMENDED FOR THE MIDWEST

Seeding Rate per Acre in Pounds							
Rotation Pastures:							
1. Alfalfa	6–8	Smooth brome	5–10				
2. Alfalfa	6–8	Meadow fescue	6–8				
3. Alfalfa	6–8	Smooth brome	6				
4. Alfalfa	6–8	Orchard grass	6	Meadow fescue	2		
5. Alfalfa	7–8	Red clover	3	Orchard grass	6		
6. Alfalfa	5–6	Ladino	½	Smooth brome	8		
7. Alfalfa	3	Ladino	½	Red clover	3	Smooth brome	8
8. Alsike	3	Ladino	½	Redtop	2–4	Timothy	2–3
9. Red clover	6–8	Timothy	4–6				
Permanent Pastures:							
1. Birdsfoot trefoil	4–6	Bluegrass	3–4	Smooth brome	6–8	Timothy	2
2. Birdsfoot trefoil	4–6	Alfalfa	3–4	Smooth brome	6–8	Timothy	1–2
Emergency Pastures:							
1. Sudan grass	15–25						
2. Sudan grass	5–10	Soybeans	60–80				

of the season will need to be supplemented with considerable amounts of additional forage and grain at other times. Young pasture grass will average from 18 to 22 percent protein and 70 to 80 percent total digestible nutrients on a dry-matter basis (water removed), compared to about 5 to 9 percent protein and 40 to 50 percent total digestible nutrients for matured grasses. Young pasture grass contains more protein and is highly digestible.

Management of Pastures and Grazing Systems

Turning cattle into the entire pasture area early in the spring and leaving them there for the season is not conducive to making the most efficient use of pastures. Under free range conditions it may be the only way, but whenever possible drift fences or electric fences should be used to limit the area cattle graze during any one period.

Rotation grazing is accomplished by grazing one area for two to four weeks and then moving the cattle on to another area. This allows the presently grazed area to recover and make new growth. If it becomes evident that any areas will grow more than necessary to provide sufficient feed, they may be clipped and put up for hay or silage.

Cutting and Hauling Pasture Forage

Green-lot feeding is a term applied to the practice of cutting and hauling pasture to cattle that are confined to a dry lot. Experiments conducted at several stations show that one can expect to produce from 20 to 30 percent more beef per acre than can be produced from grazing. In addition to the pastures lasting longer and the beef cattle making faster gains, fences may be eliminated and bloat is seldom a problem.

Forage should be cut daily and in an amount that will be consumed by the cattle

FIGURE 13-5. Feeding the beef herd a ration which contains a high percentage of roughage helps to control efficiency of production as well as quantity and quality of the beef. (*Courtesy* Iowa Structures, Inc.)

FIGURE 13-6. Proper pasture practices might improve the carrying capacity of this range land. Note the unproductive areas shown here. (*Courtesy* American Hereford Association)

in a twenty-four-hour period. Larger amounts will mold and result in waste. The taller legumes and grasses have shown considerable increase in total yield from green-lot feeding. The shorter crops have shown little if any increase. The practice of green-lot feeding is more adaptable to legume and grass pastures grown in rotation than to permanent pastures. This system is more adapted to high-priced crop land than to permanent pasture areas.

While there are the advantages mentioned in the preceding paragraphs, there are also many disadvantages to green-lot feeding. Some of the more important disadvantages are: (1) the job must be done each day, creating a labor problem, (2) considerable investment is needed in machinery, wagons, and other equipment, and (3) rainy spells and wet spots hinder heavy machinery. One must carefully weigh the advantages and disadvantages before changing from a grazing system to green-lot feeding. The large operator may consider the possibility of hauling pasture to his cattle. It is doubtful that the small herd owner would find the extra investment and labor needed to be economically sound.

Feeding the Herd in Dry Lot

Dry lot feeding of the breeding herd, such as is necessary in areas of the United States too far north for winter pastures, presents more problems than pasture or summer feeding.

Danger of Nutritional Deficiencies. Deficiencies in protein, vitamin A, and minerals may occur in cattle confined to a dry lot, especially when low-quality forages are used without supplements. However, breeding cows can utilize large quantities of corncobs, cornstalks, and some other low-quality feeds, if they are properly supplemented, and still produce a good calf crop.

TABLE 13-2 RATIONS FOR WINTERING BEEF COWS

Ration	Pounds Daily
1. Legume hay	15–20
Mixed minerals	Free choice
2. Legume hay	5–10
Oat straw	10–15
Mixed minerals	Free choice
3. Corn or sorgo silage	20–30
Legume hay	5
Oat straw	5–7
Mixed minerals	Free choice
4. Ground corncobs	14–15
Dehydrated alfalfa meal	1
Purdue Cow Supplement	3.5
Mixed minerals	Free choice
5. Corn or sorgo silage	30–40
Chopped dry corn stalks	5–10
Cottonseed, linseed, or soybean meal	1
Mixed minerals	Free choice
6. Legume and grass silage	25–30
Corncobs	4–5
Mixed minerals	Free choice
7. Cornstalk silage	30–40
Legume hay	5–10
Grain	2
Mixed minerals	Free choice
8. Cornstalks in field	Graze at will
Legume-grass hay	10
Mixed minerals	Free choice
9. Dry winter range	Graze at will
30–40% protein supplement	2
Mixed minerals	Free choice

Note: Mineral mixtures are the same as those recommended for cattle on pasture; salt should be fed free choice in addition to the mineral mixture.

If the roughage used is at least one half good legume hay or legume silage and the cows are fed liberal amounts, they will go through the winter in good condition and produce a strong calf with no extra feed except for a mineral mixture and salt self-fed. Legume hay or silage is high in protein, vitamins, and most of the minerals.

When low-quality roughages, such as corncobs, cornstalks, or coarse, stemmy hay is fed as the only roughage, some high-energy feed such as molasses or grains and a complete supplement that contains the needed vitamins, minerals, and proteins, are needed to properly balance the ration.

Experiments conducted at Purdue University show that cows receiving 14.5 pounds of ground corncobs, one pound of dehydrated alfalfa meal, 3.5 pounds of Purdue Cow Supplement (fed daily), and a mineral mixture (fed free choice) wintered and produced as good a calf crop as did cows receiving 20 pounds of alfalfa-brome-timothy hay plus minerals. The Purdue Supplement used in the cow-feeding trials consisted of 636.8 pounds of soybean meal, 285.8 pounds of 45 percent molasses feed, 51.4 pounds of bone meal, 17.2 pounds of iodized salt, and 2.5 pounds of vitamin A concentrate.

Work at the Iowa station has shown that if corn is picked when it is about 30 percent moisture and the stalks are harvested and made into silage, the cornstalk silage will provide good cattle feed, if properly supplemented. Cows receiving 2 pounds of grain and 5 pounds of good legume hay daily, plus a full feed of cornstalk silage and minerals fed free choice, could be expected to winter well and produce a good calf crop.

If corn or sorghum silage (grain included) is used as the only roughage, the addition of one to 2 pounds of cottonseed, soybean, linseed, or any good 30 to 40 percent cattle supplement plus a mineral mixture will provide adequate nutrients for the herd. If 5 to 7 pounds of good legume hay is added to the silage ration, the protein concentrate may be eliminated.

Grass silage used as the only feed for wintering bred cows has not proven as economical as grass silage plus a small amount of dry roughage. The dry roughage need not be of high quality. Four to 5 pounds of ground corncobs or other low-grade roughage will produce good results. Cows receiving 40 pounds of alfalfa-oats silage and 4 pounds of corncobs, plus a mineral mixture, gained an average of 0.36 pound per head daily and produced a good calf in a recent Iowa trial.

During the winter, cows should ordinarily gain weight equal to the weight gain of their unborn calf, fluids, and membranes of advanced gestation. This would equal about ⅓ pound per day.

Cows should have a wintering ration that will range from 8 to 10 percent protein (air-dried basis) plus adequate minerals and vitamin A. Table 13-2 suggests rations for cows in dry lots.

MATING

The period of heat or estrus is the time when the female will be receptive to the bull and the act of mating will occur. The duration of the heat period will vary from 12 to 30 hours with individual cows. The average heat period is from 12 to 18 hours. The time between heat periods will vary from 17 to 26 days, with an average of about 21 days.

The time from conception until the cow calves is known as the gestation period. Most authorities agree that 283 days is about the average length of the gestation period.

Age to Breed Heifers

Heifers that have made a good growth for their breed may be bred successfully as yearlings to calve as two-year-olds. Heifers should weigh at least 850 pounds when bred. Unless they are this heavy, small, stunted cows may result from early calving. The size rather than the age should determine when heifers are bred.

Age of the Bull

Under range conditions a bull should be two years old before he is turned with the herd. Bulls past seven years old may not be depended upon to meet the strenuous conditions of the range and to breed their quota of cows. Where they may be properly cared for, many bulls will breed when 12 or more years old.

In a small herd where the number of cows a bull serves can be regulated, a young bull 15 months old can be depended upon to breed 10 to 15 cows.

Artificial Insemination

Artificial insemination offers some benefits for the cow-calf producer. The small herd owner may not be able to justify keeping a bull for 20 to 30 cows. Large herd owners may be able to breed a hundred or more cows to the same bull during one breeding season.

A. I. permits injection of new genetic material into the herd. Also, it is the best breeding method to use in preventing disease spread from the bull to the cow. Conception rate will run about the same when compared to natural service methods.

The future of A. I. in beef cattle production will largely depend upon the solving of two important problems: (1) heat detection and (2) the availability of efficient technicians.

Determining Pregnancy

Veterinarians and animal scientists are carrying on research in an attempt to find a simple, sure way of determining pregnancy during early stages. However, for the present an economical and reliable method is an internal examination of the uterus made through the rectum. The uterus and the overies lie just beneath the colon and can easily be felt through the wall of the large intestine. If the cow is pregnant, the fetus can be felt. Later in pregnancy the uterus is pulled down and cannot be felt by the examiner.

FIGURE 13-7. Very few sights are more rewarding than a newborn animal with its mother watching over it. Reproduction and mothering ability are important traits to the success of the beef industry. (*Courtesy* American Hereford Association)

Early examination for pregnancy by a veterinarian or trained operator can be carried out on small beef cow herds without too much trouble, but large herds under range conditions would make the process of examination much more difficult.

Time for Calving

When warm housing is available, early calving is recommended. Calves that are two to three months of age when the pasture season arrives will utilize more grass and the feed cost will be reduced.

Shelters

When cows calve during cold weather, it is a common practice to keep the calves shut up in pens located in barns or sheds as protection against the cold. The calves are turned with the cows morning and night so that they may nurse. This practice permits the cows to graze the cornstalk field or the winter pasture without exposing the young calves to severe cold. Most cattlemen prefer to leave the cow and calf together continuously for the first two or three days. After that the cow may be turned out and the calf transferred to a group pen.

Well lighted, well ventilated, clean, dry pens, adequately bedded and free from drafts, are essential to the successful raising of calves born during cold weather.

It is important to avoid overcrowding the pens and to group the calves according to size and age. Calf pens should be equipped with a feed box, a hay rack, and watering facilities. Feed boxes should be about 10 inches wide, 6 inches deep, and long enough to provide 2 feet of feeding space for each calf. They should be 20 inches from the floor and away from the waterer. The hay rack should be constructed so as to prevent waste and provide at least a foot of feeding space per calf.

FIGURE 13-8. A properly trained operator can pregnancy test the female by feeling the fetus with his hand.

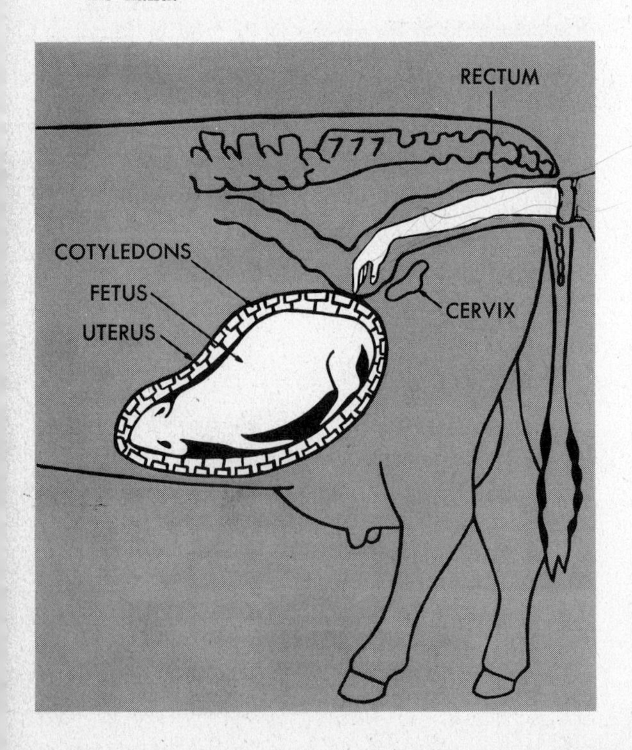

DEHORNING

While horns that have been well trained and polished add to the attractiveness of show cattle of the horned breeds, commercial beef cattlemen should remove the horns.

Advantages of Dehorned Cattle

Cattle should be dehorned for the following reasons:

1. **More room is needed for horned cattle in sheds, barns, and lots.**
2. **Cattle with horns inflict more damage on equipment.**
3. **There is more danger to the operator in handling cattle with horns.**
4. **Horned cattle that are inclined to fight will keep others away from the feed.**
5. **Cattle with horns inflict bruises on each other that may result in heavy economic losses.**

When to Dehorn

The age to dehorn calves depends upon the conditions and the method used in performing the operation. Generally, the sooner it can be done the less inconvenience and discomfort to the calf.

Methods of Dehorning

There are a number of methods that may be employed for dehorning cattle. The age of the calf to be dehorned and facilities available determine which one is best.

Chemical Method of Dehorning. The horn buttons may be prevented from growing by burning with chemicals. This method is most successful if done before the calf is ten days old. The chemicals that are commonly used are caustic potash or caustic soda. They come in a white stick about the size of blackboard chalk, or in a commercially prepared dehorning paste. Care should be used in handling them to prevent serious burns to the operator. The hair should be clipped around the horn button and a ring of heavy grease applied to the clipped area to prevent the burning action of the chemical from spreading too far. If using the caustic stick, dip the end in water to moisten it and rub with a rotary motion on the horn button. If the paste is used, it may be smeared on with a swab or flat wooden spatula. In a few days, a heavy scab forms over the horn and drops off in about ten days. The calf suffers little inconvenience, and there is no open wound to become infected.

While removing horns by chemicals is effective in eliminating the horns, there is a tendency for the face to get longer as the animal matures. The frontal bone becomes somewhat oval in outline and extends above the point where the horns would normally have been. For this reason, the use of caustics for dehorning is rejected by those who desire to show their animals at fairs and other shows.

Hot Iron. Burning the horn button with an electric dehorner or hot iron is a method used on some farms for dehorning young calves. Calves should not be over three months old if burning is to be effective in preventing horn growth. The electric dehorner has an automatic control that maintains the temperature at about 1,000° F. Applying the electric dehorner to the horn button for ten seconds is sufficient to destroy the cells and prevent growth of the horn. This is a convenient method.

Spoons and Tubes. There are a number of instruments, such as spoons and tubes, on the market for dehorning. Up to 3½ months of age, horns are only skin appendages and may be gouged or scooped out with a tube which is cylindrical in shape with a hand grip on one end and a sharpened edge on the other. After the age of 3½ months, the horns become fastened to the skull and tubes are not effective in removing them.

Dehorning tubes come in several sizes varying from ¾ of an inch to 1⅛ inches in diameter. Since the horn sizes vary with individual calves, several different sizes of tubes should be made available. In using tubes, (1) select a sharp tube of the proper size to fit over the horn base, (2) place the cutting edge straight down over the horn, (3) push and twist both ways until a cut of from ⅛ to ⅜ of an inch deep has been made. Calves nearing three months of age

FIGURE 13-9. These calves await their turn to be dehorned. Several methods can be used depending upon the age of the calf. (*Courtesy* American Hereford Association)

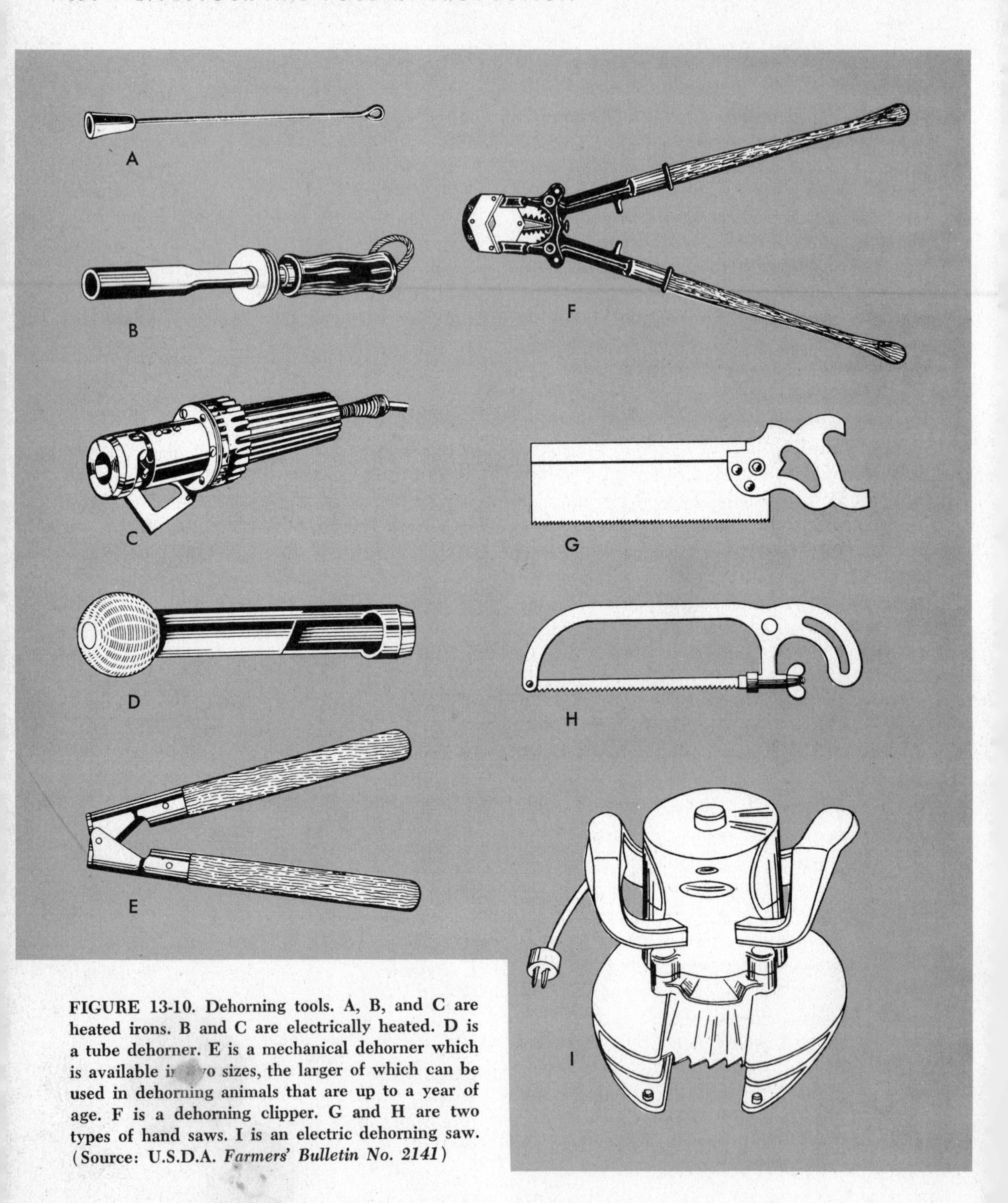

FIGURE 13-10. Dehorning tools. A, B, and C are heated irons. B and C are electrically heated. D is a tube dehorner. E is a mechanical dehorner which is available in two sizes, the larger of which can be used in dehorning animals that are up to a year of age. F is a dehorning clipper. G and H are two types of hand saws. I is an electric dehorning saw. (Source: U.S.D.A. *Farmers' Bulletin No. 2141*)

will require the deeper cut. Do not go deeper than necessary to cut through the skin, as excessive bleeding will result. (4) Turn the tube down to a 45-degree angle and lift the horn button out.

An open wound results from the use of either tubes or spoons and therefore it is better to perform the operations in cool weather. Otherwise use a good fly repellent on the wound.

Clippers and Saws. When older cattle are to be dehorned, specially designed clippers or saws are used. A considerable amount of bleeding may follow the operation. To prevent this, the main horn artery should be tied off with a cotton or silk thread. This may be done by sliding a sewing needle under the artery to pull the thread in place before tying. It is necessary, when sawing or clipping the horns, to take about ½ inch of skin in order to get the horn roots. An open wound results, so it is better to perform the operation during cool weather when there are no flies. In warm weather, a good fly repellent should be smeared over the wound.

CASTRATION

All male calves not to be used for breeding should be castrated. The operation should be done in cool weather to prevent screw worm infestation or other infections. Between the ages of one and three months is probably the ideal time to perform the operation, although many cattlemen prefer to castrate when the calves are one to two weeks of age. Generally speaking, the younger the calf is, the less inconvenience it suffers.

Methods of Castrating

Pulling downward on the scrotum and cutting off the lower third, exposing the testicles from below, is a common method of castrating. For show cattle, where a well developed cod is desired, one testicle is pulled down at a time and held firmly with the left hand so that the skin of the scrotum is tight over the testicles. An incision is then made on the outside of the scrotum next to the leg, through both the scrotum and the membrane surrounding the testicle. The pressure exerted by the left hand will expose the testicle so that it may be grasped and held in the right hand. The left hand may be used to separate it from the supporting tendons. While holding the testicle with the left hand, the tendons should be cut close to their lower attachments. The spermatic cord should now be stripped of all surrounding membranes and severed by scraping rather than cutting, as less bleeding will result. As much of the spermatic cord as possible should be removed. This may be accomplished by pulling downward on the testicle, drawing the cord out as far as possible before it is severed.

A specially constructed instrument, known as the Burdizzo or castrating pincers and designed to crush and destroy the spermatic cord and the blood vessels that supply the testicles, leaving the testicles to dry up and be absorbed, has found some favor in the South. The operation is bloodless and no open wound is left for infestation by screw worms or infection. However, unless the instrument is used by a skilled operator, the cord and blood vessels will not be completely destroyed. The result will be a calf known as a "slip," which will show stagginess when about a year old.

MARKING

Marking is essential to good management. It permits identification of animals as to ownership and breeding.

Ear notches are easy to identify and may be used for ownership as well as breeding records. One of the best marks for permanent identification of cattle is a tattoo on the inside of the ear with indelible ink.

FIGURE 13-11. (*Above*) Branding and vaccinating the modern way, demonstrated on the Phillips Ranch, Baker, Oregon. (*Courtesy* American Angus Association) FIGURE 13-12. (*Below*) Rubber ear tags like the one shown here are increasing in popularity with beef producers. A black tag with white numbers or letters is easy to read from a distance. (*Courtesy* G. A. Hormel & Co.)

Horn brands may be used as easily identifiable marks on mature horned cattle. Metal ear tags or buttons with letters and numbers may be inserted in the ear as a means of identifying calves. Leather neck straps or neck chains with a number plate attached also make identification easy.

Hide brands are probably the oldest system used in marking cattle. They are used primarily as a means of identifying cattle as to ownership on the open ranges of the west. Hide branding is usually accomplished by having irons shaped into letters, numbers, or a design to be used. The irons are heated in a fire and applied to the body of the animal. The identifying brand is seared into the hide, leaving a permanent mark.

Freeze branding of cattle is becoming more popular. Round-faced numbers made of copper or bronze alloy are best because they retain the cold better than other materials. It is important to clip the hair as short as possible and dampen the skin area where the brand is to be applied.

A mixture of Dry Ice and 95 percent alcohol has been used successfully for chilling the irons; however, liquid nitrogen is preferable since it is much colder and will chill the irons faster. Freeze brands can be altered more easily than hot brands, and this is a disadvantage for this type of marking.

FEEDING SUCKLING CALVES

Beef cattlemen usually follow the practice of letting the calves nurse the cows. Whether or not it is profitable to feed concentrates to suckling calves depends upon the pasture conditions, the age that the calves will be marketed, the cost of the supplemental feed, and whether the calves are to be sold as feeders or kept in the herd for breeding purposes. The feeding of concentrates to calves that are to be sold as yearling feeders may not be profitable where pastures are good and there are plenty of good roughages for winter feeding.

Creep-feeding Calves

The practice of creep-feeding calves has the following advantages: (1) the calves will have more weight at weaning time, (2) the cows are less suckled down, (3) calves will not miss their mothers as much at weaning time, (4) calves will be more uniform in size as creep-feeding helps to make up for the differences in milking ability of the cows, (5) calves that are to be finished will be in better condition when they go into the feed lot, and the feeding period will be shortened, (6) heifers and bulls that are to be kept or sold as breeding animals will be bigger, (7) condition is a good seller even of feeder calves. Creep-fed calves are usually larger than other calves and therefore will sell for more, either as feeders or for breeding purposes.

The disadvantages of creep-feeding are that (1) the calves' intake of pasture is reduced, (2) their efficiency is lower when put into the feed lot, and (3) the cost of the creep feed may be greater than the value of the additional weight.

Feeds for Suckling Calves

A large variety of feeds may be successfully fed to suckling calves. Availability and price should largely determine those that are selected.

Corn, oats, barley, sorghum grain, or wheat are all good grains for self-feeding suckling calves. The price per pound of digestible nutrients should be the determining factor in making a selection.

If a good mineral mixture is available for the cow herd, and is fed in a place easily accessible to the calves, it may not be necessary to provide a mineral supplement in the creep ration. However, many good cattlemen reserve a small section of the creep where a mineral supplement is placed separately for the calves.

Antibiotics for Suckling Calves

Experiments conducted at agricultural universities show a marked reduction in scouring among suckling calves, plus a growth stimulation, when Aureomycin is fed at the rate of 24 milligrams per 100 pounds of live weight.

Judging from similar results at agricultural experiment stations, it seems advisable to include Aureomycin in the creep rations.

Rations for Suckling Calves

The accompaning table lists some suggested rations for creep-feeding calves.

FEEDING REPLACEMENT HEIFERS

After replacement heifers (those which will replace old or non-productive cows) have been weaned, it is important to feed them separately from the cows. Heifers need a better quality ration than do mature cows, if they are to reach normal size at breeding time.

Whenever it is possible, during the first and second winters, breeding stock should receive at least 5 pounds of good legume-grass hay, plus additional amounts of sorgo or corn silage in the colder regions. In the South, if good winter pastures are available, additional forage is not essential.

Unless a roughage that will supply some grain like sorghum silage or corn silage is fed, grain in addition to good forage will speed up the growth and maturity of heifers. It is important to raise the replacement heifers as cheaply as possible without sacrificing growth and development. In an Iowa experiment, heifers gained satisfactorily when pastured in cornstalk fields and fed an additional 2 pounds of oats, one pound of 32 percent protein supplement, and one pound of alfalfa hay daily.

Antibiotics for Growing Heifers

Experimental results reported at Kansas State College and Purdue University indicate that antibiotics fed to calves receiving a ration consisting largely of roughage, after weaning and up to 18 to 20 months of age, will increase gains and efficiency of feed utilization.

The level of Aureomycin fed daily is important. Purdue scientists recommend a level of about 10 milligrams to 100 pounds of live weight. When greater amounts are fed, the antibiotics have a depressing effect on the appetite of the cattle.

FEEDING YOUNG BULLS

The feeding of young bull calves up to the time they will be put into service does not differ greatly from that of replacement heifers. However, because their growth is more rapid, their feed requirements are greater. The grain ration should be increased over that recommended for heifers. In addition to an unlimited amount of good cured roughage or pasture, young bulls should receive about a pound of grain per hundred pounds of live weight daily. If the roughage is low in protein ½ to one pound of protein

TABLE 13-3 SUGGESTED RATIONS FOR CREEP-FEEDING SUCKLING CALVES

With Good Pasture or Legume Hay (Pounds)		With Poor Pasture, Low-protein Forage, or Poor Milking Cows (Pounds)	
1. Ground oats	1,000	1. Ground oats	1,000
Ground corn	1,000	Ground corn	800
Antibiotics †		Linseed, soybean, or cottonseed meal	200
2. Ground oats	1,000	Antibiotics †	
Ground barley	1,000	2. Ground oats	1,000
Antibiotics †		Ground barley	850
3. Ground barley	1,000	Corn gluten meal, soybean, cottonseed, or linseed oil meal	150
Ground sorghum grain	1,000	Antibiotics †	
Antibiotics †		3. Ground sorghum grain	1,000
4. Ground oats	500	Ground barley	800
Ground wheat	500	40% commercial protein	200
Ground corn	500	Antibiotics †	
Ground sorghum grain	500		
Antibiotics †			

† Any antibiotic supplement will provide 40 grams of Aureomycin per ton of feed.

Note: Grains should not be ground too fine. Rolled or cracked grains are best unless fine grinding is necessary for mixing ingredients.

supplement should be fed daily in addition to the grains. A mineral supplement should be available free choice.

Young bulls about to be put in service should be in good flesh but not overly fat. Bulls in service will usually stay in condition on the same rations fed to the cows, unless they are used exceptionally heavily over a considerable length of time. If that is the case, an additional amount of concentrates will be required to maintain them.

BEEF CATTLE HOUSING AND HANDLING EQUIPMENT

Well-planned buildings, lots, feed bunks, and handling facilities are essential to a successful beef cattle enterprise.

Housing for Beef Cattle

Beef cattle are not especially sensitive to changes in weather conditions. Warm and expensive barns are not needed except when the cattle program calls for cows to calve during severe cold weather.

The reproducing herd may be successfully wintered without any shelter even in the Northern regions where winters are cold. However, a good windbreak will reduce the amount of feed necessary to winter the herd and provide a shelter for the cattle. Trees, a high board fence, or a natural windbreak provided by a hill or a similar wind barrier will provide ample protection except under the most severe conditions.

In the Northern areas, when the cattle production plan calls for calving during the cold months, warm housing will need to be provided during the calving period. Feeder cattle need a windbreak.

Feeder cattle protected from extremes in weather conditions gain faster and more efficiently. A shed open on the south is satisfactory for finishing cattle or those being wintered for later feeding.

Young growing cattle or cattle being finished on grass will make more rapid gains during the hot part of the summer if they have access to shade. When natural shade such as trees is not available, artificial shades will prove profitable.

Confinement Housing

Research has shown that efficient animal production can be improved through closer control over the environment. Available records indicate total-confinement cattle should gain on the average of one-half pound more per head per day in the winter, and at least

FIGURE 13-13. A workable system that includes lots, buildings, and feeding equipment is required for any program in beef feeding production. Note the design of the features being used by the Harmson brothers, Clinton County, Iowa. (*Courtesy* American Angus Association)

FIGURE 13-14. (*Above*) Profitable beef production being achieved with the use of a board fence for protection from the weather and well-drained lots providing dry resting areas. (*Courtesy* American Hereford Association) **FIGURE 13-15.** (*Left*) One of the largest environment-controlled, total confinement beef systems in the United States is located on the Rohlf Farms, Kaukauna, Wisconsin. Bill Rohlf, Manager, can feed 1,680 head in one single feeding operation by operating a control panel. (*Courtesy* Butler Manufacturing Company)

one-fourth pound more during the summer. This would depend upon the temperature for winter and summer months in your area. Total confinement systems—completely enclosed, insulated, fan-ventilated, and temperature-regulated, incorporating slatted floors and manure pits below—eliminate bedding costs and reduce the labor involved in manure handling.

Total confinement systems are expensive, with the cost of some buildings running over one hundred dollars per head. No set cost can be stated because variations in building and equipment types are too wide. If a producer is interested in such a system, he should visit different styles and decide which one, or combination of several, meets his needs.

Semi-confinement type systems with "open front" buildings are popular with some producers. Total control of the environment is not possible, but investment costs are reduced and many labor-saving features can be utilized.

Feeding and Other Equipment

Stationary bunks may be constructed along one side of the cattle lot, close to the feed supply and built of a heavy material such as concrete. If properly constructed,

FIGURE 13-16. This controlled environment, total confinement system features slatted floors, automated feeding equipment, and forced air ventilation. (*Courtesy* Butler Manufacturing Company)

FIGURE 13-17. An open-front type confinement system using concrete slatted floors with a manure pit below. (*Courtesy* Butler Manufacturing Company)

FIGURE 13-18. (*Above*) The use of self-feeders are popular with some producers. Management is an important factor in keeping the animals on full feed. (*Courtesy* Elanco Products Company) FIGURE 13-19. (*Below*) A simple but safe loading chute constructed of concrete and heavy lumber. (*Courtesy* Hawkeye Institute of Technology)

they are very durable and maintenance costs are small.

Portable feed bunks have the advantage of being easily moved from one lot to the next or onto the pastures if desired. They are usually less durable than well constructed, permanent bunks.

Hay Mangers. As with grain feed bunks, hay mangers may be portable or fixed for feeding along one side of the feed lot. Portable mangers may be of two types: those designed for feeding hay or silage directly from a stack or haybarn, and those that are designed to be moved easily from lot to lot or onto the pasture.

Self-feeders. Self-feeders reduce labor in feeding cattle, especially when feeding concentrates to cattle that are on pasture.

Mineral Feeders. Mineral feeders, built to protect the mineral mixture from rain or snow, should be placed in an area where it will be easy for the operator to fill them and where the cattle can get at the minerals conveniently, usually near the watering facilities.

Loading Chutes. Every cattleman needs a loading chute. These may be portable or fixed. If a fixed chute is used, it should be located so that both large and small trucks and trailers can reach it conveniently at any time of the year.

Restraint Equipment. It is necessary to secure cattle for hoof trimming, dehorning, and similar operations. Stocks and squeezes are used for these purposes.

Cattle Yards and Lots. Well-planned corrals and lots make handling of stock easier and save labor. Paved lots are ideal, especially in the more humid regions. They keep cattle out of the mud, make it easier to work with them, and save manure. Paved lots are easy to clean. If the lots are not paved, they should be located on well drained areas.

Fences. Feed lot fences have to be strong. A good fence can be made by combining plank and woven wire, with posts set not more than 10 feet apart. While it is more expensive, a plank fence made by bolting two-inch planks to wooden posts makes a very durable fence. Cattle confined to small lots subject fences to considerable punishment.

Mechanized Feeding Systems. One who is anticipating the building or remodeling of a cattle feeding plant will profit by making a study of various labor-saving systems that have been devised by many cattle feeders. By using properly constructed bins, mechanical augers, and feed conveyers, it is possible to mix feed and deliver it into the bunks for the cattle with very little hand labor.

Since labor is always an expensive part of any farm or feeding enterprise, labor-saving equipment that will reduce the number of man-hours necessary for carrying on any farming program should always be considered. Lack of capital may prevent the operator from completely developing all of the plant at one time. However, a complete plan adapted to the conditions should be made, and each building or piece of equipment added should fit in with the general plan.

FIGURE 13-20. This 150-foot long, automatic feeding system is designed to save labor for Jon Youngquist who ranches in southwestern Nebraska. (*Courtesy* Butler Manufacturing Company)

FIGURE 13-21. Grain storage and feed processing equipment. (*Courtesy* Butler Manufacturing Company)

SUMMARY

Cows should be fed as cheaply as possible without reducing the number and vigor of the calves or sacrificing any of the productive life of the cows.

Plenty of roughage is the key to economical production of calves, An ample amount of pasture will maintain the herd except for minerals and salt, which should be provided at all times during the pasture season. Inadequate pastures should be supplemented. Legumes grown in combination with grasses provide excellent, highly nutritious grazing. Wherever they are adapted, legumes should be grown as part of the pasture crop.

There is a danger of mineral, vitamin, and protein deficiencies when cattle are fed in dry lots unless proper precautions are taken by supplementing the ration.

Mature cattle can utilize large quantities of low-quality roughages such as cornstalk silage, corncobs, and straw, if a supplement containing protein, minerals, and vitamins is provided. When good legume hay, or legume hay and silage, is fed as the major part of the ration, bred cows can be expected to winter well and produce a calf on little else but self-fed salt and minerals. Bred cows should normally gain about ⅓ pound per day.

The heat period for cows will average from 12 to 18 hours. The time between heat periods will be from 17 to 26 days, with an average of about 21 days. The gestation period is about 283 days, with some variation among breeds and individual animals.

Heifers that are well grown out may be bred to calve as two-year-olds, but under some conditions it is recommended that they drop their first calves when three years of age.

A young bull 15 months of age may be used to breed up to 15 cows. Bulls should be at least two years old and not more than seven when turned on the range. Older bulls, when properly handled, may be used for limited service up to 12 years of age.

Calves should come as early in the spring as conditions will permit. Calves two to three months old will utilize more grass and be heavier in the fall.

During cold weather clean, dry, well lighted, and well ventilated pens should be provided for the calves. Pens should not be overcrowded and calves should be grouped according to age and size. Pens should be equipped with hay racks, feed boxes, and watering facilities. The operator will need to give more attention to sanitation when calves are confined to pens than when they are on pasture.

Commercial cattlemen should dehorn their cattle. Dehorning may be accomplished by chemicals, hot iron, dehorning spoons and tubes, saws, or clippers. The method used will depend upon the age of the animal and the experience of the operator.

Calves may be castrated any time after they are a few days old. Most experienced cattlemen prefer to perform the operation when they are from two to three months of age.

Hide brands, ear notches, and tattooing will provide permanent marks that will not be lost. Other marks include horn brands, neck chains or straps, and ear tags.

Suckling calves will grow quite satisfactorily on milk and grass or other good forage. However, growth and weight gains will be faster if a concentrate ration is provided in a creep. Antibiotics have increased the rate of gain and reduced the incidence of scours among suckling calves.

Replacement heifers need a better ration than is ordinarily necessary for mature cows. When the ration is made up primarily of roughage, antibiotics will increase the rate of growth of young stock up to 18 to 20 months of age. Young bulls may be fed the same as replacement heifers except that they will require more feed.

Well-planned equipment is essential to the successful handling of beef cattle. Housing need not be elaborate or expensive for beef cattle unless cows calve during cold weather.

Feeding equipment may consist of either fixed or portable grain bunks and hay racks or both. Mechanized feeding systems will reduce labor and should be considered before building or remodeling a cattle feeding plant.

Loading chutes and restraining equipment are an essential part of the cattleman's equipment.

Lots and corrals should be either paved or located on well drained land. Lots should be planned for convenience in sorting, loading, and handling cattle.

QUESTIONS

1. What are the important considerations in feeding the cow herd?
2. Under what conditions will it be necessary to supplement pasture?
3. Give several examples showing how, when, and what kind of additional feed you would provide for cows on pasture.
4. List some mineral mixtures that could be used under various conditions.
5. How may additional vitamin A be provided for cows on pasture?
6. How do grasses and legumes compare as pasture plants?
7. Why should grasses and legumes be grown in combination?
8. List the common pasture legumes.
9. Give some pasture mixtures suitable to your area.
10. Develop a pasture program suitable to your area.
11. Explain how the nutrient value of pastures will vary.
12. How should pastures be managed in your area for greatest production?
13. Why is there more danger of nutritional deficiencies when cattle are fed in dry lots?
14. Explain how various quality roughages can be utilized successfully.
15. Give several rations that can be depended upon to winter beef cows successfully.
16. What is the duration and length of time between heat periods?
17. What is meant by the *gestation period* and what is the average length of this period?
18. When should heifers be bred? Explain.
19. How old should bulls be before they are put into service?
20. Why isn't artificial insemination practiced as much in beef production as it is in the dairy industry?
21. When should cows calve? Why?

22 What steps would you take to provide sanitary conditions for young calves?

23 Describe the type of pens you would recommend for young calves housed in barns or sheds.

24 Why should calves be dehorned?

25 Describe the kind of protection needed for beef cattle on your farm. Describe where beef cattle are kept on the average farm in your community.

26 Make a list of equipment needed for the handling of beef cattle on your farm.

27 Make a sketch of a plan for beef cattle yards or corrals that would meet the needs of the average cattleman in your area.

28 What advantages do paved lots have?

29 Make a plan for a mechanized cattle feeding plant.

REFERENCES

Albaught, Reuben, C. F. Kelly, and H. L. Belton, *Beef Handling and Feeding Equipment,* Agricultural Experiment and Extension Service Circular 414, Berkeley, California, University of California, 1952.

Beesen, W. M., and T. W. Perry, *Chopped Forage Vs. Pasture for Feeding Cattle,* Agricultural Experiment Station Mimeo A. H. 122, Lafayette, Indiana, Purdue University, 1953.

Lindgren, H. A., *Feed Requirements and Values for Livestock,* Extension Bulletin 639, Corvallis, Oregon, Oregon State System of Higher Education, 1951.

Paynter, W. G., and L. B. Embry, *Block-Feeding Antibiotics,* from Farm and Home Research, Volume XIII, No. 3, 1962.

Progress Report in Beef Cattle Production, Lincoln, Nebraska, University of Nebraska, 1965.

Recommended Nutrient Allowances for Beef Cattle, Report of the Committee on Animal Nutrition (Revised), 1970, National Research Council, Washington, D. C.

Results of Cattle Feeding Experiments, AS-183, Ames, Iowa, Iowa State University, 1966.

Smith, Harry, and Ford C. Daugherty, *Beef Production in Colorado,* Extension Bulletin 389-A, Fort Collins, Colorado, Colorado A & M College, 1950.

Snapp, Roscoe R., and A. L. Neuman, *Beef Cattle* (Sixth Edition), John Wiley and Sons, Inc., New York, 1969.

Totusek, Robert, E. C. Hornback, T. W. Perry, and W. M. Beesen, *Corncobs for Wintering Beef Cows,* Agricultural Experiment Station, Mimeo A. H. 94, Lafayette, Indiana, Purdue University, 1952.

14 Feeding and Management of Stockers and Feeder Cattle

In addition to the initial cost of the cattle, the cattle feeder has made an investment in feed. If his feeding operations go wrong through mistakes in judgment, he will lose heavily.

SOURCES OF PROFIT FROM FEEDING CATTLE

Profits from Margin and Gain

Unless there is a general price decline in cattle, or conditions have created unusually high prices for feeder cattle, the feeder can reasonably expect to sell the finished cattle for a higher price per hundred pounds than he paid. This increase in price is known as margin. To assure himself of getting a margin, the cattle will need to be finished to the best market grade they are capable of making.

Profits result from receiving a greater total price for the increased weight than the feed cost to produce the gain.

Profits Resulting from Manure Value

The value of the manure is more difficult to determine than are margin and gain profits. Manure value must be determined in terms of increased crop production, as a result of the fertilizing value of the manure, and in the amount of commercial fertilizer replaced by the manure.

DRY ROUGHAGES FOR STOCKER AND FEEDER CATTLE

Cattle feeds are many and varied. The rations used depend upon the kind of cattle to be fed, and the cost and availability of feeds.

Roughages should make up a large part of the cattle ration. They may be economical and, when properly fed, decrease the cost of gain. Roughages vary considerably in feeding value. For example, hay wafers (hay compressed into small blocks, bite-size or smaller) showed greater feed efficiency in recent experiments than baled hay. Roughage substitutes, such as oyster shells or plastic tabs, can be used to perform much the same functions as natural roughages in the digestive tract. There is little or no nutritive value from the roughage substitutes, however.

Legume Hays. High-quality legume hay is recognized as the best roughage from a nutritional standpoint. It is high in proteins, minerals, and vitamins.

Grass Hays. Most of the grasses are excellent roughages, but low in protein and minerals compared to legumes. One pound of a 40 to 45 percent protein supplement is usually all that is needed when legume hay is used as the entire roughage. If grass hay is fed as the only roughage, an increase of ½ to one pound of protein supplement will be required to balance the ration. More attention must be given the mineral and vitamin content of the ration to make up for the roughage deficiency. Feeding a mineral mixture free choice and one or two pounds of

FIGURE 14-1. Margin and value of gain over feed cost are factors which determine profit or loss for any feedlot operation. (*Courtesy* Hawkeye Steel Products, Inc.)
FIGURE 14-2. (*Below*) Roughage is an important ingredient in rations for feeder cattle. Note the hay storage area and feeding facilities pictured here. (*Courtesy* Butler Manufacturing Company)

dehydrated alfalfa meal per head daily will compensate for the mineral and vitamin deficiencies of the grass hays.

Mixed Hays. The term mixed hay usually applies to a mixture of grasses and legumes. The food value of mixed hay is between that of legume and grass hay, depending upon the percentage of legumes in the mixture.

Beet Tops. Fresh beet tops have a very high feeding value but lose much of this if left too long in the field. They should be harvested and stored to conserve their feeding value, or they should be made into silage.

Dehydrated Hays. Dehydrated hays are the result of harvesting green plants and artificially drying them. This process saves the leaves and results in a higher protein and vitamin A content. However, since vitamin D is absorbed by the hay from the sunshine while curing in the field, sun-cured hay is higher in vitamin D. Since cattle that are exposed to direct sunshine are not likely to suffer from a vitamin D deficiency, the vitamin A content of forage is more important than the vitamin D content.

Dehydrated alfalfa meal or pellets have been a consistent gain booster when added to cattle rations that did not include legume hays, or when the legume was of poor quality. Alfalfa apparently has growth-stimulating qualities that have not been duplicated in other feeds. When the crop is cut at the proper stage and artificially dried, few nutrients are lost.

In areas where corn is raised, more total digestible nutrients (T.D.N.) may be harvested from an acre if the corn is made into silage than if only the ears or shelled corn is harvested. The shelled corn contains about 61 percent of the T.D.N., the cob, 10 percent, and the stalk, 29 percent.

Corncobs and Dry Chopped Cornstalks. Corncobs have been used successfully, when properly supplemented, for growing and finishing cattle. Cobs are high in fiber and very low in protein, vitamin, and mineral content. Cornstalks, as well as the cobs, when properly supplemented, will produce substantial gains on feeder cattle.

The bacteria that are responsible for the ability of cattle to digest high-fiber feeds must be supplied with enough energy, protein, mineral, and vitamin nutrients for their own growth. Unless these bacteria are well fed, ruminants fail to get enough food value from cobs, cornstalks, and similar feed to survive and produce gains.

Silage. Silage is produced by putting feeds containing a high percentage of moisture into one of several types of silos, or by packing the feed in a compact stack known as a silage stack.

When alfalfa, brome grass, oats, and other legumes, grasses, or small grain crops are made into silage, it is generally referred to as grass silage; actually it would be more correct to refer to silage made from legumes as legume silage, that from grasses other than small grains as grass silage, and that made from small grains as small grain silage.

These crops all vary considerably in their nutrient content. Silage made from legumes will contain more protein than that made from grasses.

The use of oxygen-free silos has been increasing in recent years. These silos permit the storage of crops at a lower moisture level than in conventional silos. When hay is stored in such silos, it is usually referred to as haylage. This product has a higher value per ton than regular silage made from the same hay. The addition of 30 to 40 pounds of liquid molasses per ton will substantially increase the energy value of haylage stored in oxygen-free silos.

Corn and the sorghums make excellent silage. The silage from these crops will be lower in protein but higher in carbohydrates than legume or legume and grass mixed silage. The addition of urea or high protein feeds made from oil seed crops will im-

FIGURE 14-3. High-quality silage and grain can provide the cattle feeder with economical nutrients to assist him in obtaining efficient gains. (*Courtesy* Iowa Structures, Inc.)

prove the protein content of the silage made from low protein crops. Urea is less expensive than oil meals, and the recommended amount is 10 pounds of urea per ton of silage.

Cornstalk silage is made from the stalks after the ears have been harvested. Water may be needed to increase the moisture as the stalks are usually too dry for silage when the corn is ready for harvest. The addition of molasses as a preservative improves the feeding value and prevents possible spoilage.

Recent feeding trials using cornstalk silage have revealed that a considerable amount of food value, especially carbohydrates, is present in cornstalks.

GRAINS AND GRAIN SUBSTITUTES FOR STOCKER AND FEEDER CATTLE

The grains are the most important concentrates used for finishing cattle. Corn, barley, sorghum grain, wheat, and rye are the major grains.

Corn. Corn is the standard finishing grain, and all other grains or grain substitutes are compared to corn in determining their value. Corn may be fed shelled, cracked, rolled, flaked, or as ground ear corn. Corn grain furnishes more nutrients per pound than does ground ear corn, and will produce faster gains when full fed. It takes more experience to feed straight corn because there is greater danger of the cattle overeating and going off feed. Although gains will be somewhat slower, ground ear corn can be fed more safely and has the advantage of utilizing the cob. Unless hogs follow the cattle to utilize undigested corn, it pays to break the grain for more complete digestion.

Recent experiments show that high moisture corn, such as corn ensiled or stored in air-tight storage facilities containing up to 32 percent moisture, produced cheaper gains than dry corn. Cattle consume from 12 to 15 percent less feed per pound of gain when high moisture corn is used. This has resulted in approximately 3 percent less cost per pound of gain.

Barley. Barley may be substituted for corn. Barley has about 85 percent of the value of corn for cattle feed. Feeding tests

TABLE 14-1 COMPARABLE VALUE OF ROUGHAGES WHEN ALFALFA IS SELLING FOR $20 PER TON

Roughage	Value	Roughage	Value
Alfalfa hay	$20.00	Legume and grass silage	$7.50
Red clover hay	18.00	Corn silage	8.00
Soybean hay	16.00	Sorghum silage	7.50
Lespedeza hay	15.00	Cornstalks (dry chopped)	6.00
Brome grass hay	13.00	Cornstalk silage	3.00
Timothy hay	12.00	Corncobs (ground)	6.50
Prairie hay	13.00	Oat straw	5.00
Sudan grass hay	12.00		

show that from 12 to 15 percent more pounds of barley are used per 100 pounds of gain. Barley should be crushed or ground for greatest feed efficiency.

Wheat. Wheat may replace all or any part of the corn fed pound for pound, but it is better when it does not exceed 50 percent of the ration. Wheat is from 2 to 5 percent higher in protein than corn.

Oats. Oats are better for young calves than for older feeder cattle. They are considered more of a growing feed than a finishing feed. Oats provide bulk when added to corn. If cattle are to be fed on shelled or cracked corn, adding oats to equal half of the ration at the start and continuing to use from 2 to 3 pounds of oats in the daily ration will eliminate part of the danger of cattle going off feed from overeating. When used as a finishing ration, they are low in food value compared to corn.

Rye. Rye is not palatable to cattle, and if fed in large quantities for a long period, it may slow down feed consumption and gains. Rye should not exceed 25 percent of the grain ration. It can be used as a corn saver if the price is reasonable.

Grain Sorghums. Grain sorghums can replace corn in the cattle finishing ration, and will produce nearly equal results. Because grain sorghums are hard, treating them in the same manner as corn improves them for cattle feed.

Molasses. Molasses is well liked by cattle and is often used as an appetizer. When unpalatable feeds, especially low-quality roughages, are fed, cattle may be induced to eat them by sprinkling molasses over the feed. Molasses is not a protein but a carbohydrate feed and a partial replacement for grains. Liquid molasses has a feeding value equal to about 70 percent of corn pound for pound. Because of its palatability, its chief value is in increasing feed consumption. Molasses may be purchased in liquid or dry form. Mixed protein supplements, designed to be fed in conjunction with low-quality forages, often contain molasses as a source of high energy feed to stimulate bacterial growth in the rumen. Rumen organisms require carbohydrates in order to digest roughages.

Animal Fats. Animal fats, such as tallow and lard, are often surplus products of the packing industry. Considerable research is now in progress to determine to what extent these products may be used in livestock feeds. Experiments at Nebraska University and the Texas Agricultural Experiment Station indicate that animal fats may be successfully used in livestock feed. At the University of Nebraska, the addition of 5

TABLE 14-2 COMPARATIVE FEED VALUE ON A PER-POUND BASIS OF GRAINS AND GRAIN SUBSTITUTES FOR FINISHING CATTLE

Feed	Bushel Weight (Pounds)	Value Compared to Corn (Percent)	Approximate Quart Weight (Pounds)
Shelled corn	56	100	1.7
Corn (ear, ground)	70	85	1.4
Barley (ground)	48	85–90	1.1
Wheat (cracked)	60	100	1.7
Oats (ground)	32	70–75	0.7
Rye (cracked)	56	70–80	1.5
Sorghum grain (ground)	50–60	95	1.4–1.5
Molasses (liquid)		70	3.0

percent tallow to a high roughage cattle ration brought cattle to market weight with about half as much corn. The energy value of one pound of fat equals 2¼ pounds of carbohydrates.

Since fats must be heated to 150° to 160° F. to be mixed with other feed substances, special mixing facilities are required. Cattlemen should follow the research and make use of new developments in the feeding of animal fats.

PROTEINS AND PROTEIN SUBSTITUTES FOR STOCKER AND FEEDER CATTLE

There are a large number of protein concentrates suitable for cattle feeding. The cost per pound of protein should be the chief consideration in making a selection.

Soybean Oil Meal. Soybean oil meal is equal to or higher in percentage of protein than other protein meals. With the increase in soybean acreage, it is often cheaper than similar feeds.

Linseed Oil Meal. Linseed oil meal has long been used as a source of protein among cattle feeders. It gives cattle a sleek hair coat and is especially popular among cattlemen who fit animals for fairs and shows. Many feeders use a mixture of protein meals, including linseed oil meal.

Cottonseed Meal. Cottonseed meal is a widely used protein supplement. It is especially popular as a cheap source of protein among cattlemen in the Cotton Belt. It is not considered as good as linseed or soybean oil meal for calves under four months of age. It may prove toxic to young calves if fed in too large a quantity.

Dehydrated Alfalfa Meal or Pellets. Dehydrated alfalfa meal or pellets will range from 17 to 20 percent protein and therefore closely approach what we may term a protein concentrate. Dehydrated alfalfa may be used to replace a part of the protein in the cattle ration. It is also a rich source of carotene and several essential minerals.

Peanut Oil Meal. Peanut oil meal is about equal to cottonseed, soybean, and linseed oil meal for finishing cattle. In some experiments, it was slightly inferior but, for all practical purposes, may be used as a protein supplement for cattle. The price per pound of protein is the deciding factor.

Soybeans. Ground soybeans provide a good source of protein. Generally, the price is too high to make them an economical feed. However, if soybeans are a cheaper source of protein than the other common protein feeds, they provide a good homegrown protein supplement.

Tankage, Fish Meal, and Other Animal Proteins. Animal proteins may be used to make up a part of the protein content of the rations, if they are more economical than the vegetable proteins. Experiments show they have no other advantage over the vegetable proteins in rations for cattle over three months of age.

Urea

Urea is a nitrogen compound. Cattle can convert a certain amount of urea to protein. This is accomplished through the bacterial action which takes place in the rumen. Protein is made by combining the urea nitrogen with carbohydrates into the correct chemical combination. It is essential that some good source of carbohydrates be fed, if urea is contained in the ration. Molasses or grains are usually recommended. It should be remembered that urea is not a protein. The feeding form of urea is known as 262 for it has a protein equivalent of 262 percent. In other words, one pound of urea will combine with the carbohydrates to make 2.62 pounds of protein. When urea is fed, one pound mixed with 6 pounds of grain will replace 7 pounds of the oil seed meals. Urea may be mixed with soybean or linseed meal and fed as a protein supplement. When mixed, 0.1 pound of urea plus 0.75 pound of linseed oil meal are equal to 1.5 pounds of linseed oil meal, or 1,760 pounds of linseed oil plus 240 pounds of urea are equal to 4,000 pounds of linseed meal in protein equivalent. If soybean oil meal is used with urea, 0.1 pound of urea plus 0.65 pound of soybean oil meal equals 1.25 pounds of soybean oil meal.

If urea is mixed with one of the common proteins, as with linseed oil meal, the following example will serve to show how the percentage of protein equivalent of the mixture may be calculated.

Precautions in Feeding Urea. It should be remembered that urea, if fed in too great quantities, will cause considerable difficulty. It is toxic if overfed; it will make cattle sick and could possibly be fatal. It is important to have urea evenly mixed with grain or protein supplement; therefore, good mixing facilities are required. Urea should not exceed 1 percent of the total dry matter in the ration or 3 percent of the concentrate mixture. The maximum safe limit is 0.3 pound of urea per day per animal over 800 pounds, 0.2 pound for animals between 500 and 800 pounds, and 0.1 pound for those between 300 and 500 pounds. Smaller animals should not be fed urea.

1,900 lbs. linseed oil meal = 35% protein = 665 lbs. protein
100 lbs. urea feed = 262% protein equivalent = 262 lbs. protein equivalent

2,000 lbs. total feed — 927 lbs. protein equivalent

927 ÷ 2,000 = 46.35% protein equivalent

Urea has no advantage, except as an economy measure, over other recommended protein concentrates. Its value should be determined in terms of replacing the protein value of other feeds. It has no energy, mineral, or vitamin value. It should not be mixed with raw soybeans or untoasted soybean oil meal, which contain an enzyme that causes urea to become toxic to the animal.

Biuret. Biochemists have known for over a century that when urea is heated, a major product, biuret, is produced. Nutritionists have been impressed with its low toxicity and adaptability as a source or protein for cattle and sheep. If the demand for natural protein oil and seed meals continues to increase, man-made products like urea and biuret will increase as a protein source for ruminant animals.

Mixed Supplements

There is a variety of mixed cattle supplements designed for various types of feeding programs. Where good-quality forage is used as the primary roughage, almost any of the protein meals that have been discussed will prove satisfactory. When low-quality roughages are used, a protein supplement reinforced with additional vitamins and minerals is recommended. If urea is to be used to replace part of the protein in the ration, it is usually fed in a mixed supplement. Following is a listing of some mixed supplements. The high urea supplements should be fed only when a high level of concentrates is included in the ration.

MIXED SUPPLEMENTS CONTAINING 30 TO 35 PERCENT PROTEIN OR PROTEIN EQUIVALENT

1. Purdue Cattle Supplement with Urea

Feed	*Pounds*
Soybean oil meal	400.5
Molasses feed (50% molasses)	280.0
Corn or its equivalent	208.0
Urea	40.0
Bone meal	52.0
Salt (mineralized)	17.0
*Vitamin A and D concentrate	2.5

2. Purdue Supplement A

Feed	*Pounds*
Soybean oil meal	650.5
Molasses	140.0
Alfalfa meal	140.0
Bone meal	52.0
**Salt with cobalt	17.0
*Vitamin A and D concentrate	.5
	1,000.0

* Vitamin A and D concentrate contained 2,250 I. U. of A and 300 I. U. of D per gram.

** One ounce of cobalt sulfate added per 100 pounds of salt.

MIXED SUPPLEMENT CONTAINING 48 TO 52 PERCENT PROTEIN EQUIVALENT

Iowa Supplement

Feed	*Pounds*
Linseed oil meal	666
Distiller's grains	666
Molasses	268
Urea feed	214
Bone meal	134
Iodized salt	36
Trace mineral mixture	8
Vitamin A and D oil (2,250–300)	8
	2,000

Purdue cattle researchers developed a supplement designed to be fed with grass silage. Since grass silage is relatively high in protein and vitamins, the Purdue Supplement G is lower in protein, and the vitamin A and D concentrate was not included.

PURDUE SUPPLEMENT G

Feed	*Pounds*
Alfalfa meal	400
Molasses	329
Dried brewer's grains	132
Bone meal	105
Salt with cobalt	34
	1,000

Experiment stations are conducting research on new protein supplements and mixtures being developed. Check with your area stations to receive the latest formulas and data.

Protein Blocks

Protein blocks, containing a mixture of the oil seed meals, salt, minerals, and molasses, and made in the shape of a salt block (approximately 10 × 10 × 12 inches), provide an easy way to supply protein to cattle on the range or grazing a corn field. A typical block weighs about 33 pounds and is made hard enough so that it must be licked and not chewed. This limits consumption to about one and a half pounds per head per day.

MINERALS AND VITAMINS FOR STOCKER AND FEEDER CATTLE

Minerals

A mineral mixture for cattle recommended for the area (see Chapter 13), fed free choice, will generally meet the mineral requirements of stocker and feeder cattle if a high-quality roughage, such as legume hays, is fed to the extent of 5 pounds or more per day. Where the roughage consists mostly of low-quality materials such as corncobs and corn stalks, a supplement containing minerals is recommended as assurance that the animal will consume enough needed mineral for proper nutrition.

Vitamins

Usually vitamins A and D are all that need concern the cattle feeder. When high-quality forage is used to make up 50 percent or more of the roughage part of the ration, and the cattle are exposed to sunshine, these vitamins may be supplied in sufficient quantities. When low-quality roughage is used almost exclusively, a supplement containing vitamin A and D concentrate is recommended. However, it is good insurance to include a vitamin A supplement in any dry lot ration.

FEED ADDITIVES FOR CATTLE

Recent experimental work at Iowa and other agricultural experiment stations have shown that a number of feed additives, when added to cattle rations, improve the rate and efficiency of gain. Many different additives have been investigated, but few have shown consistent benefits to the cattle industry.

Stilbestrol

Stilbestrol is a manufactured chemical that will produce effects similar to the hormone estrogen estradiol, which is secreted by certain glands in the animal's body. Live weight gains may be stimulated as much as 30 percent on high-grain finishing rations. Because of controversy involving possible health hazards as a result of feeding stilbestrol to animals, the cattle feeder is advised to obtain the most recent information available from the U.S.D.A. concerning the feeding or implanting of the chemical and the withdrawal period recommended before slaughter.

Implanting Hormones

Hormones may be implanted at the base of the ear with an especially designed instrument. Cattle are placed in a squeeze chute, and the hormone implanted in pellet form. Pellets are absorbed over a period of 90 to 180 days. The gains made by implanted cattle compare favorably with those made by animals which were fed hormones. The labor and equipment necessary for implanting must be considered before deciding whether to implant hormones.

Implant materials including Rapi-Gain, Synovex S., Synovex H., and Ralgro, can be used to increase gain and feed efficiency. Representatives from the experiment stations in the area can assist the cattle feeder on deciding which one might be used in this operation.

Melengestrol Acetate (MGA)

MGA is a hormone developed to supress estrus in heifers and cows as an aid in synchronizing cattle-breeding cycles. During the development of MGA it was found that small amounts stimulated the rate of gain and reduced feed costs. MGA is effective within a range of .25 to .50 milligrams per heifer per day.

MGA is effective only in open heifers and does not stimulate gains of steers or pregnant heifers. It is a feed additive which must be withdrawn from the ration 48 hours before slaughter. MGA and antibiotic combinations are not approved at this time.

Feed Additive Withdrawal Periods

Even though some feed additives have proved to be beneficial to the cattle feeder, traces of the additives remaining in the animal tissue when slaughtered can be unsafe when used as eatable food. The Food and Drug Administration has established time periods whereby the different additives must be withdrawn from the rations, thus allowing the animal enough time to rid the carcass of any trace of them. The recommended withdrawal period for the additive used is printed on the feed tag, and legal action can be taken against a feeder if it is not followed.

Tapazole

Tapazole, trade name for a product known chemically as 1-methyl 2-mercaptoimidazole, is a white powder in its pure state. It has a blocking effect upon the thyroid gland in animals.

While Tapazole shows promise as a gain booster *it has not been approved by state and Federal authorities.* Therefore, no recommendation can be made for its use until further experiments have been made and approval granted. It looks promising, however, and cattlemen should follow the research results.

ANTIBIOTICS FOR FINISHING CATTLE

Experiments have indicated some increased rate of gain and feed efficiency when antibiotics, especially Aureomycin and Terramycin, have been fed to feeder cattle. Cattle receiving a comparatively low-quality ration have shown the greatest percentage of gain, while those receiving a high-energy ration consisting of full grain feeding and high-quality roughage have shown a lower rate of gain. Most of the advantage from antibiotic feeding came during the first three or four months of the feeding period.

The addition of an antibiotic premix to the supplement, to provide 50 to 80 milligrams of Aureomycin or Terramycin daily for cattle on a high roughage ration, would be profitable in view of the results of such feeding trials.

RATIONS FOR FINISHING CATTLE

Using the feeds that have been discussed, a large variety of finishing rations may be developed. Many areas produce feeds, not included here, that may be successfully substituted. It is important to use rations that will supply needed nutrients as cheaply as possible. Homegrown feeds are usually more economical to use than feeds that are shipped in. In selecting a ration the age and quality of the cattle should be considered. Cattle being finished for high choice or prime grades are usually limited on roughages in order to induce them to consume more finishing feeds. Cattle of the plainer grades, which are unable because of quality to finish into top slaughter grades, may utilize roughages free choice. Since the selling price of plain cattle is less, cheaper gains are essential.

For years, emphasis on fat has been one of the primary goals in the breeding of beef cattle. However, the consumer wants beef with sufficient marbling to be tender, a high percentage of lean, and a limited amount of outside fat; 1/10 inch per hundred pounds of dressed carcass weight is the maximum desired amount of fat thickness over the back and loin areas or at the twelfth rib. Recent experimental work indicates that what has been considered well bred beef cattle tend to put on too much fat when maintained on a high concentrate ration, and it appears that choice feeder cattle will develop a leaner carcass if the ration is not too high in concentrates.

Lower quality feeder cattle, especially those used in dairy breeding, can utilize larger amounts of concentrates without developing excessive back and loin fat. This does not mean that low-quality feeder cattle should necessarily be promoted, but it does point up the fact that more emphasis must be placed on developing muscling in the beef breeds and not on feeding for too high a finish.

FIGURE 14-4. Feed handling equipment as shown here can be used to prepare feeder rations containing protein, carbohydrates, fats, minerals, vitamins, and feed additive pre-mixes. (*Courtesy* Butler Manufacturing Company)

Rations for Wintering Feeders

Feeders who have an abundance of pasture which they wish to utilize often buy 400- to 600-pound calves or yearlings in the fall, carry them through the winter on a growing ration, and turn them on grass in the spring. Young cattle that are wintered should be fed a ration that will produce from one to 1½ pounds gain per day, which is considered a normal growth rate. Following are some suggested rations for wintering feeders intended to be finished at a later period.

Feed	Pounds Daily
1. Legume-grass hay	Full feed
Minerals	Free choice
2. Ground corncobs	10–14
Purdue Supplement A	3–3.5
3. Legume-grass hay	5
Corn or sorghum silage	15–25
35–40% protein meal	½
Minerals	Free choice
4. Legume-grass hay	5
Legume-grass silage	15–25
Minerals	Free choice
5. Legume-grass hay	5–8
Corn and cob meal	2–3
Minerals	Free choice

Preparing Feed for Cattle

The common method of feeding dry feeds to cattle has been to grind the grains and feed the hay or fodder either chopped or long. Grinding the grains increases their digestibility and reduces the amount of undigested material that passes through the digestive system.

Pelleting Dry Feeds. Recently a great deal of interest has been shown in the use of complete pelleted rations (except for silage or other succulent feeds). When this system is used, the dry roughage, grain, supplement, and molasses (if molasses is included in the ration) are ground and run through a pelleting machine.

In experiments conducted by the Illinois Agricultural Experiment Station, cattle fed a complete pelleted ration consisting of 65 percent ground ear corn, 5 percent blackstrap molasses, 10 percent soybean meal, and 20 percent hay gained 100 pounds on

Rations for Cattle That Are Intended to Make Choice or Prime Grades

Feed	Pounds
1. Legume hay	½ lb. per cwt. live weight, daily
Shelled crimped corn	*Full feed*
Linseed, soybean, or cottonseed meal, or a 30 to 35% protein commercial supplement	1 lb. daily
2. Legume hay	½ lb. per cwt. live weight, daily
Ground ear corn	*Full feed*
Linseed, soybean, or cottonseed meal, or a 30 to 35% protein commercial supplement	1 lb. daily
3. Legume hay	½ lb. per cwt. live weight, daily
Rolled barley	*Full feed*
Any recommended protein meal or 30 to 35% protein commercial supplement	1 lb. daily
4. Legume hay	¼ lb. per cwt. live weight, daily
Legume-grass silage	2 lbs. per cwt. live weight, daily
Ground sorghum grain	*Full feed*
Protein meal or 30 to 35% protein commercial supplement	1 lb. daily
5. Legume hay	½ lb. per cwt. live weight, daily
Legume-grass silage	2 lbs. per cwt. live weight, daily
Corn	500
Sorghum grain	450
Protein meal or 30 to 35% protein commercial supplement	50
6. Legume-grass hay	½ lb. per cwt. live weight, daily
Corn	825
Oats	100
Protein meal or 30 to 35% protein commercial supplement	75
7. Legume-grass hay	½ lb. per cwt. live weight, daily
Ground ear corn	875
Protein meal or 30 to 35% protein commercial supplement	125
8. Grass hay	½ lb. per cwt. live weight, daily
Sorghum or corn silage	2 lbs. per cwt. live weight, daily
Corn or sorghum grain	850
Protein meal or 30 to 35% protein commercial supplement	150

729 pounds of pellets while cattle fed the same ration as meal required 845 pounds of feed for 100 pounds of gain and gained 22 pounds less during the feeding period.

Pelleting is especially recommended for self-feeding since the intake of grain and roughage can be controlled. However the cost of pelleting must be considered before deciding whether to feed pellets. If the feed saving and other advantages will offset the cost of pelleting, it will be more economical.

Rolling and crimping. Rolling is a method of flattening out the grain by passing it through heavy rollers. Crimping is done by a machine that cracks the grain coarsely and then flattens it out. These methods are recommended for the preparation of corn, barley, and oats that are to be fed to the older cattle, if the feed has not been pelleted.

Flaking and Roasting. Flaking is achieved by adding moisture and heat in the form of steam and then rolling the grain by passing it through heavy rollers. Roasting is done with controlled moisture and heat machinery. Both processes have advantages over whole grain, but the advantages must be weighed against the total cost of processing. Large cattle–feeding operations could justify the initial costs of such machinery.

Pastures for Feeders and Finishing Cattle

Good pastures are essential for maximum growth and finishing of cattle on grass. Like breeding cattle, feeders make greater gains on legume-grass pasture than on straight grasses. The pasture mixtures discussed in Chapter 12 are applicable for feeders and finishing cattle.

Factors Influencing Gains on Grass

Several factors, such as age, weight, and condition, affect the gains cattle will make on grass.

Age and Weight. Big, thin cattle have more frame on which to put weight and a greater capacity for grass consumption. They can be expected to improve most quickly in condition. Yearlings would rank second in total pounds gain, and calves would be last. However, in proportion to starting weight, calves will make greater gains per hundred pounds of original weight than older cattle.

Condition. Cattle that have been wintered well, and are in high condition, will need grain while on grass if they are to continue to gain. It is not economical to turn cattle carrying too good a condition on grass without including grain in the ration. Such cattle will lose weight rather than gain. Cattle that are intended for grass during the pasture season, to be further grown out and finished the following fall and winter, should not be wintered to gain more than 1½ pounds per day.

Concentrate Rations for Cattle on Grass. The kind and amount of concentrates fed cattle on grass depend upon the gains expected, the amount of pasture and the grains available, and the time cattle are to be marketed. Cattle fed grains on pasture can be expected to make greater total gains than cattle receiving pasture only. However, when concentrates are fed, less gain due to grass alone will result. It may be more economical not to feed grain to cattle on pasture when the grass is abundant and the feeder expects to finish them in a dry lot at the end of the pasture season.

Cattle receiving concentrates while on pasture may be fed the same grain mixtures, except for roughage, as those in dry lots. The roughage will be provided by grass. If the pasture is from 20 to 50 percent legumes, the amount of protein supplement may be reduced to half the amount generally recommended for dry lot rations in which the various legumes, such as hay or silage, are fed.

When pastures are not adequate to supply sufficient forage, the pasture may be supplemented with either hay or silage.

Hauling Pasture to Cattle. Green-lot feeding of feeder cattle is practiced by some cattle feeders. (The practice was described in Chapter 12.) The advantages and disadvantages of green-lot feeding of feeder cattle are the same as for the breeding herd.

Finishing Cattle on Grass

Cattle feeders who have a considerable amount of pasture available may wish to utilize this grass to replace roughage in the ration. When cattle are pastured, they are usually bought in the fall, wintered at the rate of one to 1½ pounds of gain daily, and turned on pasture when the grass has made sufficient growth in the spring. They are fed a concentrate ration while on grass. Some feeders finish them entirely while on grass; others will put them in dry lots for a few weeks or months for the final finish. The kind of cattle, market conditions, and available feed largely determine the system followed.

Advantages and Disadvantages of Pasture Feeding. Some advantages of feeding cattle on grass are:

1. **Pasture gains are cheaper because less grain and protein is used in the finishing process.**
2. **The manure is dropped on the pasture (unless green-lot feeding is practiced), saving labor in hauling the manure onto the fields.**
3. **Little or no roughage other than grass is required, reducing labor in feeding and preserving roughages.**

Some disadvantages of feeding cattle on grass are:

1. **Cattle take longer to reach a desirable finish.**
2. **They have a lower carcass yield and usually sell for less than the same quality cattle dry-lot fed.**
3. **Summer heat and flies may cause slower gains.**
4. **Cattle to be fed on grass have to be purchased in the spring, when feeders are scarce, or in the fall, when they must be wintered, thus increasing the time they must be kept.**

FIGURE 14-5. Grass being converted to beef by these high-quality Brahman cattle. (*Courtesy* American Brahman Breeders Association)

Starting Cattle on Feed

When cattle are shipped in from a distance, they are tired and need rest. They should have plenty of water and comfortable quarters. Until the cattle become adjusted to their new home, no attempt should be made to start grain feeding. Cattle should be given all the dry hay (preferably a legume and grass mixed hay) they can eat for from 10 to 15 days before starting them on grain. They may go on pasture until fully recovered from the effects of shipping. A pasture that has been allowed to grow with partially mature plants provides an excellent feed supply for newly purchased feeders from the range areas. After the cattle have become completely adjusted to their new environment, those intended to be immediately placed on a finishing ration may be started on whatever concentrates the feeding program calls for.

If grains are to be the principal concentrates, yearlings may be started on from 2 to 3 pounds of corn, or its equivalent in other grains, per head and continued on this amount until all the cattle are eating grain. Many Western cattle have never tasted grain and must learn to eat it. Placing grain over the roughage will help to get them started on the grain. Care must be exercised to prevent the few that start to eat quickly from overeating. The grain may be increased one pound per head daily until the cattle are consuming 1½ pounds per 100 pounds live weight for cattle which are intended to be finished on a high-grain ration. Further increases should be made at the rate of one pound every four days until they are receiving 2 pounds per 100 pounds of live weight, which is generally considered a full feed. Cattle to be finished on high roughage rations may be started on grain the same way, but only brought up to 4 to 6 pounds or the amount called for in the feeding program. Whatever protein is to be fed should be started with the grain. A ration that is 9 to 11 percent protein is sufficient for finishing cattle. If a high protein equivalent supplement such as one containing urea is used, the amount of the supplement will need to be limited in accordance with the protein content. The common oil seed meals will range from 32 to 47 percent protein. If the supplement is a 55 to 60 percent or higher protein or protein equivalent, the amount used will be from ⅓ to ½ pound.

To start calves, feed from 2 to 3 pounds of oats per head daily until all calves are eating. When all the calves are eating, add one pound of shelled corn per head. Increase the corn by ½ pound daily until the calves are consuming 2 pounds of grain per 100 pounds of live weight. The oats may be continued at from 2 to 3 pounds or gradually cut back and discontinued. Protein supplement may be fed the same as for yearlings.

When ground ear corn is fed to feeder cattle, it should be remembered that the cob makes up about ¼ of the total weight, so more will be needed than if shelled corn is used.

Pre-conditioned Feeder Cattle

The pre-conditioning concept is a program designed to prevent losses in the cattle industry by preparing animals to withstand the stress of movement from ranch or farm to feed lot. These losses are due to changed enviroment; excitement of sorting, loading, and shipping; long periods without feed or water; change of feed; exposure to disease —which add up to fatigue, stress, shrink, and lowered disease resistance.

The pre-conditioning program contains specific steps and procedures for preparing animals for the stress of moving to the feed lots. These steps include: weaning 30 days before shipment, with castration and branding done before weaning; animals put on a "trough and bunk" adjustment period of 30 days; recommended vaccination program followed; treat for grubs; worm if necessary; rest stops at specified time intervals when shipped by rail or truck; and having the preceding treatments certified by a veterinarian.

Pre-conditioning of feeder animals is a sound veterinary-medical and management approach to solving a mammoth problem. Universal adoption of the program appears to be on its way.

FEEDING EFFICIENCY FACTORS

Rate of Gain

Thin two-year-old cattle that are thrifty and healthy will make the greatest gains in

FIGURE 14-6. Some of the 119 steer and heifer calves — all pre-conditioned — being fed by Paul Sowers, a cattle feeder, of Story City, Iowa. (*Courtesy* G. A. Hormel & Company)

the feed lot, followed by yearlings and calves. Big cattle consume more feed that can be turned into beef and, therefore, make more rapid gains.

Economy of Gains. Calves will produce 100 pounds of gain on less feed than older cattle of the same quality, although the time required for the same gain will be greater. The gain in body weight of older cattle is due largely to fat, while much of the gain on calves is due to growth. The time required to put market finish on calves is greater than for older cattle. Calves digest their feed more thoroughly than do older cattle.

As cattle finish, the amount of feed required per hundred-weight gain becomes greater. Experiments show that cattle weighing approximately 600 pounds when started on a finishing ration require 10 units of feed for the first 100 pounds of gain, 13 units for the second 100 pounds, 14 to 15 units for the third, 17 to 18 for the fourth, and 22 or more for the fifth hundred pounds of gain. The difference in market price for highly finished cattle, as compared to medium finish, is the important consideration in deciding how long to feed cattle.

Self-feeding Feeder Cattle

Cattle that have been brought up to a full feed of grain may be successfully put on self-feeders. Self-feeders are especially valuable when full-feeding cattle on pasture. Self-feeders reduce labor, but they require that proteins and grains be thoroughly mixed.

Purdue researchers successfully self-fed cattle using a ratio of 8 pounds of ground

TABLE 14-3 COST OF GAINS IN CATTLE FEEDING (Good to Choice steers, dry lot, high grain ration, corn calculated at $1.10 and $1.50 per bushel) †

	400-lb. Calf		640-lb. Yearling		840-lb. 2-year-old	
Gain	$1.10	$1.50	$1.10	$1.50	$1.10	$1.50
1st 100 lbs.	9.57	13.05	11.55	15.75	11.77	16.05
2nd 100 lbs.	10.67	14.55	13.53	18.45	14.41	19.65
3rd 100 lbs.	11.99	16.35	15.73	21.45	18.48	25.20
4th 100 lbs.	13.64	18.60	19.25	26.25	25.74	35.10
5th 100 lbs.	17.16	23.40	24.64	33.60	(17.93)	(24.45)
6th 100 lbs.	19.14	26.10				
7th 100 lbs.	24.09	32.85				

† For 50 pounds of gain.
Iowa State University Extension, EC Inf. 70.

ear corn to 1 pound of Purdue Supplement A. No hay was available after the first 28 days. The only roughage the cattle received after that period was the cobs contained in the ear corn. A complete pelleted ration, consisting of 65 percent ground ear corn, 5 percent blackstrap molasses, 10 percent soybean meal, 20 percent ground hay self-fed, plus 12 pounds of corn silage, produced an average of 2.75 pounds of gain per head daily and produced 100 pounds of gain on 729 pounds of pellets and 442 pounds of corn silage in an Illinois experiment.

SUMMARY

Cattle feeding involves a considerable capital outlay and risk. Only those who carefully plan their feeding operations and thoroughly understand cattle feeding can expect to make a profit. Direct cattle feeding profits must come from margin on original weight, profit on gain, or both. Manure value and increased fertility of the land are important sources of indirect profits.

Roughages should make up a large part of the cattle ration. Preserved dried roughages may be divided into two major classes: (1) dry roughages and (2) silage. There may be further classification as high-quality and low-quality roughages. Good legume and grass hay or silage, corn or sorghum silage are recognized as high-quality roughages, while corncobs, dry cornstalks, or cornstalk silage, straw, and coarse stemmy hay are considered low-quality roughages.

All of the grains may be successfully fed to cattle; however, corn, sorghum, and barley are the more common feed grains used. Molasses may be substituted for part of the grain. Animal fats have also been successfully substituted for part of the grain in cattle finishing rations.

The oil seed meals are the most common sources of protein supplements used in cattle feeding. Urea, a nitrogen product, may be successfully substituted for a part of the protein in the ration. The amount that can be safely fed depends upon the age of the animal and the kind of ration fed.

Several complete supplements have been developed for feeding cattle, especially when low-quality roughage is used. These supplements include vitamins, minerals, and some high-energy feeds, such as molasses or corn, in addition to 30 to 50 percent protein.

Minerals and vitamins A and D need to be provided in the ration for beef cattle. When legume hay or a good quality silage is fed, vitamin A will usually be supplied. Sunshine will supply vitamin D if cattle are exposed to direct sunlight.

Antibiotics, especially Aureomycin and Terramycin, have increased gains on young cattle, especially those receiving a high roughage ration.

Pelleted rations have shown some advantage in gain and efficiency over ground feeds.

Cattle will make substantial gains on pasture, and big, thin cattle will make the greatest gains. Cattle turned on pasture without grain should not be in too high a condition if they are expected to gain in weight.

Grass may be substituted for dry roughage or silage in the finishing ration. The concentrates fed to cattle on good pasture will not need to be as high in protein as those added in dry lot feeding. Grass usually produces cheaper gains, but grass-fed cattle will usually not sell for quite as high a price.

Cattle should be started on feed slowly and brought gradually up to full feed. Caution must be exercised to prevent overeating during the start of a finishing ration. Self-feeding is practical if done according to recommendations.

Calves will gain on less feed per 100 pounds than will older cattle. As the weight increases, more feed is required for each 100 pounds of gain.

QUESTIONS

1. Where are most of the cattle that go into the feed yards produced?
2. What are the sources of profit from feeding beef cattle? Explain.
3. Discuss the comparative value of the various kinds of dry roughages.
4. Why is it essential to include some high-energy feeds in the ration when low-quality roughage is fed?
5. What kind of materials make good preservatives?
6. Compare the feeding value of the grains for finishing cattle.
7. What advantages are there in feeding molasses?
8. How do the various oil seed meals and dehydrated alfalfa meal compare as cattle supplements?
9. Under what conditions is it advisable to feed a complete supplement?
10. How may urea be used as a protein replacement?
11. What are the limitations of feeding urea?
12. Discuss the need for minerals and vitamins in the rations for stockers and fattening cattle.
13. What is MGA?
14. What advantages are there in feeding MGA?
15. What advantages are there in feeding antibiotics to cattle?
16. Make up several rations for cattle intended to make Choice grade.
17. Give several rations using low-grade roughages.
18. Give some good rations for wintering feeders to make only a growth gain.
19. How do pelleted rations compare to ground feeds?
20. What kind of cattle gain best on grass? Discuss.
21. What are the factors that affect the feed efficiency when cattle are being finished?
22. What advantages are there in feeding cattle on grass?
23. Discuss the various ways one could successfully start cattle on feed.

REFERENCES

Annual Cattle Feeders' Day Report, 1965, 1966, 1967, Agricultural Experiment Station, Purdue University, Lafayette, Indiana.

Annual Cattle Feeders' Day Report, 1970, 1971, 1972, Iowa State University, Ames, Iowa.

Baker, Guy N., and Marvel L. Baker, *The Use of Various Pastures in Producing Finished Yearling Steers,* North Platte Substation Bulletin 47, Lincoln, Nebraska, University of Nebraska, 1952.

Beef Cattle Feeding and Breeding Investigations, Kansas State University of Agriculture Experiment Station, Bulletins 472 and 483, Hays, Kansas, 1964–1965.

Beef Cattle Investigations in Texas, 1888–1950, Bulletin 724, Texas Agricultural Experiment Station, College Station, Texas.

Beesen, W. M. and T. W. Perry, *Supplementing Growing and Fattening Rations for Cattle,* Agricultural Experiment Station, Mimeo A. H. 123, Lafayette, Indiana, Purdue University, 1953.

———, *Chopped Forage vs. Pasture for Feeding Cattle,* Agricultural Experiment Station, Mimeo A. H. 122, Lafayette, Indiana, Purdue University, 1953.

———, *Antibiotics for Suckling Calves and Yearlings,* Agricultural Experiment Station, Mimeo A. H. 130, Lafayette, Indiana, Purdue University, 1954.

Beesen, W. M., T. W. Perry, and M. T. Mohler, *Fattening Cattle on Corn Silage and Grass Silage,* Agricultural Experiment Station, Mimeo A. H. 105–106, Lafayette, Indiana, Purdue University, 1953.

Burroughs, Wise, C. W. McDonald, J. M. Scholl, and Bob Zimmerman, *Grass Silage, Chopped Cornstalks and Various Supplemental Feeds for Wintering Yearling Steers,* Agricultural Experiment Station FSR-54, Ames, Iowa, Iowa State University, 1951–1952.

Cattle Feeders Day Report, Urbana, Illinois, University of Illinois, annually, 1954–1966.

Corn and Sorghum Silage, E. C. 131, Lincoln, Nebraska, University of Nebraska, 1954.

Herrick, John B., *Pre-Conditioning: Preventive Veterinary Medicine,* Agricultural Experiment Station and Extension Service Pamphlet FS—1291, 1968, Iowa State University, Ames, Iowa.

Mayo, Henry, *Pastures for Growing and Fattening Cattle,* Agricultural Experiment Station, Mimeo A. H. 128, Lafayette, Indiana, Purdue University, 1954.

Recommended Nutrient Allowances for Beef Cattle, A Report of the Committee on Animal Nutrition, No. IV (Revised), 1970, National Research Council, Washington, D. C.

Trenkle, Allen H., *Feed Additives for Beef Cattle: Which? Why? When?,* Agricultural Experiment Station and Extension Service Pamphlet FS 1234, 1967, Iowa State University, Ames, Iowa.

Urea-corn Silage for Beef and Dairy Cattle, Pamphlet 316, Ames, Iowa, Iowa State University, 1965.

Webb, R. J., and G. F. Cmarik, *Comparison of Feeding a Ration as Pellets and as Meal to Yearling Steers,* Agricultural Experiment Station PS-27, Urbana, Illinois, University of Illinois.

15 Buying and Selling Beef Cattle

The farmer, rancher, or feeder who has produced or fed cattle for the market is very much concerned about the price received for his product. The market price at the time of sale is largely responsible for determining whether the enterprise has been profitable.

MARKETING FEEDER CATTLE

The producer of feeder cattle and the grain-producing farmer are dependent on each other. The producer of feeder cattle must sell his cattle to the commercial feed lots and to the grain-producing farmers, and the grain producers must get the cattle for their feed yards from the producer of feeder cattle. How they do business with each other is important to each of them.

Direct Buying and Selling

In recent years there has been an increase in the number of sales made by ranchers directly to the feeder. Such sales are generally made on the ranch. Feeders make the purchases either by contract in advance of delivery date or take the cattle immediately after sale. Approximately 20 percent of cattle sales are made directly to farmers and ranchers.

Contract Sales. When cattle are contracted for, an agreement is made between the buyer and seller. The important provisions in contract sales are: (1) *Price,* which is the amount paid down to seal the bargain plus the final amount paid at the time the cattle are removed from the farm or the ranch. (2) *Time and place of delivery.* If the ranch is equipped with scales and loading facilities, the cattle may be taken directly from the range by the buyer; otherwise the seller usually agrees to deliver the cattle to a certain weighing and loading point. (3) *The percentage of shrinkage to be allowed.* When cattle are moved for any distance or handled extensively, a certain amount of body weight is lost, resulting largely from the feed and water the animal eliminates that is not being replaced by regular feeding and watering. The amount of shrinkage depends upon the time in transit, weather conditions, the length of time cattle are allowed to rest, feed, and drink before weighing, and the age and weight of the cattle. No figure can be used that will be accurate enough to determine shrinkage in all cases, but 3 percent is commonly used for feeder cattle.

If it is agreed to pay a certain price per hundred at the ranch, but it is necessary to transport the cattle several miles before weighing, a shrinkage percentage ranging from 1 to 5 percent (depending upon the distance moved and time in transit) may be agreed upon. The practice of adding to or deducting from actual weight is referred to as *pencil shrinkage.*

Direct Selling with Immediate Delivery. Immediate delivery or acceptance of the cattle involves no type of contract other than the price per hundred and allowable shrinkage. Ownership of the cattle is transferred from seller to buyer within a short time.

The chief advantage of direct dealing is that it eliminates the middleman and the profit he would necessarily have to make for handling the cattle. Contract sales are usually made for the following reasons: they insure the buyer the cattle he wants when he wants them; usually, both buyer and seller feel they have made a good price bargain; and both know in advance what the price will be and when delivery will be made.

Selling Through a Dealer. Cattle dealers are individuals or companies that make a business of buying cattle for the feeders. The dealers either buy on a commission basis or depend on reselling the cattle at a higher figure than the cost for their profit. Many times small operators can get their cattle more cheaply through a dealer than if they spent the time and money necessary to inspect and buy their own cattle.

Order Buyers. An order buyer is a cattle dealer who specializes in buying cattle to meet the needs of his customers. Since the order buyer makes it his business to know where the cattle are and the market conditions, he can very quickly inform the feeder of the prospects he has of filling his order.

Auction Sales

Auction sales are a common method of transferring cattle from producer to feeder. Most auctions are private or company owned. The selling service is paid for by a commission (which is usually paid by the seller) charged on a percentage of the selling price.

FIGURE 15-1. Feeder calves with high quality are in demand by some cattle feeders. Their genetic ability and the feeds they are fed will determine if their carcasses will be in demand when they are in the cooler. (*Courtesy* American Hereford Association)

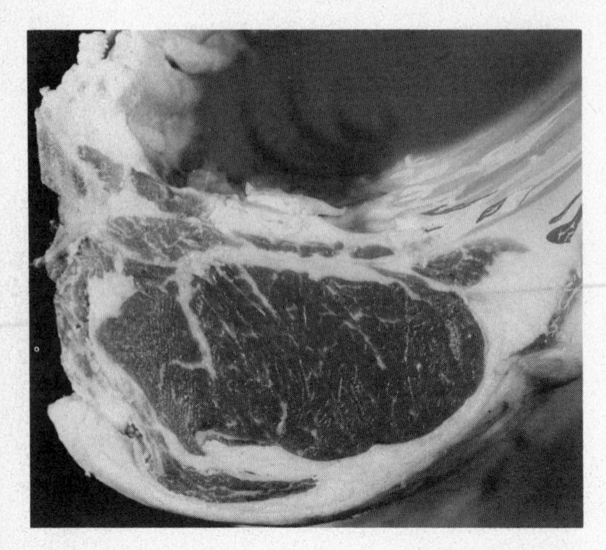

FIGURE 15-2. "We're having steak for dinner!" This cry seldom has to be repeated to bring an American family to the dinner table. (*Courtesy* American Brahman Breeders Association)

Terminal Markets

Large terminal feeder cattle markets are located in many Western and Midwestern cities. Producers who sell through these markets consign their cattle to a commission firm that will sell on a commission basis. A terminal market usually consists of a stockyard company which furnishes yards, scales, loading and unloading facilities, and feed and water for the stock. For this service, the stockyard company charges a fee. The amount will vary, but it usually ranges from 75 cents to $1.25 per head for yardage.

If the cattle are shipped to a commission firm, selling experts will handle and dispose of the cattle for the owner. They sort mixed grades into uniform lots and see that the cattle bring the best possible price obtainable at the time. The costs of yardage, feed, insurance, and selling charges are deducted, and the balance is paid to the owner. Cattle shipped to a central market must be consigned to a commission firm.

Selecting a Method of Marketing

The important factor in selecting the method of marketing is to determine which method will bring the greatest net income. The wise operator, in attempting to determine what his cattle will bring at home, will get market information from prospective markets and estimate shrinkage and other costs before finally deciding upon his method of marketing.

No one can consistently predict the time when the market will be highest. However, a knowledge of the factors discussed in the next paragraphs will be helpful in understanding the various factors.

The Total Number of Cattle. The number of cattle on the market will affect prices. When there are many many cattle to be marketed, prices will go down.

Conditions in the Grain Area. The Corn Belt farmer is the principal buyer of feeder cattle. The number he buys will depend upon his feed supply. If crop conditions are good, the producer can expect a heavy demand for feeders. The demand will be light if the crop outlook is unfavorable.

Slaughter Cattle Prices. When slaughter cattle prices are high in proportion to feed costs, it indirectly affects the price of feeder cattle. Cattle feeders are inclined to buy more cattle and pay higher prices if the outlook for slaughter cattle is good. Also, packing companies will furnish more competition for range cattle that are in good flesh.

Grass Condition on the Range. When the range areas are well supplied with moisture, there will be lots of feed. Ranchers are inclined to keep their cattle until late fall, putting on as much weight as possible before selling. Under good grazing conditions, early fall feeder markets are usually high, followed by an exceptional slump with the first general snowfall, at which time feeders pour into the market.

If a severe drouth should occur over a large area of the range country, it will usually force large numbers of cattle to market unseasonally. Such a condition will usually cause a rather sharp drop in feeder cattle prices.

Employment Conditions. Farmers and ranchers are dependent upon the consumer for their final market. What, and how much, the average American eats are the factors that determine prices of farm products. When wages are low or unemployment is high, beef and all farm products can be expected to go down in price.

Sources of Market Information

Current market information based on the previously mentioned conditions may be obtained from agricultural colleges, state and Federal market information services, newspapers, magazines, radio and television reports, and private outlook information services.

While no one can always be sure when the market conditions will be best, a careful study of the available information should enable the cattle producer to judge the market with reasonable accuracy, resulting in a more profitable beef business.

MARKETING SLAUGHTER CATTLE

Classes and Grades

The successful marketing of slaughter cattle is determined by feeding for the proper grade, methods of marketing, and the time the producer sells. Slaughter cattle are classified and graded according to sex, age, weight, and grade. Table 15-1 shows the commonly used system of classifying and grading.

The higher grades consist of cattle well developed in the areas of valuable cuts (back, loin, and hind quarters) and uni-

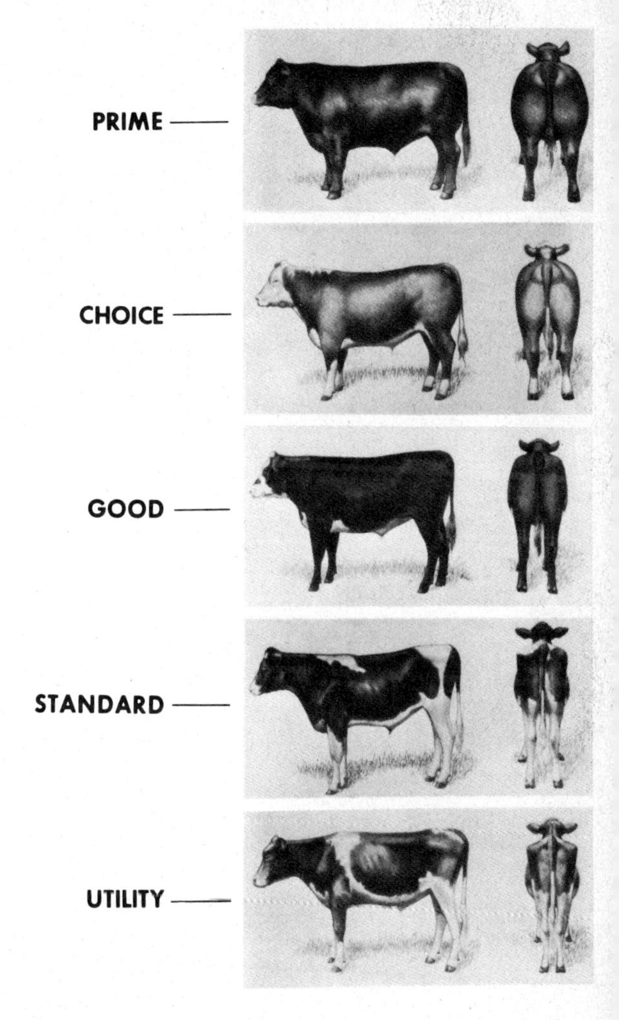

FIGURE 15-3. U. S. quality grades of slaughter cattle. (*Courtesy* U.S.D.A.)

formly covered with the right degree of finish. Choice and prime cattle usually are grain-fed cattle. They not only cut a high yielding carcass, but also one with a clear, white, firm fat that is well distributed around and through the muscle or lean meat. Calves, yearlings, and two-year-olds

TABLE 15-1 CLASSIFICATIONS AND GRADES OF LIVE SLAUGHTER CATTLE

Sex	Age	Weight	Grade
Vealers:			
Bulls Heifers	Less than 3 months	Light Medium Heavy	Prime, choice, good, standard, utility, cull
Calves:			
Bulls Heifers Steers	3 to 8 months	Light Medium Heavy	Prime, choice, good, standard, utility, cull
Steers	Yearlings	Light Medium Heavy	Prime, choice, good, standard, utility, cutter, canner
	2-year-olds	Light Medium Heavy	Prime, choice, good, standard, utility, cutter, canner
Heifers	Yearlings	Light Medium Heavy	Prime, choice, good, standard, utility, cutter, canner
	2-year-olds	Light Medium Heavy	Prime, choice, good, standard, utility, cutter, canner
Cows	All ages	All weights	Choice, good, standard, commercial, utility, cutter, canner
Bulls	Yearlings	All weights	Choice, good, commercial, utility, cutter, canner
	2-year-olds and older	Light Medium Heavy	Choice, good, commercial, utility, cutter, canner
Stags	All ages	All weights	Choice, good, commercial, utility, cutter, canner

TABLE 15-2 RELATIVE PERCENTAGES OF ALL CATTLE BY GRADE

U.S.D.A. GRADES (Quality)	Percent of All Cattle
Prime	2.9
Choice	45.6
Good	25.3
Standard, Commercial	8.5
Utility	7.0
Cutter, Canner	10.7

produce the more popular-sized cuts and are more tender and flavorful than older cattle. In contrast to older and lower grade animals, young prime cattle usually are sold at premium prices. The carcass yield is important in determining grades. Young prime cattle will dress from 62 to 67 percent; choice, from 60 to 64 percent; good, from 58 to 60 percent; standard, from 52 to 58 percent; and cutters and canners will average around 42 percent. Vealers have a flavor different from calves, yearlings, and two-year-olds. Their meat is classed as veal rather than beef.

U.S.D.A. Yield Grades for Beef. The value of a beef carcass depends chiefly upon two factors—the quality of the meat and the amount of salable meat the carcass will yield, particularly the yield of the high-value, preferred retail cuts.

U.S.D.A. yield grades for beef, which have been available for industry use only since 1965, provide a nationally uniform method of identifying *quantity* or *cutability* differences among beef carcasses. Specifically, they are based on the percentage yields of boneless, closely trimmed retail cuts from the high-value parts of the carcass—the round, loin, rib, and chuck—which account for more than 80 percent of its value. However, they also reflect differences in total yields of retail cuts.

There are five U.S.D.A. yield grades, numbered 1 through 5. Yield Grade 1 carcasses have the highest yields of retail cuts, Yield Grade 5 the lowest. A carcass which is typical of its yield grade would be expected to yield about 4.6 percent more in retail cuts than the next lower yield grade, when U.S.D.A. cutting and trimming methods are followed.

When used in conjunction with quality grades, yield grades (cutability) can be of benefit to all segments of the beef industry. To the producer they provide a means to identify breeding and slaughter cattle for

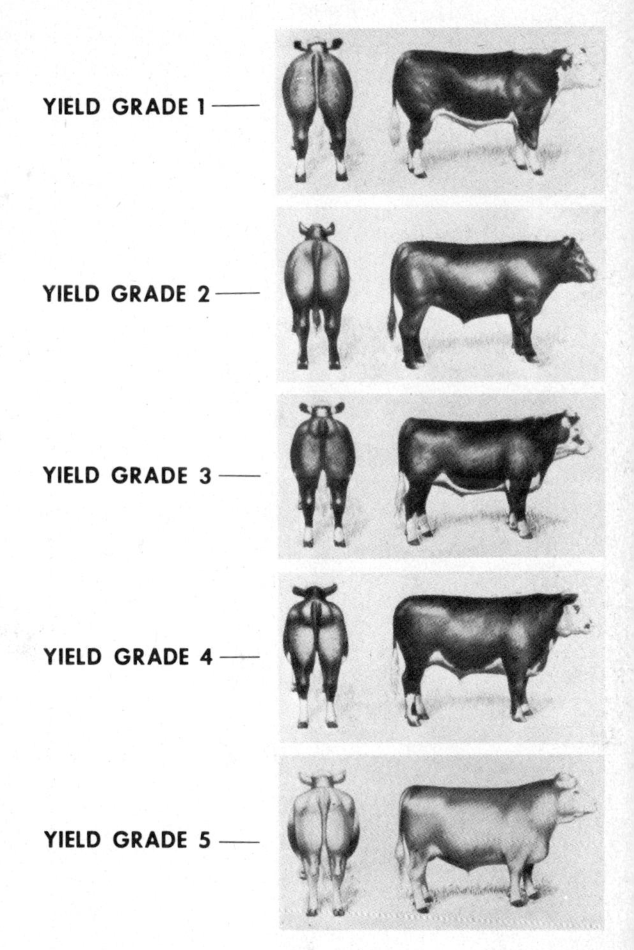

FIGURE 15-4. U. S. yield grades for slaughter cattle. (*Courtesy* U.S.D.A.)

differences in yields of salable meat. Yield grades can be a valuable tool in selecting breeding stock and in planning and operating the most efficient production, feeding, and marketing program. Properly used, yield grades could bring about more efficiency in production and marketing—and increase profits.

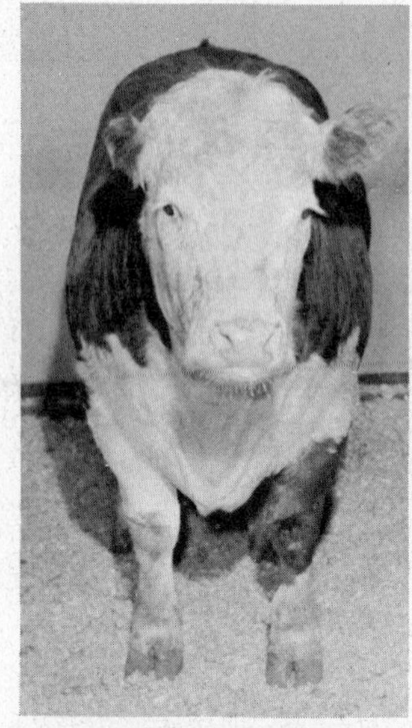

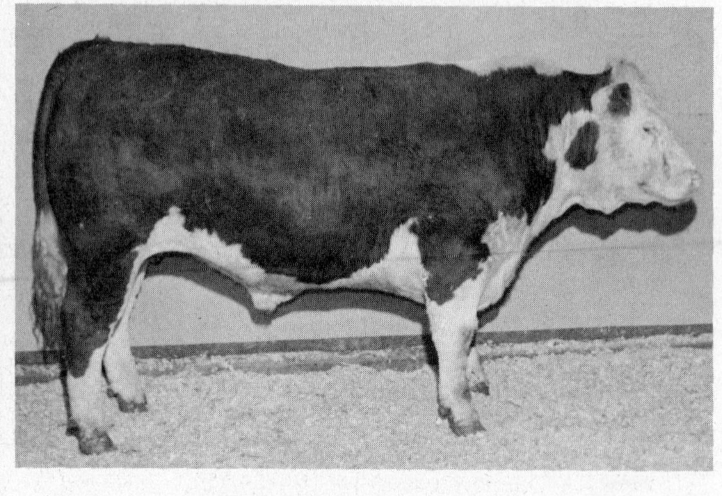

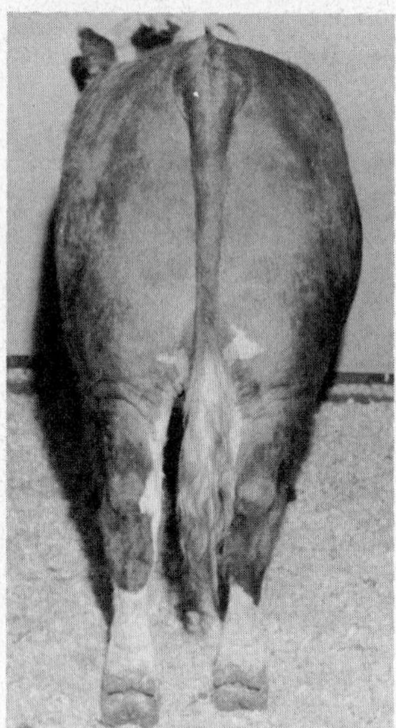

FIGURE 15-5. A front, side, and rear view, and a view of the rib area of a top-quality animal. Note the muscle expression and clean, trim appearance. This individual yielded a high percent of lean cuts with a low percent of fat. (*Courtesy* G. A. Hormel & Company)

General Guidelines for Beef Carcasses. A desirable beef carcass should fit these general standards:

1. Grade U.S.D.A. Choice or Prime.
2. Have 1.85 (or preferably 2.00) square inches of rib eye per 100 pounds of carcass weight.
3. Have less than 1/10 inch of fat per 100 pounds carcass.
4. The carcass should weigh between 550 and 750 pounds to best fit market demand.
5. U.S.D.A. Yield Grade should be 1 or 2.
6. If the actual age of the animal is known, there should be 1.4 pounds of carcass per day of age.

FIGURE 15-6. A front, side, and rear view, and a view of the rib area of an over-finished animal. Note the amount of waste in the brisket, middle, and tail head areas. The amount of fat-back is excessive and the percent of fat to lean will not be desirable. (*Courtesy* G. A. Hormel & Company)

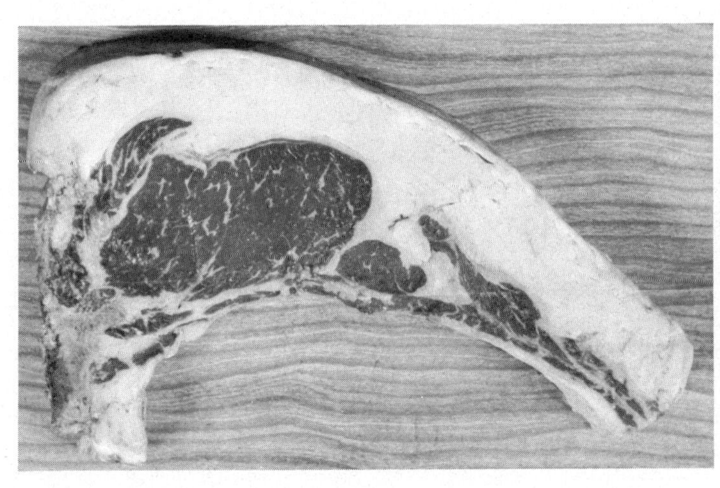

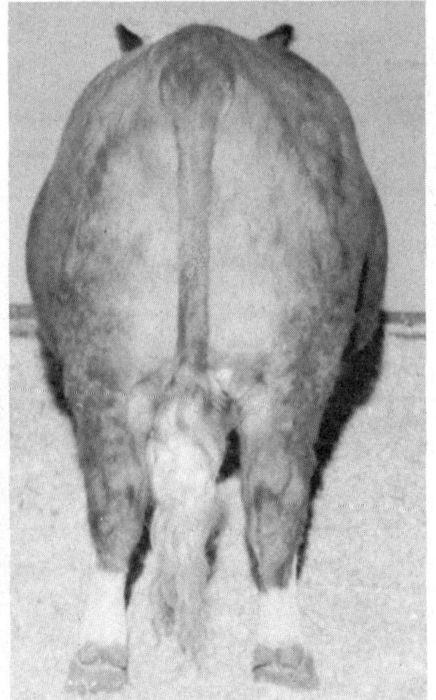

Types of Market Procedure

The cattle feeder may (1) sell cattle directly to the packer or a packer-buyer, who generally works on a salary paid by the packer he represents, (2) ship to a central market and consign to a commission firm, or (3) sell to a private buyer, who buys from the feeder and either sells directly to the packer or ships to a central market.

Selling to the Packer or Packer-buyer. Many cattle are trucked or shipped directly to the packer. Usually on the day they arrive they are weighed and paid for at the price being paid for their grade. Most packing companies have buyers in the field who will bid on the cattle at the farm; such bids may be qualified by adding a shrinkage stipulation.

Selling on Grade and Yield. During recent years there has been some interest on the part of farmers in selling their cattle to the packers on what has become known as the *grade and yield*. Under this method of sale, the live weight of the cattle has nothing to do with the price that is paid. After the cattle are slaughtered, the pounds of meat, the grade, and the yield are the determining factors in establishing the price. For example, suppose a farmer sells a 1,000-pound steer and the steer hangs up a 560-pound carcass of good beef. If the value given to this grade of beef is $.50 per pound, then 560 pounds times $.50 equals $280, the amount the farmer received. This is equivalent to $.28 per pound of live weight. Should the animal grade higher or lower than good, the price is adjusted upward or downward, depending upon the grade. If the dressing percentage is higher or lower than that given in the example (56 percent), then there will be more or less weight in the dressed carcass, and the farmer will be paid accordingly.

Selling Through a Commission Firm. When slaughter cattle are shipped to a central market, either by rail or truck, they are billed to a commission firm. The process of yarding and marketing slaughter cattle is essentially the same as that described for the marketing of feeder cattle through a central market. The chief difference lies primarily in who buys the cattle. In the large, central markets that deal essentially with slaughter cattle, the packer is the chief purchaser and there are several packing plants located near the stockyards at the central markets. They rely principally on the cattle shipped to that market for their supply. In addition to the packing plants located at the market, other plants may send buyers to purchase animals of a certain weight and grade for which they have a demand. These buyers are known as *order buyers*.

Selling to Private Buyers. Many communities have private stock buyers who buy cattle and other livestock and send them on to the packers. Such buyers offer an outlet for small lots of animals, because they generally have yard facilities where they can bunch animals according to age, sex, and grade before shipping or trucking them on to market.

Cooperative Marketing

Various types of cooperative marketing facilities have developed during recent years. Cooperative commission companies have been the principal type of cooperative that has affected beef cattle marketing. The commission companies operate much like private concerns except that the profits are paid to the members of the cooperative periodically in the form of patronage dividends.

Futures Trading in Live Beef Cattle

The cattle feeder may protect his profits by selling on a futures market. He may elect to sell as many 25,000-pound units as he wishes. For example, suppose a feeder buys cattle in November. He estimates he can bring them to choice grade or better at a weight of 1,100 pounds at a total cost of $22 per hundred. Assume that November beef futures for the following October are selling at $24 per hundred. By selling on the future market he can protect his profit. He may also buy his contract back at any time if there is an advantage to him. For further information on how to use the beef futures market as a hedge, contact any stockbroker dealing in beef futures.

Seasonal Price Trends

The marketing of slaughter cattle follows a rather definite seasonal pattern, if a period of years is used to determine that pattern. However, so many factors may affect the marketing for any one year that the long-time trends cannot always be relied upon as a means of hitting the top market in any one given year. However, for the cattleman who continues in the feeding business over a period of years, a study of these long-time trends may prove profitable.

LOSSES DUE TO DAMAGED CARCASSES

Losses running into millions of dollars are sustained by the livestock industry as a result of damaged carcasses. Most of this damage could be prevented. Producers are

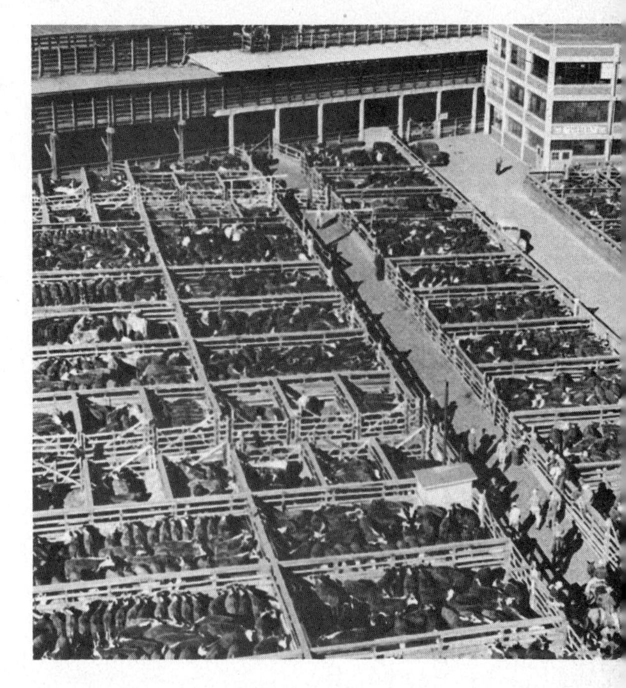

FIGURE 15-7. A view of the stockyards at a large cattle market. (*Courtesy* American Hereford Association)

FIGURE 15-8. This photo shows the location of the wholesale cuts of beef and the average percentage of each. (*Courtesy* American Brahman Breeders Association)

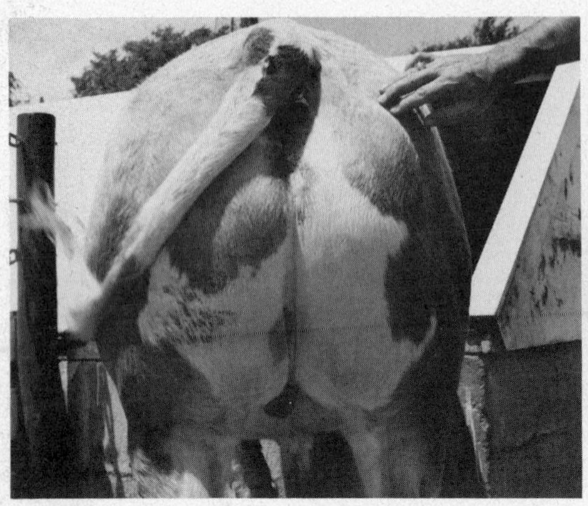

FIGURE 15-9. Rear and side view of a steer that is double-muscled. This trait is expressed when animals carrying recessive genes for double muscling are mated together. This trait is undesirable. Finished animals seldom grade very well because of the lack of marbling in the lean meat. Animals with this characteristic seldom place very high in live or carcass shows. This 1,225 pound steer, when slaughtered, had a carcass yield of 69.7 percent with a Federal grade of low good, had a fat rind of .15 inches, and a loin eye measurement of 17.9 square inches. (*Courtesy* Hawkeye Institute of Technology)

not aware of the losses caused by bruises, because such injuries can only be determined after slaughter. Usually the producer has received payment by the time the animal is slaughtered and thinks this loss is the packer's. However, packers have learned through experience the percentage of such losses, and these losses are reflected in the prices bid for live animals. Most bruised carcasses result from horn damage, careless handling, overcrowding, prodding with clubs, and feed lot obstacles. Bruised areas in the carcass must be trimmed out. The areas of the carcas removed cannot be sold for human consumption. They usually go into the manufacture of livestock feed where they are worth less money. Most of the bruises occurring on beef animals are in the regions of the most valuable cuts—loins, rumps, rounds, and ribs.

SUMMARY

Feeder cattle may be marketed by direct selling either on contract for future delivery or immediate transfer from seller to buyer. Other methods of selling feeder cattle are through auctions, dealers, and central markets.

The shipping cost, shrinkage, and prices are important considerations in determining the methods of marketing.

Information that is helpful in deciding when to market feeder cattle consists of the supply of cattle, feed conditions, prices of slaughter cattle, and the employment situation. This information may be obtained from agricultural colleges, state and Federal market information services, newspapers, magazines, radio and television reports, and private outlook information services.

Slaughter cattle are classified according to sex, age, weight, and grade. They may be sold directly to a packer or a packer-buyer on the basis of live weight, grade and yield, through a commission company at a central market, or to a private buyer.

Information helpful to the successful marketing of slaughter cattle consists of knowing the seasonal price trends for the various grades, national income and employment conditions, supply of cattle, and the feed situation.

Damage to valuable parts of the carcass resulting from unwise handling of the live animals costs the cattle industry millions of dollars yearly, an unnecessary loss.

QUESTIONS

1. List the different methods by which feeder cattle may be marketed and describe each method.
2. What is meant by *contract sales?*
3. What is *shrinkage*? How does the shrinkage affect the prices received?
4. What are the important factors in selecting a method of marketing feeder cattle?
5. What information is helpful in determining when to sell and how may it be secured?
6. What are the classes and grades of slaughter cattle?
7. What is meant by *yield grade* (cutability)?
8. List the different yield grades.
9. How may slaughter cattle be marketed? Explain.
10. What information is needed to market slaughter cattle successfully?
11. Explain the price trends for different grades of cattle.
12. Discuss the losses due to damaged carcasses and how they may be averted.

REFERENCES

Hamilton, Eugene, *Seasonal Market Variations and Their Importance to the Iowa Farmer,* Agricultural Experiment Station and Extension Service, Bulletin P 5, Ames, Iowa, Iowa State University.

Malone, Carl C., and Lucile H., *Decision Making and Management for Farm and Home,* Iowa State University Press, Ames, Iowa, 1958.

Marketing Feeder Cattle and Sheep in the North Central Region, Agricultural Experiment Station, Bulletin 410, Lincoln, Nebraska, University of Nebraska.

Potter, E. L., *The Marketing of Oregon Livestock,* Agricultural Experiment Station, Bulletin 514, Corvallis, Oregon, Oregon State College.

Rust, Robert E., *Beef Carcass Evaluation,* Agricultural Experiment Station and Extension Service, Bulletin AS-288, Iowa State University, Ames, Iowa, 1968.

Stevens, I. M., R. T. Burdick, H. G. Mason and H. P. Gazaway, *Marketing Western Feeder Cattle,* Agricultural Experiment Station, Bulletin 317, Laramie, Wyoming, University of Wyoming.

The Futures Market in Live Beef Cattle, M-1021, Ames, Iowa, Iowa State University, 1966.

DAIRY PRODUCTION

16 The Dairy Industry

The production of dairy products in this nation is big business. Income from dairy products in 1972 for farmers amounted to 7.2 billion dollars. From 40 to 50 percent of total farm marketings in New York, Pennsylvania, and Wisconsin and about 10 percent of farm marketings in the Corn Belt were dairy products. Nationally, dairy sales receipts amounted to about 10.2 percent of the total farm income. The income from the sale of dairy products was larger than the income from hog sales or from poultry and sheep sales combined.

Milk has an important place in the American diet. It is palatable and nutritious. It contains most of the food nutrients needed by human beings and by young animals.

CONSUMPTION OF DAIRY PRODUCTS

In the days of our grandparents, cows were kept to supply the family with milk and butter. Occasionally, the milk and cream were used in making ice cream and cheese. The development of refrigeration, pasteurization, homogenization, and scientific methods of producing and processing milk and milk products have made it possible for us to make more effective use of these products.

In 1924 the average consumption per person was 17.8 pounds of butter, 4.5 pounds of cheese, and 6.8 pounds of ice cream per year. In 1972 the average annual consumption per person was 4.9 pounds of butter, 13.2 pounds of cheese, 17.5 pounds of ice cream, and 17.1 pounds of dry, condensed, and evaporated milk. The average consumption of butter has decreased yearly since 1924, and the consumption of cheese and ice cream has more than doubled.

Fluid Milk and Cream

In 1925 the per capita annual consumption of fluid milk and cream was 353.5 pounds, whereas the consumption at the

TABLE 16-1 U.S. PER CAPITA CONSUMPTION OF DAIRY PRODUCTS (Pounds)

Year	Butter	Cheese	Ice Cream	Evaporated and Condensed Milk	Nonfat Dry Milk	Fluid Milk and Cream[1]
1930	17.6	4.7	9.8	13.0		337
1940	17.0	6.9	11.4	19.3		331
1950	10.7	7.7	17.2	20.1	3.7	348
1960	6.7	8.3	18.3	13.9	6.2	322
1965	5.8	9.1	18.8	15.8	5.6	302
1970	4.4	11.1	17.8	11.4	4.6	264
1972	4.9	13.2	17.5	11.9	4.6	258

[1] Whole milk equivalent.

Source: *Dairy Statistics,* U.S.D.A.

close of World War II in 1945 was 425.0 pounds. The consumption of milk and cream in 1972 was equivalent to 258 pounds of whole milk.

Evaporated and Condensed Milk and Dry Milk Solids

The consumption of these products has not been large and there has been some decrease during recent years. Some processors are marketing a concentrated liquid milk which may greatly influence the methods used in processing and transporting fluid milk. During the 1935–1939 period, the per capita consumption in this country was 15.2 pounds of evaporated whole milk, 1.6 pounds of condensed milk, and 1.9 pounds of nonfat milk solids. The 1972 consumption was 6.4 pounds, 1.2 pounds, and 4.7 pounds of these whole-milk products per person.

Butter Versus Margarine

We consume only one-fourth as much butter today as we consumed before World War II. We use fewer fats in our diets, and margarine is replacing part of the butter consumption. From 1935 through 1939, the per capita margarine consumption was 2.9 pounds, but the average in 1957 was 8.6 pounds. In 1971 we consumed an average of 11.1 pounds per person, which was more than twice the average amount of butter consumed.

Oleomargarine production has increased from 614 million pounds in 1945 to nearly 2,290 million pounds in 1971. Margarine sold in 1973 for about one-third the price of butter.

UTILIZATION OF MILK

The changes in consumption of dairy products have influenced the methods used in processing milk and the emphasis placed on the butterfat content of milk. The butterfat content is becoming less important in milk production.

Shown in Table 16-2 are the changes that have come about in the use of the milk produced in this country since 1925.

PRODUCTION OF DAIRY PRODUCTS

With the exception of a few drought years (from 1937 to 1940), the number of cows increased in this country until 1944.

TABLE 16-2 USE OF MILK SOLD BY FARMERS (In Percentages of Total)

Year	Fluid Milk	Butter	Cheese	Evaporated Milk	Condensed Whole Milk	Dry Whole Milk	Frozen Dairy Products
1930	36.5	43.4	6.9	4.1	0.9	0.2	4.0
1940	34.3	42.6	9.1	6.1	0.7	0.3	4.4
1950	43.1	28.3	12.0	6.3	0.8	1.0	7.0
1960	47.6	25.8	10.9	3.5	0.9	0.6	7.7
1965	55.4	28.5	15.8	5.3*	—	0.6	9.8
1970	51.5	24.2	21.0	3.6*	—	0.4	9.9
1972	44.7	19.6	19.8	3.1*	—	0.5	9.6

* Evaporated, condensed and dry whole milk.

Source: *Dairy Situation*, May 1973.

Since that time, the number has decreased from nearly 26 million in 1944 to 12.4 million by January, 1973. However, production of dairy products has increased more rapidly than the number of cows has declined, because of increased production per cow. Shown in Figure 16-2 are the changes in the number of cows and in production per cow from 1947 to 1965.

Leading States in Numbers of Dairy Cows

The states with the most dairy cows in January 1973 were:

1. Wisconsin	1,823,000
2. Minnesota	918,000
3. New York	907,000
4. California	790,000
5. Pennsylvania	670,000
6. Iowa	449,000
7. Ohio	430,000
8. Michigan	419,000
9. Texas	359,000
10. Missouri	318,000

Nearly 48 percent of the dairy cows in the nation in January 1973 were in the North Central States. Fourteen percent were in the South Central States, 18 percent in the North Atlantic States, 8 percent in the South Atlantic States, and 13 percent in the Western States.

In 1924 there was roughly one cow to every five people in this nation. There was only one cow to 16 people in 1971. We had only 56 percent as many cows in 1970 as we had in 1950, but they produced 101 percent of the 1950 milk production.

The prospects for the demand for milk and milk products in the future are bright. Our national population is increasing at the rate of about 2.8 million persons per year. We will need 125 million gallons of milk each year just to feed the increase in population.

Average Milk Production Per Cow

According to the U.S.D.A. Statistical Reporting Service, the average production per cow in the United States in 1972 was 10,271 pounds of milk. The high states in

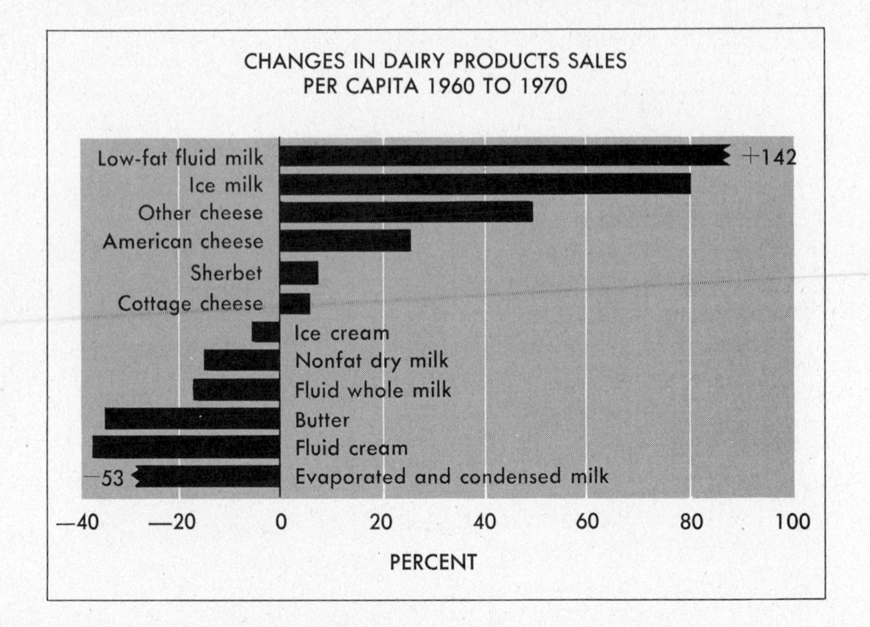

FIGURES 16-1. Changes in per capita sales of dairy products from 1960 to 1970. (U.S.D.A. Economic Research Service)

FIGURE 16-2. The number of cows in the nation decreased 33 percent between 1958 and 1971, but total milk production only decreased 3.7 percent. Milk production per cow increased 45.4 percent during this period. (U.S.D.A. Economic Research Service)

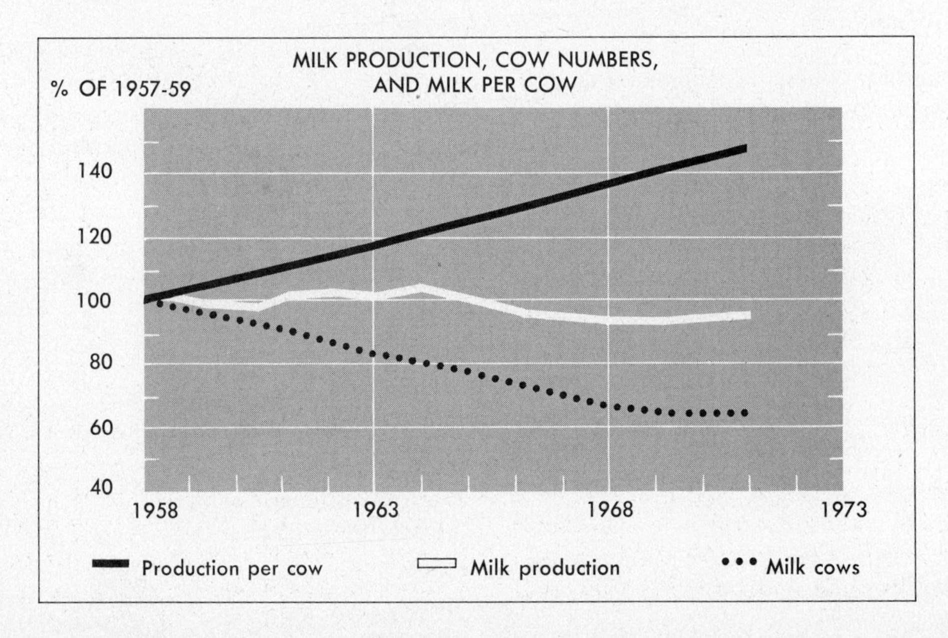

TABLE 16-3 AVERAGE PRODUCTION OF COWS IN DAIRY HERD IMPROVEMENT ASSOCIATIONS, 1906–1972

Year	Cows on Test (Number)	Average Milk (Pounds)	Milkfat Production (Pounds)
1906	239	5,430	215
1910	25,000*	5,730*	227*
1920	203,472	6,241	247
1930	507,549	7,642	303
1940	676,141	8,133	331
1950	1,088,872	9,172	370
1960	1,867,469	10,796	418
1970	3,191,457	12,839	480
1972	2,246,685	13,226	496

* Estimated

Source: U.S.D.A. *Dairy Herd Improvement Letter,* June 1973.

average production per cow in 1972 were: California, 13,406; Arizona, 12,800; Washington, 12,538; Colorado, 11,633; Michigan, 11,513; and Utah, 11,351.

Production of Cows in Dairy Herd Improvement Association (DHIA) Herds

On January 1, 1973, there were 56,698 herds with 3,338,491 cows on test in the 1,231 associations in the United States. These cows produced an average of 13,226 pounds of milk and 496 pounds of milkfat during the 1972 test year. They represented 26.12 percent of all dairy cows in the nation.

Table 16-3 presents a summary of the production of cows in the dairy herd improvement associations (DHIA) in the United States from 1906 to 1972.

Creamery Butter Production

Although only about 23 percent of the milk produced in this country is used in making butter, the butter industry is very important in some states. In 1972 Minnesota produced 262 million pounds of butter, Wisconsin 198 million pounds, California 112 million pounds, and Iowa 85 million pounds. These four states yielded 59 percent of the 1,102 million pounds of butter produced in 1972 in the United States.

Cheese Production

About 23 percent of the milk produced in 1972 was used in the production of cheese. Wisconsin was responsible for about 41 percent of the 1972 cheese production, 1,063 million pounds. Minnesota produced 235 million pounds, New York 198 million pounds, and Iowa 147 million pounds. The nation's total in 1972 was 2,604 million pounds.

Ice-Cream Production

The consumption of ice cream has nearly doubled since 1930. Approximately 11.2 percent of the milk produced in this country in 1972 was used in ice-cream production. California led all states in ice-cream production in 1972 with 74 million gallons.

Pennsylvania ranked second with 73 million; New York, third with 66 million; Ohio, fourth with 45 million; and Illinois, fifth with 42 million gallons produced.

ADVANTAGES AND DISADVANTAGES OF DAIRYING

Dairy cattle are maintained on 40 to 50 percent of the farms in some states, but the number of dairy farms is decreasing, indicating that there are certain disadvantages in dairy farming.

Advantages of Dairying

1. *Dairying fits well in diversified farming programs.* Diversification is highly recommended on many farms to permit efficient use of farm labor and economical use of buildings and equipment, and to reduce the risk involved in having but one or two sources of income.
2. *Cows are efficient consumers of roughages.* Cows make effective use of large quantities of roughages which might otherwise be wasted.
3. *Dairying provides a stable income.* Beef and sheep prices are less stable than the prices of dairy products.
4. *Income is distributed throughout the year.* Most farm income, such as that from beef, lamb, corn, wheat, and other crops, is seasonal. Dairy production and income may be distributed throughout the year.
5. *Dairy production improves the family diet and reduces food costs.* Milk is a basic food and an important item in the family food budget. A small dairy enterprise can be justified on some low-income, marginal farms for the production of milk products for family consumption. This is especially true where large families are concerned.
6. *Skim milk is of high value as poultry and swine feed.* Farmers who sell milkfat make effective use of skim milk in feeding pigs and poultry. Skim milk is an excellent source of protein, minerals, and vitamins.
7. *Dairying aids in maintaining soil fertility.* Dairy farming fits in well with grassland farming. Legumes and grasses are grown for hay and pasture. These crops are soil-conserving or soil-building crops. The manure produced is distributed on the land and returns plant food nutrients to the soil.

FIGURE 16-3. A Guernsey dairy herd in Georgia. (*Courtesy Hoard's Dairyman*)

Disadvantages of Dairying

1. *Dairying has a high labor requirement.* Dairying is a full-time job. Cows must be fed and milked at least twice each day, and time is consumed in managing the enterprise and in marketing the products, whereas other types of livestock require less labor.
2. *Considerable capital is required.* The production of high-quality dairy products requires that certain sanitary housing and equipment standards be met. These standards have made the investment in the dairy enterprise more expensive. According to U.S.D.A. data, the average value of a dairy cow in the nation in January 1971 was $317. Cows in New Jersey were valued

at $375, in Wisconsin at $340, and in Texas at $340. A herd of 50 good cows requires an investment of from $17,000 to $20,000.

The investment in feed for a herd of 50 cows is quite large. The average cost of feeding a cow in Minnesota in 1972, according to DHIA records, was $217. A herd of 50 cows will cost $10,000 to $12,000 to feed.

3. *There are many hazards in dairy production.* Dairy cows may become infected with brucellosis, tuberculosis, and other diseases. Losses are serious when valuable animals become diseased.

   Breeding, nutrition, housing, and market problems also may cause losses, but these losses are usually no more serious with dairy cattle than with other types of livestock.

4. *Substitutes for dairy products are materially affecting the dairy enterprise.* The change in status of butter due to butter substitutes has affected the profitableness of the dairy enterprise. The quality of butter must be improved, and better marketing practices are needed. The cost of producing butter must be reduced if it is to compete with butter substitutes.

FACTORS IN PROFITABLE PRODUCTION

Dairying usually can be expected to produce a rather high return on investment in feed, when labor costs are low. Where family labor can be used, dairying is quite profitable. It may not be so profitable, however, if hired labor must be used. Labor and feed costs are the major factors in dairy production. Building and equipment costs are secondary in most cases.

Data shown in Table 16-4 indicate the returns per $100 feed fed to dairy cattle and to other classes of livestock.

Effect of Production Quantity on Cost

According to Missouri DHIA records, the best way to increase the profit from the dairy herd is to increase production. Records show that the income over feed costs increases as the level of milk production

FIGURE 16-4. A Jersey dairy herd maximizing pasture in Oregon. (*Courtesy* American Jersey Cattle Club)

TABLE 16-4 RETURNS PER $100 WORTH OF FEED FED TO DIFFERENT CLASSES OF LIVESTOCK

Classes of livestock	Returns Per $100 Feed		
	1971	3-yr. average 1967–69	15-yr. average 1956–70
Beef cow herds	$180	$152	$137
Dairy cow herds	$200	$205	$191
Feeder cattle bought	$156	$138	$126
Native sheep raised	$122	$132	$124
Feeder pigs	$122	$143	$133
Hogs	$150	$179	$162
Poultry	$135	$166	$151

Source: *Summary of Illinois Farm Business Records,* 1971, University of Illinois.

is increased. In 1970 it cost about $35 more per year to feed a cow which produced 11,473 pounds of milk than it cost to feed a cow which produced 9,139 pounds. The income over the feed cost was $278 for the cow which produced 9,149 pounds, and $388 for the cow which produced 11,473 pounds. A herd of 50 cows, producing an average of 11,500 pounds of milk, will net the owner $5,500 more per year than will the same number of cows which produce an average of only 9,100 pounds, even though the feed bill for the former will be higher.

METHODS OF INCREASING DAIRY PROFITS

The average production of the cows in the dairy herd improvement associations of this nation in 1970 was about one and a half times as much milk and milkfat as the average cow in the nation. Production can be increased on most farms by carefully analyzing production methods and making the changes necessary to bring the herd up to the most profitable level of production.

The following are essential in developing a profitable dairy enterprise:

1. **Select production and breeding stock carefully.**
2. **Follow a forward-looking breeding program.**
3. **Provide adequate but economical housing.**
4. **Feed cows a balanced ration according to their maintenance and production needs.**
5. **Make efficient and economical use of high-quality hay, pasture, and silage crops.**
6. **Provide an adequate water supply.**
7. **Use good management practices in growing out herd replacement stock.**
8. **Follow production practices conducive to the production of high-quality milk.**
9. **Safeguard the herd from disease and parasites.**
10. **Keep production records of individual cows, and cull animals whenever necessary.**
11. **Select the best method of marketing and the best outlet.**
12. **Adjust the scope of the dairy enterprise to the capital, feed, and labor available, and to the market outlook.**

SUMMARY

Dairy cattle provided 10.2 percent of our nation's cash farm income in 1971. The number of dairy cows by January 1973 was 12.4 million, a reduction of 4.9 million from the number in 1966. Wisconsin, Minnesota, and New York lead the states in numbers of milk cows.

TABLE 16-5 HIGH-PRODUCING HOLSTEIN COWS IN DHIA PRODUCE MILK EFFICIENTLY (U.S. DHIA 1971–1972)

Level of Milk Production (Pounds)	Value of Product (Dollars)	Feed Cost (Dollars)	Income Over Feed Cost (Dollars)	Feed Cost Per 100 lbs. Milk (Dollars)
8,500– 9,499	$539	$238	$301	$2.62
9,500–10,499	$590	$249	$341	$2.47
10,500–11,499	$654	$266	$388	$2.41
11,500–12,499	$708	$275	$433	$2.29
12,500–13,499	$769	$289	$480	$2.22
13,500–14,499	$835	$307	$528	$2.19
14,500–15,499	$899	$324	$575	$2.17
15,500–16,499	$962	$340	$622	$2.14

Source: U.S.D.A. *Dairy Herd Improvement Letter,* August 1972.

The average production per cow in 1972 was 10,271 pounds of milk.

Minnesota, Wisconsin, California, and Iowa produced 59 percent of the nation's butter in 1972, but only about 23 percent of the nation's milk production was used in making butter. Twenty-three percent of our total milk production is used to make cheese. Wisconsin produced about 41 percent of the nation's cheese in 1972. About 11.2 percent of the milk produced in this country is used in the manufacture of ice cream. California, Pennsylvania, and New York are the largest ice cream-producing states.

In 1972 the average per capita consumption in the United States was 4.9 pounds of butter, 13.2 pounds of cheese, and 17.5 pounds of ice cream. Butter consumption is declining and margarine consumption is increasing.

Dairying fits well into diversified farming programs. The income from dairying is quite stable and well distributed throughout the year. Dairy production is less risky than beef production, encourages, soil conservation, and aids in reducing family living costs.

FIGURE 16-5. Mechanized equipment makes it possible for one man to maintain a large herd of dairy cows. (*Courtesy* Holstein-Friesian Association of America)

The income per hour of labor is not great in dairy production because of the high labor requirement. Labor and feed are the two major factors in milk production. Dairy income may be increased by careful selection of production and breeding animals, by feeding better rations, by making efficient use of hay and pasture crops, by controlling diseases, by improving the quality of the product marketed, and by adjusting the scope of the enterprise to the market outlook and to the capital, housing, feed, and labor available. All factors must be carefully considered.

QUESTIONS

1 How does the income from the dairy enterprise in your community compare with the income from the other livestock enterprises?

2 What changes have come about in the consumption of dairy products in this country during the past 25 years?

3 Which states lead in milk production, and how do you account for their large production?

4 What changes have come about in average production of milk and milkfat per cow during the past 20 years?

5 How does the production of the average cow in your herd, or in the herd with which you are most familiar, compare with the average for cows in the nation? in Dairy Herd Improvement Associations?

6 Which provides the best market for milk in your community: fluid milk distributors, the cheese factory, the creamery, or the condensed milk plant? Explain.

7 Explain the chief advantages and disadvantages of dairy production.

8 How much milk and butterfat must the cows on your farm produce to net a profit?

9 What factors must you consider in improving dairy production on your farm?

REFERENCES

Davis, Richard F., *Modern Dairy Cattle Management*, Prentice-Hall, Inc., Englewood Cliffs, New Jersey, 1962.

Diggins, R. V. and C. E. Bundy, *Dairy Production*, 3rd Edition, Prentice-Hall, Inc., Englewood Cliffs, New Jersey, 1969.

Ensminger, M. Eugene, *Dairy Cattle Science*, The Interstate Printers & Publishers, Inc., 1971.

Lascelles, H. R., *Western Dairying*, The Interstate Printers & Publishers, Inc., 1965.

Reaves, Paul M. and C. W. Pegram, *Southern Dairy Farming*, The Interstate Printers & Publishers, Inc., 1956.

U. S. Department of Agriculture, *Agricultural Statistics*, Washington, D. C., 1972.

———, *Production of Manufactured Dairy Products*, Washington, D. C., 1972.

———, *Dairy Situation*, Washington, D. C., May 1973.

17 Selection of Breeding Stock

Good seed stock is essential for profitable production of both crops and livestock. Fertilizers applied to high-yielding hybrid strains of corn net more profit than those applied to low-yielding strains. Heavy feeding of concentrates to choice feeder cattle may be very profitable, while the same feeding program applied to common steers usually is unprofitable. The same applies to dairy production: good stock and good feeding are essential for more profit.

Dairy cows are maintained for their production of milk and offspring, and they vary in their productive capacity. In 1969, the herds in the Pennsylvania DHIA produced an average of 12,258 pounds of milk. The state average for all cows was 9,705 pounds. The cows on test produced an average of 2,553 pounds of milk above the state average. At $5.46 per hundredweight, this would bring in $139 additional income per cow, or $5,560 from a herd of 40 cows. The differences in production were due to several factors, but most important were the type and inherent productiveness of the cows, and the feeding programs which were followed.

It is not unusual to discover cows in individual herds which produce one and one-half to two times as much milk and milkfat as other cows in the same herd which were fed the same rations. In one state nearly 9 percent of all cows in DHIA herds were sold because of low production. It pays to use good judgment in selecting the cows for use in starting a dairy enterprise, and it pays to cull the unprofitable cows in established herds.

FACTORS IN SELECTING DAIRY ANIMALS

Four criteria may be used to select dairy animals: (1) breed, (2) pedigree, (3) production records, and (4) physical appearance.

Most dairies that maintain breeding herds use one breed. By using artificial insemination, however, it is possible to keep cows of more than one breed and provide satisfactory bull service. Pedigree information is available for purebred dairy animals but is not always available for grades.

Production records are available for animals in DHIA herds and in many other herds that have been tested by vocational agriculture students, 4-H Club members, or the individual breeders. Almost 28 percent of the milk cows of the nation by January 1973 were on test in DHIA herds.

Quite often the selection of dairy animals is based largely on the general physical appearance and type of the individual. When the pedigree and production-record information are not available, selection must be based on the characteristics of the individual animals.

The ideal method of selecting dairy animals is to have available and to consider carefully all four criteria—breed, pedigree, production record, and physical appearance.

PARTS OF A COW

It is important that we know the parts of a cow and understand the terminology applied to those parts, in order to follow the discussion related to the selection of dairy animals. Figure 17-1 is a picture of a dairy cow with the various parts named.

BREEDS OF DAIRY CATTLE

There is no one best breed of dairy cattle, but there may be a best breed for an individual farmer. The type and quality of breeding stock available in the community, the climatic conditions, the availability of

FIGURE 17-1. Parts of the dairy cow. (*Courtesy* American Jersey Cattle Club)

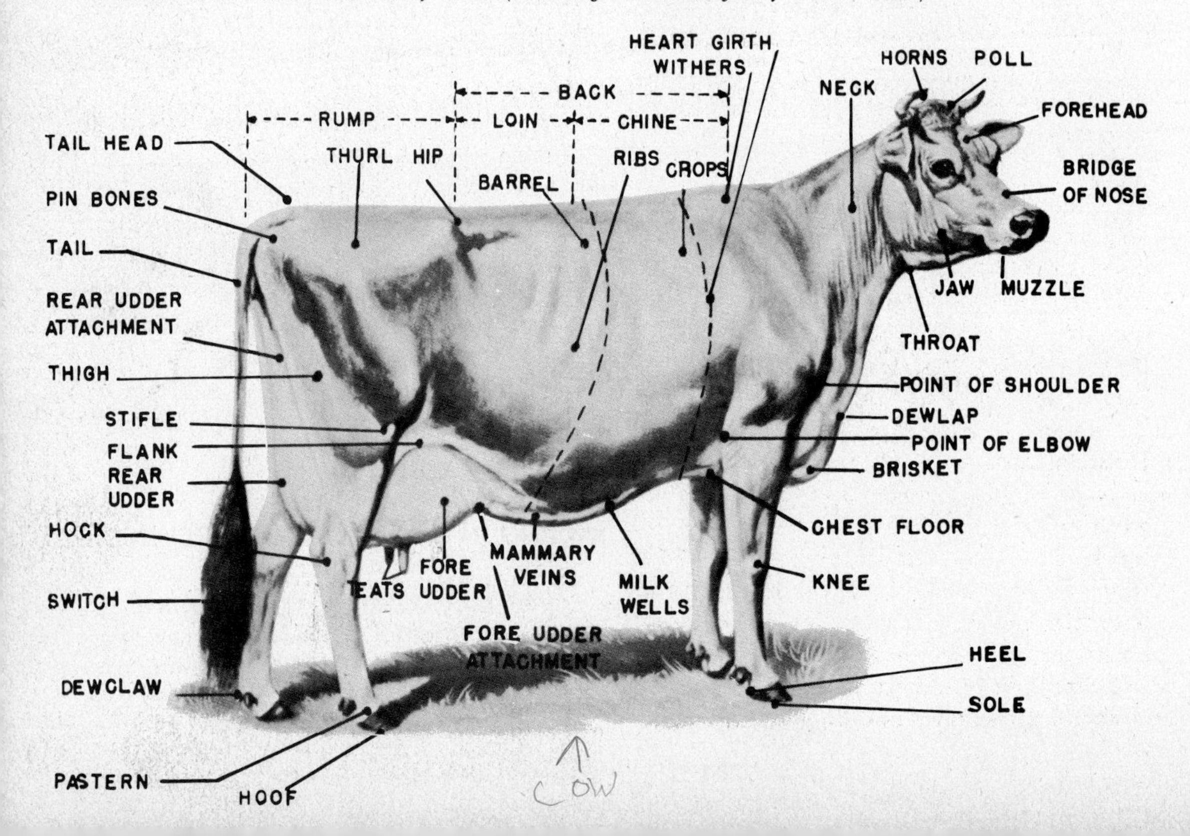

markets for milk and milkfat, the types of pasture and forage crops available, and the personal preference of the individual dairyman are the determining factors in selecting a breed.

Some breeds are more rugged than others and, as a result, are better grazers and can stand colder climates. Some breeds consume larger amounts of roughages and concentrates. The breeds differ somewhat in the fat content of the milk.

There are good individual cows in every breed, and the differences among breeds are of less importance than the differences within the individual breeds. The breed best suited to a production program should be carefully selected, and major emphasis placed on breeding or selecting the cows which will be most productive.

The five common breeds of dairy cattle are the Holstein-Friesian, the Guernsey, the Jersey, the Ayrshire, and the Brown Swiss. Two other dairy breeds are grown in smaller numbers: the Red Danish and the Dutch Belted.

Holstein-Friesian

Holstein cattle originated in Holland, and were imported to this country beginning about 1625. In Holland they were called Friesians and were bred to produce large quantities of milk for use in making cheese. Most of the importations of Holsteins were made between 1875 and 1905.

Color. To be eligible for registration, a Holstein must be black and white, and the amounts of black and white may vary from white with a few black spots to almost all black. The switch must always be white. Animals that are all white or all black, or have black on their legs below the knees or hocks, are not recorded.

Size. The standard weight for Holstein cows is 1,250 pounds and bulls should weigh at least 1,800 pounds. Many cows weigh 1,300 to 1,600 pounds and some bulls weigh over a ton. Holsteins are larger than most of the animals of other breeds.

Conformation. Animals of this breed are ruggedly built and possess large feeding capacity and udders. The head is long, narrow, and straight. Straight thighs and slightly rounded withers are desired.

Disposition. The cows are quiet and docile, but the bulls may be vicious.

Grazing Ability. Holstein cows are excellent grazers, especially on good pastures. They have large middles and can consume

FIGURE 17-2. (*Above*) The ideal type Holstein bull. (*Courtesy* Holstein-Friesian Association of America) FIGURE 17-3. (*Below*) The ideal type Holstein cow. (*Courtesy* Holstein-Friesian Association of America)

FIGURE 17-4. Skagvale Graceful Hattie 6069325, milk champion of the Holstein breed. She produced 44,091 pounds of milk and 1,505 pounds of milkfat in 1971. She is owned by Tenneson Brothers of Sedro Woolley, Washington. (*Courtesy* Holstein-Friesian Association of America)

large amounts of forage. They do not thrive as well on poor pastures as some of the smaller breeds.

Production. Holstein milk is lower in fat than is milk from any of the other dairy breeds. It averages about 3.67 percent milkfat but varies from about 2.5 to 4.3. Holsteins produce large quantities of milk and rate high in total milkfat production.

The all-time top producer of milkfat of all breeds is a Holstein, Gladell Governess Bess. She is owned by Allen and Sara Rearick of Millheim, Pennsylvania. Her official 359-day record, made on two milkings per day and completed in 1973, totaled 36,969 pounds of milk and 1,913 pounds of milkfat.

The milk champion of the Holstein breed is Skagvale Graceful Hattie 6069325, owned by Tenneson Brothers of Sedro Woolley, Washington. She completed in 1971 a record of 44,091 pounds of milk containing 1,505 pounds of fat in 365 days. She was milked twice a day.

The Holstein-Friesian Association of America in 1969 approved a plan to provide a second herd book for the registration of red-and-white offspring of purebred Holstein cattle.

Official production testing is completed through the Dairy Herd Improvement Registry or the Herd Improvement Registry. The Advanced Registry program was terminated in 1965. In 1967 the Dairy Herd Improvement Registry became the only official testing program for the breed.

The 17,436 Holstein herds in the 1971–1972 standard DHIA program in the United States produced an average of 13,365 pounds of milk and 489 pounds of milkfat. The average test was 3.66 percent milkfat.

Holstein calves are large and vigorous, weighing an average of about 90 pounds. Steers of this breed feed out well and produce lighter-colored fat than steers of the other dairy breeds.

Availability of Breeding Stock. More Holsteins were recorded in 1972 than cows of all other dairy breeds. A total of 277,851 animals was recorded with the Holstein-Friesian Association of America, Brattleboro, Vermont. Of the 2,514 dairy sires used by the artificial insemination associations in this country in 1971, 1,682 were Holsteins. According to DHIA records, more Holsteins were reported in some states than cows of all the other dairy breeds combined.

Guernsey

This breed has been imported from the small Island of Guernsey, which is located off the coast of France. Approximately 113,000 head of Guernseys have been imported.

Color. The color of the Guernsey varies from a light fawn to almost red, with white markings on the face, legs, switch, and flank. Some white may appear on the body. The nose should be cream or buff-colored; how-

FIGURE 17-5. (*Left*) Lockshore Sultan Rebel, National Grand Champion Guernsey bull. Owned by J. A. and A. B. Fish, Hickory Corners, Michigan. (Danny Weaver photo. *Courtesy* American Guernsey Cattle Club) FIGURE 17-6. (*Right*) Paula's Viscount Paulette, National Grand Champion and Best Uddered Guernsey Cow. Owned by Henry Leffler and Sons, Marion, Indiana. (*Courtesy* American Guernsey Cattle Club)

ever, smoky color is permitted. The skin of the Guernsey is yellow.

Size. Guernsey cows average about 1,100 pounds but vary from 800 to 1,300 pounds. The bulls average about 1,700 pounds.

Conformation. The Guernsey is less rugged than the Holstein, but more rugged than the Jersey. The cows may be inclined to be rough over the rump and weak in the loin. The udders are less symmetrical than those of the Jersey. The face of the Guernsey is double-dished but longer than that of the Jersey.

Disposition. Cows of this breed are alert and active but not nervous. They are easily managed.

Grazing Ability. The Guernsey is a good grazer but not equal to the Holstein on good pasture nor to the Jersey on poor pasture.

Production. Guernsey milk has a golden color and is popular on the consumer market. It contains nearly 5 percent fat. "Golden Guernsey" milk quite often sells at a premium. Guernseys usually produce less milk than Holsteins, but it is higher in fat content.

Of the 2,514 sires in artificial breeding service in the United States in 1971, 167 were Guernseys. The average production of milk and milkfat of cows sired by bulls in artificial breeding service in 1970 was 9,954 pounds of milk and 467 pounds of milkfat.

The 1,045 Guernsey herds entered in the 1971–1972 DHIA program produced an average of 9,758 pounds of milk and 459 pounds of milkfat. The average test was 4.70.

Fox Run AFC Faye 2253871 (Excellent) is both the champion milk and milkfat producer of the breed. In 1970 she produced 32,110 pounds of milk and 1,397 pounds of fat. In 1971 she produced 31,040 pounds of milk and 1,736 pounds of milkfat, the new fat record of the breed. Both records were made in 365 days on twice-a-day milking. She is owned by Lee and John Housely of Riceville, Tennessee.

Availability of Breeding Stock. Guernseys are grown in all areas of the nation, and in 1972 ranked third among the breeds in number of purebred animal recordings. A total of 34,451 animals was recorded by the

FIGURE 17-7. (*Above*) Fox Run A.F.C. Faye, 1969 All-American Aged Cow and new all-time Guernsey Milk and Fat Champion. In 1971 she produced 31,040 pounds of milk and 1,736 pounds of fat. Owned by Lee and John Housley, Riceville, Tennessee. (Danny Weaver photo. *Courtesy* American Guernsey Cattle Club) FIGURE 17-8. (*Below left*) Toreador the Ambassador, Reserve Grand Champion Bull at the 1971 All-American Jersey Show. Owned by Young's Jersey Dairy, Inc., Yellow Springs, Ohio, and Richard Wilt and Family, Springfield, Ohio. (*Courtesy* American Jersey Cattle Club) FIGURE 17-9. (*Below right*) Etta Master Babe, Grand Champion Cow at the Minnesota and Wisconsin Fairs, and at the 1971 All-American Jersey Show. Owned by Heaven Hill Farm, Lake Placid Club, New York. (Danny Weaver photo. *Courtesy* American Jersey Cattle Club)

American Guernsey Cattle Club of Peterborough, New Hampshire.

Jersey

This breed was developed on the Island of Jersey in the English Channel. Importations of Jerseys to this country began around 1850. About 37,000 animals have been imported, of which 4,800 head have been brought in since 1935.

Color. Jerseys vary in color from light fawn to black and from white-spotted to solid in marking. The tongue and switch may be black or white. The muzzle is black with a light encircling ring.

Size. The Jersey is the smallest of the dairy breeds. The cows range in weight from 800 to 1,100 pounds and the bulls from 1,200 to 1,600 pounds. Imported animals are usually smaller than those produced in this country.

Conformation. The Jersey approaches the true dairy type. The cows have straight toplines, level rumps, and sharp withers. Their heads have a double dish, and they show dairy temperament in eyes and neck. Jerseys have excellent udders, both in shape and in fore and rear attachment.

Disposition. Jerseys are inclined to be nervous and sensitive. They can be pets under good management or mean under poor management. The bulls of this breed are often quite vicious.

Grazing Ability. No dairy animal excels the Jersey in grazing ability on medium to poor pastures. They are small and active, so their maintenance requirements are lower than those of cows of the larger breeds.

Production. Jersey calves are small at birth and more difficult to raise than calves of some of the larger breeds. Calves usually weigh from 50 to 60 pounds. Jerseys mature, however, in 24 to 26 months.

Jersey milk averages about 5.3 percent fat and contains almost 15 percent solids. It sells at a premium because of its superior quality. It is yellow in color and is often sold as "Jersey Cream Line" milk.

Jerseys do not produce large quantities of milk, but they produce it economically on little feed.

There were 966 herds of Jerseys in the United States entered in the 1971–1972 DHIA (Standard) testing program. The cows in these herds averaged 8,877 pounds of milk and 445 pounds of milkfat. The average milkfat test was 5.02 percent.

Of the 2,514 sires used in artificial insemination associations in 1971, 167 were Jerseys.

The champion milk and milkfat producer of the breed is The Trademarks Sable Fashion owned by Victory Jersey Farm of Tulia, Texas. In 1970 she produced, on twice-a-day milking, 29,320 pounds of milk and 1,550 pounds of fat. She is officially classified as Excellent, with a score of 94.

Availability of Breeding Stock. Jerseys are most numerous in the South, East, and Pacific Coast areas. In 1972, 39,396 animals were recorded. The breed ranks second in number of animals recorded and in the number of bulls used in artificial insemination associations, and second in the number of cows in DHIA herds.

FIGURE 17-10. The Trademarks Sable Fashion, the Champion Jersey milk and milkfat producer. In 1970 she produced 29,320 pounds of milk and 1,550 pounds of fat. Owned by Victory Jersey Farm, Tulia, Texas. (*Courtesy* American Jersey Cattle Club)

Ayrshire

This breed was developed in Scotland and was first imported to this country in 1822. Since 1920 Ayrshire importations have been largely from Canada.

Color. Ayrshires are red with white markings or white with red markings. The red may be very light or almost black.

Size. The Ayrshire is between the Guernsey and the Holstein in size. Cows average about 1,250 pounds, and the bulls weigh from 1,600 to 2,300 pounds.

Conformation. The Ayrshire is considered by many as the most beautiful dairy breed. The animals have straight toplines, level rumps, and good udders. Ayrshires have long horns which are trained upward. Ayrshires may be shorter and thicker in the neck than some of the other dairy breeds.

Disposition. Ayrshires are very active and may be nervous and hard to manage. They are good rustlers.

Grazing Ability. Ayrshires are excellent grazers because of their ruggedness, stamina, and activity.

FIGURE 17-11. (*Left*) **Vista Grande King, a Grand Champion Ayrshire Bull, classified Excellent. (Strohmeyer and Carpenter photo. Van Devort Studio.** ***Courtesy*** **Ayrshire Breeders Association)** **FIGURE 17-12.** (*Right*) **Oak Ridge Classy Heiress, Grand Champion Ayrshire Cow at the 1971 Grand National Livestock Exposition, San Francisco, California. (Bill Serpa photo.** ***Courtesy*** **Ayrshire Breeders Association)**

Production. Ayrshires do not produce as much milk or milkfat as some of the other breeds. The milk usually averages 4 percent milkfat and about 12.75 percent total solids.

There were 335 herds of Ayrshire cattle participating in the 1971–1972 DHIA (Standard) program in the United States. The cows produced an average of 11,020 pounds of milk and 437 pounds of fat. The average fat test was 3.97 percent.

The average production of cows tested on the Ayrshire Herd Test Plan up to 1948 was 9,016 pounds of milk and 367.8 pounds of milkfat. The average milkfat test was 4.08 percent.

Fairdale Betty Gem, owned by Meredith Farm of Topsfield, Massachusetts, is the current milk-production champion of the breed. In 1970 she produced in 305 days 32,250 pounds of milk and 1,109 pounds of fat. She was milked twice daily. The milkfat test was 3.4 percent.

The champion milkfat producer is Bob's Pansy Girl, owned by Douglas Shores, Skowhegan, Maine. She produced 20,240 pounds of milk and 1,213 pounds of fat in 305 days in 1963. She was milked twice daily.

Ayrshire calves weigh 70 to 80 pounds and make good vealers. The meat from Ayrshire cattle has better color than that from Jerseys or Guernseys. Ayrshires mature in 26 to 28 months.

Availability of Breeding Stock. Ayrshires are more prevalent in the New England States, Pennsylvania, and New York. A total of 13,883 animals was recorded by the Ayrshire Breeders Association of Brandon, Vermont, in 1972, and only 44 of the 2,514 bulls used by the artificial insemination associations in 1971 were of this breed.

Brown Swiss

Brown Swiss cattle were developed in the mountainous areas of Switzerland and the first importation to this country took place in 1869. A total of about 185 animals has been imported.

FIGURE 17-13. (*Left*) L-J Label 156109, Grand Champion Brown Swiss Bull at the Eastern National Show. Owned by L-J Farm, Inc., Hilliards, Ohio, and L. E. Harrison and J. H. Brown, Wooster, Ohio. (Agri-Graphic Services photo. *Courtesy* Brown Swiss Cattle Breeders Association) FIGURE 17-14. (*Right*) Mable's Tamarind Violet 325683. Classified "4 Excellent." In 1962 this Grand Champion Brown Swiss Cow produced 24,499 pounds of milk and 1,158 pounds of milkfat. (Strohmeyer photo. *Courtesy* Brown Swiss Cattle Breeders Association)

Color. The color varies from a light fawn to almost black. The muzzle and a stripe along the backbone are light in color. The nose, tongue, switch, and horn tips are black.

Size. The Brown Swiss breed is the largest, most rugged, and meatiest breed of dairy cattle. Mature cows weigh from 1,200 to 1,400 pounds, and bulls weigh from 1,600 to 2,400 pounds.

Conformation. In Switzerland, oxen of the breed are used for plowing and for pulling carts, as well as for milk and beef production. As a result, the breed is not as refined as the other dairy breeds. However, the type and conformation of Brown Swiss cattle have been greatly improved during the last 25 years. Cows of this breed have large bones, large heads which are usually dished, and thick, loose skin. They are not as angular as the other dairy breeds.

Disposition. This breed is quiet, docile, and easily managed.

Grazing Ability. Brown Swiss cattle are good grazers because they are rugged and active.

Production. Calves of this breed weigh from 90 to 100 pounds at birth. Their fat is white, and they make good vealers. Steers and cows put on gains rapidly, producing good beef.

Brown Swiss milk is white in color and contains about 4 percent fat. Registry of Production and Herd Improvement Tests are supervised by the Brown Swiss Breeders Association of Beloit, Wisconsin.

To October, 1947, 11,171 cows had completed 305-day records, averaging 9,446 pounds of milk and 365 pounds of milkfat on twice-a-day milking.

The 338 Brown Swiss herds participating in the 1971–1972 Standard DHIA testing program produced an average of 11,392 pounds of milk and 463 pounds of fat. The average production was 4.07 percent milkfat.

Lee's Hill Keeper's Raven, owned by Lee's Hill Farm, New Vernon, New Jersey, is the champion milk producer of the breed. On three milkings daily for 365 days in 1958, she produced 34,850 pounds of milk and 1,579 pounds of milkfat. The champion fat producer of the breed is Letha Irene Pride, owned by White Cloud Farm, Princeton, New Jersey. In 1959, on three milkings daily for 365 days, she produced 34,810 pounds of milk and 1,733 pounds of fat.

Availability of Breeding Stock. Brown Swiss cattle are most numerous in Wisconsin, Illinois, New York, and Michigan. Of the 2,307 proven bulls used in artificial insemination associations in the nation in 1970, 64 were Brown Swiss. In 1970–1971, 13,743 ani-

CARNATION MILLION HEIR 1507983
Plus Proven Sire Classified "Excellent" 91
First 10 daughters in seven herds averaged 17,166 pounds of milk, 3.54 percent fat, 607 pounds of fat. Has a predicted difference of +394 pounds of milk and +16 pounds of fat with a 31 percent repeatability.
Bred and owned by
Carnation Farms
Carnation, Washington

FIGURE 17-15. Carnation Million Heir 1507983. (*Courtesy* Carnation Farms)

OSBORNDALE IVANHOE

Excellent - Gold Medal **1189870**
USDA May '67

10,989 Daus. Avg.	14,818M	3.7%	554F
Pred. Diff.	+630M		+23F

Predictability 99%
HFA 9/70 TQ 9/70
5,499 Cl. Daus. Avg. 82.3 — 10% BAA
Diff./Exp. +1.65
Honor List Sire
1964, 1965, 1966, 1967, 1968, & 1969
Daughters Include:
Daniella Farm O I Debbie
6 Yrs. 2X 365 35,712M 4.1% 1,463F
Fultonway Ivanhoe Rae (VG-89)
7 Yrs. 2X 365 33,693M 4.8% 1,615F
Sinking Springs Ivan Bright (VG)
4 Yrs. 2X 365 31,820M 3.8% 1,216F
Choptank O I Nell
6 Yrs. 2X 353 29,570M 4.2% 1,249F
Bayfield Ty Grawin Johanna (EX)
10 Yrs. 2X 365 29,408M 4.3% 1,260F
Collins Crest Ivanhoe Triune J. (EX - 95)

TEXAL RICH HERD LILY

Excellent - 94 - 4E **4519246**

5 Yrs.	2X	363	16,956M	3.9%	663F
6 Yrs.	2X	365	20,018M	4.1%	814F
7 Yrs.	3X	365	22,094M	4.1%	898F
9 Yrs.	3X	365	24,696M	3.8%	944F
11 Yrs.	3X	365	21,008M	4.2%	889F
Lifetime Prod.			122,720M	4.0%	4,870F

All-American Aged Cow 1959
Grand Champion National 1959
First 4-E cow of the breed
Daughters Include:
Carnation Sally Texal Lily (VG)
6-5 3X 365 24,030M 3.8% 907F
Maple Lily Burke (EX)
6 Yrs. 2X 365 19,824M 3.5% 690F
Carnation Fobes Lily (VG-87)
4-6 2X 365 23,816M 3.8% 893F

OSBORNDALE TY VIC

Excellent - 92 - Gold Medal **848777**
30 Daus. Avg. 12,339M 4.1% 502F
33 Classified Daughters Avg. 83.3
Diff./Exp. +3.15
1st Sr. Get E.S.E. 1948
Daughters Include:
Osborndale Musetta Ormsby (VG)
6 Yrs. 3X 365 21,981M 4.4% 966F

QUALITY FOBES ABBEKERK GAY

Excellent - Gold Medal **2471271**

2 Yrs.	3X	365	18,679M	3.9%	732F
4 Yrs.	3X	365	24,413M	3.8%	933F
6 Yrs.	3X	365	29,905M	3.7%	1,119F
8 Yrs.	3X	365	30,056M	3.7%	1,110F
9 Yrs.	3X	365	26,857M	3.7%	999F
10 Yrs.	3X	365	26,223M	3.6%	952F
11 Yrs.	3X	365	22,039M	3.7%	807F
14 Yrs.	2X	365	16,205M	3.7%	604F
Lifetime Prod.			221,366M	3.7%	8,155F

BOND HAVEN RAG APPLE MAPLE

Very Good - Extra **218036C**
USDA 9/70

481 Daus. Avg.	13,750M	3.6%	491F
Pred. Diff.	+4M		−5F

Repeatability 96%
HFA 9/70
580 Cl. Daus. Avg. 82.7 — 101.8% BAA
Diff./Exp. +.45

PCN HARTOG HIGHTEX

Good Plus **743946C**

2 Yrs.	2X	365	14,091M	3.6%	510F
7 Yrs.	2X	304	11,859M	3.7%	442F

Daughter with 4 records over 20,000M

mals were registered. Offspring of Brown Swiss bulls used in beef-cattle crossbreeding programs can be recorded and registered.

Red and White

The Red and White Dairy Cattle Association was recently organized, largely to provide for the registration of red-and-white progeny of the Holstein breed. The association has headquarters at Elgin, Illinois. Many Red and White cattle are registered, both in the Holstein-Friesian Red Book and in the Red and White Dairy Cattle Association Herd Book.

Red Danish

This breed was imported from Denmark by the U.S. Department of Agriculture, and animals were placed with some state colleges for experimental purposes. There were two males and 20 females in the original shipment.

Michigan State College and a group of Michigan farmers have been proving sires for the U.S. Department of Agriculture, and as a result the American Red Danish Cattle Association was formed, with headquarters at Fairview, Michigan.

Production. The milkfat test averages slightly over 4 percent. Mature cows in average condition weigh from 1,300 to 1,500 pounds.

Dutch Belted

Dutch Belted cattle are native to The Netherlands and were imported in 1840 by P. T. Barnum of the Barnum and Bailey Circus. A number of herds were developed in the New York and Pennsylvania area.

Mature cows weigh from 900 to 1,500 pounds and bulls average about 2,000 pounds. The milk tests about 4 percent milkfat. The highest producing Dutch Belted cow, Loraine No. 3020, produced 18,211 pounds of milk and 816 pounds of fat.

PEDIGREE

A pedigree describing the ancestors of an animal may be very helpful in selecting dairy cattle. Only a small percentage of all dairy cattle are purebreds, and the pedigrees of grade animals are not usually available. On pages 258 and 259 is a partial pedigree for an outstanding bull owned by Carnation Farms, Carnation, Washington.

PRODUCTION RECORDS

The production records of a purebred animal and of its ancestors appear on the pedigree. Increasingly large numbers of our dairy herds are being tested in dairy herd improvement association programs. During 1972, more than 2,660,000 lactation records were reported by the dairy herd improvement association supervisors in the United States. Some purebred cows were on official tests that were supervised by the various breed associations. It is not difficult to find breeding stock with production records, or from sires and dams with production records. This information takes much of the guesswork out of dairy cattle selection.

Proven Sires

The Bureau of Dairy Industry began, in 1936, a sire-proving program and in 1965 18,619 dairy sires used in DHIA herds had been evaluated. Of these, 11,245 were Holsteins, 3,389 were Guernseys, and 2,326 were Jerseys. Seventy-five percent of the sires were in natural service, averaging 13.9 daughter-herdmate pairs per sire in the comparison summary. The remaining 25 percent were sires used in artificial insemination. An average of 253 daughter-herdmate pairs was compared for each sire.

Testing the Home Herd

Only about 28 percent of our dairy cows is in DHIA herds because of the cost of the program, and because of shortages of DHIA supervisors. Some producers, too, have not been convinced of the usefulness of the testing program. What methods can the small dairyman use in making his own tests? In the main, there are two methods: (1) he can cooperate with the local DHIA in an owner-sampler testing program, or (2) he can take his own samples and do his own testing, using equipment owned privately, cooperatively, or by the local school or extension service.

On January 1, 1973, there were 56,698 herds participating in dairy herd improvement associations in the United States. A total of 33,578 herds involving 2,359,611 cows was on the official DHIA plan. Owner-sampler dairy records were being kept in 1973 by 20,895 owners of 806,270 cows.

Owner-Sampler Testing. County-wide dairy herd improvement cooperatives with central laboratories for testing milk and keeping records are the answer to the need for more record keeping at lower costs. The owner-sampler records are based on a one-day test each month. The supervisor leaves sample bottles, plastic bags, or barn sheets and scales at several farms each day. Each dairyman records the weight and takes a composite sample of the evening and morning milk from each cow. The association supervisor picks up the samples and weight records and takes or sends them to the laboratory, where the samples are tested and the records computed for each cow. The completed records are mailed to the producer and are discussed at length with the supervisor on the next trip to the farm. This system provides accurate testing and record keeping at low

FIGURE 17-16. Generators Topsy, Reserve Grand Champion Cow, 1971 All-American Jersey Show. Owned by Briggs and Beth Cunningham, South Kent, Connecticut, and Happy Valley Farm, Danville, Kentucky. (Danny Weaver photo. *Courtesy* American Jersey Cattle Club)

FIGURE 17-17. Grandview P Babs, Classified Very Good, First Fat Producer of Guernsey breed. She produced 25,040 pounds of milk and 1,247 pounds of fat in 305 days. Owned by Maurice D. Lippely, Silver Lake, Indiana. (Jim Miller photo. *Courtesy* American Guernsey Cattle Club)

cost, since one supervisor can handle several herds in one day.

The Babcock Test. An individual dairyman can weigh the milk from each cow, each milking, during one day of each month, take a composite sample, and make his own tests, if equipment is available. Facilities may be available in the vocational agriculture department of the school or in the office of the agricultural extension director. The following suggestions are in order in making home tests by the Babcock method:

1. Weigh the milk from each cow for each milking during one day.
2. Take a small sample of milk from each milking. The same amount of milk should be taken from each milking.
3. Mix the samples to make a composite test.
4. Warm the thoroughly mixed sample to 65° or 70° F.
5. Draw milk into a pipette until it is level with the 17.6-cubic-centimeters mark.
6. Discharge the milk from the pipette to the test bottle, being careful that every drop of milk is transferred to the bottle.
7. Add slowly, as the test bottle is being tilted and rotated, 17.5 cubic centimeters of cold commercial sulfuric acid. Add about 50 percent of the acid and mix it with the milk before adding the other half.
8. Mix the acid with the milk by rotating the bottle. Do not shake the bottle or hold a finger over the opening.
9. Place the mixed samples in the centrifuge tester. Space them properly to balance the centrifuge.
10. Whirl the centrifuge for five minutes at the speed indicated on the machine.
11. Add warm, soft water to bring the contents of each bottle up to the base of the bottle neck. Centrifuge for two minutes.
12. Add water carefully, using a pipette, to bring the fat column up to the graduated scale of the bottle neck. Centrifuge for one minute.
13. Place the bottles in a water bath at 130° to 140° F. for five minutes.
14. Read the test by using dividers to get the height of the fat column, then lower them, placing one divider at 0; the other divider will indicate the butterfat test.
15. The reading is made from the extreme top of the top meniscus to the extreme bottom of the lower meniscus.
16. The weight of milk multiplied by the butterfat test, times the number of days in the month, will indicate the monthly butterfat production of the cow.

FIGURE 17-18. Green Banks Admiral Mooie, Grand Champion Female at the 1970 Western National Holstein Show. Owned by Paclamar Farm, Louisville, Colorado. (Danny Weaver photo. *Courtesy* Holstein-Friesian Association of America)

Te Sa Test. This test is a recent development and has been approved for use in connection with dairy production record-keeping programs sponsored by the American Dairy Science Association. This test does not require the use of an acid and a centrifuge. Instructions are provided by the equipment supplier—Technical Industries, Fort Lauderdale, Florida.

Milko-Tester. During the past few years many states have eliminated the use of the Babcock test in DHIA programs. The Milko-Tester is an automated device for determining fat content. The steps commonly used with the Babcock method of pipetting milk, measuring acid, adding acid, mixing, and controlling temperatures are all done by the machine through push-button operation. DHIA supervisors now do little or no testing. They forward samples to centralized laboratories. These laboratory facilities can handle a large volume. It is estimated that at least 10,000 cows must be involved to justify the use of the machine.

PHYSICAL APPEARANCE

It has been found that certain physical characteristics in cattle are associated with high production. An understanding of these characteristics makes it possible for us to judge the productive capacity of a cow for which production records are not available.

Dairy Cow Score Card

The five dairy breeds of this nation, through the medium of the Purebred Dairy Cattle Association, have prepared score cards for dairy cows and bulls. The score cards have been approved and made official by the five dairy breeds.

DAIRY COW UNIFIED SCORE CARD †
(In Order of Observation)

1. **GENERAL APPEARANCE** (30 points)
   Attractive individuality with femininity, vigor, stretch, scale, harmonious blending of all parts, and impressive style and carriage. All parts of a cow should be considered in evaluating her general appearance.
   BREED CHARACTERISTICS—*Color, size,* and *horns* characteristic of the breed.
   HEAD—Clean-cut, proportionate to the body; broad *muzzle* with large, open nostrils; strong *jaws;* large, bright *eyes;* broad, moderately dished *forehead; bridge of nose* straight; *ears* medium size and alertly carried.
   SHOULDER BLADES—Set smoothly and tightly against the body.
   BACK—Straight and strong; *loin* broad and nearly level.
   RUMP—Long, wide, and nearly level from *hook bones* to *pin bones;* clean-cut and free from patchiness; *thurls* high and wide apart; *tail head* set level with backline and free from coarseness; *tail,* slender.
   LEGS AND FEET—*Bone* flat and strong; *pasterns* short and strong; *hocks* cleanly moulded. *Feet* short, compact, and well rounded with deep heel and level sole. *Forelegs* medium in length, straight, wide apart, and squarely placed. *Hind legs* nearly perpendicular from hock to pastern from the side view, and straight from the rear view.
2. **DAIRY CHARACTER** (20 points)
   Evidence of milking ability, angularity, and general openness, without weakness; freedom from coarseness, giving due regard to period of lactation.

† Copyrighted permission to reproduce score cards granted by the American Dairy Science Association.

FIGURE 17-19. Schulte's Sunrise Pat 499920, Supreme Grand Champion Brown Swiss Cow, 1971 National Dairy Cattle Congress; and Grand Champion, 1971 Central National Show. Owned by Bernard Monson, Gowrie, Iowa. (Pete's photo. *Courtesy* Brown Swiss Cattle Breeders Association)

FIGURE 17-20. C. Carlspride Vogel Reflection 1595002. Grand Champion Holstein Bull at the 1971 Eastern and Central Holstein Shows. Owned by Lakehurst Farms, Inc., Sheboygan, Wisconsin. (Danny Weaver photo. *Courtesy* Holstein-Friesian Association of America)

NECK—Long, lean, and blending smoothly into shoulders; clean-cut throat, dewlap, and brisket.

WITHERS—Sharp. *Ribs* wide apart; rib bones wide, flat, and long. *Flanks* deep and refined. *Thighs* incurving to flat, and wide apart from the rear view, providing ample room for the udder and its rear attachment. *Skin* loose and pliable.

3. **BODY CAPACITY** (20 points)
Relatively large in proportion to size of animal, providing ample capacity, strength, and vigor.
BARREL—Strongly supported, long and deep; ribs highly and widely sprung; depth and width of barrel tending to increase toward rear.
HEART GIRTH—Large and deep, with well-sprung *foreribs* blending into the shoulders; full *crops;* full at *elbows;* wide *chest floor.*

4. **MAMMARY SYSTEM** (30 points)
A strongly attached, well-balanced, capacious udder of fine texture, indicating heavy production and a long period of usefulness.
UDDER—Symmetrical, moderately long, wide and deep, strongly attached, showing moderate cleavage between halves, no quartering on sides; soft, pliable, and well collapsed after milking; *quarters* evenly balanced.
FORE UDDER—Moderate length, uniform width from front to rear and strongly attached.
REAR UDDER—High, wide, slightly rounded, fairly uniform width from top to floor, and strongly attached.
TEATS—Uniform size, of medium length and diameter, cylindrical, squarely placed under each quarter, plumb, and well spaced from side and rear views.
MAMMARY VEINS—Large, long, tortuous, and branching.

Selecting Dairy Bulls

The same general characteristics desired in dairy cows are wanted in dairy bulls, but greater emphasis is placed upon the production records and pedigrees of their ancestors. The bull provides half the genetic makeup for all the progeny he sires, and he must be selected very carefully.

DAIRY BULL UNIFIED SCORE CARD
(In Order of Observation)

1. **GENERAL APPEARANCE** (45 points)
Attractive individuality, with masculinity, vigor, stretch, and scale; harmonious blending of all parts, and impressive style and carriage. All parts of a bull should be considered in evaluating his general appearance.
BREED CHARACTERISTICS—Color, size, and horns characteristic of the breed.
HEAD—Clean-cut, proportionate to the body; broad *muzzle* with large, open nostrils; strong *jaws;* large, bright *eyes;* broad and moderately dished *forehead; bridge of nose* straight; *ears* medium size and alertly carried.
SHOULDER BLADES—Set smoothly and tightly against the body.
BACK—Straight and strong; *loin* broad and nearly level.
RUMP—Long, wide, and nearly level from *hook bones* to *pin bones;* clean-cut and free

from patchiness; *thurls* high and wide apart; *tail head* set level with backline and free from coarseness, *tail,* slender.

LEGS AND FEET—Bone flat and strong, *pasterns* short and strong, *hocks* cleanly moulded. *Feet* short, compact, and well rounded with deep heel and level sole. *Forelegs* medium in length, straight and wide apart, squarely placed. *Hind legs* nearly perpendicular from hock to pastern from the side view, and straight from the rear view.

2. **DAIRY CHARACTER** (30 points)
   Angularity and general openness, without weakness; freedom from coarseness.
   NECK—Long, with medium crest and blending smoothly into shoulders; clean-cut *throat, dewlap,* and *brisket.*
   WITHERS—Sharp. *Ribs* wide apart; *rib bones* wide, flat, and long. *Flanks* deep and refined. *Thighs* incurving to flat, and wide apart from the rear view. *Skin* loose and pliable.
3. **BODY CAPACITY** (25 points)
   Relatively large in proportion to size of animal, providing ample capacity, strength, and vigor.
   BARREL—Strongly supported, long, and deep; ribs highly and widely sprung; depth and width of barrel tending to increase toward rear.
   HEART GIRTH—Large and deep, with well-sprung foreribs blending into the shoulders; full crops; full at elbows; wide chest floor.

DISPOSITION

Disposition is imporant in the selection of dairy animals. The tendency for an animal to be exceedingly nervous or mean is very undesirable. Dairy animals should be active, but responsive to the command and action of the caretaker. Docile and gentle animals are more likely to respond to care and to adjust themselves to the environment of the dairy herd.

HEALTH

Sterility, udder troubles, mastitis, Bang's disease, leptospirosis, and tuberculosis are the most serious causes of health problems. It is a good idea to check carefully the health history of a herd before making final selection of animals. An inspection of the entire herd and a visit with the owner or herdsman will provide much of the needed information. The reliability of the owner is very important.

The health of the herd and of individuals, the number of calves produced by the cows during the past year, the number of cows in production and the stages of production, and the milk production of the herd will give a good indication of the herd's health.

It is a good idea to buy animals only from herds that have been tested for Bang's disease, leptospirosis, and tuberculosis. Animals vaccinated for Bang's disease and leptospirosis are preferred. Insist upon a retest if the last test was made more than 30 days prior to the purchase. Sometimes it is necessary to have a veterinarian examine the animal.

The physical factors discussed in connection with the selection of dairy stock will provide a fairly good indication of the general health of the animal.

SUMMARY

Efficient selection of dairy animals requires consideration of the animal's breed, physical appearance, pedigree, production record, disposition, and health.

There is a close relationship between type and production. It is possible to select dairy animals on the basis of type and to predict to some extent their productive capacity.

The five most common breeds of dairy cattle ranked according to numbers in the United States are (1) Holstein-Friesian, (2) Guernsey, (3) Jersey, (4) Ayrshire, and (5) Brown Swiss. Holstein cows lead all breeds in milk and milkfat production, but the milk is low in milkfat content. The milk of the

Jersey and Guernsey leads in quality, color, and milkfat content. Ayrshires are more rugged than Jerseys and Guernseys, and are noted for their style and conformation. Brown Swiss and Holstein are the most rugged of the breeds, and Jersey and Guernsey cattle are the most refined. Brown Swiss cattle have been greatly improved during recent years.

There is no best breed of dairy cattle, but an individual dairyman may select a breed best for his farm and market conditions. The pedigree, production record, and physical appearance are the determining factors in selecting dairy animals. They should be angular and have sharp withers, smooth thighs, sharp hip and pin bones, and long-dished faces. Body capacity is important in selecting both cows and bulls. Width, depth, and length of body are essential. Dairy temperament is shown by angularity, leanness of the neck, absence of beefiness, and the quality of the skin and hair.

A good mammary system includes a large udder with good fore and rear attachment. The teats are normal in size and properly spaced, and the udder is mellow, possessing no meaty tissue. The milk veins are large and crooked and lead to large milk wells.

Masculinity, breed character, dairy temperament, and a strong constitution are essential in selecting dairy bulls.

Production records and pedigrees of the individual and of the ancestors should be evaluated in selecting dairy animals. A comparison should be made of the production of five daughters with their herdmates in proving a bull.

Health should be considered in the selection of dairy animals. It is best to bring into the herd only animals free from Bang's disease, tuberculosis, mastitis, udder difficulties, and sterility. It is a good practice to inspect the herd and check up on the production of calves and milk.

QUESTIONS

1. Which of the following factors should be given most consideration in selecting dairy animals: breed, pedigree, production record, or physical appearance? Why?
2. List the most common dairy breeds, and outline the advantages and disadvantages of each.
3. What factors should be considered in selecting dairy cows by their physical appearance?
4. Which is most important in a dairy cow—the general appearance, the body capacity, the dairy temperament, or the mammary system? Why?
5. Describe the ideal mammary system of a Guernsey cow.
6. Survey your home county and determine the number of herds of dairy cattle for which production records are available.
7. Of what value is a pedigree in selecting foundation animals?
8. What is the place of the Dairy Herd Improvement Association in your community?
9. Outline the plan which you will use on your farm to improve your herd through selection.
10. In selecting a bull, what factors will you consider which are not considered in selecting dairy cows?

REFERENCES

The Ayrshire Cow, Ayrshire Breeders Association, Brandon, Vermont, 1964.

Descriptive Type Classification, Holstein-Friesian Association of America, Brattleboro, Vermont, 1970.

Diggins, R. V. and C. E. Bundy, *Dairy Production,* 3rd Edition, Prentice-Hall, Inc., Englewood Cliffs, New Jersey, 1969.

How to Judge Guernseys, The American Guernsey Cattle Club, Peterborough, New Hampshire, 1971.

Jersey Judging Made Easy, The American Jersey Cattle Club, Columbus, Ohio, 1970.

Judging Registered Holsteins, Holstein-Friesian Association of America, Brattleboro, Vermont, 1971.

Trimberger, George W., *Dairy Cattle Judging Techniques,* Prentice-Hall, Inc., Englewood Cliffs, New Jersey, 1958.

U. S. Department of Agriculture, *Dairy Herd Improvement Letter,* Washington, D.C., June–July 1972; May–June 1973.

18 Feeding and Management of the Producing Herd

Profitable dairying may be achieved by (1) selecting foundation animals with inherent producing ability and (2) feeding and managing the herd in such a way as to enable the cows to reach maximum economical production. It makes little difference how well bred your herd is; if the cows are not given the proper kind and amount of nutrients for manufacturing milk, they will be inefficient and unprofitable producers.

FEED FOR THE DAIRY COW

Roughages

Dairy cattle are efficient users of roughages. When the ration is properly balanced, dairy cows will convert large quantities of relatively inexpensive roughages into milk.

Legumes. Legumes lead the field as roughages for dairy cattle, with alfalfa ranking first. Alfalfa is unexcelled as a dry roughage, being high in protein, net energy, and total digestible nutrients. Red clover, if cut before the blossoms turn brown, ranks only slightly below alfalfa as a dry roughage for dairy cows. Other clovers, lespedeza, soybeans, and many other legumes are highly regarded as roughages for dairy cattle. When legumes furnish the bulk of the roughage fed, a considerable savings in protein concentrates will result. Legumes minimize the risk of vitamin and mineral deficiencies because they are high in these nutrients.

Nonlegume Hays. Native grasses, brome, timothy, fescue, millet, orchard grass, and many others will provide forage. However, the grasses are considerably lower than legumes in protein and in some of the minerals and vitamins. When nonlegume hays are used extensively, greater care must be exercised if the ration is to be balanced.

Fodder, Corncobs, Cornstalks, and Straw. All these roughages are low in proteins, minerals, vitamins, net energy, and total digestible nutrients. Although they may be used under certain conditions for cattle feeding, their value is doubtful as a feed for dairy cattle. The good milk cow must consume large amounts of highly digestible feed if she is to maintain production. Most dairymen agree that high-quality forage, such as legumes, and good silage are the cheapest feeds.

Corn or Sorghum Silage. Corn or sorghum silage makes very good roughage for dairy cattle. If well eared corn or grain sorghum silage is fed, less of other grains will be required because a considerable amount of grain will be contained in the silage. However, the protein content of these silage crops is low, requiring additional amounts of protein concentrates to balance the ration.

Legume and Grass Silage. More dairymen are using legume, grass, or a mixture of the two as a silage feed for dairy cattle. Very little protein is lost when these crops are made into silage. When the silage is made from legumes, or a mixture with a high percentage of legumes, the feed is very nutritious.

Citrus Pulp

In the citrus fruit growing areas of the United States, dried citrus pulp is becoming increasingly popular as a dairy cattle feed. The pulp is largely a by-product of the canneries. The bulk of the dried pulp comes from grapefruit and oranges; however, lemons, limes, and tangerines are entering the dried pulp industry. The feeding value varies to some extent, depending upon methods of manufacture and the content of the pulp. It is fairly high in carbohydrates, but low in protein and vitamin A. It contains from 70 to 76 percent digestible nutrients. Since calcium carbonate is added in the processing, it is fairly high in calcium but low in phosphorus. Owing to its bulky nature, it is used more as a roughage than as a concentrate, but as a nutrient it is more like the grains.

FIGURE 18-1. Dairy cows can convert large quantities of relatively inexpensive roughages into milk. (*Courtesy* American Guernsey Cattle Club)

Beet Pulp

Beet pulp is the residue of the sugar beets after the sugar has been removed. Beet pulp is often referred to as a roughage, but in nutrient composition it is more nearly like the grain concentrates. Beet pulp carries about 88 percent as many total digestible nutrients as corn but is lower in protein. Its value should be figured on the cost per pound of digestible nutrients.

Pastures

The natural feed for dairy cattle is pasture. The pasture season should be made as long as possible. Fall-seeded rye provides early spring pasture, in areas where it is adapted, and may be followed by native grasses, or a legume and grass mixture. Sudan grass, or a Sudan grass and soybean combination, makes an excellent summer and early fall pasture. In general, a legume and grass combination provides more grazing per acre of highly nutritious forage than any other common pasture crop. Pastures recommended for beef cattle are suitable for dairy cattle.

Concentrates

The grains and by-products of grain or oil seed, which are used extensively in dairy cattle rations, make up the group of feeds commonly referred to as concentrates.

Corn. Corn is palatable and very nutritious as a feed for dairy cattle. Throughout the Corn Belt, it is considered the most important grain for dairy cattle production. Corn, like most other grains, is fed primarily for its energy value, but it does supply some protein, minerals, and vitamins.

Grain Sorghums. Grain sorghums may be substituted for corn, pound for pound, in the dairy ration, if the price per pound is equal to or less than that of corn. The nutrient value of sorghum grains is very similar to that of corn.

Wheat. Wheat is equal to corn in food value, but because of its gluten content, it becomes pasty when eaten unless mixed with other feeds. Wheat should not make up more than half of the grain ration. Wheat is 3 to 5 percent higher in protein than is corn.

Ground Ear Corn. Ground ear corn has only about 90 percent of the feed value of ground shelled corn, but many dairymen prefer it. Probably the chief advantage lies in saving labor, as it does not have to be shelled.

Oats. Oats, an excellent feed for dairy cows, are somewhat less digestible than corn, but they have a higher protein content. Pound for pound, quality oats have about 90 percent of the value of ground shelled corn, or about the same value as ground ear corn, for dairy cattle. Oats vary probably more than most other grains in feed value. Light oats, containing a high proportion of hulls to berry, may not have more than 50 to 60 percent of the value of corn.

Barley. Barley, which may be used to replace corn, is the chief grain fed in many dairy herds. Good quality barley has about 95 percent of the value of corn on a pound-for-pound basis for dairy cows.

FIGURE 18-2. Sudan grass makes an excellent emergency pasture. (*Courtesy* Holstein-Friesian Association of America)

Rye. Rye is unpalatable to dairy cows. In food value it is about equal to barley. It should not be used to make up more than one-fourth of the grain ration.

Wheat Bran. Wheat bran is a by-product resulting from the processing of wheat into flour. It is composed primarily of wheat hulls and is a very common dairy feed. It is higher in protein than the common grains and is somewhat laxative. Bran compares favorably with oats in total digestible nutrients.

Molasses. Both cane and beet molasses are fed to dairy cattle. Molasses is low in protein and has about 60 to 70 percent as much total digestible nutrients as corn. Molasses is very palatable and some authorities believe it speeds up the fermentation process in the digestive system. Under most conditions, it does not pay to buy molasses unless it is as cheap as or cheaper than corn per pound of digestible nutrient.

Soybean Oil Meal. Soybean oil meal ranges from 40 to 47 percent protein and is an excellent source of protein for dairy cattle. It is palatable, and in the soybean growing areas it is often the cheapest protein concentrate.

Linseed Oil Meal. Linseed oil meal is an old standby and is well liked by dairymen as a protein supplement. It is an excellent feed and may well be used when its price per pound of digestible protein is in line with that of other high protein feeds.

Cottonseed Meal. Cottonseed meal is a common high protein feed used for dairy cattle. Experiments show that it is a very good source of protein. The price per pound of protein, compared to that of other feeds, should be the principal consideration in its use for dairy feed.

Soybeans. Ground soybeans provide an excellent source of protein. Generally, the price is too high to make them an economical feed. However, when soybeans are cheap, they provide a good homegrown protein supplement.

Other Protein Feeds. Corn gluten meal, corn gluten, and distiller's grains are good sources of protein for cattle, but are generally lower in percentage of protein than meals made from the oil seed crops.

Animal proteins, such as meat and bone meal, fish meal, and tankage, may be used to make up part of the protein content of the ration if they are more economical than the vegetable proteins. Otherwise, experiments show that they have no advantage over the vegetable proteins in a ration designed for cows in production.

Urea

Urea is a nitrogen compound. Dairy cattle can convert a certain amount of urea into protein. Except for high producing animals, urea could replace the protein concentrates in a ration that provides a full feed of good legume hay or silage, plus sufficient grains to meet the production requirements of the cow.

Urea has no advantage, except as an economy measure, over the oil seed meals and other common protein supplements. Its value lies in replacing the protein value of other feeds. It is toxic if overfed; it will make cattle sick and could possibly be fatal. For safety, urea should not exceed 1 percent of the total dry matter in the ration, 3 percent of the concentrate mixture, or 5 percent of a high-protein supplement. The maximum safety limit is 0.3 pound of urea per day per cow.

Commercial protein supplements containing urea have no advantage over feeds similar in protein equivalent, unless they can be purchased more cheaply.

The rules for feeding urea are the same for both beef and dairy cattle. (See Chapter 14.)

Vitamins

When dairy cows have access to good pasture during the summer and are fed liberally on good roughage during the winter, they seldom need vitamin supplements. The group of vitamins known as the B-complex, or water-soluble vitamins, are manufactured in the cow's digestive system. Vitamin A is found in yellow corn, good-quality cured roughage, pasture, and many other feeds. Vitamin D is supplied to animals by sunshine and is contained in sun-cured forages.

Very inferior rations could cause a deficiency in vitamins A and D. Fish liver oils are high in vitamins A and D, but they may reduce the fat content of the milk and cause other injurious effects. If a need for these vitamins exists, a prepared vitamin concentrate for dairy cattle should be used.

Minerals

The mineral requirements of dairy cattle vary considerably, depending upon the soil that produces the feed they are eating and the amount and kind of ration provided. Soils vary greatly in content of the various minerals. Cattle are known to require more than a dozen mineral elements for growth and production. If the plants they eat are high in any one or several of these minerals, it may not be necessary to provide a supplement that contains all the required minerals.

It is generally necessary to provide salt, calcium, and phosphorus in larger amounts than would ordinarily be provided in the ration. If mineralized salt is used, trace minerals will most likely be provided. Many dairymen feed a pound of mineralized salt and a pound of steamed bone meal for every hundred pounds of concentrate mixture. Bone meal is high in calcium and phosphorus, and the mineralized salt provides the salt and trace minerals. The recommendations of the agricultural college in your area should be followed regarding mineral supplements. The mineral mixtures recommended for beef cattle (see Chapter 13) may be used for dairy cattle.

FEEDING THE DAIRY COW

The dairy cow uses feed for maintenance, for developing her unborn calf, and for milk production. About half of the daily ration consumed by the cow is used to maintain her body and to nourish her unborn calf. No matter what their milk flow may be, the maintenance costs will be the same for cows of equal weight. That is why high producing cows are so important for profitable production. The cow that produces only 200 pounds of butterfat in a year will consume about two thirds as much feed as the cow that produces 400 pounds of butterfat. It will take the entire amount received for the milk, under average conditions, to pay the feed bill for the cow producing 200 pounds. The cow producing 400 pounds will pay her feed bill and leave the owner from 100 to 150 pounds of butterfat as profit. Table 18-1 shows the protein, total digestible nutrients, estimated net energy (ENE), calcium, and phosphorus requirements for maintenance of cows of various weights. Table 18-2 gives the digestible protein, estimated net energy, and the total digestible nutrients required for each pound of milk in addition to the maintenance requirements for the cow. The nutrient requirements increase as the butterfat percentage increases.

Recently more consideration has been given to total milk solids than to fat, and tests have been developed to determine total solids in milk. While the emphasis in the past has been on total milk and fat production, it appears that in the future nonfat solids will be the measure of a cow's productive ability.

As a measure of milk quality—and thus price—butterfat content may be replaced by proteins, minerals, vitamins, and other milk solids, or solids non fat (SNF).

The unfavorable price of butter in competition with vegetable fats and the importance of SNF in the human diet may be

TABLE 18-1 DAILY MAINTENANCE REQUIREMENTS OF DAIRY COWS FOR PROTEIN, TDN, CALCIUM, PHOSPHORUS AND ENE

Weight of Cow	Digestible Protein	TDN—Total Digestible Nutrients	ENE—Estimated Net Energy	Calcium	Phosphorus
	POUNDS		THERMS	GRAMS*	
700	0.48	5.8	4.6	14.0	7.00
750	0.51	6.2	4.9	15.0	7.50
800	0.54	6.5	5.2	16.0	8.00
850	0.56	6.9	5.5	17.0	8.50
900	0.59	7.2	5.8	18.0	9.00
950	0.62	7.6	6.1	19.0	9.50
1,000	0.65	7.9	6.3	20.0	10.00
1,050	0.68	8.3	6.6	21.0	10.50
1,100	0.71	8.6	6.9	22.0	11.00
1,150	0.73	9.0	7.2	23.0	11.50
1,200	0.76	9.3	7.4	24.0	12.00
1,250	0.79	9.6	7.7	25.0	12.50
1,300	0.82	10.0	8.0	26.0	13.00
1,350	0.84	10.3	8.2	27.0	13.50
1,400	0.87	10.6	8.5	28.0	14.00
1,450	0.90	11.0	8.8	29.0	14.50
1,500	0.92	11.3	9.0	30.0	15.00
1,550	0.95	11.6	9.3	31.0	15.50
1,600	0.98	11.9	9.6	32.0	16.00
1,650	1.00	12.3	9.8	33.0	16.50
1,700	1.03	12.6	10.1	34.0	17.00
1,750	1.06	12.9	10.3	35.0	17.50

* 28.35 grams equal one ounce.
Adapted from Extension Bulletin 218 (Revised), University of Minnesota, St. Paul, 1964.

responsible for this reevaluation of milk quality.

COMPUTING RATIONS FOR DAIRY COWS

Before the proteins and total digestible nutrients required for any given cow can be determined, it is necessary to know the weight of the cow, daily milk production, and butterfat percentage of the milk.

Computing T. D. N. and Protein Requirements

Suppose we have a cow weighing 950 pounds and producing 40 pounds of 4.5 percent milk. If we look at Table 18-1, we note that she requires about 0.62 pound of protein and 7.6 pounds of total digestible nutrients per day for maintenance. Table 18-2 tells us we will need to add approximately 0.05 pound of protein and 0.36 pound of total digestible nutrients for each pound of milk produced daily.

After determining the protein and total digestible nutrients required for any given cow, our next problem is to plan a ration that will provide these requirements. Many different feeds and combinations of feeds may be used. It is important that we use those feeds that will provide necessary

TABLE 18-2 PROTEIN (DIGESTIBLE) AND ENERGY (ESTIMATED NET ENERGY—ENE OR TOTAL DIGESTIBLE NUTRIENTS—TDN) REQUIRED FOR EACH POUND OF MILK, AS RELATED TO LEVEL OF PRODUCTION AND FAT CONTENT OF THE MILK

Fat Content of Milk, Percent	Pounds of Milk Produced Per Cow Per Day																				
	30			40			50			60			70			80			90		
	1 *	2 †	3 ‡	1 *	2 †	3 ‡	1 *	2 †	3 ‡	1 *	2 †	3 ‡	1 *	2 †	3 ‡	1 *	2 †	3 ‡	1 *	2 †	3 ‡
3.0	.040	.27	.29	.041	.28	.30	.043	.29	.31	.045	.30	.32	.046	.31	.33	.048	.32	.35	.051	.34	.37
3.5	.043	.29	.31	.044	.30	.32	.046	.31	.33	.048	.31	.34	.049	.32	.35	.051	.34	.37	.054	.36	.39
4.0	.046	.31	.33	.047	.31	.34	.049	.32	.35	.051	.33	.36	.052	.34	.37	.054	.36	.39	.057	.38	.41
4.5	.049	.32	.35	.050	.33	.36	.052	.34	.37	.054	.35	.38	.055	.36	.39	.057	.38	.41	.060	.40	.43
5.0	.051	.34	.37	.052	.35	.38	.054	.36	.39	.056	.37	.40	.057	.38	.41	.059	.40	.43	.062	.41	.45
5.5	.053	.36	.39	.054	.37	.40	.056	.38	.41	.058	.39	.42	.059	.40	.43	.061	.41	.45	.064	.43	.47
6.0	.055	.39	.42	.056	.40	.43	.058	.41	.44	.060	.41	.45	.061	.42	.46	.063	.44	.48	.066	.46	.50

* lb. digestible protein
† therms net energy
‡ lb. TDN
Source: Extension Bulletin 218 (Revised), University of Minnesota, St. Paul, 1964.

Requirements	Digestible Protein	Nutrients Total Digestible
For maintenance	0.62	7.60
For milk production	40 × 0.05 = 2.0	40 × 0.36 = 14.40
Total	2.62	22.00

amounts of protein and total digestible nutrients at the lowest possible price. By using the tables in Chapter 2 that show the average composition of common feeds, we can select the kind and amount of feeds best suited to our conditions and prepare a ration for the cow just described.

In Example 2, when timothy hay, which is low in protein, was substituted for alfalfa, which is a high protein roughage, it was necessary to add a protein concentrate in order to bring the protein content up to the desirable level. This increased the total digestible nutrients slightly above the required level. However, the cow should receive sufficient protein if her production is to be maintained.

Under average conditions the farmer would not need to compute a ration for each individual cow, as shown in the example, but could select a typical cow in the herd and plan a concentrate mixture that would fit the roughage available. Farmers should follow one of the practical rules described later in this chapter.

Roughages Are the Basis of Dairy Rations

Cows can maintain themselves and produce a calf and a limited amount of milk on a full ration of good-quality roughage, mineral mixture, and water. Roughages vary greatly in feeding value. A high-quality

Example 1	Digestible Proteins	Total Digestible Nutrients
20 pounds corn silage	0.26	4.00
17 pounds average alfalfa hay	1.78	8.55
8 pounds ground ear corn	0.42	5.85
4 pounds oats	0.38	2.80
Total	2.84	21.20

Example 2	Digestible Proteins	Total Digestible Nutrients
20 pounds corn silage	0.26	4.00
17 pounds good timothy hay	0.61	8.64
8 pounds ground ear corn	0.42	5.85
4 pounds oats	0.38	2.80
2 pounds soybean oil meal	0.91	1.52
Total	2.58	22.81

roughage is one that has approximately the same total digestible nutrients and digestible protein as the second or third cutting of alfalfa harvested in the early bloom stage. Grain and protein supplement are added to the ration in proportion to the amount of milk a cow produces, over and above what can be produced from the roughage part of the ration.

Amount of Roughages Consumed by Cows. When dairy cows are fed all the dry roughages they will consume, they will eat approximately 2½ pounds per 100 pounds of live weight. Thus, an 800-pound cow will consume about 20 pounds, and a 1,200-pound cow will consume about 30 pounds of roughage. When fed high-moisture roughages, such as silage, they will consume about 2½ to 3 times as much by weight as when dry roughages are used, the difference being in the water that is taken in with the feed. In food value, 3 pounds of silage equals about 1 pound of dry roughage.

Concentrates for Milk Production

The larger breeds of dairy cattle (Holstein, Brown Swiss, Ayrshires) can be expected to maintain their body weight and produce from 15 to 20 pounds of milk daily from roughages. The smaller breeds (Jerseys, Guernseys) can be expected to produce only 10 to 15 pounds of milk daily from roughages alone. Concentrates must be provided if the cow is to produce all the milk over these amounts that she is capable of producing.

Feeding Concentrates According to Profits. Except for cows on test for the purpose of making a high production record, it is seldom profitable to feed for maximum production. When the dairyman reaches the point where the value of the increased production is less than the cost of the feed, he is not getting economical production. The concentrates should be increased only until the cow has reached the maximum point of profitable production.

Rules for Feeding Producing Cows Not on Pasture

Several rules have been worked out to govern the amount and kind of concentrates for cows in production. A concentrate mixture to fit the protein content of the roughage should be selected for a herd that is not on pasture but is getting all the roughage it will consume. When cows are receiving all the high-quality legume hay or legume silage they will consume, a concentrate mixture containing from 9 to 10 percent protein will be sufficient except for very high producing cows. If the roughage is average-quality legumes, the protein level of the concentrate mixture should be about 13 percent. When the roughage is half grass hay, fodder, corn or sorghum silage, or any other similar feed, the concentrate mixture should contain 15 percent protein. If the roughage contains no legumes and is entirely of a low protein nature, a concentrate ration of 17 to 20 percent should be fed.

How Much Concentrates to Feed. When a concentrate ration has been selected that will fit the roughage, one of the following rules may be used to govern the daily allowance of each cow:

1. **A pound of concentrate mixture for each 3½ pounds of milk testing below 4 percent, and a pound for each 3 pounds of milk testing above 4 percent.**
2. **The pounds of butterfat produced monthly divided by four equals the pounds of grain mixture to feed daily.**
3. **When legume roughages are fed, a pound of grain for each 2 pounds of milk a Holstein, Brown Swiss, Ayrshire, or Shorthorn produces over 20 pounds daily, a Guernsey, over 15 pounds daily, and a Jersey, over 12 pounds daily.**

Table 18-3 may serve as a guide in planning rations when high-quality roughages are used.

TABLE 18-3 AMOUNT OF CONCENTRATE MIXTURE REQUIRED FOR DAIRY COWS THAT ARE FULL-FED HIGH-QUALITY ROUGHAGE

Breed	Amount of Milk Cows Can Produce on Roughage Alone (Pounds)	Pounds of Concentrates for Each Extra Pound of Milk
Holsteins	20	0.40
Brown Swiss	18	0.45
Milking Shorthorns	16	0.45
Ayrshires	16	0.45
Guernseys	12	0.55
Jerseys	10	0.60

Following are some suggested rations of various protein levels that may be used in areas where they are economical:

9 to 10 percent digestible protein

No. 1

Ground Ear Corn	550 Pounds
Ground Oats	400 "
Soybean or Linseed Oil Meal	50 "
	1,000 Pounds

No. 2

Ground Grain Sorghum	550 Pounds
Ground Barley	425 "
Cottonseed Meal	25 "
	1,000 Pounds

No. 3

Ground Shelled Corn	450 Pounds
Ground Barley	200 "
Ground Oats	325 "
40 Percent Protein Meal	25 "
	1,000 Pounds

No. 4

Ground Shelled Corn or Grain Sorghum	700 Pounds
Wheat Bran	300 "
	1,000 Pounds

12 to 14 percent digestible protein

No. 5

Ground Ear Corn	600 Pounds
Ground Oats	200 "
Soybean, Linseed, or Cottonseed Meal	200 "
	1,000 Pounds

No. 6

Ground Barley	300 Pounds
Ground Soybean Grain	300 "
Wheat Bran	200 "
Soybean Meal	200 "
	1,000 Pounds

15 to 16 percent digestible protein

No. 7

Ground Ear Corn	500 Pounds
Ground Oats	100 "
Wheat Bran	200 "
Corn Gluten Meal	200 "
	1,000 Pounds

No. 8

Ground Shelled Corn	500 Pounds
Ground Oats	300 "
Soybean Meal	100 "
Cottonseed Meal	100 "
	1,000 Pounds

17 to 20 percent digestible protein

No. 9	
Ground Shelled Corn	200 Pounds
Wheat Bran	200 "
Ground Oats	200 "
Cottonseed Meal	200 "
Soybean Meal	100 "
Dried Brewer's Grains	100 "
	1,000 Pounds

No. 10	
Ground Ear Corn	500 Pounds
Ground Oats	200 "
Soybean Meal	200 "
Linseed Meal	100 "
	1,000 Pounds

Urea may be used to replace part or all of the proteins, if fed according to directions. Biuret, being low in toxicity, may be used in place of urea, as a source of protein.

From 10 to 15 pounds of a mineral mixture recommended for the area should be added to each of the suggested rations. Cattle should have free access to salt.

Preparing Feeds for Dairy Cows

The digestibility of grains is usually increased by grinding them. However, too finely ground grains become pasty in the mouths of cattle and are not palatable. Coarsely ground or rolled feed is better. Little, if any, extra feeding value results from grinding forages.

Feeding Producing Cows That Are on Pasture

Pasture is the most nearly perfect dairy feed available. The producer who has given careful attention to the amount and quality of pasture crops is able to maintain high production at relatively low cost. Feed specialists have sometimes likened pastures to concentrates. Young pasture grasses and legumes are high in protein, low in fiber, and high in moisture. The fiber content is the chief factor that determines whether a feed is a roughage or a concentrate. Young pasture grasses contain about 18 percent fiber, whereas good alfalfa hay contains from 23 to 28 percent and grains from 2 to 12 percent.

Pasture crops are usually high in minerals, although the mineral content varies with the fertility of the soil. Phosphorus is most likely to be lacking in pastures. Occasionally, however, other essential minerals may be insufficient to meet the needs of animals. When soils deficient in minerals are properly fertilized, the forage they produce will generally provide adequate minerals. Vitamins are seldom a problem for cattle on good pasture. Vitamin A is found abundantly in green forage, and the sun will provide vitamin D.

Limitations of Pasture

Although good pastures are important in the economical production of dairy products, they have certain limitations. The limitations of pasture must be understood by the producer, if the best use is to be made of pasture crops.

Too High Water Content. In the more humid areas, pasture grasses may run about 80 percent water, especially in the spring. Since the cow's capacity for water is limited, she may not be able to consume enough of these grasses to provide the nutrients needed for milk production. Successful operators usually provide some good dry forage and grain for high producing cows on pasture.

Nutrient Value of Pasture Varies. The nutrient value of pasture crops does not remain the same during the season. As plants become more mature, the fiber content increases and the protein and vitamin content decreases. The same is true when the growth of plants is retarded because of dry weather.

Not only does young pasture grass contain more protein, but it is of a highly digestible nature. Any experienced dairyman knows that when cows are turned on fresh young pasture they usually respond with an increased production commonly referred to as a spring flush. This spring flush is the result of the high nutrient content and palatability of the young pasture grasses. For the best results, young pastures should be maintained by grazing down or mowing so that new fresh grass is continually produced.

Feeding Concentrates to Cows on Pasture. When pastures are good and cows can get their fill in two to four hours of grazing daily, very little grain is necessary for cows of average production. Jerseys and Guernseys will produce up to 30 pounds of milk; Holsteins, Brown Swiss, Ayrshires, Milking Shorthorns, and other breeds of nearly equal size will produce 40 pounds of milk daily on pasture alone. There are several rules followed by different producers for feeding concentrates to cows on pasture. One common rule is to feed one pound of concentrate for each 5 to 7 pounds of milk to cows producing one pound or more of fat daily. Kind and quality of pasture determine the protein percentage of the concentrate.

Feeding the Dry Cow

The dairy cow needs from 45 to 60 days of rest between lactation periods, depending upon the body weight lost. While she is dry, the cow should replace the body weight lost during the previous lactation period. Unless she has at least six weeks of rest and a proper ration during the dry period, she will not be in condition for best production during the next lactation. Cows that are put into the best possible condition by calving time generally respond by increased production during the following lactation period.

When the drying-off process (which will be discussed later) has been accomplished, the conditioning ration should be started. The stored fat, which the cow puts on during the dry period, serves as a reserve supply to be drawn upon for milk production. Protein cannot be stored, so a ration with a protein content in excess of what the cow needs to put on flesh is not necessary. The ration should contain minerals to put back into her bones the calcium and phosphorus that have been utilized in the milk production. Experiments show that cattle can store vitamins A and D. Cows that are dry during the summer will probably not need any special attention paid to the vitamins contained in the ration if they are on good pasture. During the winter, greater care needs to be exercised to insure ample supplies of important vitamins.

Feeds for the Dry Cow. The amount and kind of feed necessary for the dry cow depends upon her ability to produce and upon her condition when she is dried up. High producing cows are usually thin and need to be full-fed grain during the rest period. Average to low producers will not lose a great deal of weight, and pasture or cured roughages are sufficient to recondition them.

Feeding the Cow Just Before Calving

A week before calving, grain may be partially replaced with more bulky feeds. Do not reduce the feed too much before and after calving, for ketosis may develop in high producing cows from insufficient carbohydrate foods. The ration should be mildly laxative. Beet pulp is used by many dairymen as part of the ration during the last week of the rest period. Following are two examples of rations that may be used two days before calving.

Ration 1

200 pounds ground oats
100 pounds wheat bran
25 pounds linseed oil meal

TABLE 18-4 RATIONS FOR CONDITIONING DRY COWS †

Kind of Roughage	Concentrate Mixture (Pounds)
Good-quality legume forage or good pasture	1. 100 ground oats 400 ground ear corn 2. 200 ground barley 200 ground shelled corn 100 ground oats 3. 300 ground grain sorghum 200 ground oats
Mature grass pasture or low protein forage	1. 300 ground shelled corn or sorghum grain 100 ground oats 50 wheat bran 50 linseed, soybean, or cottonseed meal 2. 200 ground barley 200 ground shelled corn 75 wheat bran 25 soybean meal

† Five to 7 pounds of a mineral mixture used for cows in production should be added to each of these rations.

Ration 2

200 pounds ground oats
100 pounds beet pulp
25 pounds soybean oil meal

Feeding the Cow Just After Calving

The cow should be given all the fresh, reasonably warm water she will drink after calving. For the next three days she may continue on the same ration as the last week before calving. Some dairymen add 2 to 3 pounds of molasses to the ration to keep the energy value high as a precaution against ketosis. Concentrates may be moistened with water and sprinkled with a pinch of salt. The regular dairy ration may be started on the fourth day. The concentrates should be increased as rapidly as the cow's appetite will permit.

MANAGING THE DAIRY HERD

The successful management of the dairy herd depends upon an intimate knowledge and understanding of the animals.

Exercise

Contrary to popular belief, dairy cows need only limited exercise. Too much movement will result in food nutrients being used for body maintenance that would otherwise go into milk production. However, it seems advisable, during cold weather when the cows are confined, to take advantage of any warm days to turn the cows out for an hour or so. When the loose housing system is used, cows have room for sufficient free movement to provide the needed exercise. During summer months, cows receive adequate exercise, especially when they are on pasture.

Cleaning and Grooming

Cows in the milking herd should be kept clean, not only for the production of clean milk but for the health of the animal as well. Daily brushing will remove dirt and loose hair. Regular grooming will help to keep the hide pliable and the hair will develop a sheen that improves the appearance of the cows. The clipping of the long hair from the

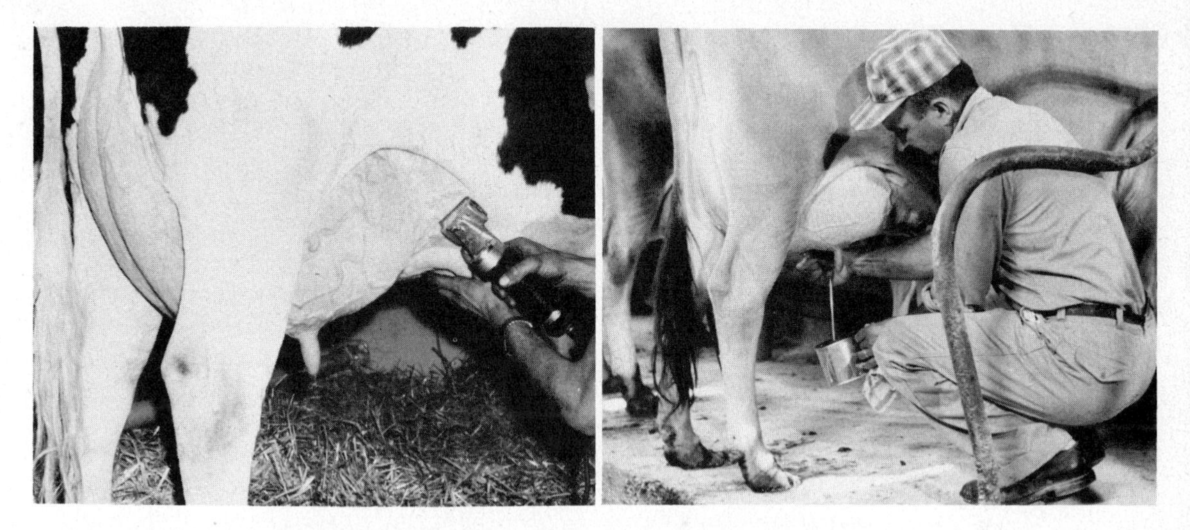

FIGURE 18-3. (*Left*) **Clipping the underline, udder, and rear flank is important in keeping cows clean. FIGURE 18-4.** (*Right*) **The use of a strip cup will help detect udder infections and clean the teat canal. A few streams should be milked from each quarter before applying the milker. (Figures 18-3 and 18-4** *Courtesy* **Hoard's Dairyman)**

udder, hind legs, and rear flanks will help to prevent accumulation of filth.

Milking a Cow

Research studies have provided us with much information concerning successful milking practices. Cows will respond favorably to certain stimuli and conditions, and unfavorably to others. Individual dairymen vary in the routine they follow during the milking period. It is important that the same procedure be carried out in the same order at each milking period.

Preparing the Milking Area and Equipment. The milking area should be clean and free of odors before the cows enter. Milking equipment should be set up and made ready for milking. The operator must be prepared to give his undivided attention to the job at hand.

Feeding. Some dairymen feed the cows during milking, others before or after. The cow will adjust herself to the time of feeding, provided the same practice is followed for each milking.

Washing the Udder and Preparing the Cow for Milking. Washing the udder has two chief advantages: (1) it stimulates the cow to let down her milk, and (2) it decreases the danger of contaminating the product. When the cow is stimulated by washing her udder with warm water and massaging it, a hormone is released into the bloodstream and reaches the udder in about one minute. This hormone causes a contraction of the muscles surrounding the milk cistern. If the cow is frightened or made nervous by any change in routine, another hormone is secreted, causing a constriction of the blood vessels that prevents the milk-secreting hormone from entering the bloodstream. Even though the cow is not disturbed in any way, the milk let-down hormone is effective for only about seven minutes after it enters the udder. Milking should be accomplished within this period.

Strip Cup. Use a strip cup and draw two or three streams of milk from each quarter before attaching the milking machine or hand milking. This removes any dirt from the teat canal and gives the operator a chance to see if the milk is normal. If any abnormality appears, the cow should be milked last and that milk kept separately.

Drying Off Cows

Frequently dairymen have considerable difficulty in drying off high producing cows. Three methods are used to dry up cows: (1) intermittent milking, (2) incomplete milking, and (3) abrupt cessation of milking.

Intermittent Milking. Under this procedure, the cow that is to be dried off will be milked once a day for a while, then once every other day, and finally milking will be stopped altogether.

Incomplete Milking. Dairymen who follow this system start by not extracting all the milk from the udder at milking time for the first few days after the drying-off period has begun. Later they milk intermittently but never completely. After the production decreases to only a few pounds daily, milking is stopped.

Abrupt Cessation of Milking. Of the various systems, experiments have proven that for cows producing 25 pounds of milk or less abrupt cessation of milking is the best, if the udder is sound and there has been no mastitis infection. Three days before milking is stopped, all concentrates should be removed from the ration. The hay should be reduced to about ½ to ⅔ of the normal ration. The reduction in feed will reduce the milk flow. It has been shown that milk will build up to a certain pressure, secretion will stop, and reabsorption of the milk into the bloodstream will begin. It is recommended that after the last milking, the teats be washed and dipped in collodion, which will seal the ends and prevent infectious organisms from entering the udder.

Care of the Cow at Calving Time

A few days before the cow is expected to calve, she should be placed in a well bedded

FIGURE 18-5. This cow has delivered her calf in a clean, well-bedded box stall. (*Courtesy* Washington State University)

box stall with ample room for calving. The stall should be isolated insofar as possible from any disturbances. Occasionally, in high producing cows, the udder may become swollen to such an extent that milking before calving will be necessary. If such is the case, the milk should be frozen and saved for the newborn calf. The practice of milking before calving is not recommended except in extreme cases.

Provide the cow with plenty of water to drink. In cold weather, many dairymen prefer the water to be warm. Remove the afterbirth as soon as the cow has eliminated it to prevent her from eating it. (Cows will eat the afterbirth, probably owing to instinct developed by their wild ancestors to prevent the odor from attracting marauding animals.) If the cow does not expel the afterbirth completely within 48 hours, call a veterinarian. Cows should be observed closely for signs of milk fever or ketosis. (See Chapter 22.) Should either of these ailments develop, treatment should start immediately. As soon as the cow shows a desire to eat, she should be given feed according to the recommendations for feeding the fresh cow. If the cow has not been milked before calving, allow the calf to nurse or save the colostrum and hand feed it. Start regular milking to relieve the pressure on any quarter not nursed by the calf. The mature cow should not be milked dry for at least four milkings, especially those cows known to be subject to milk fever.

Heat Periods

For cows, the average time from one heat period to the next, known as the oestrus cycle, is 21 days, but it varies from 17 to 26 days. The heat period lasts from 6 to 36 hours, with an average duration of 18 hours for cows and 15 for heifers.

Breeding the Cows

Cows may be bred naturally (direct service by a bull) or artificially. The best method depends upon the size of the herd, production level of the herd, whether or not the sale of bulls is part of the dairy program, and whether artificial insemination service is available in the community.

Natural Breeding. Natural breeding consists of direct service of the cow during her heat period by a bull. In some localities, this is the only means by which a cow may be bred, since artificial insemination service is not available in all localities. The natural breeding method has the following advantages: (1) It is easy to detect when the cows are in heat. Many cows have rather feeble or silent heat periods, making it extremely difficult to know when they should be bred. When the time for their heat period approaches, they may be turned in with the bull daily until they have been bred. (2) Dairymen who have developed very high producing herds may not be able to get the services of a well-proven bull for all their cows unless they own or have a share in a proven bull. (3) Breeders who sell breeding stock as a major part of their dairy business will often want bloodlines not available through artificial insemination. (4) Cows may be bred more than once during the heat period. There is some evidence that breeding the cow twice, six to eight hours apart, has some advantage in getting hard-to-settle cows to conceive.

Artificial Breeding. Artificial breeding or insemination is a practice in which the semen of the bull is transferred to the cows by a person generally known as an inseminator or technician. The semen is often produced at bull studs or semen-producing centers, and it is shipped to local associations as needed. In many localities dairymen have formed cooperative breeding associations. The usual practice is to charge a membership fee, plus a service charge for each cow serviced. The money is used to purchase bulls and equipment, and for the hiring of technicians.

Advantages of Artificial Insemination. (1) Artificial insemination, when properly administered, will give conception rates as high as those with the natural breeding method. (2) Dairymen can secure the service of proven sires or those from a high producing line at a moderate cost. (3) The bother and danger of keeping a bull is eliminated. (4) The services of outstanding sires are greatly extended. (5) The necessity of locating and purchasing a new herd sire every two years is eliminated.

How to Insure Good Results When Breeding Artificially. As previously stated, it is possible to obtain conception rates from artificial insemination equal to those from natural breeding if the proper procedure is followed. Probably the greatest disadvantage of artificial breeding is the inability of some operators to detect when the cows are in heat. To insure results, the following rules should be observed: (1) Don't breed cows until at least 60 days after calving. It usually takes a cow approximately 60 days to become normal after calving and in proper condition for breeding. Although many cows will conceive during the first heat period following calving, the conception rate is generally higher if they are bred during the second or third heat period. (2) Know when the cow first appeared in heat and tell the technician. Ovulation does not take place until about 14 hours after the end of the heat period. A study made at the Nebraska Experiment Station showed that best conception rates were obtained when cows were bred more than six hours but less than 24 hours before ovulation. Cows first noticed to be in heat during the early morning should be bred in the afternoon of the same day. Cows noticed to be in heat between nine in the morning and noon should be bred early the next day; those that come in heat during the afternoon should be bred about noon the next day. (3) Put the cow in the barn and leave her for four hours after breeding. (4) Keep accurate records of heat periods, breeding, and calving dates for all cows. (5) Have hard-to-settle cows examined by a veterinarian. (6) Have all cows examined by a veterinarian for pregnancy 12 weeks following breeding. Many cows will not come in heat owing to retained corpus luteum (yellow body) or some other organic trouble. Such conditions can often be corrected by a veterinarian. If cows are not examined for pregnancy and no heat period is evident, the operator may not notice that they are not with calf for several months.

Time to Breed

Dairymen who are retailing milk and must produce about the same amount each month find it necessary to have cows freshening throughout the year to prevent a serious drop in production during any one period. However, those who sell milk or butterfat wholesale usually prefer to have their cows freshen in the fall months (September, October, and November). Fall freshening has the following advantages: (1) Milk production is greater. Cows freshening in the fall will hold up on milk production longer than will cows freshening when it is spring or summer. The cow that was freshened in the fall will have a spring flush almost equal to a second freshening when she goes on grass. (2) Butterfat and milk prices are usually higher during the winter. (3) The farmer has more time during the winter, when he does not have to give so much attention to crop production. (4) Fall calves are often stronger, owing to the excellent ration provided the cow by good pasture, and thus there is less danger of a nutritional deficiency.

Gestation Periods

The period from the time a cow conceives until she gives birth to a calf is known as the gestation period. It varies with individual cows and with breeds. First-calf heifers will average approximately two days less than older cows of the same breed. Ayrshires, Holsteins, and Jerseys have an average gestation period of about 278 to 279 days, Brown Swiss, 288 to 290 days, and Guernseys, 283 to 284 days.

HOUSING AND EQUIPMENT FOR DAIRY COWS

The dairy cow requires about 150 man-hours of labor per year. Labor cost is second only to feed cost in dairy cattle expenses and every effort should be made to reduce labor cost. Housing and equipment should be planned with a view to reducing the time required in caring for the cow. There are three common systems for housing dairy cattle: (1) stanchion barn, (2) loose housing, and (3) free stall.

Stanchion Barn

There are two general types of stanchion barns: (1) the two-story barn, commonly used in the northern regions, and (2) the single-story barn, frequently found in the southern regions and becoming more popular in the North.

The Two-story Barn. In the two-story barn, feed, especially roughage, is stored on the second story. The lower story is used for housing cattle. The advantages of the two-story barn are: (1) it is convenient to work in because feed may be stored on the second story and moved down by gravity, and (2) it provides more cubic feet of storage space in relation to ground area covered by the building.

The Single-story Barn. With the increased use of baled or chopped hay and silage, the one-story barn has gained in popularity. Less storage room per ton of feed is required when hay is baled or chopped. Silage is not commonly stored in the barn. The weight per cubic foot of baled or chopped hay is so great that considerable bracing and heavy timbers are required in the barn when the roughage is stored in the second story. The advantages of the single-story barn are: (1) less windstorm hazard, (2) reduced fire risk, (3) easier and cheaper to build, and (4) easier repairs.

FIGURE 18-6. A modern stanchion barn equipped with an automatic barn cleaner. (*Courtesy* James Manufacturing Company)

The Stanchion-barn Plan. Space will not permit a detailed discussion of the many types of stanchion barns one may decide to build. The agricultural colleges of each state will furnish plans, free or for a small charge, from which one may select a type that will best fit individual conditions. Cows spend a great deal of time in the stanchion when this type of barn is used. They generally feed and, in the more modern type, drink from the stanchion area. It is important that the stanchion area be of a size that will be comfortable for the cattle and easy to clean. Gutters should be constructed to provide for the installation of barn cleaners.

Insulation and Ventilation. Since cows spend much time in the stanchion of the stanchion-type barn, especially in northern

TABLE 18-5 RECOMMENDED DIMENSIONS OF COW STALLS

Weight of Cow	Width of Stall	Length of Stall
800	3′ 4″	4′ 6″
1,000	3′ 8″	4′ 8″
1,200	4′ 0″	5′ 0″
1,400	4′ 4″	5′ 4″
1,600	4′ 8″	5′ 8″

areas, it is important that the building be well insulated and ventilated. Floors must be easy to clean because the cows are milked, housed, and fed in the same area. The floors should be made of cement or some other easy-to-clean material which is likely to be cold unless the barn is properly insulated. When the cow's udder is subjected to cold floors and drafty conditions, there is more danger of udder diseases developing.

Successful insulation depends upon two elements: (1) a double wall filled with a recommended insulating material, and (2) a vapor barrier applied to the inner wall. The purpose of the insulating material is to break up the air circulation between the walls, which prevents the transfer of heat or cold through them. The vapor barrier prevents moisture from penetrating the inner wall where it would condense. There are a number of good commercial insulating products from which to choose. The vapor barrier may consist of two coats of aluminum paint applied to the inner wall. Waterproof glazed surface paper or other waterproof material that will prevent moisture from penetrating the inside wall may be used. Warm air will hold more moisture than will cold air. When a barn is full of livestock, considerable moisture is picked up by the air as it is warmed from the heat given off by the animals. When the air is cooled, it can no longer hold as much water, and condensation occurs. When the air that circulates inside a building strikes the cold walls and ceiling, the moisture condenses and the walls or ceiling become wet. If the walls and ceiling are properly insulated, they remain at about the same temperature as the interior of the building and no condensation takes place.

Ventilation is the control of air movements by the proper installation of intakes and outlets. Cool, dry air from the outside is brought in. As the air warms, it picks up moisture given off by the livestock. The outlets afford a place for the warm, moist air to escape, and the process is repeated. Tight walls and a ceiling that permits the control of air circulation are necessary for good ventilation.

Temperature. Cows do not need especially warm barns in order to produce well. A stanchion barn maintained at a temperature of about 50 degrees is sufficiently warm. Sudden changes of temperature will cause a drop in milk production. A well-insulated barn is one in which ventilation may be controlled; the dairyman can prevent the radical changes in temperature that will occur in a poorly constructed building.

Gutters. Properly constructed gutters located behind the stall platform are important for sanitary purposes and for greater ease in cleaning the barn. The gutter serves as a receptacle for catching manure and urine. The standard width of the gutter is 16 inches and it should be 10 inches deep. When mechanical barn cleaners are installed, a gutter 14 inches wide may be used.

The mechanical barn cleaner will remove the manure from the gutter and load it on the spreader ready to be hauled out to

TABLE 18-6 SILAGE, HAY, AND BEDDING STORAGE SPACE REQUIREMENTS

Material	Wt. Per Cu. Ft.	Cu. Ft. Per Ton
Hay (Loose)	4.0 to 4.5 lbs.	400 to 512
Hay (Baled)	15.0 to 20.0 lbs.	100 to 133
Hay (Chopped)	10.0 to 12.0 lbs.	165 to 200
Straw (Loose)	4.0 lbs.	500
Straw (Baled)	12.0 lbs.	167
Silage (Silos up to 30′ deep)	40.0 lbs.	50
Silage (Silos of more than 30′)	50.0 lbs.	40

the fields. Such a device will save a great many hours of hand labor and will aid in maintaining a clean, sanitary barn.

Feed Storage and Watering Facilities. The cows are usually fed in the stanchions when the stall-type barn is used. It is important that both hay and grain are stored near the feeding area. Carts may be used for transporting grain and silage, reducing the amount of time and labor required for feeding.

The stalls should be equipped with automatic watering cups. Cows that have water available at all times usually produce better than those watered only twice daily.

THE LOOSE HOUSING SYSTEM

This type of housing system provides for a flexible housing arrangement that may be enlarged easily as the herd increases in size. The cows are free and have access to fresh air and sunlight. Ventilating problems are held to a minimum. The loose housing system provides low-cost housing arranged to reduce labor to a minimum.

The loose housing system consists of four units: (1) a feeding and feed storage area, (2) a paved lot, (3) the loafing and bedding storage area, and (4) the milking parlor. The four units may be under the same roof or in separate buildings.

Feeding and Feed Storage Area. The feeding area is where the feed is stored and where the cows are fed their daily ration of hay. The feed may be stored in the rear of the shed and a movable feed rack constructed. As the cows eat into the hay, the rack is moved, thereby saving the labor of handling the hay. The feeding area should be paved and the manger constructed so as to prevent the cows from pulling out the hay and lying on it.

The Loafing Area. This is the area where the cows rest. The manure pack is allowed to accumulate during cold weather. The fermentation of the manure creates heat which keeps the cows warm and comfortable. The area must be bedded heavily if the cows are to be kept clean. About 12 pounds of bedding will be required per cow per day. At least 60 square feet of space per cow should be provided. The area does not have to be kept especially warm, but drafts should be avoided. The calf pens and calving stalls may be constructed in this area. In building the loafing shed, the need for cleaning with a power loader should be kept in mind. By spring, the manure may be from 2 to 4 feet deep and trampled solid.

The Paved Area. Cows make good use of sunshine every day, even in very cold weather, if a windbreak is provided. The building may be constructed so as to give protection on the north and west. The paved lot should extend east and south of the buildings. Silage may be fed in bunks on the paved lot.

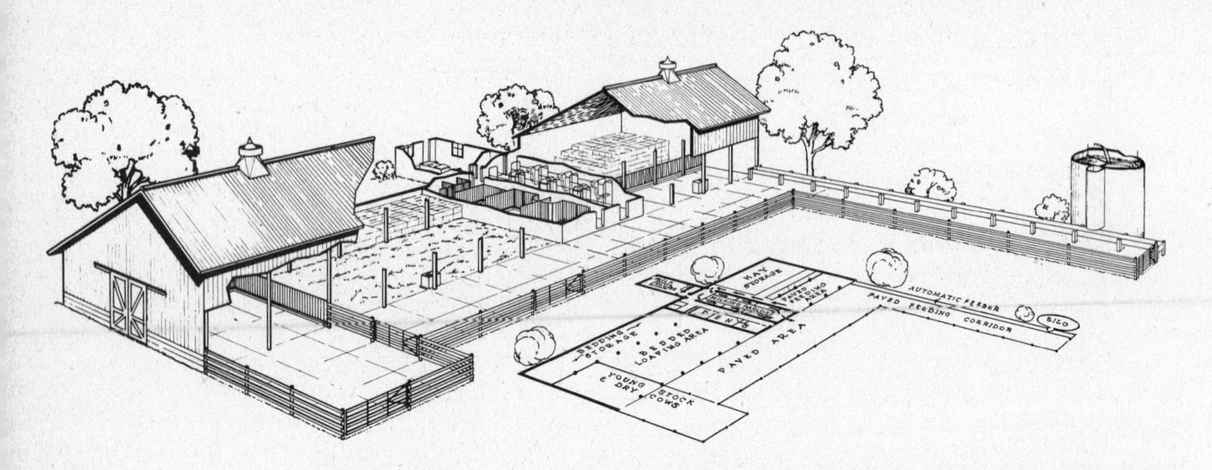

FIGURE 18-7. (*Above*) **A practical loose-housing system. The hay storage, silo, and feeding areas are separated from the loafing area by the milk room, milking parlor, and maternity pen area.** (*Courtesy* **James Manufacturing Company)** **FIGURE 18-8.** (*Below*) **Floor-level type combination milkhouse and milking parlor.** (*Courtesy* **University of Missouri)**

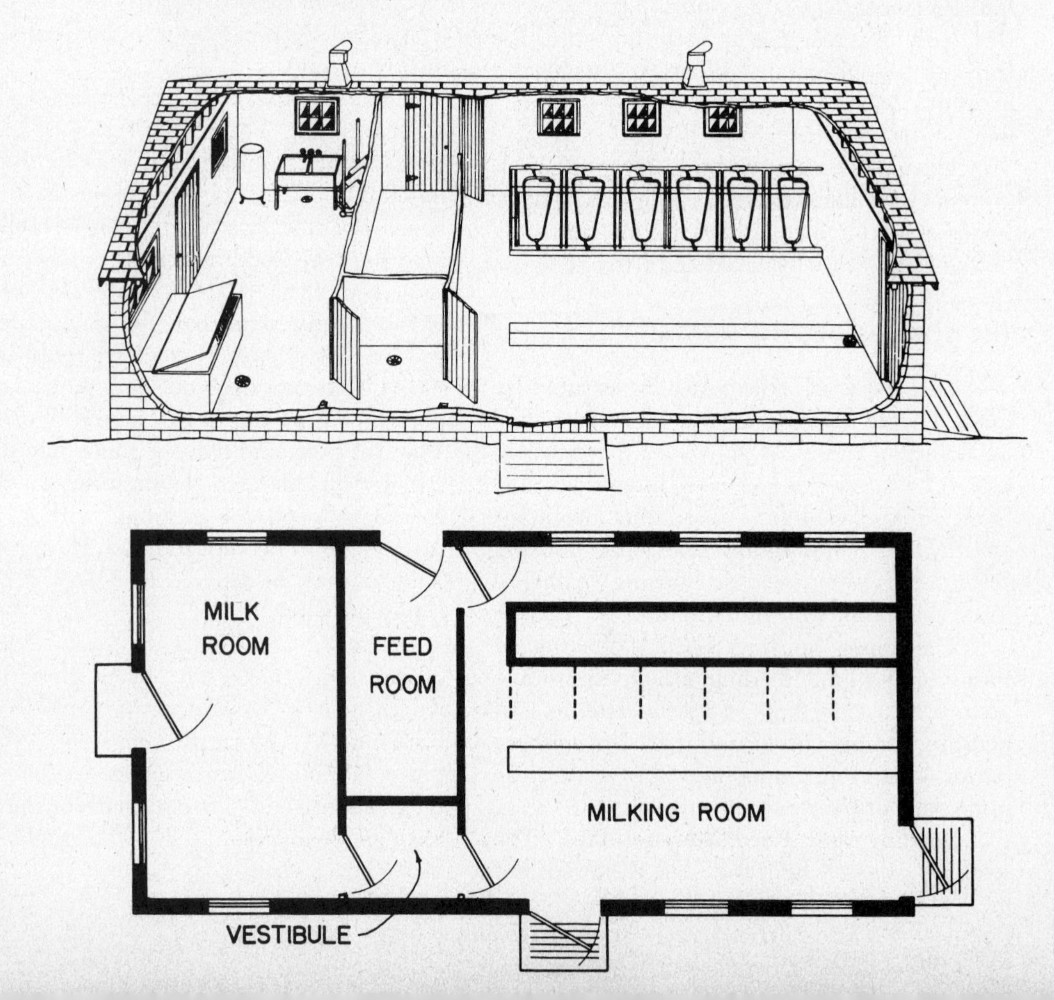

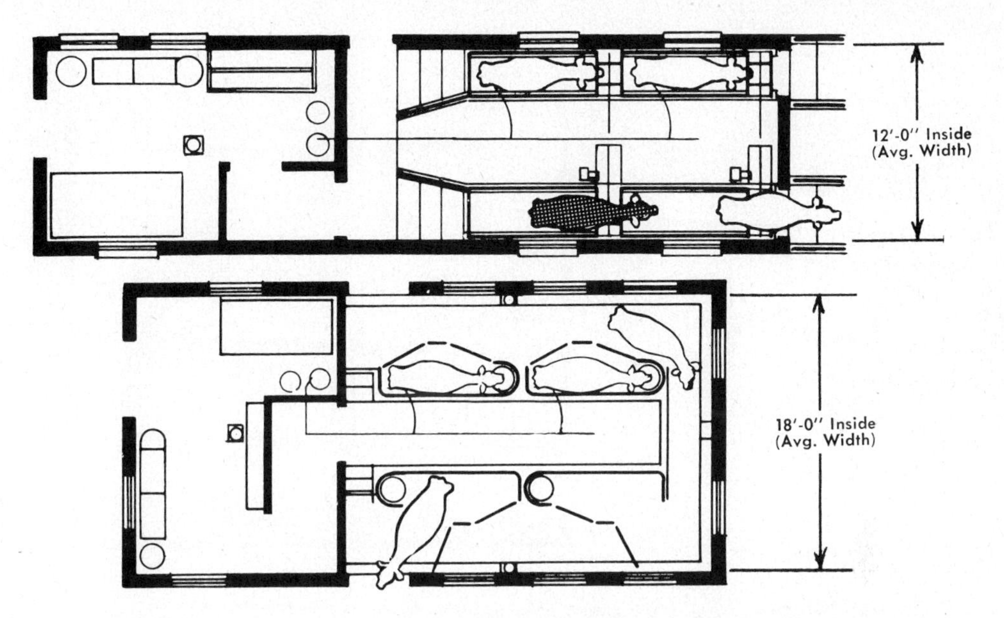

FIGURE 18-9. (*Above*) Walk-through parlor with four stalls. This can be built narrower than one with gate-type stalls. Two cows on a side are milked while two opposite are brought in and prepared. They are handled in pairs, since the slower milker on a side determines when a change can be made. **FIGURE 18-10.** (*Below*) Milking parlor with gate-type stalls. Cows are brought in from each side and released at doors ahead of them. Two are milked at a time; a slow milker does not delay the procedure, since any cow can be released and another brought in.

FIGURE 18-11. Four-stall parlor with gate-type stalls. Cows enter through door at top and move around milking pit to exit at door opposite the entrance. Two milking units are used.

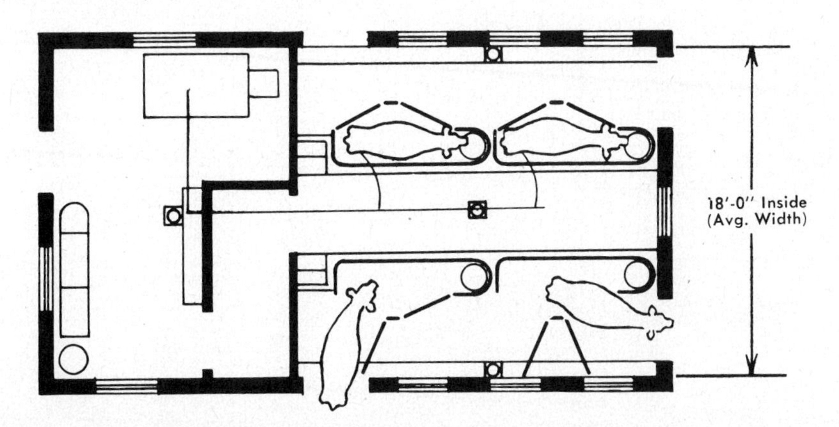

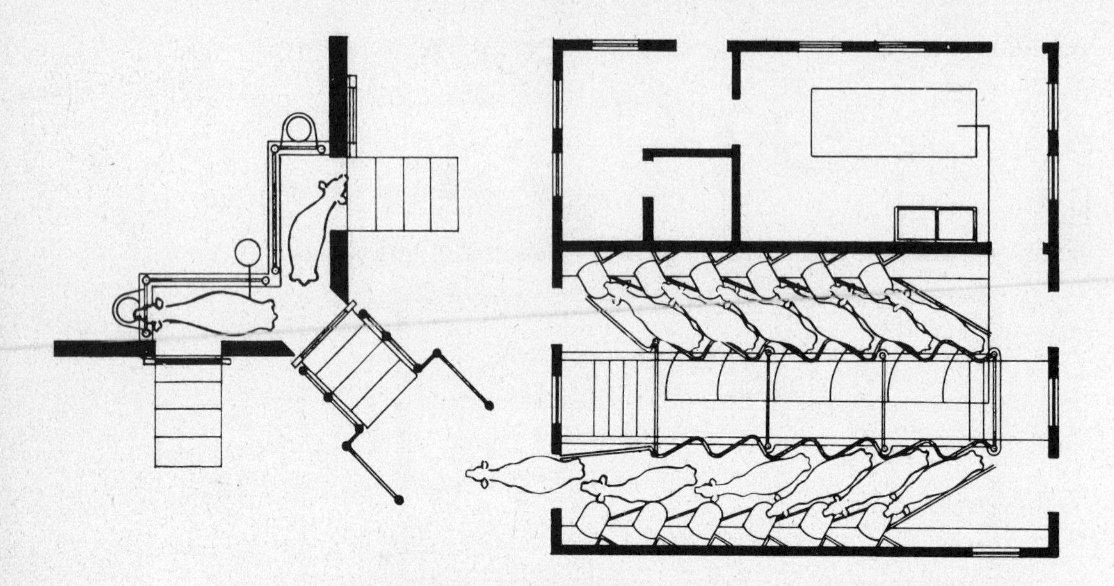

FIGURE 18-12. (*Above left*) **Cow-to-can milking with elevated stalls. The simplest milking parlor can be built in a corner of the barn. Cows come in at the corner and exit by a side door. Single milking unit mounted on the can is used for one cow at a time. FIGURE 18-13.** (*Above right*) **The herringbone milking parlor. Cows enter both sides from the right and are packed in for feeding and milking. Cows in the upper line are being milked as those below are released. Each cow is fed as she enters and goes to her trough. A milking unit is supplied for each.** (*Courtesy* **Barn Equipment Association) FIGURE 18-14.** (*Below*) **Diagram of individual stalls for loose housing. (Iowa State Extension Pamphlet 300)**

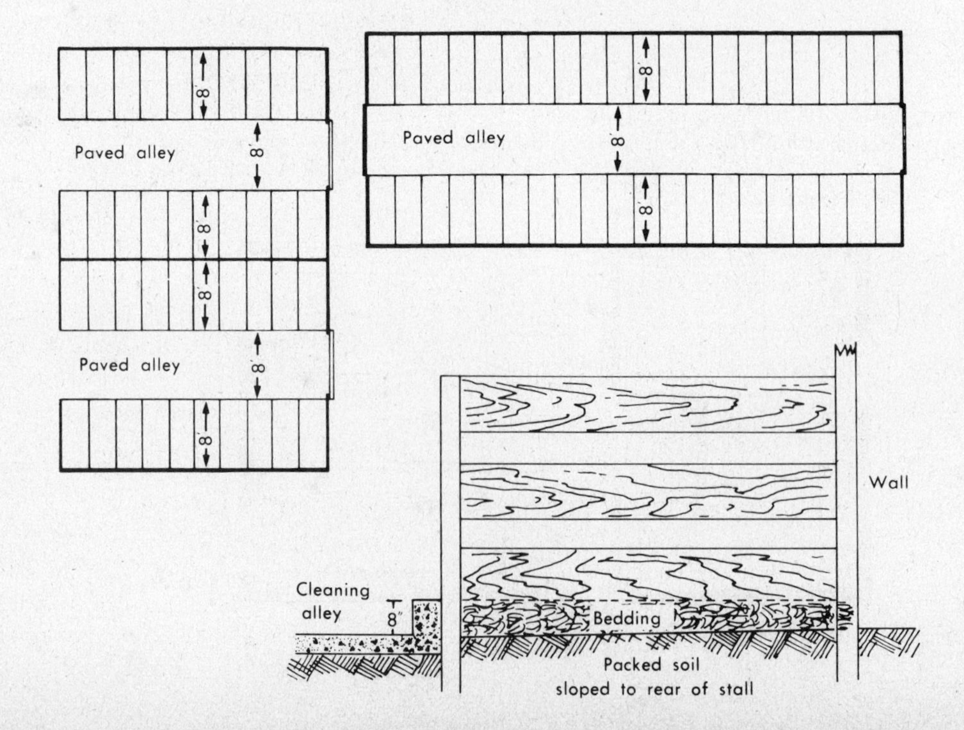

The Milking Parlor. The milking parlor is where the cows stand in either stanchions or elevated stalls for milking.

The Floor Level System consists of stanchions for confining a few cows at a time for milking. The operator stands on the same level as the cows. They are usually fed their grain ration while in the stanchion. The cost is low and construction is easy with this type of milking parlor. Usually four stanchions are required for from eight to ten cows. Most operators of floor-level milking parlors prefer to fill the stalls, milk, and refill, rather than to bring the cows in one or two at a time.

The Elevated Stall System has the cow elevated in a stall 30 inches above the level of the floor on which the operator stands. The self-closing doors may be opened by rope and pulley. One cow is changed at a time because turnover is easy. The operator can do his work with a minimum of stooping, which lessens fatigue.

FIGURE 18-15. A single-row sawtooth milking parlor. (*Courtesy* Babson Bros. Co.)

Free Stalls for Dairy Cows

Free stalls have replaced the typical loose housing system on many dairy farms. A stall is provided for each cow. Bedding costs may be reduced up to 75 percent, and cows are cleaner. There are a large number of plans for free stall housing. One example is shown in the diagram in Figure 18-14.

FEEDING AND MANAGEMENT OF THE HERD SIRE

It is desirable that outstanding dairy sires be kept in active service for a number of years. The service of bulls that have proven their ability to transmit high production qualities to their offspring is sought by all progressive dairymen. A bull is usually six to seven years of age or older by the time he has sired a sufficient number of daughters with records to establish his transmitting ability. Many bulls become sterile by the time they have proven themselves unless they have had careful feeding and management.

The herd sire should be fed in such a way as to insure his development in size to the fullest extent of his inheritance. Although a small or stunted bull may be as good a breeder as a larger one, dairymen who wish to buy breeding stock usually look upon such a sire with disfavor. Therefore, a bull of at least average size for the breed is desirable. The mature bull should be fed according to his size and to the extent he is being used for breeding.

Feeding the Mature Bull

Exclusively fat bulls may lack a desire to breed and become impotent. However, a bull should not become too thin. During the season of heavy breeding, the bull should have an increase in the daily concentrate ration.

During periods when the sire is not in heavy service, feed may be reduced. The roughage ration should be controlled to keep the paunch to minimum size, or bulls will have difficulty performing the act of breeding.

Management of the Bull

The management of the bull calf is similar to that of heifer calves until he is about five months of age. By that time, the bull may have become sexually mature enough to breed and, therefore, should be removed from the heifers.

The bull should be taught to lead while young. At first, a halter made of rope or some other inexpensive material is all that is needed. Later, a ring should be placed in his nose, and a staff should be used for leading. The bull is more easily controlled when led by a staff and ring, and there is less danger to the operator.

Ringing the Bull. When the bull calf is about six months of age, he should have a small ring placed in his nose. This may be accomplished by tying him up in such a way as to hold his head firmly. Find the soft part in the nose, force the ring through, and clamp or fasten it together with a screw, depending upon the type of ring used. A trocar may be used to puncture the nose before inserting the ring. On young bull calves, the trocar is not essential if a self-piercing ring is used. As the bull grows older, the cartilage between the nostrils becomes tougher, and a trocar is essential for ringing. The first ring should be of lightweight, nonrusting material about 1½ inches in diameter. When the bull is ten to twelve months old, the first ring should be replaced by a strong 3-inch brass or cannon metal ring.

Exercise. Plenty of exercise is essential if the bull is to be kept in good breeding condition and in active service over a long period of time. One of the best methods of providing exercise is to construct a pen adjacent to the shed or housing quarters. The bull should not be tied up for long periods at a time or he will become sore and stiff.

Housing

The dairy bull should have his own shelter located on a well drained area so that it will stay dry and built tightly enough to provide protection from cold weather and shade during warm weather. A feed alley, from which the bull may be fed without endangering the operator, may be located along one side of the shed. An overhead storage area for roughage and bedding will decrease the amount of labor that would otherwise be necessary in caring for the bull.

Exercise Lot

The safest and easiest way to provide exercise is to have a yard adjoining the bull shed. The yard need not be wide, but it should be at least 80 feet long. The bull will have a tendency to walk from one end of the yard to the other. A block of wood that is suspended about four feet from the ground and swings back and forth as the bull plays with it will encourage him to exercise.

The pen should be constructed of heavy plank or pipe fastened to substantial posts. A breeding chute located on one side of the pen is essential for the safety of the operator. The cow is placed in the chute, and the bull is allowed to reach the cow by means of a swinging gate. This eliminates the necessity of the operator's entering the bull yard. It is desirable to have the bull pen located where the cows pass frequently. The bull will be inclined to exercise more if the cows come near his pen, and cows may be more easily placed in the chute for breeding.

Service Age

A young, well grown bull may be used occasionally for breeding when ten months of age. He should not be used regularly before he is a year to 15 months old. Until the bull is 18 months old, the number of services should be limited to two per week. Mature bulls may be used for four services per week for short periods. Such heavy service should not extend for more than two weeks for any one period. One bull for 50 to 60 cows is sufficient if the breeding is spread out over the year. One bull to 30 cows is necessary if the cows are bred to freshen largely during one season of the year—that is, if the entire herd must be bred during a two- or three-month period.

SUMMARY

Profitable dairying is achieved when cattle with high inherent producing ability are fed according to their productive ability.

Roughages are the basis of dairy cattle rations. Legumes are recognized as being the best source of roughage, for they are high in minerals, protein, and vitamins. Many native grasses, brome, timothy, fescue, millet, and orchard grass are some of the more common grasses used as forage. The grasses make very acceptable roughage when properly balanced with protein, mineral, and vitamin supplements. Low-quality roughages have little value in rations for producing cows. Fodders from corn and sorghums maye be used successfully if properly supplemented, but they are not considered ideal forages for milking cows. Silage made from legumes, grasses, corn, and sorghums makes excellent roughage. However, it is low in protein unless half or more of it consists of legumes. In certain areas of the United States, citrus pulp is used as a roughage. Beet pulp has a rather wide usage among dairymen. These two feeds are low in protein, but they are higher in digestible nutrients than most roughages.

Pastures are the natural feed for dairy cows and should be extended over the longest possible time.

Corn is generally considered the best grain for dairy cattle, but combinations of other grains also make very good rations.

Soybean oil meal, linseed oil meal, soybeans, cottonseed meal, corn gluten feed, corn gluten meal, distiller's grains, and dried brewer's grains are common high protein dairy feeds. Urea or biuret may be used to replace part or all of the protein in the ration, but they should never exceed 0.3 pound per cow per day. They should be used only in conjunction with high-energy feeds, such as the grains or molasses.

Vitamin deficiencies are seldom a problem when cows have access to good pasture or any high-quality forage. Under some conditions, it is advisable to feed a supplement containing vitamins A and D. Mineral mixtures containing salt, calcium, and phosphorus are essential in most areas. Mineral mixtures recommended for the area should be followed.

The proper feeding of the dairy cow requires that her weight, her milk production,

FIGURE 18-16. A mature bull can be mated to about sixty cows by natural service if the breeding season is spread out over the entire year. By the use of artificial insemination, the same bull could be mated to several hundred cows during the same length of time. (Mann photo. *Courtesy* Farms Breeding Service)

and the butterfat percentage of her milk be known. One of the acceptable rules governing the amount of concentrates to feed should be followed.

Grinding grains reduces waste and facilitates mixing. Grains should be coarsely ground, as fine grinding reduces palatability.

Good pastures minimize feeding problems. However, some forage and grain should be fed to high producing cows on pasture. The amount and kind of forage and grain to feed depend upon the quality and quantity of pasture available.

The cow needs a dry or rest period of from six to eight weeks for replacing body weight lost during her last lactation period. She should be fed a ration in the amount and kind that will put her in the best possible condition for the next milking period.

Just before freshening, the concentrate part of the ration may be partly replaced with more bulky feeds. In the first three days following calving, a mildly laxative ration that is bulky in nature is advisable. Feed should not be reduced severely following calving, for ketosis may develop. The cow should be brought up to full feed as quickly as her appetite will permit after calving.

Managed milking is recommended. A milking routine should be planned and followed. Barns and equipment should be kept clean and free from offensive odors.

The drying-off process may be accomplished by intermittent milking, incomplete milking, or abrupt cessation of milking. Cows with sound udders will generally respond best to abrupt cessation of milking.

At calving time, the cow should be placed in a well-bedded, well-ventilated, clean stall with adequate room for calving. Occasionally, high producing cows may require milking before calving, but the practice is not generally recommended.

Cows may be bred naturally or artificially. Artificial breeding is common, with a conception rate equal to that of natural breeding. Artificial insemination provides the dairyman with the service of good bulls at low cost and eliminates the problems of securing and maintaining a bull.

Fall freshening is most profitable except when the dairyman is retailing milk and must maintain a constant supply.

There are three systems of housing—the loose housing system, the stanchion system, and the free stall method.

A moderate amount of exercise should be provided for cows but excessive exercise may reduce milk production. Cleaning and grooming the animals is essential to their health and a clean product.

The paunch of breeding bulls should be held at a minimum size to facilitate breeding. The size of the paunch may be controlled by the amount of roughage.

Ringing the bull and using a staff when leading him will minimize dangers.

The bull needs plenty of exercise, which can best be provided by constructing an exercise lot adjoining the bull shed. A bull shed that will provide shelter from the cold and shade during warm weather is essential. A breeding chute should be constructed along one side of the exercise pen. Bulls may be used for limited service when 10 to 15 months of age. Mature bulls can service 50 to 60 cows if the breeding is spread out evenly over the year.

QUESTIONS

1. List the common grain concentrates fed to dairy cows.
2. Compare the other common grains with corn as a feed for dairy cattle.
3. List some common protein concentrates used for dairy cow feed.

4 Explain how urea may be used as a protein substitute.
5 What are the precautions that should be taken when urea is being fed?
6 Why are roughages considered the basis of diary cow rations?
7 What are the advantages of legume roughages over other kinds of forage crops?
8 How much milk can a dairy cow produce on good roughages alone?
9 What information is necessary to balance a dairy cow's ration?
10 Show by example how you would determine the amount of protein and total digestible nutrients for any given cow.
11 Show by example how you would determine the protein percentage of a ration.
12 List some concentrate mixtures to be fed with various kinds of roughages.
13 What rule would you follow to determine the amount of concentrate mixture to feed a cow on pasture?
14 Why is pasture so important to the dairyman?
15 Give examples of a good concentrate mixture to fit various kinds of pasture.
16 Why is it usually necessary to provide additional forage for high producing cows on pasture?
17 What are some of the pasture limitations?
18 Give a good rule for feeding concentrates to cows on pasture.
19 Why is it usually necessary to feed concentrates to the dry cow?
20 What and how much would you feed the dry cow four or five days before freshening?
21 What and how much would you feed the cow one to four days after freshening?
22 Why should the same procedure be followed for each milking?
23 Discuss the three common methods of drying off cows.
24 Explain how you would care for the cow at calving time.
25 What are the advantages of natural breeding? of artificial breeding?
26 When should cows calve? Why?
27 What is the average gestation period for different breeds?
28 Why is it unwise to milk a cow dry shortly after freshening?
29 At what time of year is it generally best to have cows freshen? Why?
30 How can you help to insure good results from artificial insemination?
31 How should cows be prepared for milking?
32 Explain the three common systems of housing dairy cows.
33 What are the advantages of each housing system?
34 Give a good ration for the dairy bull.
35 Why should a ring be placed in the nose of the bull?
36 Explain how to ring a bull.
37 How would you feed the mature bull?

38 Why is it important to control the paunch size?

39 Describe a good bull shed and exercise lot.

40 List ways and means by which you would reduce the dangers in handling a dairy bull.

REFERENCES

A Comparison of Free Stall and Loose Housing Systems, Experimental Station Bulletin B-3, Purdue University, Lafayette, Indiana, 1966.

Diggins, R. V., and C. E. Bundy, *Dairy Production,* Englewood Cliffs, New Jersey, Prentice-Hall, Inc., 1969.

Feeding and Management of Dairy Cattle, Extension Service Pamphlet 249 (Revised), Ames, Iowa, Iowa State University, 1963.

Nibler, C. W., *Dairy Herd Management,* Agricultural Extension Bulletin EC 631, Lincoln, Nebraska, University of Nebraska, 1953.

Planning Your Dairy Production Facilities, Extension Service Bulletin A.E.-955 (Revised), Ames, Iowa, Iowa State University, 1964.

Rations for Dairy Cattle and Heifers, Extension Service Pamphlet 318, Ames, Iowa, Iowa State University, 1965.

Reaves, Paul M., *Dairy Cattle Feeding and Management,* Fifth Edition, New York, John Wiley and Sons, Inc., 1963.

Searles, H. R., R. W. Wayne, T. W. Gullickson, and R. D. Leighton, *Feeding the Dairy Herd,* Extension Bulletin 218 (Revised), St. Paul, Minnesota, University of Minnesota, 1952.

19 Feeding and Management of Young Dairy Stock

The first step in the production of dairy calves is to have strong calves at birth. To insure strong calves, the cows should have at least six weeks of rest between lactation, and they should receive a well balanced ration.

FEEDING THE CALF FROM BIRTH TO SIX MONTHS

Since the rumen of the calf does not develop and begin to function until several days after birth, the young calf is unable to manufacture vitamins and protein amino acids. Therefore, young calves are dependent on being fed a ration which contains all the essential food nutrients. The problems in feeding newborn calves are very similar to those in feeding swine and other simple-stomached animals. In addition to vitamins A and D, the B-complex vitamins and vitamin E will need to be contained in the ration. Animal proteins are necessary to insure a complete protein balance.

Antibiotics in Dairy Calf Rations

Recent experiments indicate that Aureomycin and Terramycin will increase the growth rate of dairy heifers without any apparent effect upon reproduction or milk production. Aureomycin fed in the milk at the rate of 80 milligrams of Aureomycin hydrochloride per calf per day from 4 to 116 days of age produced the following results:

1. **Increased the growth gain from 10 to 30 percent.**
2. **Improved the appetite.**
3. **Decreased the number of cases of calf scours.**
4. **Produced smoother hair coats.**

Experiments show that when Aureomycin is fed at the rate of 40 to 50 milligrams daily to calves, it reduces materially the number of cases of scours and other digestive troubles.

Feeding the Calf for the First Three Days

The first milk produced by the cow is known as colostrum. It is produced only for the first few days after calving and is rich in protein, vitamins, and minerals, and contains these food nutrients in the proper balance necessary to meet the needs of the calf. Colostrum contains antibodies that help to protect the calf from diseases and science has been unable, up to the present time, to duplicate it. Calves that fail to get the colostrum are difficult to raise, and losses are generally heavy. Dr. Karl Paul Link, of the University of Wisconsin, has developed a product called *Kafmalak* which is mixed with warm milk and fed to young calves. This product, produced by the Wisconsin Alumni Research Foundation, has reduced scours among dairy calves.

Dairymen are not all agreed as to the best way to provide the colostrum to the calves, but most seem to favor leaving the calf with the cow for the first three days. This practice insures the calf of getting the colostrum at body temperature and free from contamination.

FIGURE 19-1. A strong healthy calf is the result of good feeding and management. (*Courtesy* Holstein-Friesian Association of America)

Any additional colostrum produced by the cow in excess of what the calf needs may be fed to older calves. Some dairymen freeze enough colostrum to feed a calf the first four days following birth in case it is needed. It is a waste of highly nutritious feed to throw it away.

Systems of Feeding Dairy Calves

There are four common systems of feeding calves from the third day until they are nine weeks of age. The systems are : (1) liberal milk feeding, (2) using milk replacers, (3) limited milk feeding plus a dry calf starter, and (4) the nurse cow method.

Liberal Milk Feeding. Maximum growth can be obtained when the calf is fed liberally on milk. The common procedure, when this system is followed, is to feed whole milk at the rate of a pound to each 8 to 10 pounds of body weight for the first three to four weeks, then gradually to switch to skim milk (the same amount in proportion to body weight) until the calf is getting from 14 to 18 pounds of milk daily. No more than this amount should be fed. It is important for calves not to be overfed on milk. The milk should be fed at from 90° to 100° F. Clean pails should be used. If scouring is observed, the amount of milk should be reduced. Fifty milligrams of Aureomycin may be added to the daily milk feed. When calves are kept inside and are not exposed to direct sunlight, there is some danger that a vitamin D deficiency will develop. Whole milk will generally provide adequate amounts of vitamin A. Since the vitamin A content of milk is contained in the butterfat, additions of a vitamin A supplement when skim milk feeding is started will provide insurance against vitamin A deficiencies. Many dairymen follow the practice of adding a few drops of fish liver oil or some other vitamin A and D concentrate to the daily milk ration to prevent deficiencies of these vitamins.

Using Milk Replacers. On dairy farms where production costs are high and the product is sold as whole milk, many dairymen are turning to milk replacers as a means of lowering the cost of raising young stock. Milk replacers usually consist of a dry feed mixture that is reconstituted with warm water and fed as a replacement for milk. The successful raising of calves on milk replacers depends very largely on how nutritionally complete the product is. Many dairymen are reporting excellent success with milk replacers such as those below.

When a milk replacement similar to those listed is to be used, the calves should be fed the cow's milk for the first five days. Table 19-1 will serve as a guide for using milk replacers starting the sixth day.

Calves should be fed twice daily. After the calves are 50 days old, the milk replacer may be gradually reduced, and it may be discontinued when the calves are two months of age. Several feed companies manufacture milk replacement mixtures. If a commercial replacement is used, the manufacturer's directions should be followed.

Dried skim milk mixed with water at the rate of a pound of powder to 9 pounds of water will give good results if the change from whole to powdered skim milk is made gradually over a ten-day period. Dried buttermilk may be used instead of skim milk. When calves are fed either of these products, they should not receive more than a pound

Milk Replacers

Ingredients	1 (%)	2 (%)	3 (%)	4 (%)	5 (%)	6 (%)
Dried skim milk	48.08	42.73	98.73	40.08	86.73	61.73
Dried buttermilk						25.00
Dried whey	12.50	22.00		8.50		
Distillers dried solubles	15.00	15.00		15.00		
Oat flour	5.15			5.15		
Blood flour	10.00			10.00		
Dextrose	7.00			7.00		
Fat		18.00		12.00	12.00	12.00
Vitamin A supplement *	.70	.70	.70	.70	.70	.70
Irradiated yeast **	.02	.02	.02	.02	.02	.02
Minerals						
Dicalcium phosphate (or other suitable source of calcium and phosphorus)	1.00	1.00		1.00		
Trace	.15	.15	.15	.15	.15	.15
Antibiotic feed supplement ***	.40	.40	.40	.40	.40	.40

* 5,000 IU/gram, 15,900 IU/lb. replacer.

** 4 million IU Vitamin D/lb., 800 IU/lb. replacer.

*** Should supply at least 20 mg. of antibiotic/lb. replacer. Aureomycin (chlortetracycline) and Terramycin (oxytetracycline) seem to be the most satisfactory for milk replacers for calves.

Source: Extension Service pamphlet 253, Iowa State University.

TABLE 19-1 DAILY FEEDING RATE OF MILK REPLACEMENT

	Milk (Pounds Daily)		Water Added to Milk (90° to 100°) (Pounds Daily)		Milk Replacement (Pounds Daily)	
Age (Days)	Large Breeds	Small Breeds	Large Breeds	Small Breeds	Large Breeds	Small Breeds
6–8	5	4	4	2	0.6	0.4
8–10	3	2	7	6	0.8	0.6
11–20	0	0	9	8	1.0	0.8
21–40	0	0	12	10	1.2	1.0
40–50	0	0	14	12	1.4	1.2

of reconstituted milk for each 10 pounds of body weight daily. Antibiotic and vitamin supplements recommended for skim milk feeding will also apply when dried skim milk or dried buttermilk is fed.

Limited Milk Feeding Plus a Dry Calf Starter. When calves are approximately ten days old, they will start to eat dry feeds. On many dairy farms where the whole milk is sold, calves will be started on whole milk and gradually shifted to a calf starter. Good calf starters are nutritionally adequate and will supply a sufficient amount of the various nutrients for the growth and development of the calf. When this system is followed, as little as 250 to 350 pounds of whole milk fed over a period of six to seven weeks is used. The calves should be fed whole milk at the rate of a pound for each 10 pounds of body weight for the first three weeks following the colostrum feeding period. The dry calf starter should be made available to the calf at ten days of age. The calf can be induced to eat at an earlier age if some of the starter is rubbed on its nose and mouth after each milk feeding.

FIGURE 19-2. A calf can be induced to eat starter at an early age if some of it is rubbed on the nose and mouth. (*Courtesy* University of Wisconsin)

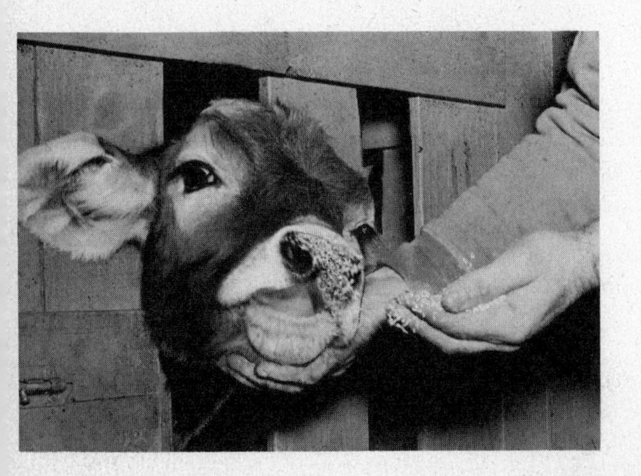

When the calf is four weeks old, it should be eating from 0.5 to one pound of starter daily, and no additional increase in the milk feeding is necessary. During the fifth week, the milk should be reduced by 2 pounds and the calf given free access to the starter. A reduction of the milk by another 2 pounds may be made during the sixth week, and the milk may be discontinued when the calf is seven weeks old. The calf should be given free access to up to 5 pounds of the starter daily. When the 5 pound level has been reached, a less expensive feed may be substituted. On the following page are four suggested calf starters:

Nurse Cow Method. Calves that are to be grown out rapidly, such as veal calves or

Ingredients *	Starters 1 (%)	2 (%)	3 (%)	4 (%)
Corn (coarse ground)	40.0	28.0	16.5	20.0
Oats (crushed)	30.0	30.0	20.0	22.0
Soybean meal	28.0	30.0	17.0	20.0
Wheat bran	—	—	15.0	10.0
Linseed meal	—	—	10.0	—
Distillers dried solubles	—	—	—	15.0
Dried whey	—	—	10.0	—
Molasses (blackstrap)	—	10.0	5.0	10.0
Dehydrated alfalfa meal	—	—	5.0	—
Steamed bone meal, dicalcium phosphate or other suitable source of calcium and phosphorus	1.0	1.0	1.0	2.0
Salt	1.0	1.0	0.5	1.0

* In some cases the following additions also may be desirable:
Vitamin A supplement to supply 5,000 IU vitamin A per lb. starter.
Vitamin D supplement to supply 400 IU vitamin D per lb. starter.
Antibiotic supplement to supply at least 10 mg. chlortetracycline or oxytetracycline per lb. starter.

Source: Extension pamphlet 253, Iowa State University.

purebred animals intended for early sale, may be put on a nurse cow. This method is more expensive but reduces the danger of calves' digestive disturbances. Sometimes two, or even three, calves are placed on one cow. The usual practice is to turn the cow and calves together morning and evening so the calves may nurse.

Purebred breeders often retain old foundation cows in the herd to produce calves long after their usefulness as milk producers is over. Many times one or more quarters of the udder no longer function. Such cows often make good nurse cows.

Grain Feeding. When a calf starter is used, no other grain is necessary for the first six to seven weeks. It is generally recommended that calves on nurse cows, and on methods featuring a liberal supply of milk or milk replacers, be fed a grain mixture starting when they are ten days old. Some recommended mixtures with good legume forage and lower quality roughage that may be fed to growing calves follow:

When Good Legume Roughage Is Fed:

1. 50 pounds ground shelled corn or ground sorghum grain
   48 pounds ground oats
   1 pound iodized salt
   1 pound steamed bone meal
   1 gram Aureomycin †
2. 25 pounds ground shelled corn
   25 pounds ground barley
   25 pounds ground sorghum grain
   23 pounds ground oats
   1 pound iodized salt
   1 pound steamed bone meal
   1 gram Aureomycin †

When Low-Quality Roughage Is Fed:

1. 30 pounds ground shelled corn or ground sorghum grain
   20 pounds ground oats
   10 pounds dehydrated alfalfa meal
   20 pounds wheat bran

TABLE 19-2 GUIDE FOR FEEDING CONCENTRATES TO DAIRY CALVES

Weight of Animals	Concentrates Daily (Pounds) with Good Legume Roughage	Concentrates Daily (Pounds) with Fair Quality Roughage
100	Free Choice	Free Choice
150	2	3
200	3	4
250	3.5	4.5
300	4	5
350	4	5.5
400	4	6

† May be any recommended antibiotic supplement for calves, fed to provide 1 gm. of antibiotic per 100 lbs. of total feed.

When Low-Quality Roughage Is Fed (cont.):

17 pounds linseed or soybean oil meal
1 pound salt
1.5 pounds steamed bone meal
0.5 pound trace minerals
2 ounces irradiated yeast
1 gram Aureomycin †

2. **30 pounds ground barley**
20 pounds ground sorghum grain
20 pounds wheat bran
15 pounds dehydrated alfalfa meal
12 pounds soybean or linseed oil meal
1.5 pounds steamed bone meal
0.5 pound trace minerals
1 pound salt
2 ounces irradiated yeast
1 gram Aureomycin †

After the fifth week, the concentrate mixture should be fed according to the weight of the growing calf.

Feeding Roughages to Calves. Calves will start to eat hay at about ten days of age. Choice green, leafy legume, or legume-grass mixed hay is best. Remove any leftovers from the rack and provide new hay each day. Silage is good feed for calves after they reach the age of four months. Calves like silage, but it should be limited to 3 pounds daily per 100 pounds of body weight until they are 4 months old.

† Any recommended antibiotic supplement for calves. The amount should be sufficient to furnish approximately 1 gm. of antibiotic per 100 lbs. of feed mixture.

Pasture. Calves under six months of age will generally do better on dry lot than on pasture since they do not have the capacity to secure the needed nutrients from pasture alone. If the calves are turned on pasture, they should receive some good quality dry forage and the recommended amount of concentrate for their body weight.

Water. Calves should have free access to clean, fresh water after they are three weeks of age. Water is especially important when calves are fed on a dry starter as the milk is reduced. Young, growing animals have high water requirements.

Minerals. If the growing calf is fed on legume hays and a complete concentrate mixture, it will seldom suffer from a shortage of minerals. As a precaution, equal parts of salt and steamed bone meals should be kept before the calves.

FEEDING HEIFERS FROM SIX MONTHS TO FRESHENING AGE

Although it is important to feed heifers sufficiently to produce normal growth, it is

TABLE 19-3 AGE, WEIGHT, HEIGHT, AND CHEST GIRTH STANDARDS FOR GROWING DAIRY HEIFERS

Age	Holstein			Ayrshire			Guernsey			Jersey		
	Weight (Pounds)	Height at Withers (Inches)	Chest Girth (Inches)	Weight (Pounds)	Height at Withers (Inches)	Chest Girth (Inches)	Weight (Pounds)	Height at Withers (Inches)	Chest Girth (Inches)	Weight (Pounds)	Height at Withers (Inches)	Chest Girth (Inches)
Birth	92	29.5	30.0	72	27.5	29.0	67	27.5	29.0	54	25.5	26.0
1st month	102	30.0	32.0	82	28.0	31.0	76	28.5	30.0	66	26.5	28.0
2nd month	138	32.0	35.0	114	30.5	35.0	98	30.0	33.0	92	29.0	31.0
3rd month	186	34.0	38.0	157	32.5	38.0	138	32.0	37.0	130	31.0	35.0
4th month	251	36.5	43.0	218	35.0	42.0	182	34.0	40.0	181	33.5	39.0
5th month	307	38.5	46.0	280	37.0	46.0	234	35.5	43.0	230	35.5	42.0
6th month	369	40.5	49.0	328	38.5	48.0	289	37.5	46.0	274	36.5	44.0
7th month	429	41.5	52.0	389	40.0	51.0	338	39.0	49.0	324	38.0	47.0
8th month	492	43.0	54.0	441	41.0	53.0	390	40.0	51.0	365	39.0	49.0
9th month	553	44.5	57.0	486	42.0	55.0	437	41.0	53.0	407	40.5	51.0
10th month	613	45.5	59.0	512	42.5	56.0	468	42.0	54.0	447	41.5	53.0
11th month	645	46.0	60.0	556	43.5	58.0	513	43.0	56.0	491	42.5	55.0
12th month	701	47.0	62.0	587	44.0	59.0	566	44.0	58.0	515	43.0	56.0
13th month	762	48.0	64.0	643	45.0	61.0	592	44.5	59.0	560	43.5	58.0
14th month												
15th month	829	49.0	66.0	700	46.0	63.0	667	46.0	62.0	613	44.5	60.0
16th month												
17th month	898	50.0	68.0	754	47.0	65.0	727	46.5	64.0	667	45.5	62.0
18th month												
19th month	968	51.0	70.0	824	48.0	67.0	761	47.0	65.0	703	46.5	63.0
20th month												
21st month	1,044	52.0	72.0	871	48.5	68.0	827	48.0	67.0	763	48.0	65.0
22nd month												
23rd month	1,122	53.0	74.0	930	49.5	70.0	901	49.0	69.0	801	48.5	66.0

Extension Circular E.C. 622, Lincoln, Nebraska, University of Nebraska.

not economical, nor is it in the best interests of producing good cows, to overfeed them. Table 19-3 shows the normal growth of dairy heifers of four common breeds.

Feeding Dairy Heifers on Pasture

Dairy heifers six months of age or older may be turned on pasture. The shift from dry feeds to pasture should be made gradually, and some good dry forages as well as grain should be provided during the first two weeks. Dry forage may be discontinued after two weeks if the pasture is good, but some grain should be provided. The amount of grain need not exceed 0.5 pound per 100 pounds of body weight with a maximum of 4 pounds. Any of the common grains or grain mixtures used for dairy cattle feeding are sufficient. When the pasture contains legumes, it is not necessary to feed protein concentrates, unless the pasture plants become dry and mature. Below are examples of concentrate mixtures that may be fed successfully to dairy heifers on pasture.

With Good Legume Pasture:

	Pounds
1. Oats (ground)	**300**
Corn (cracked)	**200**
2. Barley (ground)	**250**
Grain sorghum (ground)	**250**
3. Oats (ground)	**200**
Wheat bran	**50**
Corn (cracked)	**250**

With Dry Matured Pasture:

	Pounds
1. Oats (ground)	**275**
Corn (cracked)	**200**
Soybean, linseed, cottonseed, or peanut oil meal	**25**
2. Oats (ground)	**200**
Corn (cracked)	**200**
Wheat bran	**100**
3. Barley (ground)	**250**
Grain sorghum (ground)	**225**
Cottonseed, soybean, linseed, or peanut oil meal	**25**

Feeding Dairy Heifers in Dry Lot

Dairy heifers that are receiving all the high-quality legume hay they will consume, or legume hay plus good silage, need very little grain for normal growth. Concentrate mixtures recommended for heifers on pasture may be used for those in dry lot. When high protein forage is fed, protein concentrates are not necessary. If the forage is low in protein content, the concentrate should contain from 12 to 15 percent protein. Citrus or beet pulp may replace all or part of the oats and barley. Citrus and beet pulp are lower in protein than oats, so some additional protein concentrate should be added to the mixture.

Feeding the Dairy Heifers Two Months Before Freshening

If the dairy heifer is to give a good account of herself during the first lactation period, she will need to be properly conditioned. Usually from 6 to 10 pounds of fitting ration—used for conditioning dry cows (see Chapter 18)—is fed daily to heifers for about two months before freshening. They should have plenty of high-quality forage.

MANAGEMENT OF DAIRY CALVES

Losses range from 13 to 17 percent of live calves during the first four months after birth, and most of these result from pneumonia, calf scours, and naval infections. Proper management will eliminate a large percentage of calf deaths due to these ailments.

Care of the Newborn Calf

When the calf is born, the mucus and phlegm should be cleaned from the nose and mouth. If the calf does not start to breathe, it should be held by the rear legs and lifted from the floor with the head down. This procedure is a type of artificial respiration and will often produce results if repeated several times. Alternate compression and relaxation of the chest will often start the calf breathing when other methods fail.

As soon as the calf is breathing properly, the navel should be disinfected by squeezing out the navel cord and painting the navel with iodine. If the cow does not lick her calf dry, or if the weather is cold, the dairyman should wipe the calf to hasten drying.

The next step is to remove all cleanings and wet bedding from the pen and to wash the cow's udder with a chlorine solution to prevent any infectious organisms from being taken in through the calf's mouth during the first feeding. If the calf is normal, it will stand and suck within 30 minutes. If it fails to do so, it should be given assistance in getting a feeding of the first, or colostrum, milk.

Within two hours after the calf has had its first feed, its bowels should move to eliminate the material accumulated in the digestive tract before it was born. If this does not pass in due time, an enema consisting of ½ teaspoonful of soda in a quart of warm water should be given.

FIGURE 19-3. (*Above*) A healthy group of calves in good quarters. (*Courtesy* Holstein-Friesian Association of America) **FIGURE 19-4.** (*Below*) Disinfecting the navel of a newborn calf. (*Courtesy* Michigan State University)

Calf Pens

Well-lighted, well-ventilated, clean, dry pens, adequately bedded and free from drafts, are essential to successful calf raising.

If adequate room is available, individual solid-wall pens for calves up to eight to ten weeks of age have many advantages. The solid wall helps to prevent pneumonia, for it decreases drafts. Calves in individual pens cannot suck each other, a habit common among calves raised in groups. A recommended size is 24 square feet per pen with removable 8 × 10 inch feed boxes 6 inches deep.

When it is not possible to provide individual pens for calves, they may be raised in groups in pens equipped with stanchions for each calf. The calves may be fed milk in the stanchion and left tied up for 15 to 20 minutes after each feeding. This period immediately after milk feeding is the time when sucking of each other is common. By keeping them tied up for a few minutes, the sucking will be prevented and there will be less danger of accumulating hair balls in their stomachs. Feed boxes should be placed 20

FIGURE 19-5. Calves in individual pens cannot suck each other, a habit common among calves raised in groups. (*Courtesy* Babson Bros. Co.)

FIGURE 19-6. Weighing the milk prevents over-feeding of calves. (*Courtesy* University of Wisconsin)

inches from the floor, away from the waterer, and should be long enough to provide 2 feet of space per calf. Boxes that may be removed for cleaning are desirable.

Dehorning

Horns that have been well trained and polished add to the attractiveness of show cattle of the horned breeds. However, it is recommended that commercial dairymen remove the horns. Calves that are from three to ten days old can be dehorned more easily and with less danger of loss than can older cattle. The horn button does not become attached to the skull until the calf is past ten days of age. Several methods may be used to remove the horn; those recommended for beef cattle are suitable for dairy cattle. (See Chapter 13.)

Marking Calves

The marking of calves for identification is an important management procedure. For registration of purebreds, calves must be positively identified. (See Chapter 13.)

Removing Extra Teats

Dairy heifers often have extra teats in addition to the four normal ones. These extra teats should be removed when the calf is between one and two months old. The best method for removing an extra teat is to disinfect the area around the teat and clip it off with a pair of scissors. Usually it will bleed very little. If bleeding is considerable, holding a cotton pack over the wound for a few minutes will stop it.

Breeding and Calving Records

Breeding and calving records are essential to good dairy management. The dairyman should be able at any time to determine

BREEDING AND CALVING RECORD

Name or Number of Cow	Sire	Date Bred	Date Due	Date Born	Sex	Identification of Calf	Disposition
Molly No. 27	Chief 41683	Oct. 1, 1970	July 11, 1971	July 12, 1971	F	MA 42	Sold Jan. 17, 1973 to John Smith

the sire and dam of a calf. He should also know when the calf was born. Breeding dates of heifers and cows and expected calving dates are important. There are many different types of forms one may use for this purpose. The preceding form is an example of a simple record adequate for many herds.

Age to Breed

Size rather than age is more important in determining the best breeding time for a dairy heifer. Breeding undersized animals is never a profitable practice; they may be stunted or slow to reach maximum size. Small heifers are more apt to have difficulty in calving. Table 19-4 will serve as a guide for determining the size and age to breed heifers.

SUMMARY

Since the rumen of the calf does not develop and begin to function until several days after birth, the problems of feeding the calf are similar to those of feeding simple-stomached animals. Most of the vitamins and proteins have to be provided in the ration.

Antibiotics have proven beneficial in reducing calf scours and increasing growth rates when fed in small amounts daily.

FIGURE 19-7. (*Left*) Identification by neck straps and numbers. (*Right*) Ear tagging. (*Courtesy* University of Wisconsin)

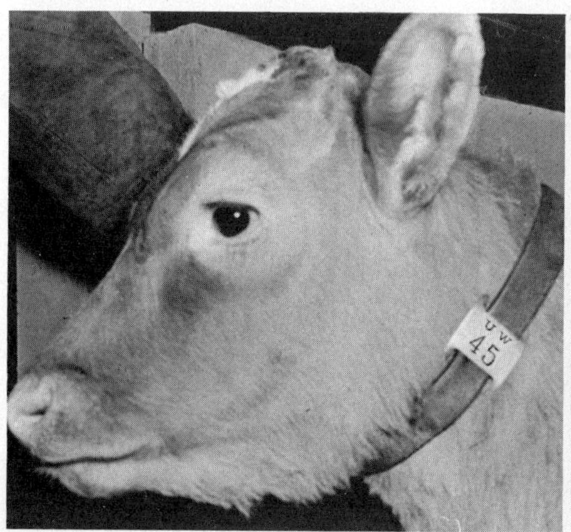

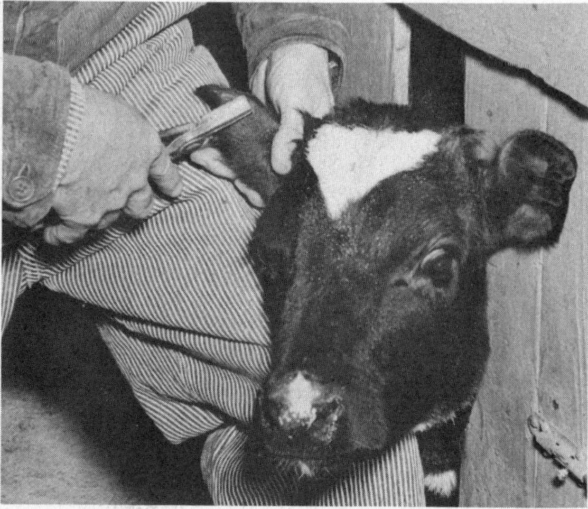

TABLE 19-4 AGE AND SIZE TO BREED DAIRY HEIFERS

Breed	Age (Months)	Weight	Heart Girth (Inches)
Holsteins	18	850–900	67
Brown Swiss	18	800–850	65
Ayshires	18	750–800	64
Guernseys	17	700–750	62
Jerseys	16	600–650	59

The calf should receive the colostrum milk for the first three days. After this, any one of several systems may be successfully followed. They are: (1) the liberal milk system, (2) using milk replacements, (3) limited milk feeding plus a calf starter, and (4) the nurse cow method. When a calf starter is used, no additional grain is necessary for the first six to seven weeks. Calves being raised on the liberal milk, milk replacement, or nurse cow methods are generally started on a concentrate mixture and fed according to body weight until they are five to six months old.

It is important to provide good roughage for young calves. Until calves are six months old, they do better in dry lot.

Water is very important to calf growth and development.

Dairy heifers six months old or older need very little grain if plenty of good pasture or other high-quality roughage is used.

The newborn calf should have the phlegm and mucus cleaned from its mouth and nose, be dried off, have its navel disinfected, and be allowed to nurse. The pen should be clean, warm, well lighted, and well ventilated.

Dehorning should be done at an early age. Caustics, hot irons, or spoons may be used successfully if calves are dehorned early. Clippers or saws, especially prepared for that purpose, are commonly used to dehorn older cattle. Calves should be marked for identification. Extra teats should be clipped off with a pair of scissors when heifers are between one and two months of age. Complete breeding, calving, and disposal records should be kept for the entire herd.

Normal healthy calves should gain from a pound to 1½ pounds daily up to six months of age. The rate of gain will vary with the breed. Size rather than age should be considered when breeding dairy heifers.

QUESTIONS

1. What are the principal differences in the digestive system of calves and older cattle?
2. Why are animal proteins and most of the vitamins necessary in the ration of young calves?
3. What effect do antibiotics have on the growth and development of calves?
4. What is the recommended level for feeding antibiotics?

5 How should the calf be fed the first three days?
6 Describe the common systems of feeding dairy calves until they are five to six months old.
7 What are the advantages of each system listed in Question 6?
8 Give a good formula for a milk replacer, a milk substitute, and a regular concentrate ration for calves getting liberal amounts of milk.
9 At what age will calves start to eat grain and hay?
10 What kind of hays would you recommend for calves?
11 At what age would you recommend turning calves on pasture?
12 Give a concentrate mixture for calves on good pasture and one for calves on fair pasture.
13 What quantity of concentrates should growing heifers receive daily?
14 What percentage of calves die during the first four months after birth?
15 How can calf losses be reduced?
16 What advantages do individual pens for calves have over group pens?
17 When group pens are used for calves, how can sucking of each other be prevented?
18 Give the methods of dehorning calves.
19 Why is sanitation important in calf raising?
20 Give a sanitation program that you would recommend.
21 What are the normal growth rates for heifer calves?
22 Explain the different systems of marking calves for identification.
23 Why should breeding and calving records be kept?
24 What information would breeding and calving records give?

REFERENCES

Dairy Facts For 1966, Extension Service of Kansas State University, Manhattan, Kansas, 1966.

Feeding Dairy Heifers, Extension Service Pamphlet 302, Ames, Iowa, Iowa State University, 1965.

Raising Dairy Calves To Breeding Age, Circular 174, University of Missouri, Columbia, Missouri, 1966.

Raising Dairy Calves, Extension Service Leaflet 302, Rutgers University, New Brunswick, New Jersey.

20 Marketing Dairy Products

The profit from dairy production is determined by the economy of production and by the selling price. Feed makes up about 50 percent of the cost of production; labor makes up about 33 percent. Efficient production practices alone, however, will not insure a profitable enterprise. The dairy products must be of such quality and produced at such a time that the consumer market will absorb the products at a reasonably high price. The form in which the products are sold may also affect the net income from the enterprise. Milk may be sold as Grade A, Grade B, or Grade C, or as ungraded milk. The milk may be separated on the farm, the milkfat being sold and the skim milk being used on the farm as feed.

A farmer can increase the income from the dairy enterprise through improved marketing practices by (1) planning production so that large quantities of milk or milkfat can be marketed during the seasons of high prices, (2) selecting the form in which to sell the products so as to yield the largest net return, and (3) improving the quality of the products to be sold.

SEASONAL PRICE TRENDS

Milk Prices

A common practice on many farms is to have the cows freshen in the spring, so that they make maximum use of pastures. This system minimizes the need for feed, labor, and housing during the winter months, but it encourages heavy production of milk and milkfat in the spring and early summer. Since a number of the dairy cows in this nation are kept by farmers with small herds, we have more production of milk in the spring and light production during the winter months. As a result, the price of milk is low in April, May, and June and high in October, November, and December.

A summary of milk production in the nation by months during the 1970–1972 period

TABLE 20-1 MILK PRODUCTION BY MONTHS, U.S., 1970–1972

Month	1970	1971	1972	Average
		Billion Pounds		
January	9.5	9.6	9.7	9.6
February	8.9	9.0	9.4	9.1
March	10.1	10.2	10.4	10.2
April	10.3	10.4	10.7	10.5
May	11.1	11.2	11.3	11.2
June	10.8	10.8	11.0	10.9
July	10.2	10.3	10.5	10.3
August	9.8	9.9	10.0	9.9
September	9.3	9.4	9.4	9.4
October	9.3	9.4	9.5	9.4
November	8.8	8.9	9.0	8.9
December	9.4	9.4	9.4	9.4
TOTAL	117.4	118.5	120.2	118.7

Source: U.S.D.A., *Dairy Situation*, May 1973.

FIGURE 20-1. Average prices received by dairies for milk, 1970–1972 period. (*Courtesy Dairy Situation*, 1973)

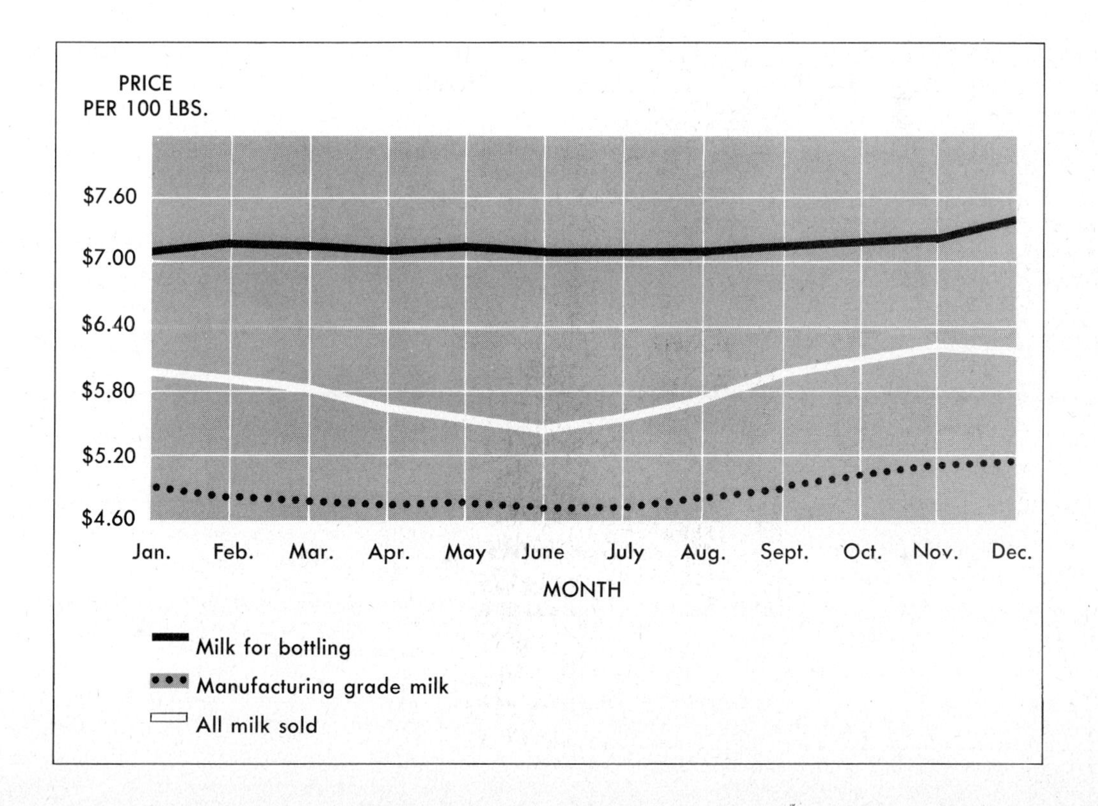

is shown in Table 20-1. Heaviest production was during April, May, and June. Lightest production was during September, October, and November. Thirty-seven percent of the milk was produced during the April through July period. The price of milk, as shown in Figure 20-1, did not vary greatly, due to the price support program of the federal government. Federal milk orders and government purchases of dairy products have eliminated much of the seasonal variation in price paid for milk.

Usually about 40 percent of a cow's yearly production takes place in the first four months after calving. It is a good idea to breed the cows so that 40 percent of the total production can be sold during the months of high prices. As shown in Figure 20-1, milk sold for bottling in May during the period from 1970–1972 brought an average of $7.10 per hundredweight, whereas milk sold in December for the same period averaged $7.21. The difference in the selling price of milk produced in one month by a herd of 40 cows producing an average of 40 pounds per day amounted to $52.80. The difference in selling price was 11 cents per hundredweight.

Milkfat Prices

A number of factors have affected the price of butter since 1940. The competition of oleomargarine has lowered the demand for butter and its price; the governmental price stabilization programs have also affected butter prices.

During the 1968–1970 period, there was only a one-cent variation in price of milkfat sold in cream during the early summer and early winter months. Milkfat sold for an average of 69 cents during May, June, and July, and 70 cents during October, November, and December. As shown in Figure 20-2, the market price was slightly above the government support price, with the exception of 1970. A rise in support price was made by the government in 1971.

Milk-Feed Price Ratios

Another method of determining the best time to produce and sell milk is to compare the milk-feed price ratios. Milk-feed price ratios compare the cost of the feed necessary to produce one pound of milk to the market value of one pound of wholesale milk.

According to the information in Table 20-2, the highest ratios during the period from 1970 through 1972 were during October and November. The lowest ratios were during the months of May, June, and July. The table indicates the advisability of planning the heaviest milk production for the fall and winter months.

FORM IN WHICH MILK SHOULD BE SOLD

Milk disposal methods vary from area to area of the country and from community to community. In 1972, 95.1 percent of the

TABLE 20-2 AVERAGE MILK-FEED PRICE RATIOS BY MONTHS, 1970–1972

Month	Ratio	Month	Ratio
January	1.79	July	1.67
February	1.77	August	1.70
March	1.73	September	1.76
April	1.70	October	1.82
May	1.66	November	1.83
June	1.63	December	1.75

Source: U.S.D.A., *Dairy Situation,* March 1970; March 1973.

milk marketed in the nation was sold as whole milk, less than 1 percent was sold as farm-skimmed cream, and 1.1 percent was retailed by farmers as milk or cream. About 2.9 percent of the nation's total milk production was used on the farms where it was produced.

Only 0.1 percent of the milk produced in the Atlantic States in 1971 was sold as farm-skimmed cream; more than 95 percent was sold as fluid milk.

In the North Central States, more than 90 percent of the 1971 milk production was sold as whole milk, and 1.5 percent was sold as cream.

Each dairyman must make a careful analysis of the market available in his community and calculate the probable income from the use of various marketing programs. At one time, skim milk was indispensable as a feed for calves, chickens, and pigs. We now have feeds which can take its place, and farmers are more willing now to sell whole milk than they were 20 years ago.

Here are the calculations made by one dairyman to determine the best method of disposing of the milk produced by his herd.

Example:

The herd consisted of 30 cows, which produced an average of 400 pounds of milkfat and 10,000 pounds of milk in a year. The milk could be sold as Grade A milk for human consumption for an average of $6.00 per hundredweight. It could also be sold as manufacturing grade milk

FIGURE 20-2. Monthly market and support prices for butterfat in cream, U.S. 1962–1970. (U.S.D.A. *Dairy Situation,* May 1970)

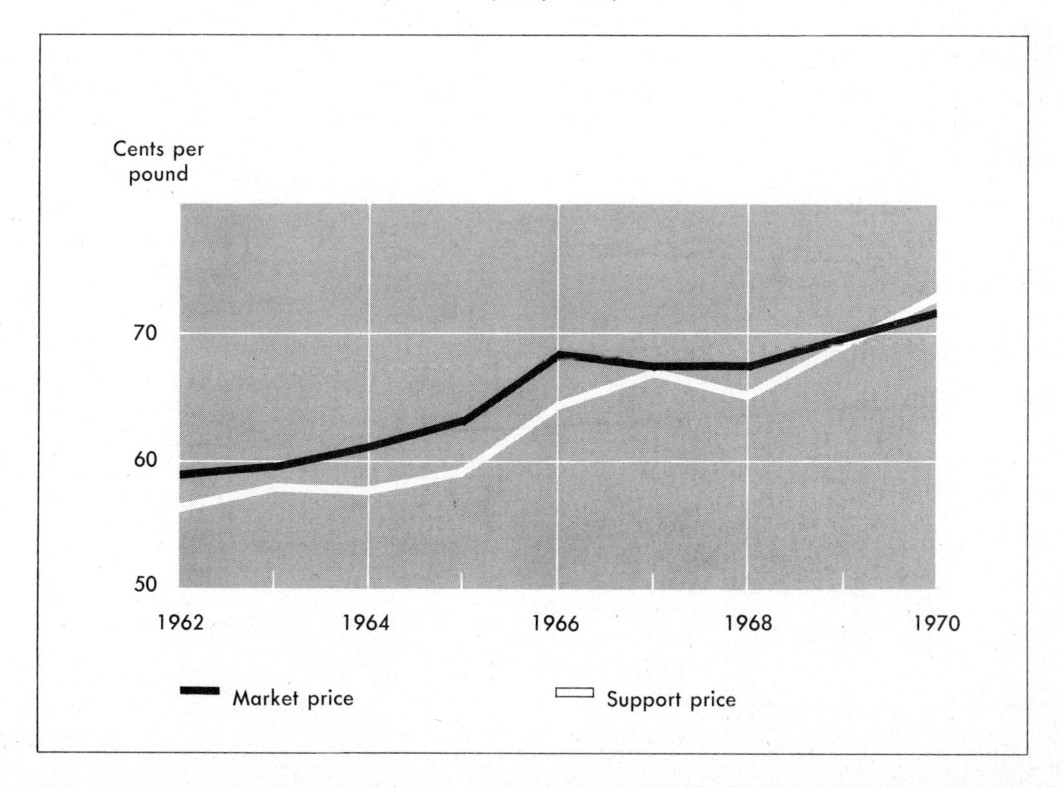

FIGURE 20-3. A pipeline milking system used in producing high-quality milk. (*Courtesy* Babson Bros. Co.)

for making cheese and milk concentrates at an average price of $5.00 per hundredweight. It could be separated and sold as milkfat for 70¢ per pound.

Income:

Plan 1. Sell Grade A milk.
300,000 pounds (10,000 × 30) @ $6.00 per cwt. = $18,000.

Plan 2. Sell as manufacturing grade milk.
300,000 pounds of milk at $5.00 per cwt. = $15,000.

Plan 3. Sell as milkfat and feed skim milk.
12,000 pounds of milkfat
@ $.70 per pound = $8,400.
288,000 pounds of skim milk
@ $.40 per cwt. = $1,152.

Total income from Plan 3 $9,552.

Difference in income between
Plan 1 and Plan 2 $3,000.
Difference in income between
Plan 1 and Plan 3 $8,448.

Under Plan 1, it would be necessary for the dairyman to provide the sanitary quarters, milk room, and milk-cooling facilities required in producing Grade A milk. Under Plans 2 and 3, these facilities are desirable but not required. Under Plan 3, a separator and cooling tank for cream must be provided and pigs, calves, or chickens must be available to consume the skim milk.

The example presents the following important points for consideration in deciding the form in which milk should be sold:

1. **The value of the milkfat must be compared with the values of Grade A and manufacturing-grade milk.**
2. **The use and value of the skim milk must be determined.**
3. **An estimate must be made of the increase in costs involved in producing Grade A milk.**
4. **A study must be made of the availability, cost, and dependability of transportation to market.**

In the case presented, the milkfat price was low in comparison to the price of Grade A milk, and the skim milk value as feed was less than its value on the commercial market for drying or condensing purposes. The price of Grade A milk was sufficiently high to pay the cost of the equipment necessary to produce it.

MARKETS AVAILABLE

The number and types of markets available in a given area are determined by the density of population in the area, the quantity of milk produced, and the cost of transporting the dairy products to market. Table 20-3 summarizes the locations and number of milk-processing plants in the United States in 1970.

IMPROVING THE QUALITY OF DAIRY PRODUCTS

The quality of milk and cream to be used for pasteurization and bottling and for

TABLE 20-3 NUMBER OF MILK-PROCESSING PLANTS IN THE U.S. IN 1970

Region	Butter	Ice Cream	Cheese	Dry Skim and Buttermilk
New England and Middle Atlantic	55	402	134	52
East North Central	125	326	572	68
West North Central	282	185	111	128
South Atlantic	2	120	—	—
South Central	25	192	47	3
Mountain	61	103	35	17
Pacific	57	287	25	40
All other states *	12	12	39	37
Total United States	619	1,627	963	345

* States having fewer than 3 plants not included in regional data, but included in total for the nation.

Source: U.S.D.A., *Production of Manufactured Dairy Products,* 1970

FIGURE 20-4. A large-scale Brown Swiss herd in New Mexico. Climatic factors are favorable for the production of high-quality milk. (Kruse photo. *Courtesy* Brown Swiss Cattle Breeders Association)

manufacturing has been given considerable attention by city health authorities and by milk processors. Most ordinances are based on Grade A milk standards recommended by the U.S. Department of Health, Education, and Welfare.

Milk retailers have found that increased quality stimulates consumption, and that consumers are willing to pay higher prices for milk of high quality. As a result, milk processors and dairies are paying the producers premium prices for high-quality milk. In some cases the premium amounts to as much as $1.50 per hundredweight. The added income from the sale of high-quality milk from a herd of 30 cows producing an average of 10,000 pounds of milk may range from $3,000 to nearly $4,500 per year.

Most creameries pay several cents more per pound for milkfat in sweet cream of superior quality than for milkfat in sour cream of poor quality. Sweet cream butter sells better on the consumer market and can better compete with butter substitutes.

GRADE A MILK

Definition

According to the U.S. Public Health Service Ordinance, Grade A raw milk for pasteurization is raw milk from producer dairies conforming to certain standards of sanitation. The bacterial plate count, or the microscopic clump count of the milk, as it is delivered from the farm, must not exceed 200,000 bacteria per milliliter. (A milliliter is approximately one thousandth of a gallon.) The ordinance provides for the inspection of the cow herd, the milking barn, the milk house, the utensils and equipment, and the milking of the cows, by an authorized public health officer.

Requirements

In order to produce and market high-quality Grade A milk successfully, a number of basic, very important requirements must be met. The paragraphs that follow describe these points in detail.

Healthy Cows. Since disease may spread through the cows to human beings and to other animals, the herd must be accredited or tested for tuberculosis at least once every six years, according to the modified accredited area plan of the U.S.D.A. The herd must also be certified as free from Bang's disease, according to procedures approved by the U.S.D.A. Bloody, stringy, and other abnormal milk produced by cows with mastitis must be excluded.

Sanitation. The presence of bacteria in milk will reduce the length of time it may be kept and may even make it unfit for human consumption.

The cows, especially their udders and flanks, should be cleaned at milking time. The yards should be well drained and clean. The barn floor should be made of concrete and kept clean.

The utensils and milking equipment should be made of tinned iron or stainless steel and kept in good condition.

The milk house should have a concrete floor, sloped for drainage. It should be separated from the barn and have wash and rinse vats, hot water, storage racks for utensils, and a tank for cooling and storing milk. The milk house should be screened and well ventilated.

The water supply for washing utensils and cooling the milk should be free from pollution. Wells should be protected from pollution from septic tanks, sewer lines, and surface water.

Cooling Facilities. Bacteria multiply very slowly in cold milk. Milk should be cooled to 50°F or lower, immediately after milking.

Milk Free from Residues. Pesticide residue in milk is not tolerated and can be costly. Some dairymen have been forced to dump milk for several months because small amounts of pesticides were found in it.

There are at least seven ways in which pesticides can get into the milk: from feed, by direct body contact with insecticides, by inhalation, from contaminated replacement animals, from bedding, from water, and from improper pesticide storage. Dairymen should check carefully to ensure that all pesticides are used properly and that none contaminate the feed produced on the farm. If there is any evidence of pesticide residue contamination, the feed and bedding should be tested immediately. The Wisconsin Alumni Research Foundation at Madison, Wisconsin provides a feed-testing service for that area.

PRODUCING GRADE A MILK

The methods recommended in this section are applicable to the production of both quality cream and Grade A milk.

Maintaining a Healthy Herd

The herd should be tested regularly, according to the ordinance requirements, for tuberculosis, leptospirosis, and Bang's disease. A veterinarian's statement certifying that the herd is free from these diseases is necessary. Cows that have swollen or sore udders, or those that produce abnormal milk, should be removed from the herd.

Housing Facilities

If a stanchion barn is used, it should be clean, adequately lighted, and well ventilated, and have a concrete floor. From 400 to 500 cubic feet of air space is needed for each cow in this kind of barn.

The interior of the barns should be kept free from dust and cobwebs. It should be whitewashed once each year or painted every two years. Manure should be removed from the barn daily. Lime should be used in gutters and on the passageways.

Under the loose housing system, 50 to 75 square feet of floor space should be provided for each cow. Adequate bedding is essential. A milking parlor with a cement floor and proper drainage should be provided.

Milk House

The milk house should be located near the milking parlor or barn, but should not open directly into the barn. The doors and windows should be screened and the doors should be self-closing. The floors should be of concrete and be well drained. The window area should be at least 10 percent of the floor area. The walls and ceiling should be finished and painted. A house 10 × 12 or 12 × 16 feet is recommended, depending on the size of the herd.

Cooling Facilities

Bacteria that cause souring and undesirable flavors in milk grow 10 times as quickly at 90°F as at 50°F. The milk should be cooled as soon as possible after milking. A cooling tank, adequate in size to take care of the volume of milk produced, should be provided. A mechanical cooler is preferred.

The cooler should lower the temperature of the milk to 50° or 55° within an hour.

FIGURE 20-5. A free-stall dairy barn. (*Courtesy* Babson Bros. Co.)

FIGURE 20-6. (*Above left*) This milk house, equipped with a bulk tank, is nicely located. It is separated from the dairy herd and milking parlor and has easy access for the milk hauler. (*Courtesy* U.S.D.A.) **FIGURE 20-7.** (*Above right*) The bulk tank must be large enough to hold at least four peak milkings. This 800-gallon tank is equipped with a cleaning device. (*Courtesy* C. P. Division of St. Regis Co.) **FIGURE 20-8.** (*Below*) A modern milk room with milk-handling and cleaning equipment. (*Courtesy* Babson Bros. Co.)

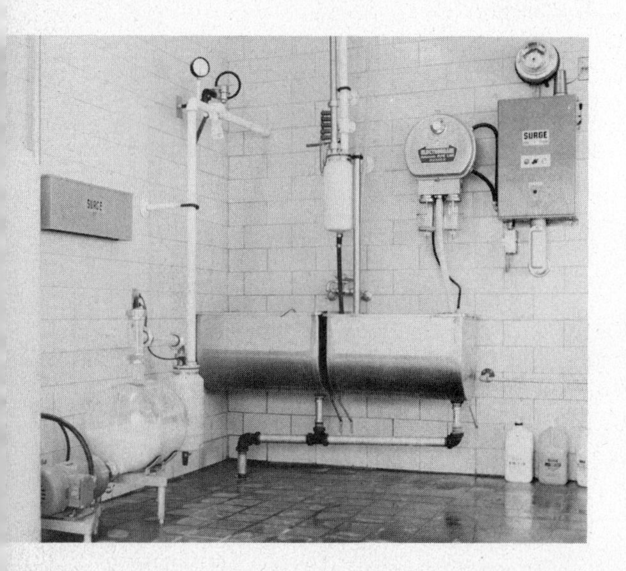

Bulk Tanks

Many farmers have installed bulk milk tanks in the past few years. On January 1, 1970, a high percentage of all producers selling milk in Iowa were using bulk tanks. With bulk equipment, milk is cooled immediately and is held under refrigeration until it is picked up and transported by bulk-tank trucks to the processing plant.

In some states, 90 percent of the market milk is now bulk-handled. On January 1, 1964, only 10,855 bulk tanks were in use in Iowa.

Capacity. In a study of the tanks in use in 39 market milk areas, it was found that the majority ranged from 200 to 400 gallons in size. Most dairymen prefer tanks which will hold two to four peak milkings.

Cost. Bulk coolers vary in cost according to size and quality of materials and workmanship. Prices of tanks, not including housing, range from $1,200 to $5,000. A herd of 30 to 40 cows is usually necessary to make a bulk tank pay.

Utensils

Seamless or electric-welded pails made of noncorrodible metals should be used. Milk strainers with single-service filtering pads and of a type approved by state departments of health or agriculture are recommended. When utensils are not in use, they should be inverted and stored on racks in the milk house.

Cleaning Utensils and Milking Machines

The milking machine teat cups, tubes, pails, and strainer should be thoroughly rinsed immediately after each milking to remove milk film. Following the rinsing, the utensils should be washed with warm water containing a suitable washing compound or wetting agent. Soap should not be used because it leaves a greasy film. The surfaces of the utensils should be brushed with a stiff brush. After washing, the utensils should be rinsed with hot or boiling water. Steam or chemical sterilizers may also be used.

Cleaning the Milking Machine. Be certain that all milking machine parts are accessible for cleaning and are in good repair. Taking the machine apart and brushing all parts during washing is the best way to get them clean. After the teat-cup assembly and tubes have been washed, they may be rinsed with very hot water and hung up to dry, or they may be immersed in a germicidal solution. All parts treated with the solution should be thoroughly rinsed with clean water before they are used again.

Cleaning the Bulk Tank

Stainless steel bulk tanks will last indefinitely if housed in a good, well ventilated milk house and if given proper care. Adequate warm water facilities for cleaning are a necessity.

FIGURE 20-9. Milk equipment cleaning tanks in the milk room. (*Courtesy* Iowa State University)

FIGURE 20-10. The bulk tank is washed, disinfected, and rinsed after milk is removed. Note the cleanliness of this milk room. (*Courtesy* Paul Mueller Co.)

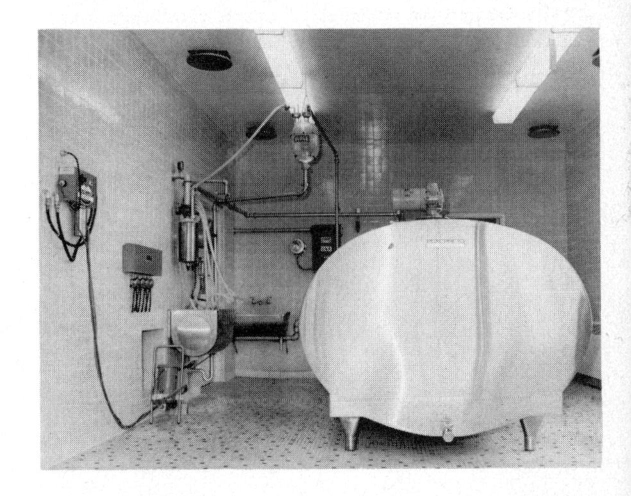

The bulk tank should be rinsed with tap water immediately after use, to remove all milk before it dries on the surface. The rinsing should be followed by cleaning with hot water and a commercial dairy cleaner. The cleaner should be used according to the manufacturer's recommendations, with the cleaning solution brushed over all the surface.

If solid deposits or milkstones appear, they should be removed with a brush and an acid cleaner or a commercial milkstone remover. After cleaning, the tank and equipment should be rinsed with warm water to remove the detergent. A hot water rinse should follow.

A germicidal solution may be applied just before the equipment is to be used again. It should be used according to directions and the equipment rinsed thoroughly before milking begins.

Cleaning the Cows

The udders of the cows should be washed and wiped with a germicidal sterilizer solution just before milking. A dirty udder provides a source of sediment and bacteria. Some dairymen brush their cows daily and clip the long hair from the legs, flank, and udder.

Milking

If milking is done by hand, the hands should be clean and dry. Wet-hand milking is not a good practice. A teat cup should be used before milking, to see whether the milk is normal. Milk produced during the first four to six days after calving should not be marketed.

FIGURE 20-11. Cleaning the cow's udder before milking. (*Courtesy* Babson Bros. Co.)

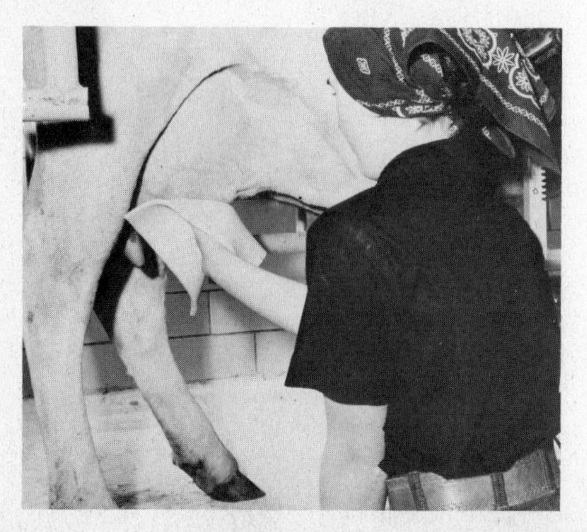

IMPROVING THE QUALITY OF CREAM

In general, the methods recommended in the production of Grade A milk are applicable to the production of cream. The milk should be separated as soon as possible after milking, and the cream should be cooled immediately in the cooling tank or refrigerator. Dust and any undesirable aroma should be kept out of the cream can, but a tight lid should not be used. For maximum profit, cream should be sold as sweet cream.

The cream separator should be housed in the milk house and given the same treatment in cleaning and sterilization that was recommended for the milking machine.

SUMMARY

Dairy income can be increased by: (1) planning production so that milk and milkfat can be marketed during the seasons of high prices, (2) selling the products in the form that will yield the largest net return, and (3) improving the quality of the products to be sold.

About 40 percent of a cow's production is in the first four months after calving. Milk and milkfat prices are highest in October, November, and December and lowest in April, May, and June.

Each dairyman must analyze his production and marketing system to determine whether to sell milk or cream. Most dairymen report greater profits from the sale of Grade A milk.

Considerable differences may exist in the prices offered by various milk processors in the same community for milk of the same quality. The local market must be analyzed carefully. To improve the quality of dairy products, the herd should be free from disease and the barn should be sanitary, ventilated, and adequate in size. The milk house should be sanitary and separate from the barn, and should provide an abundant supply of water. The separator, milking machine, and utensils should be thoroughly washed and disinfected after each milking. The cows' udders should be washed and disinfected and strip cups used before milking.

A milk and cream cooler or a cooling tank should be provided. Dairy products should be cooled to 50°F within one hour after milking.

The bulk cooling and storage of milk are recommended. The cold-wall bulk cooler should be of sufficient size to allow for the expansion of the herd and to hold two to four peak milkings, depending on whether truck pickups are made daily or every other day. The cooler should be conveniently located and kept clean. Bulk coolers vary in cost from about $1,200 to $5,000, depending on size.

The production of Grade A milk requires an additional investment in housing and equipment amounting to $2,000 to $6,000. However, the added income from the sale of Grade A milk usually justifies the investment.

QUESTIONS

1. When should you have your cows freshen to be able to market milk and milkfat at the highest prices?
2. What is meant by the *milk-feed price ratio*?
3. What changes should you make in the seasonal production and marketing of dairy products on your farm?
4. Which is the best way for you to market milk on your farm—as milk or as milkfat? Why?
5. Outline the requirements for Grade A milk production.
6. What changes would you need to make in your production and marketing program to be able to market Grade A milk? What would these changes cost?
7. Outline a program for properly cleaning the milking machine, milking utensils, and cream separator on your farm.
8. Describe a desirable milk house.
9. Outline the requirements of a good bulk cooling tank installation.
10. Outline a program for improving the marketing of dairy products on your home farm.

REFERENCES

Diggins, R. V. and C. E. Bundy, *Dairy Production*, 3rd Edition, Prentice-Hall, Inc., Englewood Cliffs, New Jersey, 1969.

Porter, Arthur R., J.A. Sims, and C. F. Foreman, *Dairy Cattle in American Agriculture*, Iowa State University Press, Ames, Iowa, 1965.

Reaves, Paul M. and H. O. Anderson, eds., *Dairy Cattle Feeding and Management*, 5th Edition, John Wiley and Sons, Inc., New York, New York, 1963.

U. S. Department of Agriculture, *Dairy Situation*, Washington D. C., March 1970, March 1971, May 1973.

———, *Dairy Statistics, 1968*, Washington, D. C., 1968.

———, *Production of Manufactured Dairy Products*, Washington, D. C., 1970.

———, *Milk Production, Disposition, and Income, 1970–1971*, Washington, D. C., 1972.

DUAL-PURPOSE CATTLE

21 Breeding and Management of Dual-Purpose Cattle

Dual-purpose cattle have been bred and developed as intermediate breeds between the exclusively dairy or beef breeds. The breeds of dual-purpose cattle will not equal the dairy breeds as economical producers of milk or the beef breeds as producers of beef. However, they produce more milk than the beef breeds and more desirable carcasses than the dairy breeds.

Dual-purpose cattle are well suited to farms where a monthly income from milk or cream is desired and where there is surplus grain for finishing cattle. Many large farm operators, who are not equipped for specialized dairying and whose cattle cannot be given the care required for high producing dairy cows, have turned to using dual-purpose cattle. These cattle permit a quick shift from dairy to beef or vice versa. Animals to be slaughtered do very well in the finishing lot; experiments have shown that dual-purpose cattle will finish into good to choice slaughter animals. The breeder of dual-purpose cattle need not veal his bull calves, as many dairymen do, but can grow them out and feed them for the market. Cows that have completed their usefulness as milk animals and those culled from the milking herd can be fed a high carbohydrate ration and will bring a good price as beef.

SELECTING BREEDS OF DUAL-PURPOSE CATTLE

The same general principles of breed selection apply for dual-purpose cattle as for other breeds of livestock. They are: (1) personal likes, (2) availability of breeding stock, (3) outlet for surplus animals, and (4) environmental conditions under which the cattle will be raised.

Two breeds of dual-purpose cattle are relatively common in the United States—Milking Shorthorns and Red Poll. Another breed, the Dexter, has not developed in population very rapidly.

Milking Shorthorns

Milking Shorthorn cattle were first developed in England from much the same parent stock as were the beef Shorthorns. In developing the two strains from the original stock, the breeders of the Milking Shorthorns emphasized the development of a breed with both good milking qualities and good beef qualities. The result was a shorthorn cow possessing good milking qualities but lacking some of the excellent qualities of the beef strain.

The Milking Shorthorn is red, white, or any combination of these colors. There are two strains of Milking Shorthorns, polled and horned. The polled animals were developed from mutants—animals that were naturally polled even though they were born of parents having horns. The polled strain has the same characteristics as the horned cattle except for the poll.

Red Poll

Red Poll cattle were developed in England by merging the Norfolk and Suffolk strains of cattle. They are solid red in color, hornless, and quite compact in body form. They are good milkers, and do well in the finishing lot. The Red Poll is popular among dual-purpose cattlemen.

Dexter

Dexter cattle were developed in the southern and southwestern parts of Ireland. They are small in size compared to other breeds. Mature bulls usually do not weigh over 900 pounds, and mature cows weigh about 800 pounds. Some mature animals are less than 42 inches tall. They can be either black or red in color.

Dexter cattle have not developed very rapidly in the United States. Some have been used in crossbreeding programs with other beef breeds.

Selection of Breeding Stock

It has been explained that the objective of the breeder of dual-purpose cattle is to combine good milking and beef qualities in the same animal. In the selection of breeding

FIGURE 21-1. (*Left*) Grand Champion Milking Shorthorn bull, Pinesedge Echo King, owned by Sam Yoder, Shoemakerville, Pennsylvania. **FIGURE 21-2** (*Right*) Grand Champion Milking Shorthorn cow owned by Clampitt Farms, New Providence, Iowa. (Figures 21-1 and 21-2 *courtesy* American Milking Shorthorn Society)

or foundation stock both the beef and dairy qualities must be considered.

Using Production Records in Selecting Breeding Stock. The most reliable indicator of a cow's ability to produce is her production record or the records of her ancestors.

In selecting dual-purpose cattle, both their records of milk production and ability to produce beef should be taken into consideration. Many herds of dual-purpose cattle have been on test in dairy herd improvement associations, and it is not difficult to find animals with good milk production records.

Other Factors to Consider in Selection. Since the dual-purpose animal should economically produce both beef and milk, this ability should be apparent in the animals selected for breeding purposes. The mammary system should be well developed. The body form should be intermediate between the beef and dairy types, the legs and body should be average in bone size and above average in structure, with the top and underline straight. The covering of flesh should be uniform.

SUMMARY

Dual-purpose cattle are adapted to both milk and beef production. The breeds do not equal dairy cattle in milk production nor beef cattle in beef production. Many farmers, not wishing to specialize in either dairy or beef cattle, have found the dual-purpose breeds satisfactory.

There are two common dual-purpose breeds in the United States: Milking Shorthorns and Red Poll. Another breed, the Dexter is not so popular.

In body form, dual-purpose cattle are intermediate between the dairy and beef types.

QUESTIONS

1 Under what conditions would raising dual-purpose cattle be a wise choice?

2 Describe the general type of the dual-purpose cattle.

3 What would you consider in making a selection of foundation stock for starting a dual-purpose herd?

4 What are the common breeds of dual-purpose cattle?

REFERENCES

Baker, Marvel L., V. H. Arthaud, and C. H. Adams, *Feeding Milking Shorthorn Steers,* Agricultural Experiment Station Circular 91, Lincoln, Nebraska, University of Nebraska, 1951.

Peters, W. H., J. B. Fitch, H. R. Searles, and W. E. Morris, *Dual-Purpose Cattle,* Extension Bulletin 203, Minneapolis, Minnesota, University of Minnesota, 1943.

CATTLE DISEASES AND PARASITES

22 Keeping Cattle Healthy

Cattle diseases, parasites, and other ailments may be divided into four groups: (1) external parasites, (2) internal parasites, (3) infectious diseases, and (4) noninfectious ailments. The average cattleman is not a veterinarian; therefore, his chief duties in maintaining a healthy herd are good sanitation, the proper use of sprays and disinfectants, and the vaccination of his herd against those diseases for which vaccines have been perfected, when recommended by a reliable veterinarian.

Most infections and diseases to which cattle are subject are brought into the herd as a result of poor management. The interchange of cattle from place to place grows each year, and this means that livestock health laws will become increasingly strict in the years ahead.

A PROGRAM OF DISEASE AND PARASITE PREVENTION

The following steps are important if a disease- and parasite-free herd is to be maintained.

1. **Bring only clean animals into the herd. Many serious diseases, such as brucellosis, can be detected through a test for the disease. A veterinarian may prevent serious losses if called to examine animals before they go into the herd.**
2. **Drain lots so that they will remain dry and free of stagnant water. Paved lots will aid in keeping cattle out of the mud and filth.**
3. **Isolate all animals that are known to have contagious infections. Newly acquired animals should be isolated until it is reasonably certain that they are free of disease.**
4. **Have the breeding herd tested regularly for brucellosis, tuberculosis, and other diseases for which tests have been developed.**
5. **Vaccinate for diseases common in the community, if a successful vaccine exists.**
6. **Disinfect housing and equipment regularly.**
7. **Treat open wounds and the navel of newborn calves with a reliable disinfectant.**

8. Provide plenty of exercise for the breeding herd.
9. Dust for external parasites, such as flies, and eliminate manure piles and other filth accumulations where flies breed.
10. If cows calve in places other than clean pastures, be sure the area is well bedded and disinfected.
11. Avoid cold floors and drafty housing quarters for young calves dropped during cold weather.
12. Provide clean fresh water at all times. Have your water supply tested for impurities every three years or oftener if necessary.

FIGURE 22-1. (*Above*) This calf is a picture of health. A complete program of disease and parasite control will help to keep him from becoming infected. (*Courtesy* Elanco Products Company)
FIGURE 22-2. (*Below*) Navel wound—a common site of attack by the screwworm—is treated with a smear. (*Courtesy* U.S.D.A.)

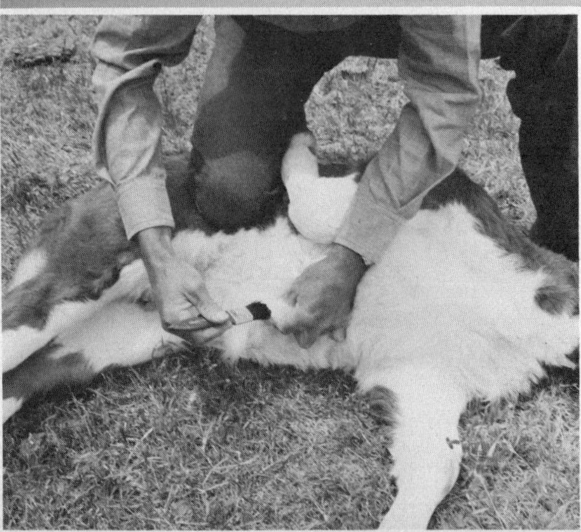

EXTERNAL PARASITES AND PESTS

The Screwworm

Screwworms are most prevalent in the Southern and Southwestern states, where they are one of the foremost causes of cattle losses.

Life History. The screwworm fly is bluish green in color, with orange shading below the eyes. Three prominent dark stripes along its back distinguish it from similar insects. The flies lay their eggs in masses along the edges of open wounds. The eggs hatch into tiny maggots that burrow into the flesh, where they feed from four to seven days.

When the worms have reached their full growth, they drop to the ground and burrow into the soil. A few days later they emerge from the pupa, or dormant stage, as adult flies. The entire life cycle may be completed within 21 days under favorable conditions.

Prevention and Control. The U.S.D.A. has developed several smears that will kill the worms or prevent their infecting wounds. One of these smears is known as Smear EQ 335. CO-Ral used as described for grub control is also effective in the control of screwworms. Due to rigid control, the screwworm has been nearly eliminated in the United States.

Grubs

Cattle grubs, or heel flies, seldom cause death in cattle, but the loss resulting from the uncomfortable effect they produce upon the animals is greater than is generally realized by herd owners. Milk production may be severely reduced and the growth rate of young animals slowed down.

Life History. The heel fly lays its eggs on the lower parts of the animal during the spring and early summer. The eggs hatch, and the tiny grubs bore through the skin and migrate through the animal's body, feeding on the tissues for about nine months. They appear as lumps along the back from early winter until spring. The mature grubs perforate the skin and come out through the openings in the hide, drop to the ground, and soon develop into adult flies. These flies seldom travel more than a mile, so if a cooperative effort is made they can be controlled in any area.

There are two species of cattle grubs, one known as the *common cattle grub* and the other as the *Northern cattle grub*. The life cycles of the two are similar except that the duration is longer for the Northern grub.

Prevention and Control. The timing of grub applications is important because there is a chance that late treatment may involve side reactions such as bloat or possible death. The time to apply treatment varies with the locality and may be learned from the vocational agriculture instructor or the county extension director. A number of products are available, and the manufacturers' directions for application and discontinuation should be carefully followed. Proper administration will destroy the grub soon after it enters the body, thus preventing extensive losses.

Systemic insecticides, although relatively new, have been widely tested and approved for grub control on beef cattle. These chemicals may be applied (1) as dilute sprays over a large area of the body, (2) as more concentrated liquids poured on a strip of the back, or (3) as feed additives. The newest and one of the most effective sprays for grub control is a product known by the trade name of CO-Ral. This material not only controls grubs but also gives effective control for most of the external parasites that affect beef cattle.

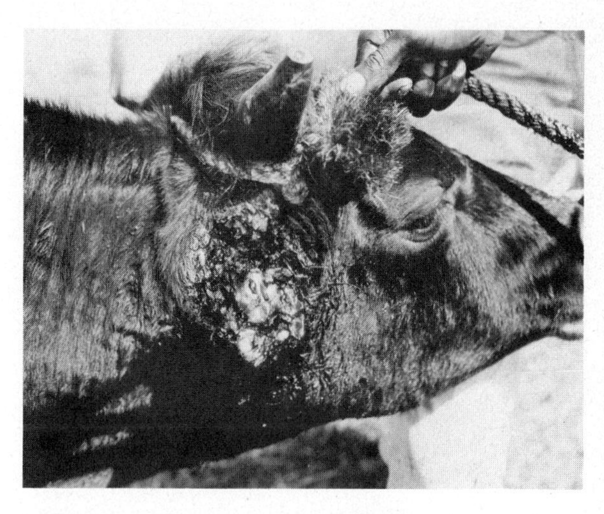

FIGURE 22-3. Screwworm infestation in the ear of a steer. An untreated, grown animal may be killed in 10 days by thousands of maggots feeding in a single wound. (*Courtesy* U.S.D.A.)

A product known as Et-57, or by such trade names as Trolene, Ectoral, and others, is very effective as a control for grubs when given as a bolus shortly after the heel-fly activity has ceased. The dosage is 5 grams of Et-57 for each hundred pounds of body weight. Treatment must not be used within 60 days of slaughter.

Rotenone, applied as a dust or spray, will destroy the grubs in the backs of cattle after the grubs have appeared. The spray is made by using 7½ pounds of 5 percent rotenone wettable powder plus 2 pounds of detergent per 100 gallons of water. Use one gallon per animal and wet the backs thoroughly. One to 1.5 percent rotenone dust is effective if rubbed onto the backs. However, the disadvantage of rotenone is that the grubs have already done most of the damage before they are destroyed.

Warbex (famphur) is another new pour-on insecticide for control of cattle grubs and lice infestations.

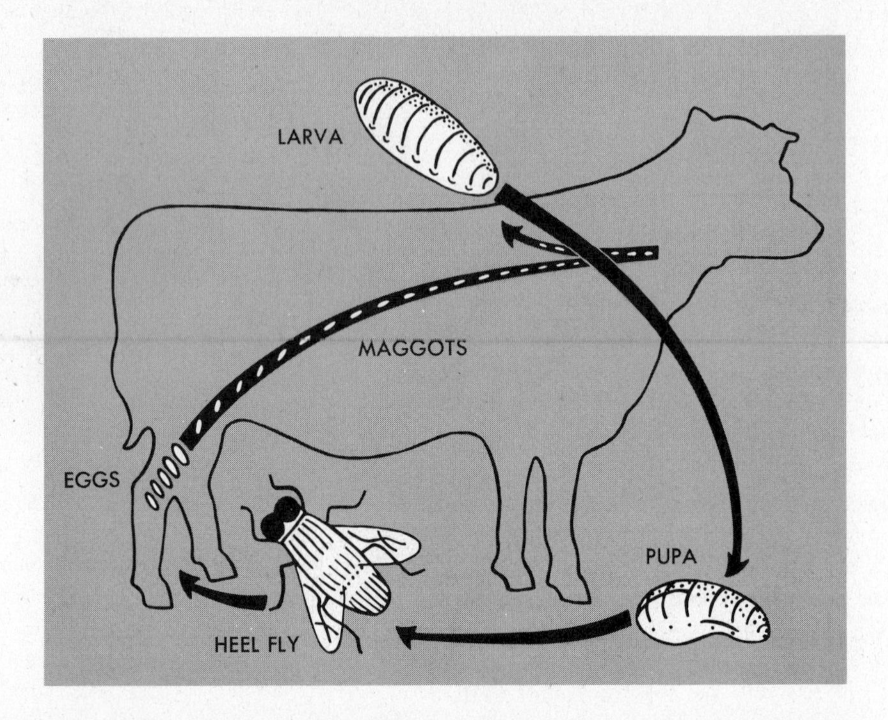

FIGURE 22-4. Life cycle of the heel fly. Heel flies begin to chase cattle during the first warm days of spring. The cattle run frantically to escape these insects, which cannot bite or sting. The female flies lay their eggs on the short hairs of the animal's heels. In a few days the eggs hatch, and the larvae or maggots—the cattle grubs—burrow into the animal's skin. For about eight months these grubs migrate through the various body organs and muscular tissue of the infested animal. Finally they reach the underside of the skin on the animal's back, where they stay for 35 to 60 days, breathing through holes they make in the hide. As the grubs mature, they fall to the ground and burrow into the soil to pupate. About 3 to 11 weeks later, depending on the temperature, they emerge as heel flies. When the female flies lay their eggs, the cycle begins again. (*Courtesy* U.S.D.A.)

Feed additives should be carefully mixed and applied to the top of the ration, following manufacturer's directions. Have all animals feeding at one time and "on feed" during the 7 to 14 days when you add the grubicide. This method of grub control has the advantage of not requiring a corral or other holding facilities. Table 22-1 lists recommended dosages of the various grub controls.

Cattle Lice

These pests are most abundant during the winter. Cattle infested with lice rub along fences and feed bunks in an effort to relieve irritation. The hair appears dry and dead. There will be bare places on the shoulders, neck, topline, and flanks where the hair has been rubbed off. Milk production may decline, and young animals may slow up in their development.

Prevention and Control. Spraying with CO-Ral or rotenone (see grub control) is very effective for the control of lice. Lindane, used at the rate of 2 pounds of 25 percent wettable powder plus 2 pounds of detergent per 100 gallons of water and sprayed thor-

TABLE 22-1 INSECTICIDES FOR FARM GRUB CONTROL †

Chemical	Spray ‡	Pour On	Feed	Days before Slaughter
Coumaphos (CO-Ral)	12–16 lbs. 25% [1]	4% undiluted solution		0
Neguvon S.P.	10 lbs. 80%	8% undiluted solution [2]		14
Ruelene	2 gals. 25% e.c.	½ gal. 25% e.c. in 1.5 gals. water [3]		28
Rotenone	7.5 lbs. 5% w.p. + 2 lbs. detergent [4]			0
Ronnel			Follow directions	21–60

† The following abbreviations are used for various insecticide formulations:
e.c. = emulsifiable concentrate; w.e. = water emulsion
w.p. = wettable powder; o.s. = water emulsion
d. = dust; g. = granules

‡ All ingredients are mixed in 100 gallons of water, unless otherwise noted.

[1] For grub *and* fly control, use 2 applications within 3 months: 8 lbs. of 25% Coumaphos w.p. in 100 gals. water.

[2] 0.5 fl. oz./100 lbs. body weight; maximum 4 oz. per animal.

[3] One fl. oz./100 lbs. body weight; maximum 8 oz. per animal.

[4] May be applied as dust: 1–1.5% rotenone applied 25 to 30 days after first grubs appear; repeat in 30 days. One pound of dust will treat 12 to 16 head.

oughly on the animals is also very effective. Spray must not be applied within 60 days of slaughter. A spray made by using 57 percent malathion emulsifiable concentrate and 2 pounds of detergent per 100 gallons of water is also very effective for the control of lice. Malathion must not be applied within 30 days of slaughter.

Horse Fly

Several species of blood-sucking horse flies attack cattle. They all breed in or on the edge of swampy areas, and some species require several years to complete one generation.

Housefly

The housefly does not suck blood. it breeds in a wide variety of decaying organic matter, and can be a severe barn and milk room pest. Sanitation and residual wall sprays are the most effective controls.

Horn Fly

This blood-sucking fly is smaller than the housefly. It begins to appear on cattle in early spring and stays on the backs and sides of the animals. It is easily controlled with back rubbers and sprays. It breeds in fresh cattle droppings in the pasture.

Stable Fly

The stable fly is about the size of the housefly, and also breeds in decaying organic matter. The adult fly feeds primarily on the legs and bellies of the cattle. After a blood meal, the flies rest on building walls, fences, and bushes. Because the flies are on the ani-

FIGURE 22-5. Dipping cattle in an insecticide kills ticks and protects the animals against cattle fever. This practice helped eradicate fever ticks from the South. The result: upgrading of the South's cattle stock and growth of a profitable livestock industry. (*Courtesy* U.S.D.A.)

FIGURE 22-6. Spraying cattle with approved insecticides to destroy insect pests and use of other methods of controlling insects affecting livestock are saving livestock producers millions of dollars a year. These cattle are being sprayed with insecticide for horn fly control. (*Courtesy* U.S.D.A.)

mal only a short time, and because any spray applied to the legs is easily rubbed off in tall grass, stable flies are difficult to control. Back rubbers offer very little protection, but residual sprays and sanitation are excellent controls.

Face Fly

The face fly looks much like the housefly, and, like it, has lapping mouth parts. During the day it clusters around the eyes, nostrils, and mouth of cattle, lapping up secretions. It leaves the animal at night. Face flies breed in fresh cattle droppings, just as the horn fly does. Face flies do not come into barns with the cattle. Back rubbers are only moderately successful against these flies since the animals seldom rub their faces on the device. Water sprays are only effective for a few days; daily sprays give the best protection.

Control on Beef Cattle. Many devices can be put up in doorways, above salt boxes, water tanks, or feed bunks, and saturated with 1% Ciodrin, 2% Malathion, or 5% Toxaphene in fuel oil. Cattle can rub their faces against the device to obtain some relief from face flies.

Control on Dairy Cattle. Syrup baits containing 0.5% dichlorvos (DDVP, Vapona) can be applied to the faces of dairy cattle every day during May and June, when face flies are most numerous. If used in combination with sanitation and residual spraying on outside barn walls, the number of face flies will be kept low.

Mosquitoes

Several species of these blood-sucking pests may be annoying to cattle. They breed in standing water, so are more severe in wet years.

Ticks

There are several species of ticks that affect cattle, one of which may carry the

TABLE 22-2 INSECTICIDES FOR FARM FLY CONTROL

	Beef Cattle			Dairy Cattle	
Chemical †	Spray	Cable-type [1] Back Rubber	Days before Slaughter	Spray	Building Spray
Ciodrin	0.15–0.3% w.e.[4]	1% o.s.[2]	0	1–2% o.s.[3]	
Ciodrin-dichlorvos			0	1% Ciodrin + 0.25% dichlorvos [3]	
Coumaphos (CO-Ral)	0.15% w.p.[5]	1.0% o.s.	0		
Diazinon					0.5–1% w.p.
Dichlorvos (DDVP, Vapona)				1% o.s.[3]	0.5–1% fog
Dimethoate (Cygon)					1% w.e.
Dioxathion (Delnav)	0.15% w.e.	1.5% o.s.	0		
Fenthion (Baytex)					0.75–1.5% w.e.[6]
Malathion	0.5% w.e. *or* w.p.	2% o.s.	0		
Pyrethrins				0.1% [3]	0.025–0.1% fog
With: synergist and repellant				+ 1.0% + 1–5%	
Ronnel (Korlan)	0.5% w.e. *or* w.p.	1% o.s.	56		0.5–2% w.e. *or* w.p.
Toxaphene [7]	0.5% w.e.	5% o.s.	28		

† Follow manufacturers' directions for mixing chemicals.
[1] Horn flies primarily.
[2] May also be used in back rubber for dairy cows.
[3] Apply 1 oz. morning and night, or 2 oz. in the morning as a fine mist with automatic or hand sprayer.
[4] Use 2 to 4 qts. finished spray per adult animal not oftener than once a week.
[5] See directions on Coumaphos for grub control.
[6] Do not use fenthion in dairy barns, milk houses, or poultry houses.
[7] Use only Toxaphene formulation labelled for use on livestock.
IC-328 (Revised), Ames, Iowa, Iowa State University, 1966.

disease known as Texas fever, which, in the past, resulted in heavy losses in the South. However, as a result of the work of the U. S. Department of Agriculture and the state experiment stations, Texas fever has all but been eliminated from the United States.

Prevention and Control. A lindane spray is an effective control measure for ticks. Lindane should not be used within 60 days of slaughter. Calves under three months of age should not be sprayed with lindane. CO-Ral as used for grubs will control ticks.

TABLE 22-3 CONTROLS FOR LICE, MANGE, AND TICKS

Pest & Chemicals	Back Rubber	Spray †	Dip	Days before Slaughter
Lice:	As in fly control [1]			
Ciodrin		1 gal. 2% e.c.[2]		0
Coumaphos (CO-Ral)		2 lbs. 25% *or* 2 qts. 11.6% e.c.		0
Dioxathion		1 gal. 15% e.c., *or* 0.5 gal. 30% e.c., + 2 lbs. detergent [3]		0
Lindane		1.5 pt. 20% *or* 1 qt. 12.4% e.c.[4]		30 [5]
Malathion		1 gal. 57% e.c. + 2 lbs. detergent		0
Ronnel (Korlan)		1 gal. 24% e.c., *or* 8 lbs. 25% w.p., + 2 lbs. detergent		56
Toxaphene [6]		3 qts. 60% e.c. + 2 lbs. detergent		28
Mange:				
Lindane		As for lice control [7]		
Toxaphene [8]				30 [5]
			6.325 lbs. Toxaphene and 12.5% emulsifier per gal.	28
Ticks:				
Ciodrin		As for lice control [2]		
Coumaphos (CO-Ral)		4 lbs. 25% w.p. *or* 1 gal. 11.6% e.c.[4]		0

† All ingredients are mixed in 100 gallons of water, unless otherwise noted.

[1] Back rubbers for lice are particularly necessary in the fall.

[2] Apply 2 to 4 qts. per animal.

[3] Spray thoroughly, but not oftener than every 2 weeks.

[4] Apply one gal. per animal.

[5] Do not use on calves less than 4 months old.

[6] Use only Toxaphene that has been prepared specifically for livestock.

[7] A second application will probably be needed 14 days after the first.

[8] Dip or "spray-dip" as for lice.

Mites

Cattle mites produce what is known as sarcoptic mange or cattle scab. The mite spends its entire life cycle on the body of the animal, piercing the skin and feeding on the lymph, producing a thickened, tough, wrinkled skin.

Prevention and Control. The same lindane spray as that recommended for ticks sprayed at 300 to 400 pounds of pressure is very effective in the control of mites.

INTERNAL PARASITES

There are several different types of stomach and intestinal worms that affect cattle, but the common stomach worm is the greatest menace.

Common Stomach Worm

Worms similar to those found in sheep may infest young cattle, particularly calves under six months of age. Adult cattle are seldom affected. Heavily infected animals lose weight and become thin and weak. The hair becomes rough, and the membranes of the mouth are pale. A soft swelling, known as *bottle jaw*, may develop under the jaw.

Prevention and Control. Sanitation is the most important means of prevention. Worm eggs pass in the droppings and hatch into tiny worms, which are picked up by other animals. Rotating pastures and placing water and feed where they will not be contaminated by droppings are effective controls.

Infested animals may be given a fluid drench containing 20 grams of phenothiazine powder per 100 pounds of body weight. The maximum dose should not exceed 60 grams, regardless of the animal's weight.

Doses should be repeated every 20 days where heavy infestation occurs. The drench may be given with a syringe placed in the side of the mouth and discharged near the base of the tongue. Usually after the first dose, low-level feeding of 2 grams per animal of phenothiazine powder will keep it free of worms. To accomplish this, the drug may be mixed with the mineral or supplement. A mixture of one part phenothiazine powder to nine parts salt and mineral mixture may be fed. A new product known as Thiabenzole has been successful when used to control stomach worms.

Coccidiosis

This disease is the result of poor management and unsanitary conditions. It is caused by the entrance of protozoa into the digestive systems of young calves through contaminated feed and water.

Symptoms. Common symptoms are loss of appetite, bloody diarrhea, and weakness.

Prevention and Control. Sanitation, the drainage of pastures and yards, rotation of pastures, and general good management are the best preventive measures. Veterinarians may successfully treat infected animals in the early stages of the disease with sulfa and other drugs.

Anaplasmosis

Anaplasmosis is caused by a protazoan parasite that destroys the red blood corpuscles. It is thought that various insects such as fleas, ticks, and mosquitoes are responsible for transmitting the disease from one animal to another. High fever, anemia, loss of weight and appetite are the common symptoms. The disease may be present in either the chronic or acute form. In the acute form death often occurs within five days after the first symptoms are noted. In the chronic form animals often recover but may be carriers of the disease.

Prevention and Control. No cure has been perfected for the disease. The U. S. Department of Agriculture has a test that will

FIGURE 22-7. A truck loaded with livestock pulls up at the livestock inspection station at Separ, New Mexico. With the cooperation of shippers, thousands of animals are unloaded, inspected, and treated, if necessary, to prevent spread of pests and diseases. (*Courtesy* U.S.D.A.)

FIGURE 22-8. Well-planned vaccination programs can prevent some diseases. Calfhood vaccination has helped reduce brucellosis infection to less than one-half of one percent. (*Courtesy* American Hereford Association)

aid in detecting animals that are carriers of the disease. The elimination of such animals from the herd is the best control measure.

INFECTIOUS DISEASES

There are many infectious diseases of cattle, but only those that cause the greatest economic losses will be discussed.

Brucellosis

Brucellosis, or Bang's disease, is first in importance, not only because of the economic losses resulting from the disease, but also because a disease known as undulant fever, which affects human beings, may be contracted from animals affected with it.

Symptoms. Infected animals may abort or give birth to a dead or weak calf; this is the most commonly observed symptom of brucellosis. On the other hand, the birth may be normal, but the cow may fail to clean (expel the afterbirth). Animals that are infected often have higher than normal temperatures at calving time. Milk production is reduced. Heavily infected herds may have 50 percent or more aborted or dead calves among heifers. Great care must be taken by persons coming in contact with animals infected with brucellosis to avoid contracting undulant fever. The disease may be contracted from the consumption of non-pasteurized milk produced by infected animals, and it is dangerous to handle newborn calves or aborted fetuses from infected herds.

Prevention and Control. Brucellosis finds its way into a herd through any of the following: (1) the purchase of infected or exposed animals, (2) contact with a neighbor's herd over a line fence, (3) exposure at livestock shows where an infected animal may be on exhibit, (4) livestock trucks that go from farm to farm handling animals and have not been properly cleaned and disinfected,

(5) public livestock auctions where proper sanitation is not practiced, (6) aborted calves that are dragged into the herd by carnivorous animals such as foxes or dogs.

Brucellosis may be detected by having a veterinarian blood test the herd. If the disease is found, the recommended plans for eradication are:

1. **Test all the cows and heifers, removing any reactors from the herd and vaccinating all calves between the ages of six and eight months that are not infected. This is the only recommended plan for dairy cattle.**
2. **Test and keep all reactors separate from the herd and vaccinate the calves. Since cows will generally produce normal calves after the second calving, reactors may be kept, but should not be in the same area with heifers since they are carriers of the disease. The plan should be to eliminate the reactors as quickly as it is economically possible. This plan may be used for beef herds.**

Since brucellosis may crop up in any herd at any time, it is advisable for any cow herd owner to start a calf vaccination program as a preventive measure.

Tuberculosis

Tuberculosis is a serious disease of cattle, but, because of state and Federal eradication programs, it is steadily declining in the United States.

Symptoms. Many times animals will show no outward sign of the disease. There may be a gradual loss of weight, swelling of the joints, and labored breathing. The part of the animal affected has much to do with the outward symptoms.

Prevention and Control. Tests have been perfected for determining the presence of tuberculosis in the herd. Periodic testing and elimination of reactors constitutes a reliable control program.

Blackleg

Blackleg is one of the most infectious diseases of young cattle and generally proves fatal.

Symptoms. Blackleg is usually accompanied by high fever, loss of appetite, and labored breathing. Rapidly developing tumors under the skin that make a crackling sound (because of gas) when subjected to pressure are one of the best indications of the disease.

Prevention and Control. If blackleg infection is found in the area or in the herd, vaccination of animals not affected is the only prevention. Affected animals will not generally respond to vaccination, but if the disease is diagnosed during its early stages, penicillin treatment by a veterinarian may cure some animals. Carcasses of animals killed by blackleg should be burned.

Anthrax

Anthrax is caused by an organism that may live in the soil for many years. For this reason, some areas have more outbreaks of the disease than others. Outbreaks usually occur during dry spells, when pastures are short and cattle tend to pick up soil while grazing. They may also occur following a flood, when pastures are overflowed.

Symptoms. High temperatures and bloody discharges from natural body openings are symptoms. However, the sudden death of cattle without these symptoms and without apparent cause may also indicate the presence of anthrax in the herd.

Prevention and Control. The prevention of anthrax lies mainly in following a vaccination program, especially in areas where outbreaks of the disease occur quite regularly. Vaccination provides immunity for a season but will not permanently immunize the animals. Infected animals may be cured by penicillin treatment, if it is given during the early stages of the disease. If the cattleman suspects that anthrax infection is present, he should immediately call a veterinarian.

Shipping Fever

Shipping fever is a blood disease chiefly affecting young cattle. A weakening of resistance, because of exposure when being shipped from one point to another, often leads to infection by the germs that cause shipping fever.

Symptoms. High temperature, coughing, and watery discharges from the nostrils and eyes are common symptoms.

Prevention and Control. Overexposure and drafty means of transportation should be avoided. Cattle should have plenty of water and should not be overcrowded. Infected animals should be separated from the rest of the herd. They should be kept dry and in protected places for a few days after shipping. Vaccines can be used, but should be given ten days before shipping. Serum should be used if animals are to be shipped immediately or are exposed by coming in contact with animals suffering from the disease. Antibiotics, especially penicillin, are effective in the treatment and prevention of shipping fever.

FIGURE 22-9. Typical appearance of the carcass of an animal that has died of anthrax. Note the bloated condition which occurs soon after death due to rapid decomposition. In warm weather, flies that are attracted to the carcass are capable of carrying virulent anthrax germs. Cremation or deep burial is recommended for such carcasses in order to check the spread of anthrax to other stock or to other farms. (*Courtesy* U.S.D.A.)

Rhinotracheitis or Red Nose

This is a serious disease in cattle. Although the actual death loss may not be large, cattle, especially feeder cattle, may lose from two to three hundred pounds of weight during the course of the disease. It is caused by a virus.

Symptoms. The symptoms are much like those of shipping fever. Animals have difficulty in breathing and usually have a discharge from the eyes and nose. They cough, have an excessive flow of saliva, and have temperatures from 103 to 108 degrees.

Prevention and Control. A number of vaccines are on the market, and veterinarians use them as a preventative measure and as a treatment. If the disease is suspected, a veterinarian should be called immediately in order to reduce losses.

Virus Diarrhea

This is a comparatively new disease in the United States. The loss in weight and the death rate will be high unless cattle are promptly treated.

Symptoms. The initial signs are nasal discharge and mouth sores. At first the feces are very hard, but later the animal will scour, and mucus and blood may appear in the feces.

Prevention and Control. No positive prevention has been found. Veterinarians have had some success treating the disease with various drugs.

Foot Rot

Foot rot is an infection of the feet of cattle and is a serious disease in many beef herds. It usually occurs when cattle are confined for long periods in muddy lots. It is caused by a germ that invades the tissues of the foot from the soil. An injury resulting in a break in the skin of the foot provides a means by which the germs may enter.

Symptoms. Affected animals become sorefooted, and the infected foot swells and is foul-smelling. There may not be any visible sore at the start. Later, the skin cracks open and a dirty yellowish material is present.

Prevention and Control. Keeping lots clean and well drained, and barns free from manure or well bedded are good preventive measures. A box 4 inches deep filled with hydrated lime and placed where the cattle will be forced to walk through it will aid in preventing the disease. Infected animals may be treated by scraping away all dead tissue and saturating the foot with Lysol or a one percent solution of copper sulfate. Veterinarians have successfully treated foot rot with sulfa and other drugs.

Pinkeye

Pinkeye attacks animals of any age, but seldom occurs during the cool season.

Symptoms. The first indication of pinkeye is a flow of tears and a tendency to keep the eyes closed. There will be a swelling of the eyelids and a general inflammation of the eye.

Prevention and Control. Affected animals should be segregated. Blowing sulfa powders into the eyes is helpful in many cases, but most animals will recover without treatment.

Mastitis

Mastitis is one of the worst diseases of dairy cattle but is less common in beef cattle. It is caused by types of germs that find their way into the udder, often as a result of an injury.

Symptoms. There are two forms of mastitis, one known as the acute and the other as the chronic form. In the acute form, the teat and quarter are swollen and painful. This form is often accompanied by fever and loss of appetite. The milk will be stringy and sometimes bloody. The chronic form is milder and may go unnoticed by the operator until it flares up in the acute form. In chronic mastitis, there may be no swelling. The milk, however, will generally show clots and watery consistency under close inspection. Usually only one quarter is involved at a time. Milk production drops and the udder may be ruined unless treatment is started during the early stages of the disease.

Prevention and Control. Eliminating sources of udder injury, such as obstacles that may cause bruises or cuts, is important in mastitis prevention. Poorly bedded cement floors are invitations for udder injuries. Milking a few streams from each quarter will give the operator an opportunity to observe any stringiness or clots in the milk. The Brome-Thymol-Blue test, although not 100 percent

FIGURE 22-10. Steer with shipping fever. (*Courtesy* U.S.D.A.)

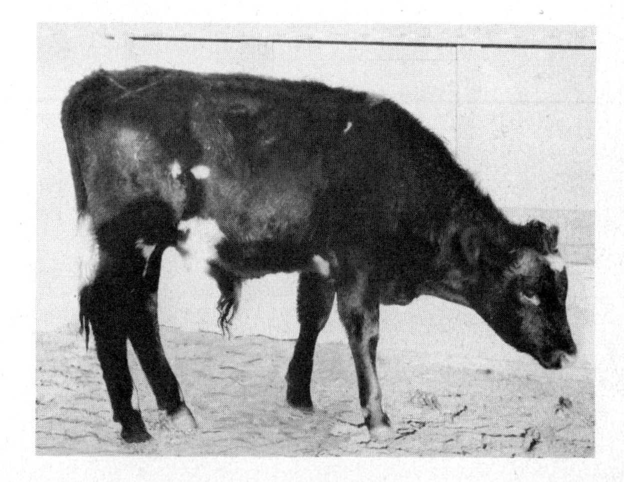

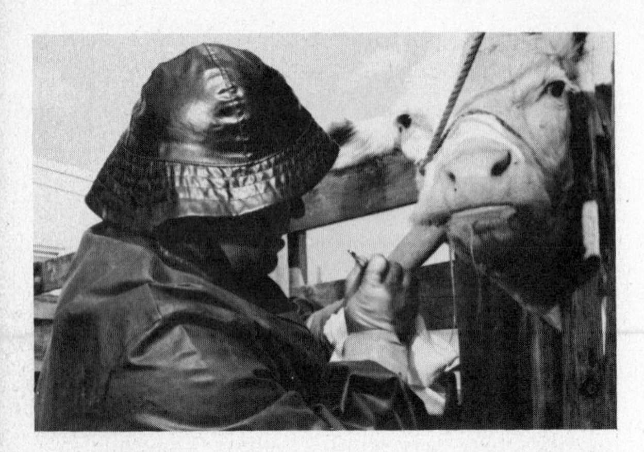

FIGURE 22-11. A Federal veterinarian scrapes an animal's tongue to get a specimen for laboratory analysis. Foot-and-mouth disease is highly communicable among cloven-hoofed animals, and is caused by a virus so small it can be seen only under an electronic microscope. Six outbreaks of foot-and-mouth disease between 1900 and 1929 cost the U.S. taxpayers $500 million. (*Courtesy* U.S.D.A. Photo)

accurate, will give a good indication of the presence or absence of mastitis. To conduct the test, place a drop of Brome-Thymol-Blue on a blotter with an eyedropper, using one drop in a separate spot for each quarter of the udder. Place a few drops of milk on the same spot. If the color changes to a bluish-green, mastitis infection may be suspected.

In mild cases, successful treatment may result by administering antibiotic ointment that comes in a small tube fitted with a spout. The ointment is injected into the affected quarter through the teat canal. In the most serious cases, or in cases where there is no response to this treatment, the cows should be placed under the care of a veterinarian.

Foot-and-Mouth Disease

Foot-and-mouth disease is highly infectious, but the few outbreaks that have occurred in the United States have been quickly brought under control.

Symptoms. The disease is characterized by blisters that form on the tongue, the lips, the cheeks, and the skin around the claws of the feet, and on the teats and udder. The blisters cause a heavy flow of saliva that hangs from the lips in strings. Infected animals smack their lips and, owing to tenderness of the feet, sway from one hind foot to the other.

Prevention and Control. To control the disease, infected animals should be destroyed and the premises disinfected with a lye solution. A vaccine has been developed that is used in areas outside the United States. Authorities in the United States have not recommended vaccination, for it is not considered conducive to complete eradication. The policy followed in the United States has been to destroy infected animals, and disinfect and quarantine the area where the outbreak occurred until the disease has been eradicated. If foot-and-mouth disease is suspected, Federal authorities should be notified, because quick control is essential.

Warts

Warts on calves are caused by a virus. The condition spreads from one animal to another by contact.

Symptoms. The warts appear around the neck and head, often near the eyes.

Prevention and Control. Wart-infected calves should be placed away from other calves. Painting the warts with iodine once or twice during a two-day period, followed by application of caster oil, will often eliminate them. A vaccine has been developed that is very effective in controlling warts.

Calf Scours

Calf scours is one of the worst diseases of young calves. It has been reported that from 10 to 15 percent of calf deaths are the result of calf scours.

Symptoms. Calf scours is characterized by pasty white scours, and usually affects calves under three weeks of age. The animals become weak and lose their appetite.

Prevention and Control. The first step in prevention is good management. Calves should be kept in clean, well-bedded stalls when they are dropped during cold weather.

Antibiotics fed in small daily amounts, particularly Aureomycin or Terramycin, have been effective in the prevention of calf scours. A dose of 3 milligrams of Aureomycin per pound of body weight for the first three or four weeks and 1½ milligrams per pound of body weight from three weeks to three to six months of age is recommended as a prevention. Calves that have the disease should have concentrated doses administered under the direction of a veterinarian.

FIGURE 22-12. Warts on a calf. (*Courtesy* U.S.D.A.)

Ringworm

Ringworm is a contagious skin disease of cattle. Young stock are more susceptible to the disease than older cattle.

Symptoms. Oval, scaly areas around the sides, neck, eyes, and back are signs of ringworm.

Prevention and Control. Infected animals should be separated from the rest of the herd and treated by applying phemerol to the areas affected. Washing with soap and water and painting the infected area with a tincture of iodine will often control ringworms. A new product known as Griseofulvim is also very effective.

Leptospirosis

Leptospirosis is a disease threat to cattle that was first discovered in the United States in 1944. The disease is one of the most serious of those affecting cattle. The death rate will average about 5 percent for those animals infected with the disease, but it may run up to 25 percent among young stock. About 25 percent of the cows infected will abort.

Symptoms. Fever, depression, loss of appetite, and loss of production are common symptoms. The milk may take on a yellow color and become thickened in consistency. Sometimes it will be bloody. There may be blood in the urine.

Prevention and Control. Great care should be taken to eliminate the possibility of bringing infected animals into the herd. Animals that are questionable should be given a blood test by a veterinarian. Antibiotics, especially penicillin, are effective in treating animals when treatment is started during the early stage of the disease. Vaccines have been perfected that are effective for a year or more. In areas where the disease is common, the vaccination of the herd is recommended.

Pneumonia

Pneumonia is a serious threat to young calves. The disease often occurs in calves weakened by scours. The highly infectious type may invade healthy calves and result in death unless treated in the early stage.

Symptoms. Labored, rapid breathing, coughing, high body temperature, nasal discharge, weakness, and loss of appetite are common symptoms.

Prevention and Control. Clean, well bedded pens, free from drafts and maintained at even temperatures, will help to prevent the disease. Infected calves may be successfully treated with sulfa drugs and antibiotics administered under the supervision of a veterinarian.

Vibriosis

This disease may attack mature female cattle. It resembles brucellosis and can be differentiated only by a laboratory test.

Symptoms. Like brucellosis, abortions, low conception rates, retained placentas, and weak calves are the common symptoms. Bulls may become affected and spread the disease to cows at the time of service.

Prevention and Control. Sanitation and careful selection of disease-free bulls are the best preventative measures. The use of streptomycin under the direction of a veterinarian is a recommended treatment.

NONINFECTIOUS AILMENTS

Milk Fever

Milk fever affects many high-producing cows. Cows are usually affected shortly after calving, but sometimes the condition will develop just prior to calving. The condition is caused by a rapid change in the balance of the blood calcium because of the inability of the blood calcium regulatory mechanism to meet the sudden demand made by milk production. Milk fever seldom affects cows later than five days after freshening.

Symptoms. Cows become dull and difficult to move, stagger, and lose their appetite during the first stages of milk fever. Constipation is very common as an early symptom. Later, the animal goes into a coma. When the coma is complete, the head will be drawn back against the chest or toward the udder.

Prevention and Control. Cows should have a mineral mixture including calcium, phosphorus, and salt available at all times. Massive doses of vitamin D (30 million units per day) for from five to seven days before the calf is born will help to ward off milk fever without apparent injury to the cow. Irradiated ergosterol, or type 142 F irradiated dry yeast, may be used for vitamin D.

Cows subject to milk fever should not be milked completely dry for the first few milkings after they have freshened. If milk fever does attack the cow, intravenous injections of calcium borogluconate will generally check the difficulty. Unless the dairyman has had considerable experience in giving the injections, a veterinarian should be called.

Ketosis

Ketosis is thought to be caused by a shortage or faulty utilization of sugar in the body. It usually affects either cattle fed for long periods on poor roughage without sufficient concentrates to balance the rations, or high-producing cows during the first month of lactation.

Symptoms. Animals affected by ketosis are nervous, lose their appetite, and drop in milk production. High-producing cows are the most susceptible.

Prevention and Control. Good rations with plenty of high-quality hay and silage are the best preventive measures. Some dairymen increase the carbohydrates for fresh cows by adding sugar or molasses to the ration. Several treatments have been administered by veterinarians to cows affected by the disease, with varying results. Excellent results have been reported in the treatment of ketosis by a single injection of 1.5 grams of cortisone. Giving affected animals ¼ pound of sodium propionate daily in two doses for three to ten days is another method of treatment that some dairymen have reported to be effective. Other ketosis control drugs made available to veterinarians include a combination of ACTH and penicillin, and a combination of ACTH and glucose.

X Disease

X disease, which is usually fatal, has appeared in many areas. The cause of the disease, according to most authorities, is chlorinated naphthalene, which is present in some types of grease used on farm and feed-mixing machinery.

Symptoms. The disease has a rather slow course, often as long as three months, before death occurs. It generally begins with a loss of weight and watery discharges from the nose and eyes. Animals slobber at the mouth, and sores are often found in the mouth.

Prevention and Control. No known treatment exists. Close inspection of animals for running eyes, slobbering, and sore mouths should be made before purchasing them. Keep the cattle away from any machinery or stored grease that may contain chlorinated naphthalene.

Bloat

Bloat is a condition in which the rumen becomes filled with gas that the animal is unable to expel. The condition may be caused by a growth or other obstruction in the esophagus, but is more often caused by feeds that ferment rapidly in the rumen, causing large amounts of gas to form. Feeding on legume pastures is one of the most common causes of bloat. Saponins (plant materials that produce a soapy lather) are thought to be the principal ingredients in legumes responsible for bloat.

Symptoms. The chief symptom is a great distention of the upper left side of the abdomen. Rapid breathing and uneasiness occur. If the animal is not relieved, it will stagger and fall. The pressure becomes great enough to prevent lung action, and suffocation is the final cause of death.

Prevention and Control. If the gas pressure is not too great, walking the animal will induce belching and provide relief. Tying a rope or a stick in the mouth will often help. If immediate relief is required, a puncture of the rumen becomes necessary. The puncture should be made in the center of the triangle formed by the last rib, the hip bone, and the transverse processes of the backbone on the left side. Unless poloxalene can be fed daily as described in the section on beef cattle, the following recommendations should be applied.

Pastures should not contain more than 50 percent legumes. Giving the cattle a fill of dry hay before they go on pasture or making dry hay available to cattle on legume pastures will help to prevent bloat. Cattle that have easily accessible water are less apt to bloat than are those that drink only in the morning and in the evening when they come from the pasture. Research work to find an effective method of controlling bloat is in progress. The most promising, at this time, is the administration of water-dispersible oil by adding it to the drinking water, and the sprinkling of crude soybean oil over fresh cut alfalfa at the rate of ¼ pound per animal daily. Antibiotics, especially penicillin, administered at rather short intervals have given some promise of bloat control.

Poisonous Plants

Many plants are poisonous to cattle. A large group of plants will develop what is known as *prussic acid*, especially under conditions that retard growth, such as extreme dry weather or frost. This group of plants includes sudan grass and other plants belonging to the sorghum family, Johnson grass, choke-cherry, black cherry, arrow grass, velvet grass, and Christmas berry. Prussic acid is very poisonous and often proves fatal within a few minutes after consumption of the plant containing it.

Sweet clover hay or silage not properly preserved often proves toxic to cattle. Certain weeds, such as snakeroot, larkspur, loco weed, lupine weed, water hemlock, milkweed, and cocklebur, will poison cattle if consumed in large enough quantities.

SUMMARY

The first steps in the prevention of disease and parasite infestation of the herd are sanitation, clean and well drained lots, clean and regularly disinfected buildings, and precautions against the purchase of sick animals or animals that have been exposed to infections. Many diseases can be prevented with vaccines.

FIGURE 22-13. This calf shows the crooked calf syndrome, a crippling disorder caused by pregnant cows eating lupine weed during early pregnancy. (*Courtesy* U.S.D.A.)

The most effective control measure for external parasites, except screwworms, is the use of proper sprays and dusts. Applying Smear EQ 335 on all wounds will prevent screwworm infestation. The common stomach worm may be controlled with phenothiazine or thiabenzole.

Brucellosis is best eradicated by a testing and vaccinating program, whereas tuberculosis is controlled by testing and eliminating reactors. Blackleg and anthrax can be prevented by vaccination.

Foot rot may be successfully treated by using disinfectants. If the case is advanced, a veterinarian should be called.

Mastitis can often be detected by the Brome-Thymol-Blue test. Antibiotics injected into the udder through the teat canal will help in many cases.

Warts may be treated with iodine followed by applications of castor oil. A special vaccine is also effective. Ringworms are best treated by applying phemerol, Griseofulvim, or iodine to the affected area.

Prevention of milk fever consists of supplying animals with adequate minerals and massive doses of vitamin D for from five to seven days before calving. Treatment consists of intravenous injections of calcium borogluconate.

Ketosis prevention consists of maintaining adequate carbohydrates in the ration. Cortisone, sodium propionate, and combinations of ACTH and other drugs have been successfully used in treating affected animals.

Bloat is caused by the formation of gas that the animal is unable to expel. It is often caused by certain feeds, such as young legume pasture. Treatment that will induce belching should be used, such as feeding poloxalene. If such methods fail, making a slit through the animal's side into the rumen will allow the gas to escape.

Poisonous plants cause death among cattle. The dairy farmer must recognize the poisonous plants and prevent their consumption to prevent poisoning the cattle.

QUESTIONS

1 List the steps of a good disease- and parasite-prevention program.

2 Give the control program for brucellosis and explain the circumstances under which you would recommend each method.

3 Name the common external parasites of cattle and give the methods of control of each.

4 How would you control the common stomach worm?

5 What diseases can be successfully controlled by vaccines?

6 Give the home treatments for warts and ringworms.

7 How would you control calf scours?

REFERENCES

Controlling Livestock Pests, IC-328 (Revised), Agricultural Extension Service, Ames, Iowa, 1966.

Herrick, John B., *Keep Feeder Cattle Healthy,* Agricultural Extension Service, Pamphlet FS-1099, 1964, Iowa State University, Ames, Iowa.

Seiden, Rudolph, *Livestock Health Encyclopedia* (Revised Third Edition), Springer Publishing Company, Inc., New York, 1968.

Snapp, Roscoe R., and A. L. Neuman, *Beef Cattle* (Sixth Edition), John Wiley and Sons, Inc., New York, 1969.

Stockdale, H. J., *Beef Cattle Insects and Their Control,* Agricultural Extension Service, Pamphlet 294, 1963, Iowa State University, Ames, Iowa.

U. S. Department of Agriculture, *Animal Diseases,* The Yearbook of Agriculture, Washington, D. C., 1956.

HORSE PRODUCTION

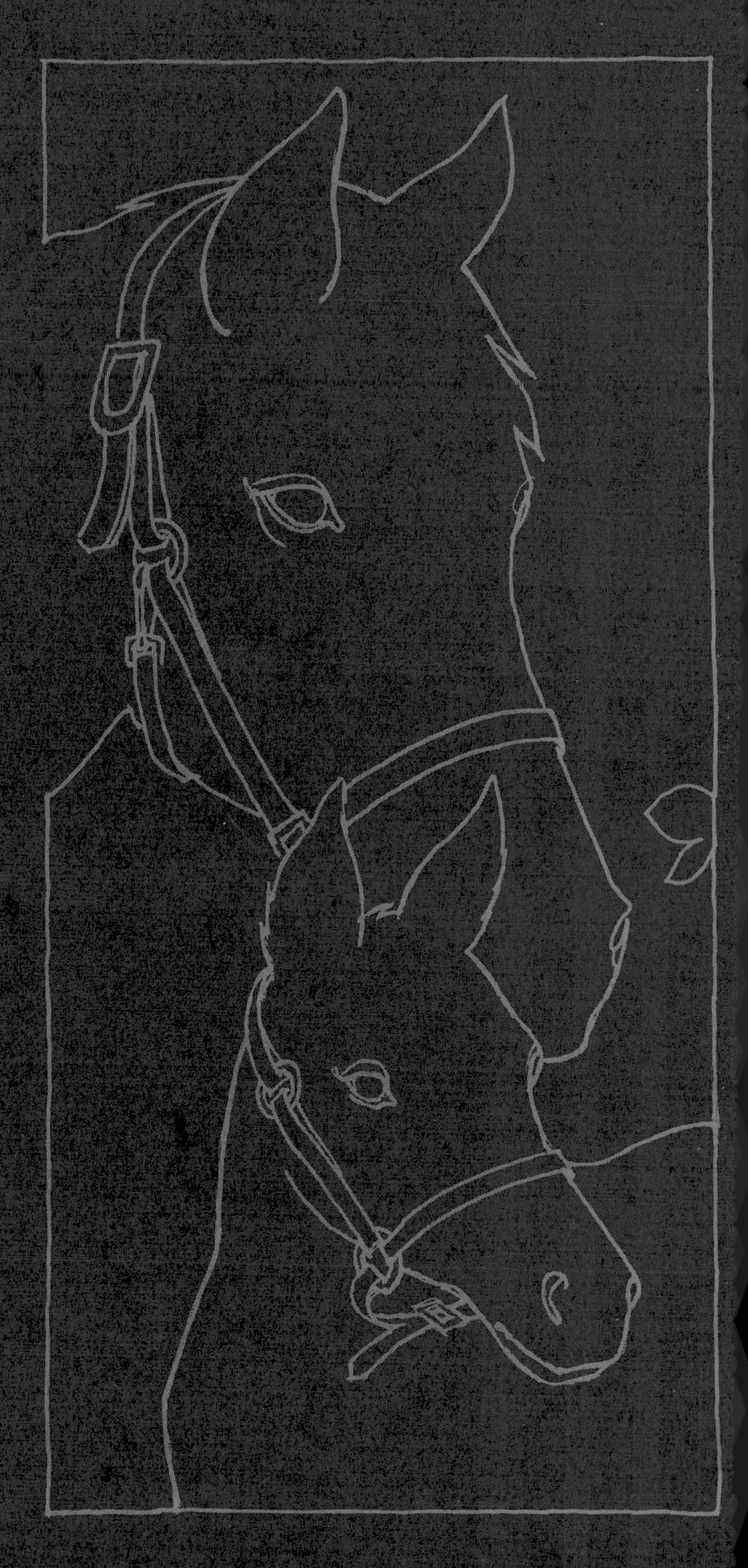

23 The Horse Industry

When Thomas Jefferson brought his bride to Monticello in Virginia, they came by horse and carriage. After traveling more than forty miles, a heavy snowstorm made it necessary to abandon the carriage and ride the horses the remainder of the way. Horses were the primary mode of travel in the early history of this country.

Horses and mules gradually replaced oxen as the source of power in the production and harvesting of crops and in heavy work in the forests. Large numbers of mules were used in the cotton fields of southern states because they had demonstrated the ability to withstand heat. Horsepower was used in building roads and railroads; small horses, donkeys, and mules were used extensively in the mining of coal.

The stagecoach and the pony express were forerunners of horse-drawn freight lines, delivery wagons, and spring buggies. Most communities had livery stables where travelers could hire horses and carriages or stable their own horses overnight.

Two types of horses were in common use: heavy draft horses or mules, used in field work and in pulling heavy loads, and light horses that were bred and maintained for riding and carriage use.

HORSE NUMBERS

Horse and mule numbers reached a peak of about 26 million in 1915. However, the introduction of the automobile in the early 1900's and the tractor in the 1920's minimized the need for horses, and their numbers began to decrease. As shown in Table 23-1, horse numbers decreased to about 18¾ million in 1930 and 14 million in 1940.

In 1929 nearly 67 million acres of land was needed to produce feed for horses and mules. By 1940 only 43 million acres was needed. Approximately 19 million acres was needed in 1950, and 5 million acres was needed to produce feed for the estimated 3

FIGURE 23-1. A common sight in the spring is a mare and her foal. Note the leather halters. (*Courtesy* United States Trotting Association)

FIGURE 23-2. The Hackney is noted for its springy action and style. (*Courtesy* American Hackney Horse Society)

million horses and mules in the United States in 1960.

The U. S. Agricultural Census discontinued reporting horse numbers in 1961. In 1971 U.S.D.A. officials estimated about 150,000 horses in each state in the eastern half of the nation, and nearly 500,000 in California. There are nearly 500,000 horses in Texas, and from 100,000 to 200,000 in several other western states. Estimates set a range of 6 to 7 million horses in the entire nation in 1971. In that year nearly 2.8 million horses were vaccinated for Venezuelan equine encephalomyelitis in 19 states.

Most of the horses in the nation in 1971 were of the light horse and pony classes. Very few draft horses and mules were to be found.

THE HORSE INDUSTRY A BIG BUSINESS

In 1972 horses were the basis for a 7 or 8 billion dollar annual business. For many people, owning a horse has become a status symbol. Large numbers of horses are owned by city dwellers. Many of these horses are pedigreed, highly trained animals, and many specialists are employed by urban owners in the breeding, grooming, feeding, and training of their horses.

Horse Racing

Horse racing is a national sport, attracting large numbers of people. More than 10,493 horses raced in American Quarter Horse running races in 1970, and several of these horses earned more than $100,000 that year. The Jockey Club of New York registered 24,628 Thoroughbred foals in 1970, to be used largely in racing and in polo.

The running horse industry in Kentucky is dominated by several corporate breeding farms. Kentucky produces about one-third of the Thoroughbreds in the nation and provides the site for the famous Kentucky Derby.

TABLE 23-1 HORSES, TRACTORS, AND AUTOMOBILES ON U. S. FARMS, 1930–70

Year	Farms in U. S.	Horses (thousands)	Tractors (thousands)	Automobiles (thousands)
1930	6,546	18,738	920	4,135
1940	6,350	13,932	1,567	4,144
1950	5,648	7,604	3,394	4,100
1960	3,949	3,089	4,684	3,629
1970	2,924	6,500 *	4,790	3,545

* Estimated
U.S.D.A. *Agricultural Statistics 1962*
U.S.D.A. *Agricultural Statistics 1971*

Trotting and pacing races are still very much a part of county and state fair sports schedules. Race tracks are often found near large metropolitan areas, attracting large numbers of sports enthusiasts to the Standardbred trotting and pacing races.

Stock Horses

Nearly a half-million horses are owned and used by ranch personnel, largely in the southern and western states. The stock horse has been an important part of ranching throughout the history of our nation. At one time, ranchers and ranch hands relied entirely upon horses for transportation about the ranches to check on calving, water and grass supplies, fences, and the health of the herd. The Jeep and the automobile have all but replaced the stock horse in providing most types of transportation, although the horse is still needed and is much used on most ranches.

The rising popularity of rodeos has accounted for increased numbers of stock horses and increased numbers of skilled horsemen and horsewomen. Many 4-H and FFA youth groups participate in and sponsor rodeos. The breeding and training of stock horses is to some people an avocational interest; to others it is a profession. With cattle numbers at an all-time high, it is likely that the demand for good stock horses will continue to be brisk in the years ahead.

U. S. Investment in Horses

According to the U.S.D.A. Statistical Reporting Service, the average horse in the nation in 1960, was valued at $113. The 3.09 million horses in the country were valued at $347 million. Since the assessors do not obtain complete information on the number and valuation of horses, ponies, and mules, it is difficult to make an accurate estimate of numbers or value. Some Shetland ponies sell for as little as $15 whereas others bring $500 or more. Good riding horses sell for $250 to $2,500. Show horses are more expensive, and racing horses and their sires and dams are often sold for hundreds of thousands of dollars each.

The U.S.D.A. estimated that 6.2 million horses, ponies, and mules in the nation in 1971 represented an investment of 620 million dollars if the average value per head was $100. At $200 per head, the investment would be $1.2 billion. The actual value probably falls somewhere between the two figures.

Pleasure Horses

Shorter working hours and early retirement have greatly increased the interest in pleasure horses. The trend in urban families

moving to rural home sites has made it possible for more people to enjoy the ownership and use of pleasure horses and ponies. These horses and ponies are used for recreational purposes rather than for monetary gain from racing, breeding, or exhibition.

At least one-third of the nation's population devotes some leisure time to horse-related activities. Thousands of people are spectators at horse racing events. A greater number, however, own one or more riding horses or ponies. A study in Minnesota indicated that 60 percent of the horseowners were women and 40 percent owned but one horse. Nearly 30 percent owned two or three horses, whereas 30 percent owned four or more.

More than 50 percent of the horseowners studied in 24 counties in Minnesota in 1970 were members of a saddle club. Many of the younger owners were members of 4-H

FIGURE 23-3. (*Above*) **Horses vary in value. This American Albino is very valuable as a show horse.** (*Courtesy* **American Albino Association, Inc.**) **FIGURE 23-4.** (*Below*) **Many youngsters participate in the youth programs of the various breed associations.** (**Guy Kassel photo.** *Courtesy* **American Quarter Horse Association**)

clubs and had one or more horses as their 4-H project.

Youth Activities

The number of 4-H horse and pony projects exceeds the number of dairy and beef cattle projects in many states. A horse or pony makes an excellent project for a town or city youngster. Many FFA members also have horse or pony projects. Some develop horse breeding and training programs. Many FFA and 4-H members participate in trail-ride and rodeo activities.

The Appaloosa Horse Club and the American Quarter Horse Association have developed excellent youth programs. A total of 114,968 youths exhibited in one or more of the 1,246 shows sponsored by the American Quarter Horse Association in 1970.

CAREER OPPORTUNITIES

Horse Production

There are many opportunities for professional specialists in the horse industry. Large horse-breeding stables need managerial, technical, and service personnel. Managers must be knowledgeable in finance and personnel management as well as horsemanship. The responsibility in the production and training of horses should be assumed by trained technicians. These persons are specialists in breeding, nutrition, health, training, racing, exhibition, and public relations. Much of the physical contact with horses is the responsibility of the service personnel. These persons do the feeding, watering, exercising, and grooming of the animals. Small breeders must have resident personnel qualified in all of these areas, or rely upon outside technicians and consultant help.

Some stable operators keep a few horses of their own as well as providing the services described previously for horseowners who do not have sufficient numbers of horses to justify maintaining their own stables.

Training of Horses

Large numbers of persons are skilled in training horses for various uses. Stock and cattle horses, trotters, pacers, jumpers, polo horses, carriage horses, gaited show horses, and running horses must be thoroughly trained. The trainer is particularly important to the racing phase of the horse industry.

Health

Many stables have a resident veterinarian, although some stable owners contract for outside veterinary services. Many persons who are not qualified as veterinarians but are the equivalent of a veterinary technician, provide supplementary veterinary services. Usually these supplementary services are rendered by horseowners who have become skilled in specific phases of equine health.

Rodeo Careers

Some horse enthusiasts make a career of organizing and managing rodeos. They travel to various sections of the country managing rodeos at state fairs, regional livestock shows, and other exhibitions.

Rodeo participants usually have another career, but some make rodeo participation a full-time job, and do well financially. Persons who breed and care for rodeo horses must be specialists in this field.

Racing

The number of people employed in the horse racing industry is not large when compared to horse production, but many excellent career opportunities exist. Race track officials such as stewards, paddockmen, starters, timers and judges must have years of

FIGURE 23-5. Bueno Bandit, a Champion Paint Stallion with a 96.3 speed and 97.4 performance rating with the Computer Horse Breeders Association. Owned by Murray Miller, Plano, Texas and Ralph Russell, McKinney, Texas. (Margie Spence photo. *Courtesy* American Paint Horse Association)

training and experience to manage races within the strict laws that govern racing. The physical specifications for jockeys and drivers are well-defined, their profession is highly esteemed, and they command high salaries.

Horse Shows

Comparatively few individuals make a full-time career of horse show activity. Many, however, supplement other careers in the horse industry with seasonal employment related to horse shows. A large number of "society horse shows" are held and many are scheduled on a circuit.

Part-time career opportunities exist for many persons who participate in or assist with the large number of horse shows conducted each year by local saddle clubs, county and district fairs, and by horse breed associations. There are several thousand of these shows conducted each year in the United States. The American Quarter Horse Association approved nearly 1,500 shows in 1972.

Full-time employment is available to persons associated with horse shows. Large horse shows are held in Columbus, Ohio; Fort Worth, Houston, and Dallas, Texas; Detroit, Michigan; Syracuse, New York; St. Paul, Minnesota; Chicago, Illinois; Denver, Colorado; and Santa Barbara, California.

Public Relations

Each breed association provides registration and promotional services. There is a demand for persons who are knowledgeable about horses as well as being qualified in journalism, accounting, and public relations. Specialists in communications are also needed by fair and race track organizations.

SUMMARY

There were an estimated 6 to 7 million horses in the nation in 1971. Most are in the light horse and pony classes. Horse and pony numbers are increasing. The horse has become a status symbol for some people.

Most horses are owned by individuals as pleasure horses. At least 30 percent of the nation's population participate in horse-related activities each year.

Horses serve as the basis for a 7 to 8 billion dollar industry. Horse racing, rodeos, horse shows, and trail rides attract large numbers of people.

Excellent youth programs have been developed through 4-H Clubs, Future Farmers of America chapters, and horse breed associations.

The estimated 6.2 million horses, ponies, and mules in the nation represent an investment of approximately one billion dollars and require a large labor force.

There are many opportunities for careers in the horse industry. Manpower is needed at the professional, managerial, technical, and service levels. Large numbers of persons are needed in horse breeding, management, and training.

QUESTIONS

1. How many horses and ponies are in your county?
2. What are the chief uses made of horses in your state?
3. How does the pleasure horse relate to present-day work and recreational schedules?
4. How important is the horse industry to the economy of the nation?
5. What career opportunities exist in the horse industry?

REFERENCES

Appaloosa Horse Club, Inc., *Horsemanship Manual,* Appaloosa Youth Program, Moscow, Idaho.

American Quarter Horse Assn., *32nd Annual Convention,* Amarillo, Texas, 1971.

Ball, Charles E., *Saddle Up,* Philadelphia, Pennsylvania, The Farm Journal 1970.

Ensminger, M. E., *Horses and Horsemanship, Fourth Edition,* Danville, Illinois, Interstate Printers and Publishers, Inc., 1969.

Gorman, John A., *The Western Horse, First Edition,* Danville, Illinois, Interstate Printers and Publishers, Inc., 1958.

U.S. Department of Agriculture, *Agricultural Statistics, 1962,* Government Printing Office, Washington, D.C., 1962.

U.S. Department of Agriculture, *Agricultural Statistics 1971,* Government Printing Office, Washington, D.C., 1971.

24 Selection and Breeding

Various factors must be considered in selecting horses or ponies. Among these are (1) use of the horse, (2) breed, (3) age, (4) body conformation, (5) action, (6) soundness, (7) disposition, (8) vices, (9) pedigree, and (10) price. Horses may be kept for pleasure, for breeding, for work with stock, for show, or for sport. Some horses may serve more than one purpose. A high percentage of horseowners keep pleasure horses. These owners are primarily concerned with selection factors related to the disposition, beauty, and riding qualities of the animal.

HORSE BREEDING

Horse breeding is a highly specialized profession. Because of this, there are comparatively few horse-breeding establishments. Although each stable may produce large numbers of horses, it is not necessary to keep a stallion because artificial insemination can be used. The gestation period is comparatively long—about 11 months, although this period can vary widely (see Chapter 26, page 388). Breeding mares are not normally used as pleasure or work horses, another reason that may account for the few specialized breeding farms.

CLASSES OF HORSES

Stock Horses

Working stock horses are very much in demand in the western and southern states. These horses must be well muscled and highly trained. They must be quick in starting and stopping, have great endurance, and at the same time, they must be calm in temperament, yet competitive in attitude. The American Quarter Horse and the Appaloosa bloodlines are commonly used in breeding stock horses.

FIGURE 24-1. The Clydesdale six-horse team is an attraction at any horse show. (*Courtesy* Dick Sparrow, Zearing, Iowa)

Show Horses

Many types of horses may be exhibited at shows and fairs. Classes may be provided for Shetland and Hackney ponies, stock horses, jumpers, three- and five-gaited saddle horses, and even for draft horses. Much pleasure may be obtained through show participation and awards received. Show horses are highly trained and require specialized trainers, groomers, and exhibitors.

The American Saddle Horse is bred for show and most gaited horses in shows are of this breed. Show horses must be attractive to the eye, but must also move with smoothness, style, and speed. They lift their feet high off the ground and have a long, straight stride. The three-gaited horses walk, trot, and canter. Five-gaited horses have two additional gaits: the slow gait and the rack. The various gaits are discussed in a following section.

Horses for Sport

While amateurs may participate in competitive racing at the local level, racing has become a highly specialized sport requiring expertise, both human and equine. Much has happened since the Justin Morgan horse won the race with the imported British horses. The story of this famous horse race was made into a motion picture, "Justin Morgan Had a Horse." It was televised nationally in 1972. The likelihood of an amateur selecting and training a winner for a major race is remote. Trotting, pacing, and running horses are highly bred, and breeding stock is expensive. Most of the horses in the running races are of the Thoroughbred or American Quarter Horse breeds. The Standardbred lines are used largely in trotting and pacing races. Harness racing, involving trotting and pacing horses, is much less extensive than running horse races.

Jumpers need to be large and exceptionally well muscled and coordinated. Many jumpers are Thoroughbreds, used in steeplechase races and in horse shows and exhibitions. Rodeo horses are usually selected from stock horses. An exceedingly good stock horse makes a good rodeo horse regardless of breed.

GAITS

Gait has to do with the distinctive rhythmic movement of the feet and legs of a horse as it moves. *Action* or *way of going* are also terms that apply to the gait of a horse.

The *walk* is a slow four-beat gait in which each foot leaves and strikes the ground at separate intervals. It is the natural flat-footed gait of horses. The *trot* is the natural rapid gait of horses in which a front foot and the opposite hind foot take off and strike the ground at the same time. It is a diagonal, two-beat gait. The *canter* is a slow three-beat gait in which two diagonal legs are paired to produce a single beat that falls between the separate beats of the other two unpaired legs. The *slow* gait is a very slow canter. The horse's head is carried low in producing this very smooth gait.

FIGURE 24-2. Whoops Speckles, a Champion Appaloosa owned by Joe Miller, Oklahoma City, Oklahoma. (Johnny Johnson photo. *Courtesy* Appaloosa Horse Club, Inc.)

The *rack*, originally known as the single-foot gait, is an artificial gait in which each foot meets the ground separately at equal intervals. It is a fast, flashy, and popular gait in the show ring. The *gallop* is the most rapid natural gait of horses. It is a four-beat gait produced when each front leg is paired with the opposite hind leg, and each foot strikes the ground at a different time. The hind legs are pulled forward under the body permitting the horse to push forward at a rapid speed. One hind foot strikes the ground first, followed by the other hind foot. The front foot on the same side as the first hind foot then strikes the ground, followed by the other front foot. This is the gait of running horses in the Kentucky Derby and other similar races.

The *pace* is a fast two-beat gait primarily used in harness racing. The front and hind feet on the same side leave the ground and strike the ground simultaneously. The feet are not lifted high but move just about ground level.

BREEDS OF HORSES

There is a large number of breeds of horses. Primarily they may be classified as: (1) saddle horses, (2) light harness horses, (3) ponies, and (4) draft horses. Table 24-1 is a summary of the various breeds of horses, their origin, color, and registration figures.

Purebred animals of each breed may be registered and recorded with the breed and registry association. The names and addresses of these associations may be found at the conclusion of this chapter.

TERMS USED IN DESCRIBING HORSES

Head Markings

Star — a small white mark on the forehead.
Strip — a narrow vertical mark extending from the center of the forehead to the bridge of the nose.

TABLE 24-1 BREEDS OF HORSES AND PONIES

Breed	Origin	Color	1970–71 Registration
Saddle Horses (light horses)			
American Albino	United States	White, pink skinned	194
American Paint *	United States	Two-toned, white, and solid color	3,053
American Quarter Horse	United States	Variable	90,877
American Saddle Horse	United States	Variable	3,942
Appaloosa *	United States	Variable, white with dark spots over loin and hips.	28,700
Arabian	Arabia	Variable	8,148
Morgan	United States	Variable	
Palomino *	United States	Golden yellow, white mane and tail	1,600
Pinto *	United States	White with large spots, or colored with white spots.	3,167
Tennessee Walking Horse	United States	Variable	7,500
Thoroughbred	England	Bay, chestnut, brown.	24,682
Light Harness Horses			
Hackney	England	Chestnut, bay, brown	637
Standardbred	United States	Variable	11,654
Ponies			
Pony of the Americas	United States	Appaloosa markings	1,161
Shetland Pony	Shetland Isles	Variable	1,501
Welsh Pony	Wales	Variable	7,500
Draft Horses (heavy horses)			
Belgian	Belgium	Bay, sorrel, roan	
Clydesdale	Scotland	Bay with white markings	
Percheron	France	Black and gray	
Shire	England	Variable; bay with white common	
Suffolk	England	Chestnut	

* These horses are sometimes grouped by *color type* instead of by breed. Some persons group the Appaloosa as a color type, but these horses actually form a full breed. The coloration and markings of Paint, Pinto, and Palomino can be found throughout many breeds.

Snip — a small vertical mark on the bridge of the nose between the nostrils.

Blaze — a wide mark, especially at the eyes, covering the entire nasal area.

Bald Face — a broad mark covering most of the forehead and face, and sometimes the lips and nostrils.

Body Markings

Appaloosa — a solid color or mottled roan of foreparts with lighter color over loin and hips, containing dark round or egg-shaped spots. Some white with spots over the entire body.

Bay — Light bay (yellowish tan). Blood bay (bright mahogany). Dark bay (dark, rich shade, almost brown).

Black — black, with fine black hair on the muzzle.

Brown — a very dark shade with tan or brown hairs on the muzzle or flanks.

Buckskin — a form of dun—yellowish or gold

body color, black mane and tail, and usually black lower legs.
Chestnut — light washy yellow to dark liver color; no black mane or tail.
Dun — mouse-color to golden; black mane and tail.
Gray — a mixture of black and white hairs.
Roan — Strawberry roan (hairs of red, white, and yellow intermingled). Blue roan (black, white, and yellow hairs intermingled).
Palomino — a golden body color, varying from light yellow to bright copper with white mane and tail.
Piebald Pinto — black spots on a white coat.
Skewbald Pinto — spots of any color other than black on a white coat.

SADDLE HORSES

American Albino

This breed was developed by the Thompson family on the White Horse Ranch in Nebraska beginning in 1918. A milk-white stallion was mated to Morgan mares. The stallion was believed to be of Arabian and Morgan ancestry. The Albino stands about 15.2 hands high (a hand is four inches, the width of a man's hand; the decimal figure expresses inches) and weighs 1,100 to 1,150 pounds at maturity. The haircoat is white, fine, and silky. The eyes are either brown, dark blue, hazel, or light blue.

Horses of this breed are gentle, intelligent, and well muscled. They are widely used in circuses and as pleasure horses.

American Paint

This breed has been developed by crossing Paint mares with outstanding Quarter Horse stallions. Some Thoroughbred blood has been infused to increase the running speed.

There are two color patterns: the tobiano and the overo. Both are two-toned in that white is surrounded with solid colors of black, bay, sorrel, dun, or gray. No two of these horses are marked alike.

About four tobianos are registered for each overo. The tobiano gene is considered to be genetically dominant, whereas the overo gene is thought to be recessive. The following are the characteristics of the two color patterns:

Overo

1. **Usually white will not cross the back between the withers and the tail.**
2. **At least one leg, and often all four, will be a dark color.**
3. **Head markings are often bald, apron, or bonnet-faced.**
4. **Irregular, rather scattered and/or splashy-white markings on the body, often referred to as calico patterns.**
5. **The tail is usually one color.**
6. **The horse may be predominately dark or white.**

Tobiano

1. **Head is marked like a solid-color horse—solid, or will have a blaze, strip, star, or snip.**
2. **All four legs will be white, at least below the hocks and knees.**
3. **Spots are usually regular and distinct, often appearing in oval or round patterns that extend down over the neck and chest giving the appearance of a shield.**
4. **The horse will usually have the dark color in one or both flanks.**
5. **The horse may be predominantly dark or white.**

The Paint is noted for its speed, stamina and performance. While it is bred in practically every state, the largest number of breeders are located in Texas, Kansas, Oklahoma, Ohio, and Colorado.

American Quarter Horse

The American Quarter Horse is the most popular breed of the light horses. This breed, which originated in Virginia and the Carolinas, got its name because of superiority in running races which were a quarter-mile in length on a straight, flat track. While Quarter Horses are used in racing, they have become quite popular as stock horses in ranching, in rodeo activities, and as pleasure horses. The King Ranch in Texas has done much to develop and popularize the breed.

Horses of this breed are usually small, short-coupled, fairly lowset, and noted for their hardiness and toughness. They are quick and strong, and can run very fast for short distances.

Mature mares and stallions stand about 14.3 to 15.1 hands high and weigh 1,100 to 1,300 pounds. Their color varies, but no animal may be registered that has one or more spots of such size or location as to indicate Pinto, Albino, or Appaloosa breeding.

FIGURE 24-3. A typical American Quarter Horse. (*Courtesy* **American Quarter Horse Association**)

American Saddle Horse

The American Saddle Horse had its origins in Virginia, Kentucky, Missouri, and Tennessee. Several breeds were used in its formation: the Thoroughbred of England, the Canadian Pacer, the Morgan, and the Standardbred.

American Saddle Horses are known for their smooth gait and style. Popular as pleasure horses, they hold their heads high and show animation. They are used extensively as three- or five-gaited horses in the show ring, and can be used in harness.

They stand 15 to 16 hands high and weigh from 1,000 to 1,200 pounds. The colors range from black to gray and from golden to bay and chestnut. They are the largest of the breeds of saddle horses.

Appaloosa

The Appaloosa is the second most popular breed of saddle horse in America in terms of numbers recorded each year. The breed had its origin in horses introduced from Mexico by Spanish adventurers, and was developed in Idaho and Washington by the American Indian. Large numbers of the Appaloosa breed are to be found in the northwestern states.

The Appaloosa is a compact horse, standing 14 to 15 hands high and weighing from 950 to 1,150 pounds. In size, muscling, and disposition it is an excellent stock horse.

Horses of this breed are easily identified by unusual color patterns. Most Appaloosas are white over the rear quarters and sprinkled with dark brown or black spots. The forequarters may be white or dark-colored. Some animals have spots over the entire body. Striped hoofs, particolored skin, roan coloration, and sparse mane and tail are characteristic.

Arabian

The Arabian is the oldest of the breeds and is widely used as a parade horse. It stands about 14.2 to 15.2 hands high at the

FIGURE 24-4. (*Above*) **Arabian horses vary in color. The bay shown illustrates the characteristics of the breed.** (*Courtesy* **Arabian Horse Registry of America, Inc.**) **FIGURE 24-5.** (*Below*) **A Tennessee Walking Horse with Palomino markings.** (*Courtesy* **Tennessee Walking Horse Breeders Association of America**)

withers and will weigh 800 to 1,000 pounds at maturity. About 50 percent are bay, 30 percent are gray, and the remainder may be chestnut or brown in color. White or black animals are seldom found. Stars, strips, or blaze faces, snip noses, and one or more white feet with stockings are common. It is a gentle, very intelligent horse.

Morgan

This breed is of American origin, resulting from the mating of a Thoroughbred stallion with an Arabian mare. Most Morgans stand 14.2 to 16 hands high and weigh 1,000 to 1,200 pounds. They usually have a bay, brown, black, or chestnut haircoat. White markings are common.

The Morgan is versatile, used widely as a pleasure horse and as a stock horse. In the past it has been used as a harness horse. The U. S. Morgan Horse Farm in Vermont, which did much to promote the breed, is now owned by the Vermont Agricultural College.

Palomino

The Palomino is very popular as a pleasure horse because of its gold haircoat and blond or silver tail and mane. Two associations register Palomino horses; however, horses of this color may be found in several breeds. When Palominos are mated, the golden color does not breed true. Usually 50 percent of the foals are golden, 25 percent are chestnut, and 25 percent are white or light cream. The palomino color results only when the chestnut is crossed with the cream or white parent stock.

Stallions and mares with Palomino characteristics may be registered in the books of the two Palomino associations if they are from ancestry that is registered in one of the other horse associations. Palomino geldings may be registered regardless of ancestry. Many Palominos are of the American Quarter Horse and American Saddle Horse breeds.

Pinto

The Pinto is a spotted horse of Spanish origin. In size, the spots may be several inches across. Markings are of three types: (1) a colored horse with a white rump on which there are colored spots; (2) a colored horse with white areas extending upward from the belly and legs; or (3) a colored horse with white areas extending downward from the dorsal (back) region. Each of the three patterns may have other white markings.

Pintos range in height from 14.1 to 16.2 hands and weigh 750 to 1,200 pounds. They are similar in body conformation to the Quarter Horse. They have glassy eyes. They are popular as parade and pleasure horses, and many are used as stock horses.

FIGURE 24-6. The Pinto is popular as a parade horse. (*Courtesy* Pinto Horse Association of America, Inc.)

Tennessee Walking Horse

This breed was established in Tennessee by immigrants from the Carolinas and Virginia who brought with them Standardbred, Thoroughbred, Morgan, and American Saddle Horse stock.

The horse is noted for the running walk, one of three gaits common to the breed. In performing the running walk, a good horse will overstride from 30 to 50 inches, meaning that the back hoof will be placed ahead of the print of the forehoof. The running walk is faster than the flat walk, and provides the rider with a smooth and gentle ride. The horse nods its head with each step.

Tennessee Walkers average 15.2 hands in height and weigh from 1,000 to 1,200 pounds. Colors of sorrel, chestnut, black, roan, white, bay, brown, gray, and gold are common. White on the face, feet, and legs is common.

Thoroughbred

This breed originated in England and has been developed into the top breed of race horse in the nation. The Thoroughbred is a large horse, 15 to 17 hands high, and weighing from 900 to 1,000 pounds in racing condition. While some are gray, they are usually bay, brown, black, or chestnut in color. White markings on the face and legs are common.

Kentucky produces many horses of this breed. Large syndicates own some of the elaborate stables found in the Lexington area. Kentucky Derby winners, their sires and progeny, are sold for thousands and sometimes millions of dollars.

LIGHT HARNESS HORSES

While some horses of each breed may be used as harness horses, most light harness horses are of the Standardbred or Hackney breeds.

Standardbred

This breed originated from the crossbreeding of Thoroughbred, Arabian, Morgan, Hackney, and various lines of pacing horses. Prior to the automobile, this breed provided many buggy and carriage horses. Today the

FIGURE 24-7. (*Left*) Thoroughbreds and Quarter Horses are noted for their running ability. (*Courtesy* American Quarter Horse Association) FIGURE 24-8. (*Right*) Nevele Pride, the World's Champion Trotter of the Standardbred breed. (*Courtesy* United States Trotting Association)

breed is largely used as trotters and pacers for harness racing.

Standardbreds range in height from 15 to 16 hands and weigh from 900 to 1,300 pounds. They vary in color from bay, brown, and chestnut to black, although other colors are found. The breed has demonstrated a gentle disposition, racing speed and endurance, and a will-to-win attitude.

Hackney

The Hackney originated in England and was used in the eighteenth century to pull the English hackney coaches. Previously they were used as riding horses and sometimes to do field work. They vary in size; some, in the pony class, stand less than 14.2 hands high, while others may attain a height of 16 hands. They vary in weight from 800 to 1,200 pounds.

Hackneys are usually short-legged and compact. They have smooth coats, are graceful and stylish, and as show horses they are highly trained to use their excellent natural action. They lift their feet high and are colorful in the show ring.

They vary in color, but bay, brown, and chestnut are common. Some have white face and leg markings.

PONIES

The three most common types of ponies in the United States are the Shetland, the Pony of the Americas, and the Welsh. Several other breeds and lines of ponies have been developed and are produced in small numbers. Among them are the White Albino and the Gotland Horse.

Shetland Pony

While only about 1,500 Shetland ponies were recorded by the breed association in 1971, large numbers are to be found. The Shetland is the smallest of our horses and is a favorite of children and adults of this nation.

They are native to the Shetland Isles, north of Scotland, and importations have continued since 1850. Two distinct types are now available: the original Shetland, lowset and built like a draft horse; and the American line, bred as a road-type horse, is lighter and has longer legs.

Shetlands are less than 11.2 hands tall and are varied in color and markings. Solid colors are preferred but many are spotted. These ponies, noted for gentle disposition and faithfulness, make excellent pets for children.

Pony of the Americas

This breed has been developed since 1950, resulting from the crossing of Shetland ponies with Appaloosa, Arabian, and Quarter Horse breeds. To be registered, animals must have the markings of the Appaloosa and be from 11.2 to 13 hands high at maturity. Body conformation standards are similar to those of the Arabian and Quarter Horse but on a miniature basis.

This pony is popular with children who have outgrown the Shetland and want a larger and more rugged pony.

Welsh Pony

The Welsh pony was imported from Wales to the United States many years ago. They range from 10 to 12 hands in height and weigh less than 500 pounds. They may be of almost any solid color, or a solid color lightened by an admixture of white hairs. Welsh ponies with white markings cannot be registered. They are somewhat larger and more upstanding than the Shetland.

DRAFT HORSES

There were 16 million horses in the nation in 1925, and a high percentage of these horses in the Midwest were draft horses. Six to ten draft animals were required on the average farm. The tractor has replaced draft horses on most farms and only a few of the following breeds are to be found: Belgian, Clydesdale, Percheron, Shire and Suffolk.

FIGURE 24-9. The Pony of the Americas has the color of the Appaloosa. (*Courtesy* Pony of the Americas Club)

Belgian

The Belgian is an import from Belgium and ranks second among the draft breeds in total number of animals recorded to date. Very few are now to be found.

Mature animals weigh 1,800 to 2,200 pounds and stand 15.2 to 17 hands high. They are usually bay, chestnut, and roan in color. They are the widest, deepest, and most compact of the draft breeds.

Clydesdale

This breed is considered by many as the most stylish and beautiful of the draft breeds. They originated in Scotland and rank third in number of registrations to date. Clydesdales are more upstanding and lighter in weight than the other breeds of draft horses. They stand from 16 to 17 hands in height and weigh at maturity from 1,700 to 1,900 pounds.

Most Clydesdales are bay or brown with white on the face and legs. They have large hooves and long white hair about the fetlocks. They excell all draft breeds in their stride and springy action. They are popular in six- and eight-horse hitches.

Percheron

This breed was imported from France about 1850 and ranks first in total numbers registered among the breed associations to date. During recent years, both the Belgian and Clydesdale have enjoyed greater popularity.

The Percheron is usually black or gray in color. Mature animals weigh from 1,800 to 2,100 pounds and are 16.1 to 16.3 hands in height. Animals of this breed have excellent dispositions, good action, and are attractive. Limited numbers are now available.

Shire

Very few Shires are to be found. They originated in England and are massive animals weighing 1,900 to 2,000 pounds at maturity. They stand 16 to 17.2 hands high at the withers. They have the height of the Clydesdale but are less active.

FIGURE 24-10. (*Above*) **A Percheron stallion and a Shetland pony foal look each other over on the Iowa State University campus.** (*Courtesy* **Iowa State University**)
FIGURE 24-11. (*Below*) **This Belgian six-horse hitch demonstrates the massiveness and attractiveness of this breed of draft horse.** (**Cline photo.** *Courtesy* **Dick Sparrow, Zearing, Iowa**)

Most Shires are bay, brown or black in color. They have white markings on the face and legs.

Suffolk

The Suffolk is the smallest of the draft breeds. It also was imported from England. Only a few horses of the breed are now available.

Suffolks are 15.2 to 16.2 hands high and weigh 1,700 to 1,900 pounds at maturity. They are chestnut in color. The color may vary from light to dark. They may have some white markings.

MULES

There were nearly six million mules in the nation in 1925. Most of them were used on farms and plantations in the south. Some, however, were to be found in the Cornbelt. Very few mules are now to be found.

The mule is the result of crossing a jack (male donkey) with a mare of one of the breeds of horses. Draft horse mares were mated to jacks to produre heavy draft mules, and mares of the light horse breeds were bred to produce small mules for light work. Mules do not usually reproduce.

SELECTION FACTORS

Horses and ponies may live 20 to 24 years, but their greatest usefulness is at three to twelve years of age. Horses and ponies are classified by sex and age as follows:

Foal—a young horse of either sex up to a year old.
Filly—a young female horse under five years of age.
Colt—a young male horse up to five years of age.
Mare—a mature female horse five or more years of age.
Stallion—a mature male horse five or more years of age.
Gelding—a desexed male horse.

Age Determination

The age of a pedigreed horse may be determined from its official pedigree. The date of birth of Thoroughbred and Standardbred horses is carried in a tatoo on the inside of the lower lip. The ages of other horses may be determined by inspecting the lower incisor teeth.

A 10-month-old foal has 12 front (incisor) teeth and 12 grinders (molars) for a total of 24 teeth. A stallion has 24 molars and 12 incisors, plus 4 canine teeth (tushes) for a total of 40 teeth. The canine teeth are located in the space between the incisors and the molars, and appear only in the male. The mare has only 36 teeth.

The temporary teeth of a foal are small and white with a distinct neck. The permanent teeth are larger, stronger, and have a darker color with distinct cups, or depressions, in the crown of the tooth in horses up to 6 years of age. A foal will have all three pairs of temporary incisor teeth when 10 months of age, and the crowns of all teeth will show wear at 16 to 18 months. As the tooth wears, the cup in the crown disappears. At about 2½ years of age, the temporary central incisors loosen and permanent incisors replace them. At 3½ to 4 years the intermediate temporary incisors are replaced, and at 4½ to 5½ years the temporary corner incisors are replaced by permanent teeth.

Figure 24-12 shows the teeth of a foal at birth, at 4 to 6 weeks, and at 6 to 9 months; and the teeth of a horse at 2½, 3½, and 4½ years. It is relatively easy to determine the age of a horse up to five years of age by inspecting the teeth.

The extent to which the cups are worn in the biting surface or crown of the incisors, and the angle of the bite are the keys to de-

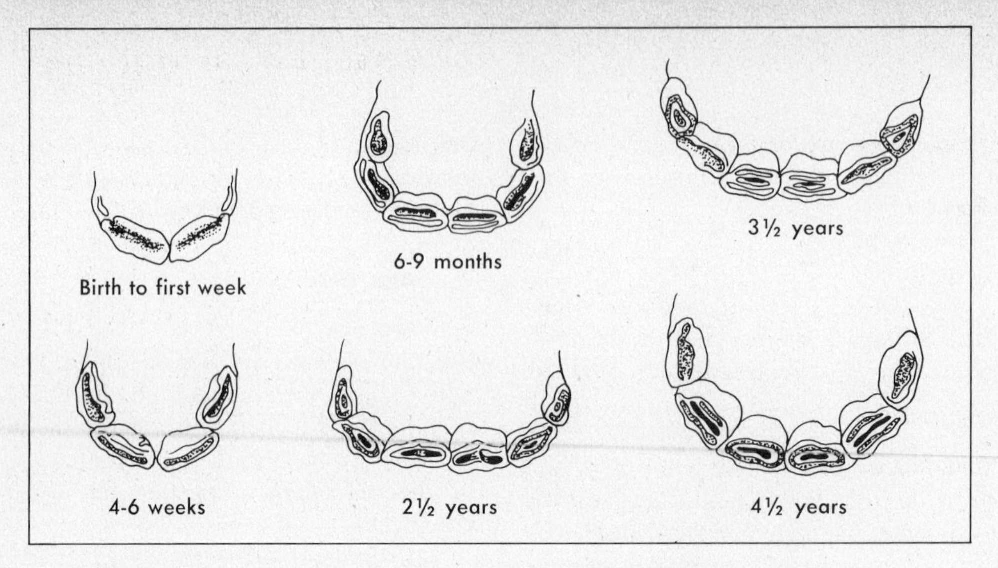

FIGURE 24-12. Note the changes in number and appearance of teeth in a horse from a foal to 4½ years of age. (*Courtesy* Iowa State University)

termining ages of horses six years old and older. The cups disappear at regular intervals beginning with the lower centrals during the sixth year. At eight years, the cups have disappeared from the lower centrals and intermediate incisors. The accuracy in determining age after nine years is more difficult. By 12 years the cups have disappeared from the upper incisors and the horse has a smooth mouth (the crowns of all teeth are smooth). The dental star or circle-like structure near the center of the wearing surface of the permanent incisors is smaller but more distinct at 15 years. The teeth become shorter and the bite more angular as the horse approaches 20 years of age.

FIGURE 24-13. The basic points of an American Quarter Horse. (*Courtesy* American Quarter Horse Association)

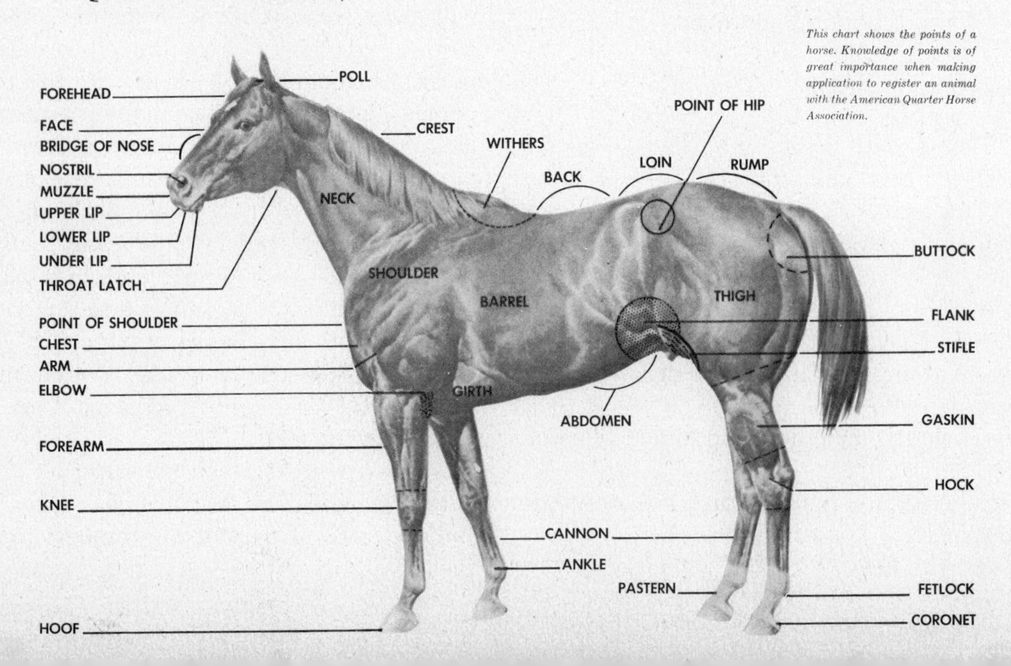

CONFIRMATION FACTORS

Parts of a Horse

It is important in the selection and judging of horses that the horse-person understands the parts of a horse and how they appear on an ideal animal. Figure 24-13 indicates the basic points of an American Quarter Horse.

Score Card for the Light Horse

The usefulness of a horse depends primarily upon its strength, stamina, and coordination. Certain body characteristics have much to do with the athletic qualities of the horse. Horseowners seek animals which meet certain standards of excellence.

A well-formed horse has long, sloping shoulders; a short, strong back; a long underline; and a long, rather level rump. A well-proportioned head and long neck are also desired. Soundness of feet and legs is essential.

Following is a score card recommended by the Appaloosa Horse Club, Inc. for use in judging light horses.

SCORE CARD FOR THE LIGHT SADDLE HORSE

Scale of Points	Standard or Perfect Score
General Appearance — 12 percent	
Height	
Weight	
Form (Close but not full made, deep but not broad, symmetrical)	4
Quality (Bone clean, dense, fine, yet indicating substance. Tendons and joints sharply defined, hide and hair fine; general refinement, finish)	4
Temperament (Active, disposition good, intelligent)	4
Head and Neck — 8 percent	
Head (Size and dimensions in proportion, clear-cut features, straight face line, wide angle in lower jaw)	1
Muzzle (Fine, nostrils large, lips thin, trim, even)	1
Eyes (Prominent orbit; large, full, bright, clear; lid thin, even curvature)	1
Forehead (Broad, full)	1
Ears (Medium-size, pointed, set close, carried alert)	1
Neck (Long, supple, well crested, not carried too high, throttle well cut out, head well set on)	3
Forehand — 22 percent	
Shoulders (Very long, sloping, yet muscular)	3
Arms (Short, muscular, carried well forward)	1
Forearm (Long, broad, muscular)	1
Knees (Straight, wide, deep, strongly supported)	2
Cannons (Short, broad, flat, tendons sharply defined, set well back)	2
Fetlocks (Wide, tendons well back, straight, well supported)	2
Pasterns (Long, oblique — 45 degrees — smooth, strong)	2
Feet (Large, round, uniform, straight slope of wall parallel to slope of pastern, sole concave, bars strong, frog large, elastic, heels wide, full, one-third height of toe, horn dense, smooth, dark color)	5

SCORE CARD FOR THE LIGHT SADDLE HORSE (Cont.)

Scale of Points	Standard or Perfect Score
Legs (Direction viewed from in front, a perpendicular line dropped from the point of the shoulder should divide the leg and foot into two lateral halves; viewed from the side, a perpendicular line dropped from the tuberosity of the scapula should pass through the center of the elbow-joint and meet the ground at the center of the foot)	4
Body — 12 percent	
Withers (High, muscular, well finished at top, extending into back)	3
Chest (Medium-wide, deep)	2
Ribs (Well sprung, long, close)	2
Back (Short, straight, strong, broad)	2
Loin (Short, broad, muscular, strongly coupled)	2
Flank (Deep, full, long, low underline)	1
Hindquarters — 31 percent	
Hips (Broad, round, smooth)	2
Croup (Long, level, round, smooth)	2
Tail (Set high, well carried)	2
Thighs (Full, muscular)	2
Stifles (Broad, full, muscular)	2
Gaskins (Broad, muscular)	2
Hocks (Straight, wide, point prominent, deep, clean cut, smooth, well supported)	5
Cannons (Short, broad, flat, tendons sharply defined, set well back)	2
Fetlocks (Wide, tendons well back, straight, well supported)	2
Pasterns (Long, obliquc — 50 degrees — smooth, strong)	2
Feet (Large, round — slightly less than in front — uniform, straight, slope of wall parallel to slope of pastern, sole concave, bars strong, frog large and elastic, heels wide, full, one-third height of toe, horn dense, smooth, dark color)	4
Legs (Direction viewed from the rear, a perpendicular line dropped from the point of the buttock should divide the leg and foot into lateral halves; viewed from the side this same line should touch the point of the hock and meet the ground some little distance back of the heel; a perpendicular line dropped from the hipjoint should meet the ground near the center of the foot)	4
Way of Going (Action) — 15 percent	
Walk (Rapid, flat footed, in line)	5
Trot (Free, straight, smooth, springy, going well off hocks, not extreme knee fold)	5
Canter (Slow, collected, either lead, no cross canter)	5
Total	100

Courtesy Appaloosa Horse Club, Inc.

ACTION

Action, or the way a horse moves, contributes 15 percentage points to the score card for light horses. Actually, the action of a horse may warrant greater importance for a horse is of little value unless it can move well.

A horse should walk in a straight line with easy, springy, rapid, and long steps. The feet should be lifted clear of the ground and placed flat when returned. At a trot the stride should be free, straight, and prompt. The action should be springy but smooth. The hocks should be carried close and there should be high flexion of the knees and hocks, but no extreme knee fold. (See Chapter 26, page 389.)

The canter should be slow, well coordinated, and smooth. The horse should be able to lead with either front foot. Figure 24-15 shows how a normal foot moves, and the action of horses with undesirable underpinning characteristics.

SOUNDNESS

The beginner must be very cautious in selecting a horse. It is often wise to obtain assistance in order to avoid selecting an unsound animal or one with blemishes. Experienced horseowners become skilled in detecting conditions which depreciate the value of an animal. Any condition which affects the action or working ability of a horse should be carefully analyzed.

Figure 24-16 is a diagram of a horse indicating locations of major unsoundnesses. Following are the most common unsoundnesses (U) and blemishes (B) of the head, neck, withers, and shoulders; the body; the front legs; and the rear limbs as outlined by the Appaloosa Horse Club, Inc.:

HEAD

1. Cataract (U) — cloudy or opaque appearance of the eye.

FIGURE 24-14. Bannock Silvertip, Grand Champion Appaloosa Gelding, National Appaloosa Show, Sacramento, California. (Johnny Johnson photo. *Courtesy* Appaloosa Horse Club, Inc.)

2. Defective Eyes (U) — impaired vision or blindness.
3. Poll Evil (U) — inflamed swelling of poll between ears.
4. Roman Nose — faulty conformation.
5. Parrot Mouth (U) — lower jaw is shorter than upper jaw.
6. Undershot Jaw (U) — upper jaw is shorter than lower jaw.

NECK

1. Ewe-Neck — faulty conformation.

WITHERS AND SHOULDERS

1. Fistula of Withers (U or B) — inflamed swelling of withers.
2. Sweeny (U) — atrophy or decrease in size of a single muscle or group of muscles, usually found in shoulder or hip.

BODY

1. Heaves (U) — difficult breathing, lung damage.
2. Roaring (U) — difficult breathing due to obstruction usually in larynx.

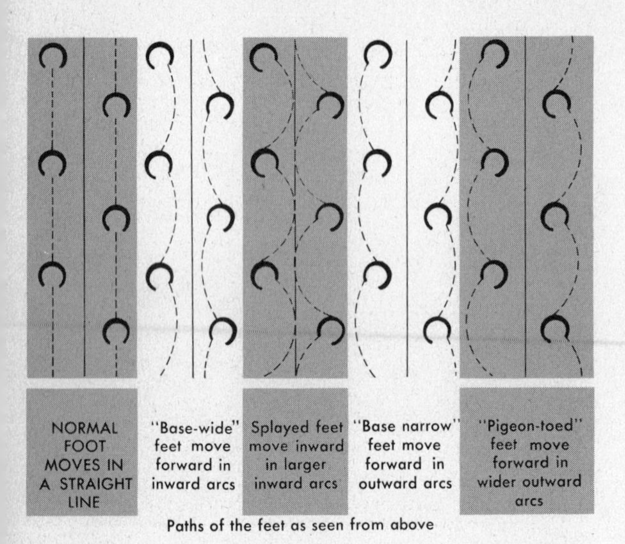

FIGURE 24-15. (*Above*) **Foot action of a normal horse and of horses with undesirable underpinning characteristics.** (*Courtesy* **Appaloosa Horse Club, Inc.**) **FIGURE 24-16.** (*Below*) **Locations of common unsoundnesses of a horse.** (*Courtesy* **University of Idaho**)

3. Rupture (U) — protrusion of internal organs through the wall (hernia) of the body. Umbilical or scrotal areas most common.
4. Sway Back — faulty conformation.
5. Hipdown (U) — fracture of prominence of hip and falling away.

FRONT LEGS

1. Shoe Boil or Capped Elbow (B) — soft, flabby swelling at the point of elbow.
2. Knee — sprung or buck knee — over on the knees. Faulty conformation.
3. Calf-kneed — back at the knees. Faulty conformation.
4. Splint (B) — capsule enlargement usually found inside upper part of front cannon.
5. Wind Puff (U) — puffy swelling occurring either side of tendons above fetlock or knee.
6. Bowed Tendons (U) — enlarged, stretched

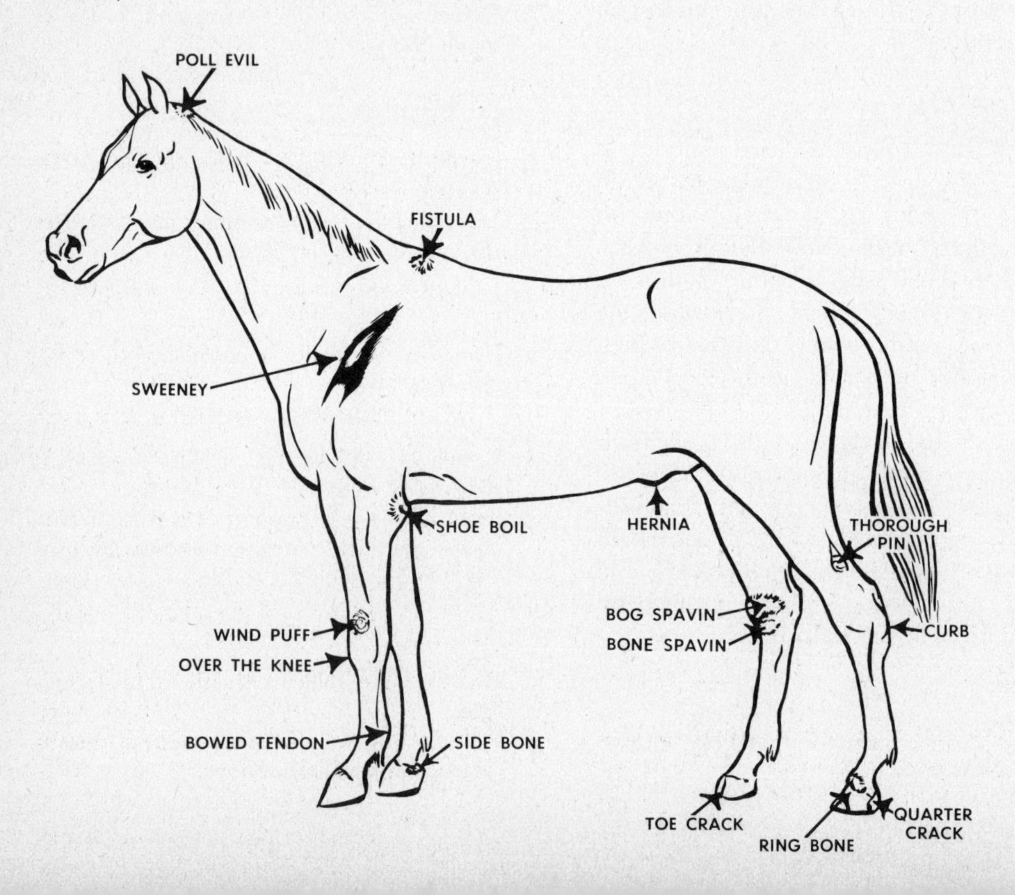

flexor tendons behind the cannon bones.
7. Ringbone (U) — bony growth on either or both sides of pastern.
8. Sidebone (U) — bony growth above and toward the rear quarter of hoofhead.
9. Quittor (U) — fistula of the hoofhead.
10. Quarter of Sand Crack (B) — vertical split in the wall of the hoof.
11. Navicular Disease (U) — inflamation of small navicular bone usually inside front foot.
12. Founder (U) — turning up of hoof and rough, deep rings in hoof wall caused by over feeding, severe concussion or disease and abnormal management.
13. Contracted Feet (B) — abnormal contraction of heel.
14. Thrush (B) — disease of the frog.

REAR LIMBS

1. Stifled (U) — displaced patella of stifle joint.
2. Stringhalt (U) — nervous disorder characterized by excessive jerking of the hind leg.
3. Thoroughpin (U) — puffy swelling which appears on upper part of hock and in front of the large tendon.
4. Capped Hock (B or U) — enlargement on point of hock. Depends on stage of development.
5. Bog Spavin (U) — meaty, soft swelling occurring on inner front part of hock.
6. Bone Spavin or Jack Spavin (U) — bony growth usually found on inside lower point of hock.
7. Curb (U) — hard swelling on back surface of rear cannon about four inches below point of hock.
8. Cocked Angle (U) — usually in hind feet, horse stands bent forward due to contracted tendons.
9. Blood Spavin (B) — swelling of vein, usually below seat of bog spavin.

DISPOSITION

The disposition of a horse helps determine the enjoyment and safety of the rider or driver. While certain qualities of disposition are inherited, others are acquired. Improper care, careless management, and undisciplined training can ruin the disposition of a good horse. A horse has an excellent memory. Poor treatment as well as good treatment will be long remembered.

Disposition is very important, particularly if the animal is to be used by children. Some breeds and lines of breeding are more docile and gentle than others.

It is wise to watch the eyes and ear movement of a horse to detect any tendency to be nervous. The reaction of a horse to noise, movement, and to other horses is good indication of its disposition. Leading, riding, or driving the horse will give an indication of its willingness to respond to commands and to its dependability. It is well to have the owner examine the horse's front and back feet and its mouth. Whenever possible, buy a horse subject to a specified trial period.

VICES

Horses may develop a number of vices (bad habits or tricks) which depreciate their usefulness and value. Some vices are dangerous to owner and horse. Usually these vices can be ascertained before a horse is purchased.

Some horses will not stand still; they may move to the side, forward, or backward. Some horses balk, shy away, or rear up on their hind legs. Some will run away. Others do not like being bridled, saddled, or harnessed.

Some horses eat or drink too fast. Others are cribbers—they bite the manger or feed box. Some pull on the tie rope or bridle strap, or weave back and forth in the stall.

It is often difficult to break older horses of vices. Most horseowners prefer to buy young animals and begin to train them before any vices are acquired.

PEDIGREE

More than 160,000 horses are recorded with the various breed associations during an average year. The pedigrees of these horses provide prospective buyers with valuable information. This is especially true if the horse or its ancestors have been raced or exhibited.

The sire and dam and the grandsires and granddams should be studied carefully to determine if the animal has in its ancestry the qualities needed for its intended use. Buyers should inspect as many horses as possible with similar bloodlines. The pedigree alone is not sufficient evidence of the prepotency of the bloodline.

PRICE

A high prices does not always reflect the true value and usefulness of a horse. Neither does a low price always indicate an inferior animal. Supply and demand usually determine price. When demand is brisk, superior animals command a good price, but weakened demand will lower prices and a horse may sell for less than it is really worth. A few years ago the demand for Shetland ponies exceeded the supply and prices rose accordingly. Currently there appears to be an oversupply of ponies and prices are comparatively lower.

It is often wise to buy when there are the fewest buyers. This may mean buying in the fall or winter rather than in the spring and summer when the buyer wants to ride the horse.

There is much satisfaction in owning and riding a good horse. The buyer should consider investing in a horse that will provide the needs and satisfaction required. It costs very little more to keep a superior horse than an ordinary one.

SUMMARY

The following factors should be considered in selecting a horse: intended use, breed, age, conformation, action, soundness, disposition, vices, pedigree, and price.

Horses may be classified as saddle horses, light harness horses, ponies, and draft horses. Most of the horses in the nation are

FIGURE 24-17. This yearling trotter, Bye Bye Sam, sold for $71,000 in Lexington, Kentucky. (*Courtesy* United States Trotting Association)

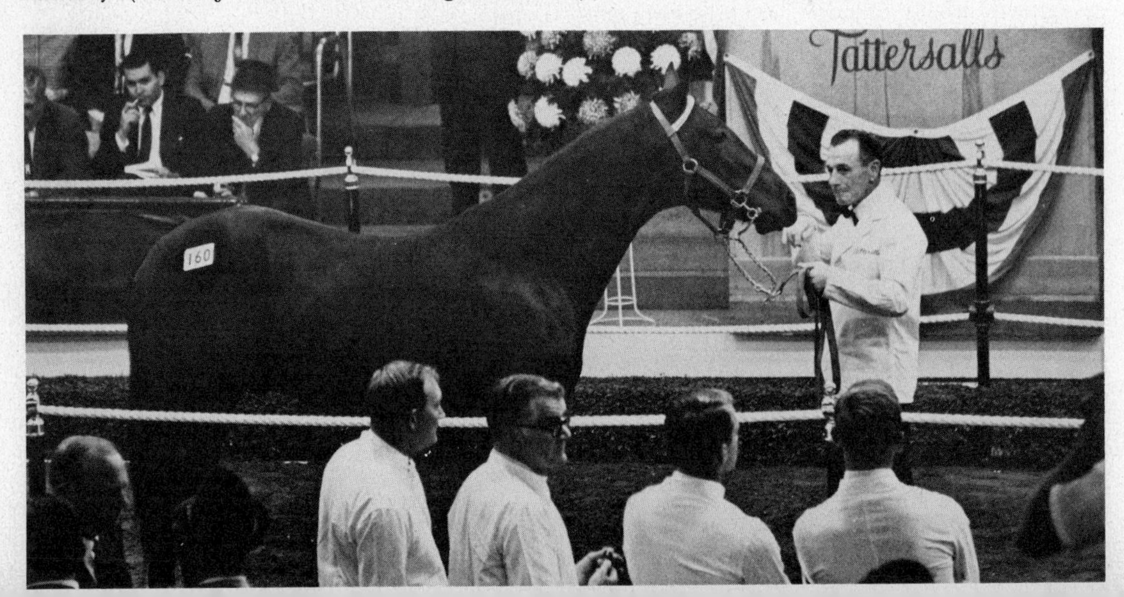

of the saddle horse class. Ponies rank second and light harness horses third in numbers. Very few draft horses are available.

The American Quarter Horse, the Appaloosa, and the Thoroughbred provide most of the saddle horses. More Quarter Horses are recorded each year than all other breeds of riding horses.

The Standardbred and Hackney breeds provide most of the light harness horses. Most running horses carry Thoroughbred or Quarter Horse breeding, whereas most trotters and pacers in harness racing are of the Standardbred breeds. The Arabian, Thoroughbred, and Standardbred breeds were used in developing many of the other breeds.

The age of a horse can be determined by examination of the teeth. The number of permanent incisor teeth, the extent that the cups are worn, and the angle of the bite indicate age. Older horses have smooth mouths. The cups are worn smooth and the bite is at an increased angle.

A well-formed horse has long, sloping shoulders; a strong, short back; a long underline; and a long, rather level croup.

In judging light horses, general appearance counts 12 percent; head and neck, 8 percent; forequarters, 22 percent; body, 12 percent; hindquarters, 31 percent; and action, 15 percent.

Other factors to be considered in selecting a horse are: unsoundnesses, blemishes, vices, disposition, pedigree, and price. It usually pays to buy the horse that will give the owner complete satisfaction.

BREED REGISTRY ASSOCIATIONS

Light Horses

American Albino Record Ass'n.
P.O. Box 79
Crabtree, Oregon 97335

American Hackney Horse Society
527 Madison Avenue
New York, NY 10022

American Morgan Horse Club, Inc.
Box 265 Westlake Moraine Rd.
Hamilton, NY 13346

American Paint Horse Ass'n.
P.O. Box 12487
Ft. Worth, Texas 76116

American Quarter Horse Ass'n.
P.O. Box 200
Amarillo, Texas

American Saddle Horse Breeders Ass'n.
929 S. 4th Street
Louisville, Kentucky 40203

Appaloosa Horse Club, Inc.
P.O. Box 403
Moscow, Idaho 83843

Arabian Horse Club Registry of America
One Executive Park
7801 E. Belleview Avenue
Englewood, Colorado 80110

Jockey Club Thoroughbred Registry
300 Park Avenue
New York, NY 10022

Palomino Horse Ass'n.
22049 Devonshire Street
Chatsworth, California 91311

Palomino Horse Breeders of America
P.O. Box 249
Mineral Wells, Texas 76067

Pinto Horse Association of America, Inc.
P.O. Box 3984
San Diego, California 92103

Tennessee Walking Horse Breeders Ass'n. of America
P.O. Box 286
Lewisburg, Tennessee 37091

United States Trotting Ass'n.
750 Michigan Avenue
Columbus, Ohio 43215

Ponies

American Shetland Pony Club
P.O. Box 2339
West Lafayette, Indiana 47906

Pony of the Americas Club, Inc.
1452 N. Federal
P.O. Box 1447
Mason City, Iowa 50401

Welsh Pony Society of America, Inc.
202 N. Church Street
West Chester, Pennsylvania 19280

Draft Horses, Jacks, and Jennets

American Suffolk-Shire Horse Ass'n.
300 East Grover Street
Lynden, Washington 98264

Belgian Draft Horse Corp. of America
P.O. Box 335
Wabash, Indiana 46992

Clydesdale Breeders Ass'n. of the U.S.
Batavia, Iowa 52533

Percheron Horse Ass'n. of America
RFD 1
Belmont, Ohio 43718

Standard Jack and Jennet Registry of America
RR #7
Rodds Road
Lexington, Kentucky 40508

QUESTIONS

1. What factors should you consider in selecting a horse?
2. List and describe the three most popular breeds of saddle horses.
3. List and describe the two most popular breeds of light harness horses.
4. Which breeds of horses and ponies are from imported bloodlines?
5. Name the breeds which have been developed in the United States.
6. Which breeds provide the horses for the Kentucky Derby and other running races?
7. What breeds provide the horses for the trotting and pacing races?
8. Describe how you can tell the age of a horse by examination of its teeth.
9. Describe the body conformation and feet and legs of a desirable saddle horse.
10. Describe desirable action of a Tennessee Walking Horse.
11. List the blemishes and unsoundnesses most likely to impair the usefulness of a riding horse.
12. How would you determine the disposition of a pony that you were selecting for a youngster?
13. Of what value is a pedigree in the selection of a saddle horse?

REFERENCES

Appaloosa Horse Club, Inc., *Horsemanship Manual,* Appaloosa Youth Program, Moscow, Idaho, 1971.

American Quarter Horse Assn., *Judging American Quarter Horses,* Amarillo, Texas, 1969.

Bradley, Melvin and Slemp, W. H., *Fundamentals of Conformation and Horse Judging,* University of Missouri, Columbia, Missouri, 1970.

Bradley, Melvin, *Unsoundnesses and Blemishes of Horses: Feet and Legs,* University of Missouri, Columbia, Missouri, 1967.

Bradley, Melvin, *Selecting Your Riding Horse,* University of Missouri, Columbia, Missouri, 1971.

Ensminger, M. E., *Horses and Horsemanship, Fourth Edition,* Danville, Illinois, Interstate Printers and Publishers, Inc., 1969.

Federal and State Cooperative Extension Service, *Horse Science,* Washington, D.C., 1969.

FS Services, Inc., *Shaping and Showing Horses,* Bloomington, Illinois, 1970.

25 Nutrition and Feeding

The amount and kind of feed to give their horses has always concerned horseowners. Knowledge about the nutritional requirements of horses is not as comprehensive as it is for other farm animals. Fortunately, considerable research on the nutrition of light horses is currently underway. It will help the industry determine such things as the energy, protein, mineral, and vitamin requirements and the feeds which best supply these nutrients. The nutritional needs of the horse will vary according to the demands which are made on him. There is no doubt that feeding horses for racing, show and performance purposes should differ from feeding during breeding, gestation, lactation, and growth.

DIGESTION IN HORSES

A horse's digestive tract is quite different from that of a cow. The cow has a rumen (large fermentation vat where much of the digestion occurs) in the forepart of the digestive tract. The horse has a relatively small, simple stomach but does have a large cecum and colon located between the small and large intestines. To a degree, the cecum and colon serve the same purpose for the horse that the rumen does for the cow. However, the cecum's location toward the end of the digestive tract reduces the horse's overall digestive efficiency. Intelligent feeding depends upon an understanding of the parts and functions of a horse's digestive system.

The Digestive Organs

The main parts of the digestive system in horses are the mouth, stomach, small intestine, and large intestine (cecum, large colon, and small colon).

Mouth. Digestion starts in the mouth. The feed is masticated by the teeth and mixed with saliva. Approximately ten gallons of saliva are secreted daily in a mature horse.

Stomach. Food taken in through the mouth moves down the esophagus into the stomach. The stomach capacity of the horse is 8 to 16 quarts which is small compared to about 200 quarts for the mature cow. The stomach secretes gastric juices by which proteins and fats are broken down. Food is passed quite readily out of the stomach and into the small intestine.

Small Intestine. The small intestine is the tube connecting the stomach to the large intestine. It is here that enzymes of the pancreas and liver assist in further breaking down the proteins, fats and sugars.

Large Intestine. The large intestine of the horse is divided into the cecum, large colon and small colon. In the cecum, digestion continues, limited vitamin synthesis occurs, and nutrients are absorbed. In the large colon there is a continuation of digestion, bacterial action and absorption of nutrients. In the small colon, the contents of the digestive tract become solid and the feces are formed.

Unlike ruminants, the cecum (the horse's fermentation vat) follows the small intestine. Thus, relatively small amounts of the cellulose of feed is digested, concluding that, in comparison to the cow, the horse should be fed less roughage, higher quality proteins and more concentrates.

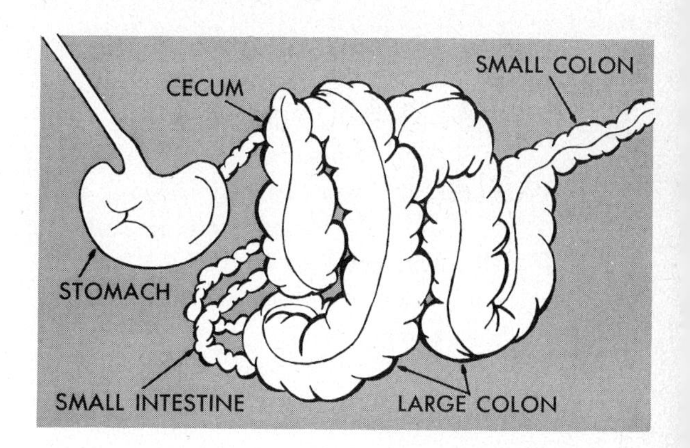

FIGURE 25-1. The organs which make up the digestive system of a horse. Note how large the cecum and large colon are compared to the other organs.

GENERAL HORSE MANAGEMENT

A good feeding program for horses depends almost as much on the feeder as it does on the feed. A well-balanced ration is the first requirement, yet there is no substitute for dependability, regularity, alertness, ability to work hard, and integrity in a successful horse feeder. There is a need for individual feeding and for a constant study of the peculiarities of each individual horse.

The level of feeding will vary with different horses and with the response being obtained. It will also depend on their condition, as well as the growth or activity desired. The quality of the ration, however, needs to be high with all horses, in order to supply them with all the essential nutrients they need to build a healthy, sound body.

Feed must be furnished on a regular schedule, especially when horses do not have access to pasture. Most horseowners feed at least two times each day. A change in ration should always be done gradually. Abrupt and rapid changes will throw animals off feed and will result in digestive disturbances. Colic (stomach-ache) can be caused by overfeeding. If the horse eats too rapidly, a larger feed box can be used or one can place a few smooth rocks, at least the size of baseballs, in the feed box. This forces the horse to eat around the rocks rather carefully. The feed boxes should be kept clean at all times. Old, dusty, moldy, or spoiled feed should not be fed to horses.

Exercise and Water

Horses should get adequate exercise. This helps them improve their appetite, digestion, and over-all condition. Pastures provide top-quality feed, exercise area, and a natural environment for developing horses.

A clean, fresh water supply should always be available. A mature horse needs 10 to 12 gallons of water a day and even greater amounts when subjected to hard work or hot weather. Horses exercised or worked very heavily should not be allowed to drink a heavy fill all at once for digestive disorders may result. Allow the horse to rest and cool down before drinking.

Droppings

The horse droppings should be observed frequently and watched for consistency. Any unusual color, odor, or abnormal characteristics observed is a good way to detect early signs that something adverse may be happening to the horse.

Teeth

A dental examination should be performed two to four times a year, depending upon the horse's age and the ration made available. At times a horse's teeth may need to be floated (sharp edges taken off). There are a number of dental occlusions or abnormalities which may affect the horse's ability to eat.

Value of Experience

There is no substitute for experience in feeding horses properly. The beginner should work with and learn from an experienced horseowner. Do not let fads, fancies, and trade secrets influence good judgment. Magic potions and tonics will decrease as more research information is obtained and horse feeding becomes more scientific.

FEED FOR HORSES

About 75 percent of the cost of raising horses is for feed. Overfeeding can be wasteful and expensive, and underfeeding or a nutritional deficiency will not permit optimum performance. Since horses depend so

FIGURE 25-2. No one feedstuff for horses is as complete in nutrients as green pasture. Pastures also provide excellent areas for exercise. (Key photo. *Courtesy* United States Trotting Association)

much on their wind, feeds should be clean and free of mold and excessive dust.

Pasture

Grass is the natural feed for horses. No one feedstuff is as complete in nutrients as green pasture grown on fertile soil and few feeds are fed under a more healthful environment. Succulent pastures may, however, have some laxative effects and produce a greater tendency for a horse to sweat. Legume pastures are excellent for horses with little danger of bloat. The specific grass or grass-legume mixture will vary from area to area, according to differences in temperature, rainfall, and soil. The state agricultural college can furnish recommendations for your area.

Grass or Nonlegume Hays

Grass hays yield less per acre, are lower in protein, calcium, and vitamins, but are less likely to be moldy and dusty than legumes.

Timothy. No other hay has attained the lasting popularity of timothy. Its wide range of climatic adaptability, ease of curing, bright color, and freedom from dust and mold make it the horseman's favorite. If it is fed as the only roughage, it should be supplemented with protein or be fed with a high-protein grain such as oats for it is low in this nutrient.

Prairie Hay. Some horseowners substitute prairie hay satisfactorily for timothy. It is lower in protein, less bright in color, and usually less palatable than timothy.

Bromegrass Hay. Bromegrass hay makes good horse hay. It is palatable when harvested in the bloom stage.

Cereal Hays. Cereals make good hays when cut early. They should be cut in the soft dough stage. They are seldom cut early enough. Oats, barley, wheat, and rye hays are preferred in this order.

Legume Hays

Legume hays are higher in protein and minerals and are more palatable than grass hays. They make excellent horse feed and should be included in the rations of young growing animals, breeding animals, and many adult working horses.

Alfalfa Hay. When properly cured, alfalfa is the best of the legumes from a nutrient standpoint. Its high protein, calcium, and vitamin content make it especially useful in balancing rations for brood mares and young growing horses. Some horseowners consider it "too laxative," and feed it as half of the roughage.

Clovers. Many varieties of clover are used alone or in combination with grass hays for horses. Red clover can be substituted for alfalfa with slightly less beneficial results. Properly cured alsike clover is a good hay for horses.

Lespedeza. When cut early, lespedeza makes an excellent hay. When cut late many leaves are lost from shattering and the stems become wiry and low in digestibility. It is higher in protein than red clover. The calcium content is about half that of alfalfa.

Silage

Various types of silage can be used to replace half of the hay ration. It must be of good quality, free of mold, and should be chopped fine. Good corn silage is preferred, but milo and grass silage can be used. Silage is too bulky for hardworking animals and foals. Legume haylage can replace silage with equal or superior results.

Grains

Horses like grains. Grains are high-energy feeds used with hay to regulate the horse's work, growth, or reproductive performance. Medium-sized, hardworking horses may need as much as 12 lbs. of grain and an equal amount of hay daily to maintain body weight; whereas idle adult horses may get fat on grass alone. In general, grain rations should be cut in half and hay increased on days that working horses are idled. Most grains are improved by grinding or rolling, but none should be ground fine.

Oats. Oats are first choice among grains by most horseowners and horses. Oats are higher in protein than other grains, which make them useful with low-protein grass hays. However, half legume hay insures a more complete ration when oats are fed as the only grain.

Corn. Corn is a good horse feed and is used extensively in the Midwest. Because of its high energy content and low fiber, corn must be fed with more care than oats to avoid colic. Corn and oats, equal parts, make an excellent grain ration. Corn can supply all the grain needed when fed according to the work that horses are performing, but large amounts of corn should not be given at one time.

Milo. Milo can substitute for corn in most rations. It is most satisfactory when used in a grain mixture. In some areas milo is often more economical than corn.

Barley. Barley is a very satisfactory feed when ground. Fifteen percent wheat bran or 25% oats fed with barley almost eliminates the risk of colic.

Wheat. Wheat is seldom fed to horses except in the Pacific Northwest. It can be fed as a part (about ⅓) of the grain ration. It should be rolled or coarsely ground. Wheat tends to be doughy when moist and creates digestive problems.

Wheat Bran. Wheat bran is highly palatable, slightly laxative and very bulky. Bran is reasonably high in protein, high in phosphorus, and like other grains, low in calcium.

Protein Feeds

The horse's need for protein is relatively low and is easy to meet with practical rations. With the exception of milking mares, most 600–1200 lb. horses need from ¾ to 1 lb. of digestable protein daily. If the roughage is half legume hay fed in adequate amounts, the protein need will be met. However, supplementing rations of milking mares and young growing horses is insurance against a deficiency and stimulates appetite.

Linseed Meal. Linseed meal produces bloom or lustre to the haircoat and is held in high esteem by many horse show personnel. Its laxative effect and palatability make it useful with roughages of poor quality.

Soybean Meal. With the exception of producing bloom, soybean meal is a preferred supplement to linseed meal. It is higher in protein and has a better balance of amino acids.

Cottonseed Meal. Cottonseed meal is used extensively for horses in the Southwest. It is not as palatable as linseed or soybean meal but is higher in phosphorus content.

Other Protein Supplements. Alfalfa meal, corn gluten meal, meat meals, dried milk products, and others can be used with horses as protein supplements.

Commercial Protein Supplements. These vary in composition, protein level, and price. They often contain needed minerals and vitamins and are convenient for horsemen who do not formulate their own horse rations. Commercial supplements are usually formulated for a specific feeding program and should be fed according to directions.

Minerals

Minerals are needed to develop sound feet and legs. Bone abnormalities, lameness and similar leg problems ruin or eliminate many horses from a useful life or from developing their full potential. Developing good, strong, sound bone, however, involves more than just calcium and phosphorus. It also involves other minerals and nutrients.

Common farm feeds provide minerals, but most horses need extra calcium, phosphorus, salt, and iodine. The daily salt requirement is about 2 to 3 ounces. The calcium requirement is about 0.5 percent of the ration and the phosphorus requirement is about 0.4 percent. A good way to furnish supplemental minerals is to offer a free-choice mixture of equal parts of dicalcium phosphate and trace-mineralized salt in a box protected from the weather.

Vitamins

Horses with access to good pasture and those receiving good quality hay, especially if half legume, will probably need no vitamin supplementation. Deficiencies are more likely to appear with horses in confinement for long periods of time on poor quality roughage. Green grasses and hays furnish carotene that the horse converts to vitamin A. Vitamin D is obtained from sunshine. A horse needs about 20,000 to 30,000. International Units (I.U.) of vitamin A and about 2,000 I.U. of vitamin D daily.

Green feeds and wheat products such as wheat bran and wheat germ oil are rich in vitamin E which is often associated with reproduction. Economical supplements of vitamins A, D, and E can be mixed in the feed, injected intramuscularly, or furnished in stabilized mineral blocks.

Antibiotics

There is practically no information available on the value of antibiotics for horses. Many horseowners are using them and are obtaining some results. Antibiotics may be helpful for young foals which suffer setbacks because of infections, digestive troubles, lack of milk, poor weather or other stress factors. The feeding of 85 milligrams of Aureomycin daily to foals up to 1 year old improves growth rate slightly.

FEEDING SUCKLING FOALS AND WEANLINGS

The first year of life is a very critical period in the development of horses. Usually, the mare's milk will supply all of the nutritional requirements of the foal for the first three or four mouths of its life, provided the mare is well fed. After this time, the foal will have to obtain a significant amount of nutrients from other sources. Before the foal is weaned, it should be familiar with the type of dry feed it will be receiving after weaning. During the suckling phase, this may be accomplished by furnishing a creep separate from the mares. At weaning time or 5 to 6 months of age, the foal should be consuming six to eight pounds of feed per day. After weaning, about 1 pound of hay per 100 pounds of body weight should be added to the daily ration. The grain portion is gradually increased up to a maximum of 10–12 pounds per day depending upon what the foal will voluntarily clean up. Continue the weanlings on a feeding program of about 1 pound of hay per 100 pounds of body weight daily plus sufficient grain mix to furnish energy to obtain the desired growth rate.

FIGURE 25-3. These foals appear to be in excellent health. (*Courtesy* American Albino Association)

FEEDING YEARLINGS AND 2-YEAR-OLDS

At this age and stage of development, it is possible to rely more extensively on hay and pasture. It is, however, very important to assure an adequate supply of protein, minerals, and vitamins so that the horses will continue the growth and maturity of their tissue and bone structure. The total amount of feed should be based upon requirements for maintenance and the growth that is desired.

Training

As a horse goes into training, it is again best to base feeding practices on maintaining the desired condition and development and growth of the horse. The work of training will increase the energy requirements; however, since the intensiveness and amount of work involved will be so variable, it is necessary to use a great deal of judgement in feeding horses at this stage. The same principles apply in feeding the performing or working horse. Each animal should be fed as an individual to maintain the desired condition and weight.

FIGURE 25-4. Nigatt, a champion Arabian stallion, owned by Bonnie and Leon Matthias, Cedar Falls, Iowa. (Knoll photo. *Courtesy* Matthias Arabians)

FEEDING THE MATURE DRY MARE OR GELDING

The mature animal which is idle or ridden as infrequently as once or twice per week can be maintained on hay or pasture alone. A calcium-phosphorus and salt supplement should be furnished to such horses. A combination of hay and grain may also be used. Oats or another cereal grain fed with good hay will be adequate.

FEEDING THE LACTATING MARE AND STALLION

The same ration fed to yearlings and two-year-olds can be used for the lactating mare and stallion. In the case of the mare which has just foaled, is lactating, and is being prepared for rebreeding, nutritional requirements increase. She must also be in a top state of health to accomplish all these functions. The mare should be in good flesh prior to foaling so that her reserves of nutrients are built up and then she should be raised to a high level of feeding as soon after foaling as possible. The actual level of feeding is based on her total energy requirement for maintenance plus lactation with an additional amount of feed which should allow her to begin to gain weight and get in the best possible breeding condition. This high level of feeding should be continued following breeding and into lactation until the mare begins to gain excess weight.

Stallions can be fed the same ration the year round. The amount, however, is based on their requirements for energy maintenance with additional allowance during the breeding season to keep him in good breeding condition.

Pelleted Rations

Good hay is often difficult to find, and storing hay is sometimes a problem. Some

TABLE 25-1 PROTEIN REQUIREMENTS OF DIFFERENT CLASSES OF HORSES

Class of Horse	Protein as % of Ration
Weanling foals	16–18%
Yearlings, 2 year olds, wet mares, show and performance horses, and stallions	12–14%
Mature, idle horses	10–12%

TABLE 25-2 FOAL CREEP AND STARTING RATION

Ingredient	Percent of Ration
Oats	45.0
Corn	18.5
Soybean meal	15.0
Dried whey	10.0
Alfalfa meal	5.0
Molasses	5.0
Limestone	0.5
Defluorinated phosphate	0.5
Salt (trace mineralized)	0.5
	100.0

TABLE 25-3 WEANLING RATION

Ingredient	Percent of Ration
Oats	45.0
Corn	23.0
Soybean meal	20.0
Alfalfa meal	5.0
Molasses	5.0
Limestone	0.5
Defluorinated phosphate	1.0
Salt (trace mineralized)	0.5
	100.0

TABLE 25-4 YEARLING, TWO-YEAR-OLD, TRAINING AND PERFORMANCE LACTATING MARE AND STALLION RATION

Ingredient	Percent of Ration
Oats	60.0
Corn	18.0
Soybean meal	10.0
Alfalfa meal	5.0
Molasses	5.0
Limestone	0.5
Defluorinated phosphate	0.5
Salt (trace mineralized)	1.0
	100.0

horseowners have tried pelleted rations (both hay and grain) and have had good results. Pelleting does not increase the digestibility of the ration but does speed its passage through the digestive tract. Thus a horse fed a pelleted ration may become bored, chew on his stall, or become a cribber. Because some vices may develop when the entire ration is pelleted, about 25 percent of the daily ration should be long hay.

SUMMARY

The amount and kind of feed to give their horses has always concerned horseowners. Nutritionists agree that the nutritional needs of the horse will vary according to the demands which are made of it.

The digestive tract of the horse is composed of the mouth, stomach, small intestine and large intestine (cecum, large colon, and small colon). Unlike ruminants, the cecum (the horse's fermentation vat) follows the small intestine. Thus, relatively small amounts of the cellulose of feed is digested, concluding that, in comparison to the cow, the horse should be fed less roughage, higher quality proteins and more concentrate.

A high-quality ration is important, but, regularity and the number of times the horse is fed and watered are also factors in sound feeding management. Grain and hay must be free of dust and mold. Feed boxes and waterers should be clean. Exercise is important for it aids in sound body structure and release of nervous tension. There is no substitute for experience in feeding horses properly. Work with experienced horseowners for much can be learned from them.

About 75 percent of the cost of raising horses is for feed. No one feedstuff is as complete in nutrients as green pasture. Both grass and legume pastures are satisfactory for horses. Several kinds of grass and legume hays are used in horse rations. Timothy, prairie grass, bromegrass, alfalfa, red clover, and alsike clover are examples of the more common ones. Silage can be fed to horses, but not over half of the roughage ration should be silage.

Several grains are fed to horses with oats being the most common. Corn, milo, barley, and wheat are other recommended grains. Most grains are improved by grinding or rolling, but none should be ground fine.

When rations call for extra protein, linseed meal, soybean meal or cottonseed meal may be added. Alfalfa meal, corn gluten meal, meat meals or dried milk products can also be used. Some owners use commercially formulated protein supplements to balance grain and roughage rations.

Minerals are needed for sound bone development and maintenance. The main minerals needed are calcium, phosphorus, salt, and iodine. Trace minerals can be furnished by the use of trace-mineralized iodized salt in loose or block form.

Very little is known about the vitamin requirements of horses. Horses with access to good pasture and those receiving good quality hay will probably need no vitamin supplementation.

The first year of life is a very critical period in the development of horses. Usually, the mare's milk will supply all the nutritional requirements of the foal for the first three or four months. Foals should be started on a dry feedstuff ration by using a creep before they are weaned. Roughage is added to the grain to complete the ration after weaning. Yearlings and two-year-olds are fed according to the desired rate of growth and development. During training, the amount of the ration is increased to meet the higher requirement of nutrients.

Mature dry mares and geldings which are idle or ridden infrequently can be maintained on hay or pasture alone. However, minerals should be furnished to such horses.

A mare should be in good flesh prior to foaling and the amount of the ration increased as she progresses into lactation and the rebreeding period. She should actually be fed enough to gain in weight.

Stallions can be fed the same ration the year round. The amount fed is determined by his breeding activity and general energy maintenance requirements.

The entire ration (both hay and grain) may be fed in pellet form. Since this form of ration passes through the digestive tract faster, some horses may become bored, chew on the stall or develop some other vice. It is recommended that some long hay be fed with pelleted rations.

QUESTIONS

1 Name the parts of the digestive system in horses.
2 How does digestion in horses differ than in ruminants?
3 What can be done to make the horse eat grain slower?
4 Why is daily exercise so important to a horse?
5 List the hays and grains available on your farm which can be satisfactorily fed to horses.
6 What is the main disadvantage of feeding complete pelleted rations to horses?

REFERENCES

Bradley, M., and W. H. Pfander, *Feeds For Light Horses,* Extension Guide 2806, Columbia, Missouri, University of Missouri, 1966.

Ensminger, M. E., *Horses and Horsemanship,* Fourth Edition, Danville, Illinois, The Interstate Printers and Publishers, 1969.

Jordan, R. M., *Horse Nutrition and Feeding,* Extension Bulletin 348 (Revised), St. Paul, Minnesota, University of Minnesota, 1972.

26 Care and Management

Most people love horses, and a renewed interest in them has taken place since World War II. The demand for good horses has increased with the increase in recreation time, riding clubs, trail rides, and horse shows. Successful horseowners know that genetic background, environment, and training are essential in every good horse. Improper management will not allow the horse to develop and perform with desired results.

REPRODUCTION

Domestication has greatly reduced the horse's reproductive efficiency. The average conception rate is about fifty percent, with few breeders reaching seventy percent. Low fertility has mainly been caused by mating under relatively artificial conditions. Wild horses normally had little difficulty in reproducing.

Age of Puberty

Most female horses and ponies reach sexual maturity when they are 12 to 15 months of age, while the colt (young male) is normally mature at about two years. The age of maturity is largely determined by how well they have been fed.

Age to Breed

Size rather than age should be the main factor in deciding the proper time for first breeding. The additional nutritional demands during gestation are not great, but the demands during lactation are heavy. Most horseowners prefer to breed when the mare is three years old. If she is properly cared for, a broodmare should produce regularly up to fifteen or so years of age.

The Estrous Cycle

Estrus (or heat) is the period during which the mare will accept the stallion. The duration of estrus averages from 5 to 7 days

FIGURE 26-1. Reproductive efficiency is improved by the use of healthy mares and sound breeding management. (Dickinson photo. *Courtesy* American Quarter Horse Association)

but may vary from 3 to 12 days. Horses will come into estrus about every 21 days. The most desirable time to breed the mare is about two days before the end of estrus. Breeding on the third day of estrus and on alternate days thereafter as long as the mare remains in heat is a common practice. Normal healthy mares usually come into estrus between the seventh and ninth day after foaling and remain in estrus for 5 to 7 days. Where it is desirable to keep the mare on schedule, she can be bred on about the tenth day after foaling. If the mare has not completely recovered from foaling or has any signs of infection in her reproductive tract, she should not be bred. Many horsebreeders will not breed the mare during this heat period.

Management at Breeding Time

Mares should be taken to the stallion, if possible, because facilities where he is stood for service are usually more desirable. Most mares are bred by stallions on halter; however, artificial insemination is practiced to a limited degree. Make sure the mare is ready before she is serviced. This is usually determined by "teasing" her with the stallion and observing the reactions.

Hand Breeding. For protection of the stallion and for ease in handling, mares should be hobbled and a twitch applied. The hobbles prevent the stallion from being injured as he approaches and dismounts. The twitch is normally removed as soon as the stallion has entered the mare. Keep control of the stallion at all times and do not allow him to mount until he is ready. It is recommended that the external parts of the mare, which are likely to come in contact with the reproductive organs of the stallion, be washed with a soap solution before and after service to reduce the chance of spreading infection. Some horsebreeders will wrap the mare's tail during breeding. Following mating, mares should be allowed to remain quiet and avoid excitement or exertion.

Pasture Breeding. Pasture breeding is accomplished by turning the stallion into a pasture with the mares he is to breed. This practice is seldom done except in the range areas of the West. Pasture breeding is discouraged due to the possibility of injury, the difficulty of accurate breeding records and the reduction in the number of mares a stallion may service during the breeding season.

Gestation

The length of the gestation period averages 336 days or a little over eleven months. The period may range from 310 to 370 days after breeding. This large variation in days is largely due to the age, size, physical condition, health, and nutritional needs of the mare. Gestating mares should, if possible, be kept separate from other horses and permitted or forced to get regular and moderate exercise. Pasture that provides high-quality feed, exercise and sunshine is an excellent environment.

FIGURE 26-2. An excellent Hackney mare and her foal. Note the wood-type constructed fence which is recommended over the wire type. (*Courtesy* American Hackney Horse Society)

CARE OF THE MARE AND FOAL

It is difficult to determine the foaling date of a mare due to the great variation in length of gestation. Frequent observation is essential. About a month before foaling, the mare will develop a distended udder and within a week of foaling the mare usually shows a marked falling away of the muscular parts of the croup. Drops of wax will appear on the end of the nipples about two or three days before foaling while milk may not appear until a few hours before the foal is born.

The best place for a mare to foal is on pasture, assuming pasture is available and the weather is moderate. If this isn't possible, the mare should be housed in a large box stall free from dirt and drafts.

The labor pains of a mare are violent and she normally foals quickly. It is usually not more than an hour from the time labor starts until the birth. Normal presentation of foal is front feet first with head between forelegs. Should the foal be in need of changing position, a veterinarian or an experienced horseowner should be called. If the afterbirth isn't expelled within 4 to 6 hours, call a veterinarian.

Dip the navel of the foal in iodine shortly after birth to minimize infection. Usually the foal will be up attempting to nurse within two hours after birth and, if not, give it assistance. If the foal fails to nurse within six hours, give it some of the mare's first milk in a bottle. Following foaling, the mare should be given small amounts of water and fed lightly with a laxative-type feed for the first few days. If bowel movements of both the mare and foal are normal, they can be turned back to pasture. Be aware that constipation and diarrhea are common in foals and should be treated when necessary.

AGE OF WEANING AND CASTRATION

Sound management practices will help in the success of weaning the foal. One such practice is providing the foal with a creep ration, thus, getting it used to a dry ration.

Age to Wean

Foals are normally weaned at four to six months of age.

The weaning date is best determined by the condition of the mare and foal, if the mare was re-bred early, if the mare is being given heavy work, or when it is desirable to develop the foal to the maximum.

Weaning is accomplished by separating the mare and foal. Place the mare in an area some distance from the foal and dry her up immediately. Rub an oil preparation on the udder and observe it at daily intervals. After about seven days, the udder should be soft and pliable. Milk out what little secretion is left. The weanling will soon adjust to the separation if allowed to do so.

Age to Castrate

Castration is best performed by an experienced veterinarian. A colt may be castrated when only a few days old but most horseowners prefer the colt be about one year of age. This allows the colt to become more muscular, bolder in features, and have better carriage of the foreparts. The time of castration is best determined by the development of the individual. Less danger from infection occurs if colts are castrated in the spring soon after they have been on pasture. Avoid fly time and hot weather when possible.

HOOF CARE AND SHOEING

The old adage, "No foot, no horse" is as true today as it ever was. The feet of most horses require attention every month. Growing horses should receive special attention in order to maintain straight legs during periods of rapid bone growth.

Hoof Care

Hoof care should start with the foal. The foal should be taught to yield its feet at an early age so that the condition of the sole, frog, and heel can be examined. The frog should be pliable and the heels should flex when squeezed if the hoof has adequate moisture. If the hoof is too dry, splitting or infection can occur.

Trimming. Abnormal growth or uneven wear of the hoof indicates the need for trimming. A foal's hooves that are properly trimmed and rasped, when walking on a flat surface, will help straighten out or prevent crooked feet. Keep in mind, however, there are some faults which are not correctable such as twisted legs, cow-hocks, sickle hocks, or other problems involving malformation of the bone structure. Generally, a turned foot, but not the leg, can be corrected. (See Chapter 24 for further discussion.)

The best way to determine whether the foot is crooked but not the leg, is to observe the foal traveling. To do this, the foal should be led directly away, then directly toward the observer. If the foot is crooked and not the leg, the travel pattern will be improper. As an example, a foal's front feet are turned out, but not the leg; the knees are comparatively straight (normal). Traveling pattern of the front feet will be swinging the feet on an inward arc during forward phase of stride (winging). A foal with front feet turned in, but knees normal, will throw or swing its feet in an outward arc during rearward and forward phase of stride (paddling). Both of these situations can be corrected by proper trimming.

Shoeing

All hooves need care, but not all horses need shoes. To determine correct shoeing for any horse, one should consider for what pur-

FIGURE 26-3. Proper hoof care and correct shoeing assist the horse in its performance. (Spence photo. *Courtesy* American Paint Horse Association)

pose the horse is going to be used; type of conformation; type of shoeing required; type of feet on the horse; and the horse's training program. The way the cutting horse is shod is not necessarily the correct way to shoe a halter horse or a pleasure horse.

Horses are shod for the following reasons:

1. **To prevent the hoof from wearing down faster than it grows out.**
2. **To keep the hoof's bearing surface flat and encourage a straight gait.**
3. **To correct faults in gaits.**
4. **To relieve pain in the foot.**

While shoeing is important, a good farrier does far more than simply nail an iron bar on the end of the leg of a horse. He examines the horse to determine whether it stands straight and moves straight. Proper trimming is the foundation of any good shoeing job.

Basically, the types of shoeing can be divided into four categories: normal, corrective, pathological, and surgical. Normal shoeing is for foot protection, to assist in traction, mobility, and utility. Corrective shoeing is constructed to alter an undesirable way of travel or change a gait. Pathological shoeing is for a condition which has evolved due to disease, injury or illness causing a change which affects the horse's feet or legs. Surgical shoeing is when a shoe is utilized in conjunction with a surgical procedure.

A good farrier knows horses, the limitations of his own skills, and through experience, can appraise the chances of success in corrective shoeing better than the average horseowner. The need for corrective shoeing stems from the conformation of the horse and affects the way he goes.

TRANSPORTING HORSES

Safety, comfort, and economics are all important in moving a horse from place to place. Transportation by truck, rail, boat, and plane is utilized, but most horses are moved about by trailer. Healthy, relaxed, and well-trained horses load, ride and unload with little difficulty. Good equipment kept in safe repair is required. Bed properly with materials such as sand covered with bright straw, sawdust or shavings. When the horses are tied, use a slip knot which can be quickly released in case of an emergency. Provide for plenty of fresh air without drafts. Horses may be blanketed if the weather is cold.

Schedule the trip with time allowed for stops and emergencies. A health certificate, statement of ownership and brand certificate may be required depending upon the purpose of the trip.

STABLE MANAGEMENT

One of the most important activities of a good horseowner is sound stable management. Carefully planned, well-designed buildings and facilities make horse care easier and more enjoyable.

Housing

Sound basic planning is required when selecting and equipping a barn to stable horses. Two of the foremost considerations in planning a facility are health and safety, not only for the horse but for the horseowner as well. Sound construction, building materials, and labor-saving conveniences are some other important considerations.

The horse barn, large or small, should provide an environment that protects horses from temperature extremes, keeps them dry and out of the wind, provides fresh air in both winter and summer, and protects them from injury. It should also contain space for hay and bedding storage, feed storage, and a tack room.

Open-Front Buildings

An open-front shed will usually provide adequate protection from the weather. Face the open side of the building away from the direction of the prevailing winds. Recommended floor space is 60 to 80 square feet per 1,000 pound animal using the shelter. Additional room may be needed for storage of hay and bedding, feed, pens, etc. Proper ventilation can be controlled by using ridge devices, panels in the back wall, or windows which open inward.

Provide an adequate supply of water. This may require a heated waterer for winter use. A single heated water bowl will serve 8 to 10 horses. Some artificial light is desirable.

Enclosed Buildings

Barn layouts are determined by the function required and the type of building construction. Most horse barns provide either box stall or tie stall type housing.

Box Stalls. Box stalls are preferred for they allow the horse more freedom while standing or lying down. The size and use of the horse will help to determine the size of the stall as shown in Table 26-1. The most popular sized box stall is 12′ by 12′ which provides enough room for large horses as well as for maternity use. Every part of the

TABLE 26-1 SPACE REQUIREMENTS FOR HORSES IN BUILDINGS

	Dimension of stalls including manger		
	Box stall size		Tie stall size
Mature Animal (mare or gelding)	10′ × 10′	Small Medium Large	 5′ × 9′ 5′ × 12′
Brood mare	12′ × 12′	or larger	
Foal to 2-year old	10′ × 10′	Average	4.5′ × 9′
	12′ × 12′	Large	5′ × 9′
Stallion	14′ × 14′	or larger	
Pony	9′ × 9′	Average	3′ × 6′

TABLE 26-2 CHARACTERISTICS OF STALLS

Item	Box Stalls	Tie Stalls
Water	In-stall	Out-of-stall
Feed	In-stall	In-stall
Bedding	More required	Less required
Manure	More carrying	Less carrying
Exercise	Limited in-stall	Out-of-stall
Space	100–300 sq. ft.	45–60 sq. ft.
Floor	Clay or plank	Clay or plank
Partitions	Strong, tight	Strong, tight
Top Guard	On partitions	Manger end

FIGURE 26-4. This U.S.D.A. barn plan features three stalls, a feed room, a tack room, and a 12′ × 36′ working alley. There is ample overhead storage for hay and bedding. The barn is 24′ × 48′. Other U.S.D.A. plans are available. (*Courtesy* U.S.D.A.)

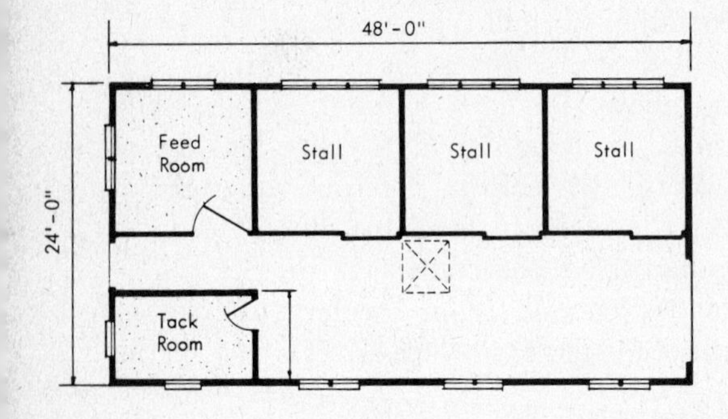

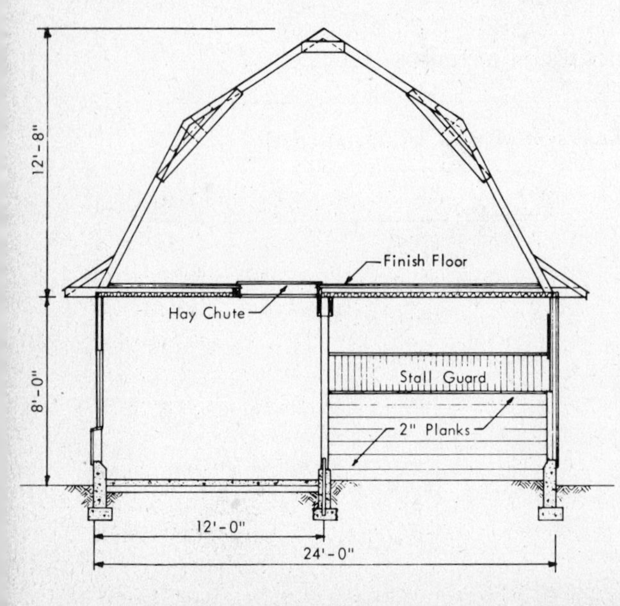

stall should be of sufficient height to prevent injury to the horse or horseowner. A minimum clearance is eight feet with nine feet preferred.

Tie Stalls. Tie stalls are usually five feet wide and nine feet long. They require less bedding and are easier to clean than box stalls. They can be constructed in existing buildings unsuitable for box stalls.

Regardless of the type of stall, there should be no projections on which the horse may injure itself. Feed boxes and hay mangers should be placed with safety and convenience in mind. Table 26-2 lists some characteristics of both types of stalls.

Alleys. For convenience, safety, and animal traffic, provide 10 feet or wider work alleys between the stalls. This allows room for moving horses, a small truck, or a tractor pulling a wagon or manure spreader. A ceiling height of at least twelve feet allows room for both the horse and rider.

The floors of the alleys and other work areas can be constructed of clay, concrete, asphalt, or brick while stall floors are usually of clay or wood. Clay floors are noiseless and springy, keep the hoofs moist, and provide firm natural footing unless wet. They are, however, more difficult to keep clean and level.

Bedding and Manure. Materials used for bedding include peat moss, straw, shavings, and sawdust. They will vary in their water-absorbing capacity and ease of handling. Manure management is an important part of caring for horses. Clean the stalls regularly and dispose of manure when possible. If storage of manure is necessary, locate the storage in an approved or safe area for convenient removal, away from any water source and out of natural drainage channels.

Feed Room. The feed room should be organized for convenience and easy housekeeping. Provide storage for feed materials, equipment and tools. Metal containers such as garbage cans and woodbins are excellent for limited bulk feed storage. Hay may be stored in an overhead loft or on the ground floor. With any type of feed storage, a good rodent-control program is essential.

Tack Room. An organized and well-maintained tack room that is dry and free of dust is important to good stable management. It can be a small room used mainly for storing riding equipment, or it can be large enough to serve as an office, a service shop for cleaning and maintaining tack, and a meeting and lounging place. The tack room may contain saddle racks, tack box, bridle and halter racks, shoeing box, first aid kit, clothes closet, office-type furniture, and heating and cooking facilities. The tack room is the center of most stable activities.

PADDOCKS AND CORRALS

Paddocks and corrals provide an area for exercise and close observation. Although paddock exercise is not as complete as that given under saddle or in harness, it will contribute to the general health of the unworked horse. Sodded paddocks that provide succulent and nutritious grass are preferable to barren areas which merely serve as exercise areas.

A well-designed and properly constructed corral serves as an exercise area as well as a training, observation, or loafing area.

PASTURE MANAGEMENT

Sound horse care and management involve the utilization of good pastures. Selecting adapted varieties of grasses and legumes when establishing a horse pasture is essential. Controlled grazing will extend the time during the year which a pasture can be utilized. The use of commercial fertilizer can add to the yield and quality of roughage grown.

Pastures should be free of holes and rocks to prevent injury to horses. Tree stumps, pits or like obstacles should be guarded. Water, shade, and minerals should be made available in all pastures.

Early pastures can be supplemented, if need be, with grass hay or straw. High-energy concentrates can also be used. Dry, mature pastures are low in energy, protein, and phosphorus which may become even more acute after a frost. Severe loss in condition of horses often follows the first fall freeze. Again, concentrates or supplements can be used to provide the nutrient requirements. Remember that no matter how good a pasture may be, it provides only roughage and should not be the entire daily ration for all kinds of horses.

FENCES

Horses require rather high, strong fences. Traditionally, fences for horses are made of wood, and it is still the most widely used material for fencing roadways, paddocks, exercise lots, and areas used for special and competitive events. However, materials such as pipe or four-inch diamond-woven wire may be used.

When wire fencing is used, select a mesh that a colt cannot get his feet through. Use a protective board on top of wire fences, and

FIGURE 26-5. A wood fence constructed of poles or boards is recommended for horses. (*Courtesy* American Quarter Horse Association)

under no conditions should barbed wire be used for horse fencing.

Use pressure-treated 5-inch diameter wood posts, or equivalent; space eight feet on center and set three feet in the ground. Pipe fences may be attached to metal posts set in concrete. Wood fences are normally made of 1 × 6 or 1 × 8 rough-sawed native lumber. The bottom board is set nine to twelve inches above the ground and additional boards spaced about eight inches apart. The normal height for most horse fences is five feet. If a higher and more rugged fence is desired, use two inch thick boards set on six inch diameter posts. Keep all fences in good repair to maintain safety and appearance.

OTHER HORSE EQUIPMENT

The design of horse equipment is most likely dominated by the fads and fancies of the horseowner. The basic needs are to provide hay, concentrates, minerals, and water to the horse without much waste or spoilage.

Feeders and Waterers

Hay Racks and Mangers. Some horseowners prefer feeding hay on the floor or ground while others use mangers or racks. Both methods have advantages and disadvantages. Feeding hay on the floor requires less equipment but is not necessarily the most economical. Feeding in racks will reduce contamination, pawing, and waste.

Grain Containers. Most grain feeding containers are made of wood, metal, hard rubber, or plastic. Regardless of type or material, the container should be easily removable for cleaning. Most horseowners will attach the container to a corner of the stall or set it in a wooden shelf or manger. Convenience to the horse and the attendant is important.

Watering. Watering facilities, either removable or stationary, should be of the type that can be drained and cleaned frequently. In cold areas, the stationary-type waterer should contain heaters. Tanks, troughs, pails, pans, etc. are examples of removable watering equipment. An adequate supply of fresh clean water must be provided for sound horse management.

Salt and Mineral Boxes. Salt and other minerals should be provided in separate boxes or compartments within one box made of wood. Metal containers are not generally used. Place the box at sufficient height to reduce contamination or rubbing upon it by the horse.

SUMMARY

The demand for good horses has increased with the increase in recreation time, riding clubs, trail rides, and horse shows. Sound management practices can improve the value of a good horse.

Domestication has greatly reduced the horse's reproductive efficiency. Sexual maturity in horses ranges from one to two years of age. Mares will usually come in heat every 21 days and remain in heat for 5 to 7 days. "Foal heat" occurs between the seventh and ninth day after foaling. Most horsebreeders will not breed their mares during this heat period.

Most mares are bred by stallions on halter, however, artificial insemination is practiced to a limited degree. Make sure the mare and stallion are ready before breeding takes place. Pasture breeding is practiced mainly in the western range country. Many good stallions have been ruined by being injured during pasture breeding. The length of gestation averages 336 days or a little over eleven months; however, the number of days may vary from 310 to 370 days.

The best place to foal a mare is on pasture. Signs of a mare preparing to foal include a distended udder and, later, wax and milk appearing on the end of the nipples. Labor pains of a mare are violent and birth should occur within an hour after labor begins. Normal presentation of foals is front feet first with head between the front legs. Should the foal be in need of changing position, a veterinarian should be called.

Most foals are weaned at four to six months of age. Drying up of the mare is done rapidly with little difficulty. Colts may be castrated at most any age, but horseowners will usually castrate when the colt is about one year old.

The old adage, "No foot, no horse" is as true today as it ever was. Hoof care should start with the foal and continue the rest of its life. Trimming the hoof does not correct all foot and leg problems. While shoeing is important, a good farrier will also observe the horse's conformation and movement. The purpose of the horse will mainly determine the kind or type of shoeing done.

Horses are usually transported by trailer. Bed the trailer floor with sand covered with straw or sawdust. Provide for plenty of fresh air, without drafts.

One of the most important activities is sound stable management. The horse barn should provide an environment that protects the horses from temperature extremes, keeps them dry, provides fresh air, and protects them from injury. Open-front buildings provide adequate protection from the weather when the open side is faced away from the prevailing winds. Enclosed buildings usually contain hay storage, box or tie stalls, a feed room, and a tack room. Box stalls offer more freedom to the horse while tie stalls require less bedding and manure-carrying. Most stalls are constructed of wood; however, metal is becoming more popular. Floors made of wood or clay are noiseless and springy.

Peat moss, straw, shavings, and sawdust are all materials used for bedding. Manure should be disposed of daily or stored in an approved or safe area for convenient removal. The feed room and tack room should be organized for convenience, and be kept dry and free of dust. Most of the training and riding equipment is maintained in the tack room.

Paddocks and corrals provide an area for exercise and close observation. Sodded paddocks that provide succulent and nutritious grass are preferrable to barren areas.

Sound horse care and management involve the utilization of good pastures. Controlled grazing will extend the time during the year in which a pasture can be utilized. Pastures should be free of holes and rocks to reduce injury. Water, shade, and minerals should be made available in all pastures. Supplement pastures with concentrates, for a good pasture only provides roughages.

Horse fences are usually constructed of wood or woven wire. Fences should be kept in good repair to maintain safety and appearance.

Hay racks and mangers, grain containers, water facilities, and mineral feeders are all essential equipment in horse care and management. Keep this equipment clean and free of contamination.

QUESTIONS

1. Why has the domestication of horses reduced their reproductive efficiency?
2. At what age should a young mare be bred?
3. How long is the gestation period in horses? Why does it vary so many days in length?
4. Why is hand breeding so popular with horseowners?
5. What factors determine the age at which a foal is weaned?
6. At what age is a colt castrated?
7. Explain "winging" and "paddling" in reference to horse movement.
8. List the reasons why horses are shod.
9. List some advantages and disadvantages for both box stalls and tie stalls.
10. Why are most stall floors constructed of wood or clay instead of concrete?
11. What is a tack room used for?
12. How can the length of utilization of pastures be extended?
13. Why isn't barbed wire used for horse fences?
14. Grain feeders are usually made of what kind of materials?

REFERENCES

Bradley, M., *Illustrated Hoof Care for Horses,* Extension Guide 2825, Columbia, Missouri, University of Missouri, 1969.

Cole, H. H., *Introduction to Livestock Production,* Second Edition, San Francisco, California, W. H. Freeman & Company, 1966.

Ensminger, M. E., *Horses and Horsemanship,* Fourth Edition, Danville, Illinois, The Interstate Printers and Publishers, 1969.

Horse Handbook—Housing and Equipment, Extension Service Bulletin MWPS-15, Urbana, Illinois, University of Illinois, 1971.

Jordan, R. M., *Horse Care and Management.* Extension Bulletin 358 (Revised), St. Paul, Minnesota, University of Minnesota, 1972.

Pinson, Alfred R., *Correct Horseshoeing,* Extension Service Technical Report Number 25, College Station, Texas, Texas A & M University, 1970.

Pinson, Alfred R., *Foot Care of the Young Growing Horse,* Extension Service Technical Report Number 25, College Station, Texas, Texas A & M University, 1970.

27 Disease and Parasite Control

The usefulness of a horse is dependent largely upon its health. Many horseowners are quite unfamiliar with the factors affecting animal health and as a result their horses are neglected. There is much that can be done to maintain animal health and to prevent losses due to diseases and parasites. The following practices should be included in a sound health program.

1. **Feed a nutritionally balanced ration in proper amounts.**
2. **Avoid feeding hay or grain that is moldy or of poor quality.**
3. **Following a heavy workout, allow the animal to become rested before watering.**
4. **Maintain open, clean, well-ventilated, no-draft quarters.**
5. **Provide the horse with daily exercise.**
6. **Avoid public facilities when watering horses.**
7. **Isolate new horses for two to three weeks until their health conditions are known.**
8. **Check housing facilities and equipment carefully for conditions which might injure the horse.**
9. **Establish and follow a systematic immunization and parasite control program with a veterinarian.**
10. **Rely upon a veterinarian for the diagnosis and treatment of diseases and injuries.**

INFECTIOUS DISEASES

There are seven common bacterial or viral diseases common to the horse. Five may be prevented or controlled by use of an immunization program. They are equine encephalomyelitis, strangles, tetanus, influenza, and infectious abortion.

Equine Encephalomyelitis

Three strains of this disease, commonly known as sleeping sickness, have been present in the United States. The eastern and western strains have been present for a long time. During 1971 the Venezuela strain was confirmed in the Brownsville, Texas area. It is now believed that this strain has been stopped. According to the U.S.D.A., in 1971 nearly 2.8 million horses were vaccinated for this disease in 19 states, involving a vaccination and mosquito-spraying program. No confirmed cases of VEE were reported in the United States in 1972.

Cause. The disease is caused by at least four distinct viruses which are transmitted by mosquitos or through birds and then to horses. This disease can be transmitted to humans.

Symptoms. The animal appears sleepy and stands with its head lowered. In the early stages, the horse may walk aimlessly and appear to be blind. The animal may become paralyzed, and about 50 percent die.

Treatment and Prevention. A vaccine, now available, that controls all three strains, administered prior to the mosquito season, is recommended.

Distemper or Strangles

This disease affects young stock and often is spread at livestock sales or in public stables.

Cause. It is caused by a bacterium called *Streptococcus equi*. The incubation period is 4 to 10 days.

Symptoms. Infected animals lose their appetite, appear depressed, and run high fevers. Pus is discharged from the nose, and the animal may develop a cough and a sore throat. The glands under the jaw enlarge the fourth or fifth day, become sensitive, and usually break open, discharging pus. This disease differs from distemper found in dogs.

Treatment and Prevention. Treatment by a veterinarian and complete rest in comfortable quarters are recommended. Antibiotics and sulfas may be used. A strangles bacterin is available as a preventative.

Tetanus or Lockjaw

This disease, which can also affect humans, usually results from a wound caused by a nail, wire, or splinter.

Cause. The disease is caused by a toxin liberated by *Clostridium tetani* bacteria. The incubation period varies from one day to two weeks.

Symptoms. The disease affects the nervous system. Infected animals chew and

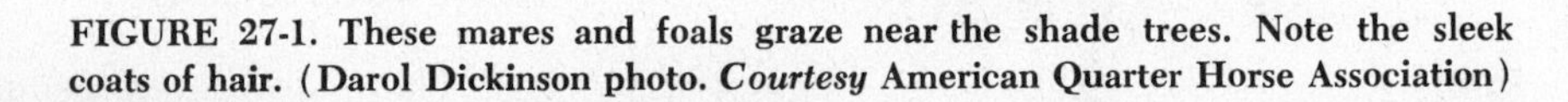

FIGURE 27-1. These mares and foals graze near the shade trees. Note the sleek coats of hair. (Darol Dickinson photo. *Courtesy* American Quarter Horse Association)

swallow awkwardly. The inner eyelid may protrude over much of the eyeball and the horse may become hyperexcited and have violent spasms.

Treatment and Prevention. A veterinarian should be called to administer a tetanus antitoxin. Treatment involves providing water and feed so the animal does not have to lower its head. A laxative may be necessary to maintain bowel function. Comfortable, quiet, and clean quarters should be provided.

Prevention consists of removing objects that may cause wounds. Active immunization results from use of tetanus toxoid.

Influenza

This respiratory disease is very contagious and can rapidly spread similar influenza strains to man. Horses may be exposed at rodeos, fairs, trailrides, while grazing, or when sudden changes in temperature occur.

Cause. Several virus strains cause the disease.

Symptoms. Younger animals are more likely to become infected than older animals; however, horses of all ages are subject to this infection. Very young foals usually receive some immunity from the dam's milk.

Infected animals usually show signs of fever, lose appetite, and become depressed. Watery discharges flow from the nostrils and eyes. Rapid breathing and a harsh cough usually follow. In later stages, pus may be discharged and there may be swelling of glands. Difficulty in breathing and pneumonia may result.

Treatment and Prevention. A veterinarian can administer vaccines for at least two of the causal viruses. Antibiotic and sulfa treatments may be of help in early stages. Infected animals should have comfortable quarters and be given complete rest as they should not exercise while running a temperature.

The isolation of infected animals, particularly from insects, birds, and rodents, and thorough sanitation measures are musts in preventing spread of the disease.

Swamp Fever

This disease (equine infectious anemia) was first discovered in the low areas of the southern states. It is found in both high altitude and low altitude states, and does not spread rapidly.

Cause. The virus causing the disease is carried in the bloodstream and is spread by insects rather than by contaminated feed or by body contact. Infected animals that appear to have been cured may still be carriers of the disease.

Symptoms. Initially infected animals suddenly appear exhausted, run high fevers, and have red or yellow coloration of the mucous membranes. There may be nose and eye discharges. Leg weakness is common as is diarrhea and frequent urination. Acute attacks are usually followed by recurrent or chronic attacks.

The disease may last a few days or several months. In chronic stages, the horse appears anemic, is listless and weak, and develops a rough haircoat.

Treatment and Prevention. Veterinarians may give the animal blood transfusions and drugs to promote red blood cell formation. No specific treatment is known and no vaccine is yet available.

Prevention consists largely of insect control, isolation of animals, and care in using injection equipment to prevent the spread of organisms.

Infectious Abortion

Abortion may be caused by physical factors, injury, or nutrition deficiency, or by one of three types of infectious organisms.

Cause. *Salmonella abortivoequina, streptococci abortion,* and *rhinopneumonitis* in-

FIGURE 27-2. Strict sanitation at breeding time, at birth, and following birth will insure a healthy foal. (*Courtesy* United States Trotting Association)

fection are the chief causes of infectious abortion. Salmonella abortivoequina enters the body through the digestive system, whereas streptococci abortion enters through the genital tract. The virus, rhinopneumonitis infection, spreads much like the common cold in humans—through the respiratory and digestive systems.

Symptoms. Usually an aborted fetus is the first indication of infection. Streptococci abortions usually occur prior to the fifth month, and the placenta is large and retained more than three hours. Abortions due to salmonella abortivoequina usually occur during or after the sixth month of pregnancy. Those due to rhinopneumonitis infection usually take place during the ninth and tenth months.

Treatment and Prevention. Streptococci abortion may be controlled by conforming to strict sanitation measures in the preparation of mares and stallions at breeding time; in the immediate removal and destroying of an aborted fetus and afterbirth; and in a general cleanup and disinfection of quarters.

Salmonella abortivoequina infection may be controlled by vaccination of mares in the fall with three injections one week apart. Rhinopneumonitis infection may be controlled by vaccinating mares with virus abortion vaccine in the fall or early winter, followed with shots during the latter stages of pregnancy.

Navel Ill

This infection may greatly retard the newborn foal, or cause crippling or death.

Cause. Navel ill may be caused by several bacteria, and other types of livestock may be affected. It may be spread by the stallion or by other animals.

Symptoms. Infected foals become listless and lose their appetites. The joints may become stiff and swelling may occur.

Treatment and Prevention. Infected foals may be given a blood transfusion, a sulfa, an antibiotic, a serum, or a bacterin. The services of a veterinarian are recommended. The best preventative is strict sanitation of the sire and dam at breeding time, and of dam and foal at the time of birth. Clean, comfortable, disinfected quarters must be provided.

NUTRITIONAL AND OTHER DISEASES

Laminitis (Foundering)

This disease is more common to ponies but affects horses of all types.

Cause. Foundering usually results from overeating grains or concentrate feeds, from over-consumption of water when the animal is hot, or from inflammation of the uterus following parturition (foaling).

Symptoms. Animals become lame due to blood accumulation and resulting pressure in their hoofs. The affected animals are reluctant to move; they suffer much pain and fever. In chronic cases, the hoof grows rapidly and becomes rigid, rough and brittle.

Treatment and Prevention. Immediate medical attention is important. Standing a horse in cool water or mud will help reduce fever. Removal of shoes and providing a laxative to eliminate toxic materials are helpful. Veterinarians may provide antihistamines, cortisone, and other drugs to relive pain. Permanent injury will result if early treatment is not provided. Prevention consists of avoiding the causes—overeating, overconsumption of water when the animal is hot, and inflammation following parturition.

FIGURE 27-3. Health is very important if the horse is involved in racing, rodeo, or horse show activities. (Dalco Film Co., Inc. photo. *Courtesy* American Quarter Horse Association, Inc.)

Heaves

Broken wind, or heaves, is a respiratory disorder common to many horses.

Cause. It may be caused by feeding dusty or moldy feed, exercising or stabling under dusty or poorly ventilated conditions, or by overexertion.

Symptoms. Affected horses develop a weak cough and a characteristic wheeze due to difficulty in breathing. Although increased exercise magnifies the difficulty, affected animals may be used for light work.

Treatment and Prevention. Temporary respiratory relief may be obtained by injections of antihistamines and other drugs. Elimination of all dusty and moldy feeds and improved environmental conditions in the stables, lots, and pastures will decrease harmful effects of the condition.

Colic

Colic is a condition involving acute abdominal pain.

Cause. There are several causes, the most common being an abnormal condition in the bowels created by the accumulation of gas. The intestines may be twisted or obstructed, preventing the movement of the gas. Over-consumption of feed, eating too fast, or eating before watering can also cause the condition.

Symptoms. Affected animals may sweat, roll, groan, strain, stretch, or turn the head around to the flank area.

Treatment and Prevention. Mineral oil, fed to establish intestinal mobility and fecal passage, is a common treatment. Antiferments, pain killers, and tranquilizers are quite effective. Prevention is best obtained by feeding high-quality feeds in proper amounts, proper watering procedures, and avoiding over- and under-exercising of the animal.

Azoturia

This condition is sometimes called "Monday Morning Sickness" or "tying up."

Cause. Azoturia is caused by continued heavy grain feeding while horses are resting and inactive.

Symptoms. Affected animals, when returned to heavy muscular activity, sweat profusely, have a stiff gait, are reluctant to move, and are nervous. The urine is usually dark in color and the back muscles become rigid.

Treatment and Prevention. Selenium, vitamin E injections, and other drugs may be administered by a veterinarian. The condition can best be prevented by providing idle horses with daily exercise and reduced rations.

INTERNAL PARASITES

A number of internal parasites cause serious health problems in horses. Ascarids, strongyles, and bots are the most serious.

Ascarids

Life History. The ascarid is a parasitic roundworm found in the upper part of the small intestine. The females deposit eggs which are passed with the feces. The eggs may live for years in a dormant state in stalls and pastures. During warm, damp weather, the eggs develop embryos and infest horses when swallowed with grass or feed. The larvae move through the intestinal wall into the bloodstream and are carried to the liver, heart, and lungs. From the lungs, the larvae are coughed up and swallowed by the animal. They return to the intestines and develop to maturity.

Damage. Infested horses usually appear weak although they eat well. They have rough haircoats, and often have extended abdomens and "tucked-in" flanks. The growth of young stock is impaired and lung damage is common.

Treatment and Prevention. The use of rotated pastures and the spreading of manure from the horse barn on nonpasture areas are recommended. Young horses should not be grazed on pastures grazed by older horses. Avoid all fecal contacts with feed, water, and equipment. A strict sanitation program should be supplemental to a treatment pro-

FIGURE 27-4. These yearling American Albino fillies are on rotated pasture. (*Courtesy* American Albino Association, Inc.)

gram using drugs recommended by a veterinarian. Piperazine, phenothiazine, carbon disulfide, trichlorophon, and thiabendazole are usually recommended separately or in combination.

Strongyles

Life History. There are nearly 40 species of strongyles infesting horses. They are small worms, less than an inch in length, living most of their lives in the large intestine and the cecum. The female deposits eggs in the large intestine which are passed with the feces. The eggs hatch in pastures in one or two days. The larvae molt twice before crawling to the tops of blades of grass where they are swallowed by horses while grazing. The larvae migrate to various internal parts of the body through the blood system. Some remain in the arteries until near maturity.

Damage. The mature worms as well as the larvae derive their nutrition by sucking blood, thus weakening blood vessels and causing blood clots. Infested horses lack appetite and show signs of anemia. They usually become dehydrated and have digestive disturbances.

Treatment and Prevention. The management practices and treatment schedules recommended for ascarid control are also recommended for the control of strongyles. A combination of phenothiazine and piperazine-carbon disulfide complex is quite effective. Thiabendazole is highly recommended and is obtainable from a veterinarian.

Bot Flies

Life History. There are several species of bot flies. The female lays eggs on the hairs of the forelegs, mane, shoulder, belly, and chin of the horse. The eggs hatch and the larvae enter the horse's mouth when it bites or licks itself. The larvae burrow into the tongue and migrate to the stomach and intestines where they become attached and remain for 10 to 12 months. The mature larvae are passed with feces and pupate in the soil. The adult fly emerges in about 16 days and the cycle is repeated.

Damage. Bot flies attacking animals cause them to toss their heads, strike the ground with their front feet, and rub their noses. Infested animals have reduced vitality, may have digestive disturbances, and are inefficient workers.

Treatment and Prevention. The methods described for ascarids will control bot fly infestation of young horses. Older horses should be treated twice each year, once just before frost and about 30 days later. Carbon disulfide, trichlorfon or other drugs recommended by a veterinarian may be used.

FIGURE 27-5. Horses can become infected with disease or infested with parasites as a result of contact at fairs and rodeos. (Darol Dickinson photo. *Courtesy* American Quarter Horse Association)

EXTERNAL PARASITES

Lice

Life History. Egg masses, or nits, are laid on long hairs close to the skin. They hatch in 11 to 20 days, and the young lice mature in 11 to 12 days. Mature lice live only a few weeks, but several generations result each year. The entire life is lived on a horse and a new deposit of eggs may result three to four weeks after the first eggs were laid.

Damage. Lice cause skin irritation resulting in restlessness and loss of condition. There is evidence of rubbing and abrasion of the skin. Infestation is usually around the tail, inside the thighs, over the fetlocks, and along the neck and shoulders.

Treatment. Sanitation in grooming and stabling is always recommended. Horses may be sprayed or dusted with coumaphos, dioxathion, malathion, or other products recommended by a veterinarian. The treatment should be repeated in two weeks and periodically thereafter.

Mange Mite

Life History. The female mite lays 15 to 20 eggs over a two-week period under the skin near the hair follicles. The eggs hatch and the mites mature in about 14 days. The cycle is then repeated. The males die after mating, and the females die after laying eggs.

Damage. The mite creates a mange condition which causes itching and scratching. The mite burrows under the skin to feed on cells and lymph. Thick, wrinkled, tough skin is evident, and loss of hair results.

Treatment. A treatment of 0.5 percent toxaphene spray is recommended and should be repeated in two weeks. Periodic treatments should be provided as necessary.

Ringworm

Life History. Ringworm is caused by a fungus organism which spreads by contact with an animal or contaminated equipment. The spores may live 15 to 20 months on an animal, on equipment, or in the housing facility. The incubation period is seven days.

Damage. Scaly areas, usually devoid of hair, appear around the eyes, ears, neck, or root of tail. The round infected patches itch, cause scratching and loss of hair, and become encrusted.

Treatment. Strict sanitation in grooming and the care of equipment and quarters is necessary. Infected areas should be washed with soap and water and treated with tincture of iodine or other drugs approved by a veterinarian. Infected animals should be isolated, and all equipment should be sprayed with an approved fungicide. Treatment should be repeated as necessary.

SUMMARY

Equine encephalomyelitis, distemper, tetanus, influenza, and swamp fever are the most serious infectious diseases of horses. Vaccines are available for all but swamp fever.

Several factors and conditions produce equine abortion. Vaccinations and injections are available.

Foundering, heaves, and colic are the most common nutritional diseases. All can be prevented by prudent practices in feeding and management. Prevention is much more effective than cures.

Ascarids, strongyles, and bots are the most serious internal parasites of horses. A systematic insecticide program should be worked out with a veterinarian and strictly adhered to. Repeated treatments are usually necessary. Complete sanitation and the use of rotated pastures are strong preventatives.

Lice, mites, and ringworm are the most serious external parasites. Repeated treatments of insecticides and fungicides as recommended by a veterinarian should be used. Federal regulations should be followed carefully in using any drugs.

QUESTIONS

1 Describe the most serious infectious diseases of horses. Indicate causal agent, symptoms, and damage.

2 Outline control methods for equine encephalomyelitis, distemper, tetanus, equine influenza, and swamp fever.

3 What nutritional diseases affect the horse and how can they be prevented and treated?

4 What are the similarities and differences in life history among ascarids, bot flies, and strongyles?

5 Outline a program for controlling internal parasites of horses in your stable.

6 How can you control lice, mites, and ringworm infestations among your horses?

7 Outline an equine immunization program adapted to conditions in your locality.

8 List sanitation and management practices necessary in the control of equine diseases and parasites.

REFERENCES

Ensminger, M. E., *Horses and Horsemanship, Fourth Edition,* Danville, Illinois, Interstate Printers and Publishers, Inc., 1969.

Dykstra, R. R., *Animal Sanitation and Disease Control,* Danville, Illinois, Interstate Printers and Publishers, Inc., 1957.

Federal and State Cooperative Extension Service, *Horse Science,* Washington, D.C., 1969.

FS Services, Inc., *Shaping and Showing Horses,* Bloomington, Illinois, 1970.

28 Training and Showing

Horses are unique among domestic animals in that their value is greatly increased with training. The horse is one of the easiest of all animals to train. Some horseowners define training as the art of forming useful habits. The correct response by the horse to the commands of the trainer signifies good training.

There are two general theories on the best way to convert an untrained, half-wild animal into one that responds. One theory is force—overpowering or breaking the horse to lead, ride, or drive. The other is the principle of taming, conditioning, and adjusting the horse's disposition. Breaking by overpowering is usually done when the horse is 3 to 5 years old. Taming and conditioning the horse takes considerably longer time and usually starts when the horse is a yearling.

TRAINER AND EQUIPMENT QUALITIES

An effective horse trainer should possess qualities of kindness, patience, firmness, consistency, courage, consideration, intelligence, and determination. When trainers develop these traits, they not only become good trainers, they also become better, more understanding persons. The horse has an instinct for reflecting the trainer's state of mind. A calm attitude, slow, easy movements, and a flow of soft-voiced talk will lessen nervousness. Discontinue a training period when the horse begins to show impatience or disinterest. About half an hour of handling daily is enough in the early stages.

Use Good Equipment

A horse, being halterbroken and tied with a halter or strap that breaks, learns that by lunging back it can escape. Use standard hackamores, bits, and similar gear which have been determined best for the horse. Avoid novelty-type equipment. A properly fitted saddle is important, but the most important part of the equipment is the bit.

TRAINING THE YOUNG HORSE

A completely acceptable horse is easily trained, regardless of what it is to be used for. This is especially true if the horse starts training at an early age.

Training Procedures

Halter Breaking. This training can't begin too early. Use a strong halter and a stout rope when the foal is first tied. To avoid entanglement in the rope, tie it slightly above the height of the withers. In a short time, the foal will realize the halter won't break and that he is confined. Generally, three or four sessions of being tied takes the fight out of the foal.

Teaching to lead. Pulling forward on the halter won't encourage the colt to move, but pulling sharply to the right and to the left will get him off balance and make him take a step. Lead him beside the dam at first. Another method used in teaching a colt to lead is to put a rope loop over the rump and have it fall just above the hocks, run the end through the halter, and apply tension. Don't expect perfection the first or second time.

With halter breaking and training to lead, a daily brushing and an examination and handling of his feet should occur. Lessons learned as a weanling will be remembered and make later training easier and safer.

Pre-bit Hackamore Training. Hackamores are used by most trainers as a pre-bit training tool. They are so designed to prevent injury to sensitive tissue in the horse's mouth while providing adequate control. The hackamore is based on the principle of the horse learning to respond to pressures on the nose and under the chin.

A hackamore, like a bridle, has a direct rein and a bearing rein. Using both reins, the rider must teach the horse to respond to different pressures. Pressure from the direct rein shows the horse the direction in which

FIGURE 28-1. A yearling being trained to lead. (*Courtesy* United States Trotting Association)

to move. Pressure from the bearing rein, applied to the horse's neck, is the beginning of training the horse to neck rein.

Bit Training. Start using a bit when the horse is over one year of age. Be gentle, tieing a straight or jointed snaffle bit to the halter and letting the yearling mouth it for short periods of time. Once accustomed to the bit, the horse can be trained to flex at the pole (bend the head down so that the forehead is perpendicular to the ground). Most horseowners prefer the head to be carried in this position.

Longeing. Longe training is accomplished by fastening a rope to the halter and, with the aid of a whip, moving the horse in a circle. With training, the horse is taught to move in both directions at the walk and trot, and should also learn verbal commands, especially "whoa."

Preventing Shyness. "Sacking out" is the training session that eliminates shyness. Using a soft rope, place a loop over the neck of the horse, then a small loop around the pastern of one of the hind legs, raising the hind foot off the ground so that it is impossible to get any weight on the hind foot; tie it up by fastening the rope to the loop around the neck. This will restrain the horse without frightening or injuring it. Next, slap the animal gently all over with a gunny sack or a pair of pants. This sacking out principle is used on almost all well-trained Quarter Horses, trail horses, and some gaited horses.

Saddling

After the horse has been trained to lead, respond to voice commands, and is used to a bit, he is ready to be saddled. Introduce the saddle blanket, slip it over his back, petting and talking to him. Place it a little forward on the withers and then pull it back so all the hairs lie straight. Then put the saddle on gently. At first, all the horse should carry is the saddle. He should be led around to grow accustomed to this weight and the flopping stirrups. Keep his head up as he moves and if he has been properly trained up to this point, he should never buck.

When the horse is used to the saddle, the trainer may mount. Some trainers quietly swing into the saddle, while others may put their weight in one stirrup for the first few times before mounting. Once mounted, the horse should be led around at a walk. Be sure he responds to the bit before being turned loose with no one at his head.

The actual gentling of a horse and getting it to ride off is the simplest part of training a good horse. The most challenging work is that of converting a saddle-broke horse to a top-performance horse. It is during this converting stage of training that many potentially good horses are lost. The trainer must know what is required and how to train the horse to respond instantly to every command.

Selecting the Bit

Bits are mainly used to control speed, direction, and head carriage. They are used to discipline the horse by applying and then releasing pressure when the horse obeys. Select the right bit and never use bits too advanced for the horse's training.

In selecting the bit, one should know the different parts and the function of them. The mouthpiece may be either straight or jointed. A straight mouthpiece applies pressure on the tongue and bars (space between the front and back teeth), while the jointed mouthpiece applies more pressure on the corners of the mouth. The sidepieces put pressure on the chin and mouth. The parts above the mouthpiece are called the cheekpieces and the parts below are called the shanks. The longer the shanks, the greater the leverage.

Snaffle Bit. The snaffle bit is the safest and least likely to damage the mouth and is particularly good for the inexperienced horse

FIGURE 28-2. The Quarter Horse at work. (*Courtesy* American Quarter Horse Association)

and rider. This type of bit puts the main pressure on the corners of the mouth and some on the cheeks and tongue. The straight snaffle is the least severe and used as one of the beginning training bridles, while the jointed snaffle has somewhat of a pinching action. There is a wide variety of snaffle bits.

Curb Bit. A curb bit provides leverage and should be used on the principle of instant pressure and immediate release when the horse responds. Curb bits put pressure on the tongue and chin. Because of the smooth stops and quick turns, this is the type of bit called for in all Western show classes. There are many versions of the curb bits. A curb bit is the all-round bit for western horses well trained first with a snafflle.

Proper fit of the bit is equally important to the correct selection. Improper fitting of bits is the biggest mistake amateur riders make. The bit is too tight if the skin is wrinkled in the corners of the mouth. If the horse mouths the bit excessively, the bit is too long and he will try to spit it out.

Hackamore. The hackamore with a noseband, curb strap, and jaw or side pieces for leverage may be used with or without a bit. The hackamore is for good riders since there is less contact between the mouth of the horse and the hands of the rider. It does offer more control than most bits for the well-trained horse will respond to every touch of the rein.

The Bridle. The main purpose of the bridle is to hold the bit in its proper place in the horse's mouth. Bridles may either be double or single. A single bridle is equipped with one bit, while a double bridle will usually have a snaffle bit and a curb bit with two pairs of reins. Only one rein is used with western bridles. A properly fitted bridle will allow the hand to pass between the throat strap and the horse's throat.

SADDLES AND BLANKETS

The English saddle and the Western saddle are the two common types in use today. The English saddle is light in construction with a flat seat. It is well adapted for use in training, pleasure riding, racing, jumping, and polo. Saddle blankets are not generally used with English saddles.

The Western saddle is the descendant of the old Spanish war saddle. Some changes have taken place over the years; however, it still has a wooden or steel tree called a horn, a leather covering, and rings to attach the cinches and stirrups. Western saddles are designed for work and to provide comfort. For example, the horn functions as a post around which a lariat can be tied or wound when handling cattle. It is recommended that an experienced rider assist the young amateur in selecting the first saddle.

It is common to use a saddle blanket with Western saddles and some riders will use one with English saddles. When kept clean and properly used, a blanket or corona will usually prevent a sore back. A corona is a blanket cut to the shape of the saddle and has a colorful roll around the edge. The blanket should be hung up to dry after being used and, when dry, brushed to remove dried sweat and hair.

Proper care of a saddle will prolong the years of use. Horse sweat, moisture, and dust are the three enemies of a saddle; therefore, wipe the saddle clean after use. Clean, oil, and polish the saddle at least twice a year.

EQUITATION

Good equitation or good horsemanship enables the horse to work better and easier and the rider to ride better and easier. Before mounting, check your equipment and stirrup length. Knees should flex slightly when riding either a flat saddle or Western saddle and a bit more when riding a hunt seat.

Methods of mounting, sitting the saddle, and dismounting are basically the same for English and Western riding.

English Equitation

To mount, stand on the left side, face the rear of the horse, take the reins in the left hand with the bight or end of the reins on the right side of the horse's neck. With left hand on the withers and left foot in the stirrup, grasp the off side of cantle with right hand. Take one or two hops on the right foot to attain momentum to mount. Swing the right leg clear of the horse's hips. Position the right foot in the stirrup before easing the body weight into the seat of the saddle. Dismounting is the same procedure in reverse.

Western Equitation

To mount a horse with Western gear, face the left side of the horse with reins in the left hand and placed on the withers. Turn the stirrup with right hand, put the left foot in the stirrup, grasp the saddle horn with right hand and mount. Here, again, dismounting is the same procedure in reverse.

Position in the Saddle

To master the basic position, sit relaxed in the center of the saddle with feet and legs close to the body. Stirrups should be adjusted to allow heels to be lower than toes. Place the balls of the feet on the treads of the stirrups. Toes should be neither extremely in or out, but should be in a comfortable position pointed in the direction of travel. Calves, knees, and thighs should maintain contact with the saddle. Sit deep but straight in the saddle without slumping or lowering chin.

Good hands go with good balance, secure seat, and consideration of the horse's mouth. Normally, keep the hands low and, in western show classes, close to the horn. Leg pressure can be applied to urge the horse forward or to indicate direction. Body weight is used in turning, stopping, and going forward. The voice of the rider is not generally used during shows for fear of distracting other horses and riders. If used, the tone of voice is more important than the words.

Confidence, control of the horse, and seat balance are three ingredients of any good rider. All are more easily attained if one has proper seat, correct hand and leg position, and proper posture.

GAITING A HORSE

A saddle horse is not really a saddle horse in a true sense if it is not gaited. The word, "gaited," signifies that it is able to go the five gaits required and do them neatly and swiftly at the proper command. Saddle horses are classified as either the three-gaited or the five-gaited. (See Chapter 24, page 356 for a description of the various gaits.)

The difference between the three-gaited and the five-gaited horse is training. Any five-gaited horse must have the walk, trot, and canter properly developed before he is taught the slow gait (also called running walk or fox trot) and the rack. A gaited horse should start to canter from a walk, not a trot. In Western show classes, horses can move from the trot into the canter. Many trainers have difficulty teaching the horse the right lead. A good rider can put a horse into the desired lead, on call, at any time, at any place, regardless of what the horse is doing.

SHOWING HORSES

It takes hours and hours of work in developing a horse and rider who can consistently place in today's competitive shows. First, one must decide on what type of class to concentrate. The horse may make the decision, for not all horses are capable of performing in all classes. Next, the trainer must work the horse several days and weeks in advance of the show if satisfactory results are to be expected. Having a well-trained performance horse doesn't mean he can never be ridden at high speeds, but it does mean that at least 25 percent of the riding should consist of training sessions suitable to the type of performance expected from the horse.

Start with a horse of good type and conformation, and condition it so that its coat shines. The horse should have a little fat on his ribs and a trim middle. Grooming means daily brushing over the horse's entire body. Brushing with a rice-straw brush will produce a fine, glossy coat. Keep long beard hairs about the nose and lower jaw trimmed. The long hair about the cannon bones, fetlocks, and pasterns should be clipped with an electric clippers.

The tail should be thinned from underneath by plucking. Don't use a clipper unless

FIGURE 28-3. Many hours are spent in training before a horse and rider are ready for competition. A Western pleasure class entry is shown here. (Knoll photo. *Courtesy* Matthias Arabians)

you are fitting a horse such as a three-gaited saddle horse. The foretop should be thinned, but not clipped, unless the entire mane is clipped. Some owners clip the mane on quarter horses, except for the foretop, and a small area of mane over the withers. Learn the established custom and groom as desired.

Showing at Halter

Showing at halter classes is common in many shows. Many people think that to show at halter is to clean the horse up a little the day of the show and lead him into the arena. This is untrue, for it requires every bit as much time, effort, and know-how to properly show a horse at halter as it does for any of the other events.

To show at halter, lead the horse from the left side, keeping just behind the horse's head. The right hand should hold the lead shank about six inches from the halter with the excess lead shank folded in the left hand.

Most horse judges prefer to get an overall view of the entire class before concentrating on individual horses. To obtain this, the horses are moved in a large circle. The horse should move with a long stride and pick up his feet and not shufflle along.

When asked to line up, stand the horse up by leading it forward until the left hind foot is in the position wanted. Then place the other three feet by use of the lead shank.

When the judge is ready to watch the horse travel, walk straight from the ring steward to the judge. Stop momentarily in front of the judge, then turn the horse on its hind legs and proceed directly away from the judge at a trot. Return to place in line and stand the horse up. When showing, dress appropriately in clean, neat western clothes. Keep an eye on the judge and steward for instructions and respond quietly.

English and Western Pleasure Classes

English pleasure is usually defined as anything that is not Western. The type of saddle and the dress of the rider are two differing factors. Most horses are trained the

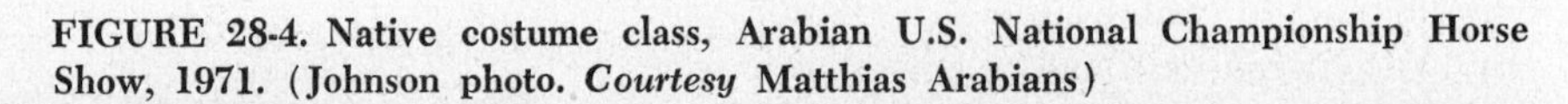

FIGURE 28-4. Native costume class, Arabian U.S. National Championship Horse Show, 1971. (Johnson photo. *Courtesy* Matthias Arabians)

same way and specialization in either English or Western styles takes place after basic training.

Training and showing the roping horse, the jumping horse, etc. takes time and experience. Performance classes are so many and varied that it is not practical to list them here. Therefore, it is recommended that the reader obtain the official rule book of the American Horse Show Association and rules governing local horse shows.

Good show people are courteous, respectful of others, and follow safety procedures. There are many tricks to make a horse show to the greatest advantage. The best way to learn important points is to observe how successful horseowners fit their horses and show them.

SUMMARY

Horses are unique in that their value is greatly increased with training. The correct response by the horse to the commands of the trainer signifies good training. Breaking a horse to lead, ride, or drive by overpowering him, is done in a relative short time and when the horse is 3 to 5 years old. Taming, conditioning, and training a horse usually starts when the horse is a yearling.

The horse has an instinct for reflecting the trainer's state of mind. A calm attitude, with slow and easy movements, will lessen nervousness. Use good equipment which is properly fitted.

A completely acceptable horse is easily trained if training starts at an early age. Halter break and train the colt to lead before weaning. A hackamore can be used in several areas of training.

Bit training should be accomplished before the horse is broke to drive or ride. The selection on fit of the bit is important.

Shyness can be prevented by a process called "sacking out." Most well-trained quarter horses, trail horses, and some gaited horses go through this phase of training.

Introduce the horse to a saddle blanket by placing it a little forward of the withers and pulling it back so all the hairs lie straight. Place the saddle on the horse and lead him around to get used to the weight before mounting.

The actual gentling of a horse and getting it to ride off is the simpliest part of training. The most challenging work is that of converting a saddle-broke horse to a top performance horse.

In selecting the bit, one should know the different parts and the functions of them. The snaffle and curb bits are the two most common. Other bits are usually variations of these two. Improper fitting of bits is the biggest mistake amateur riders make. The main purpose of the bridle is to hold the bit in its proper place in the horse's mouth.

The English and the Western are the two main types of saddles used. The English saddle is light in weight while the Western is heavier and has a horn which projects upward at the front. This can be used to hold on to or tie a rope to. Saddle blankets are commonly used with Western saddles, but seldom used with English saddles. Keep saddles and blankets clean and free of dust and moisture.

Before mounting, check the equipment and length of stirrups. Methods of mounting, sitting the saddle, and dismounting are basically the same for English and Western riding.

Good hands go with good balance, a secure seat, and consideration of the horse's mouth. Normally, keep the hands low and close to the horn. Leg pressure and body weight can be used to aid in directing the horse's movements.

A saddle horse is not really a saddle horse in a true sense if it is not gaited. Saddle horses are classified as either three-gaited or five-gaited. The main difference here is training. The five gaits are: walk, trot, canter, slow gait, and the rack. A gaited horse should start to canter from a walk, not a trot.

It takes hours of work to develop a horse rider who can consistently place in today's competitive shows. The make-up of the horse may decide the type of class on which to concentrate. Grooming means daily brushing with a rice-straw brush which will yield a fine, glossy coat. Learn the established custom and groom as desired.

Showing at halter classes is common in many shows. When showing at halter, lead the horse from the left side, walk toward the judge, stop, turn, and then proceed directly away from the judge at a trot. Keep an eye on the judge and steward and respond to their instructions.

Performance classes and show rules vary from show to show. It is recommended that one showing horses should obtain the official rule book of the American Horse Show Association and the rules printed for local horse shows. The best way to learn points in showing is to observe how successful trainers fit and show their animals.

QUESTIONS

1. Explain the two general theories of training horses.
2. List the qualities of an effective horse trainer.
3. Explain how a young colt is taught to lead.
4. What is a hackamore?
5. What is meant by the term "Longeing"?
6. How is shyness prevented?
7. Explain the steps in saddling.
8. What are the main uses of the bit? List the two main kinds of bits.
9. What is the main purpose of the bridle?
10. What is the difference between the English and Western saddle?
11. Describe the correct way to mount with western gear.
12. List the three ingredients of any good rider.
13. List the gaits of the three-gaited and the five-gaited horse.
14. Explain how a horse is shown at halter.

REFERENCES

Bradley, M., and S. D. Hardwicke, *English Equitation: Mounting, Correct Seat, Dismounting,* Extension Guide 1875, Columbia, Missouri, University of Missouri, 1965.

Bradley, M., M. L. Jones, and S. D. Hardwicke, *Western Equitation: Mounting, Correct Seat, Dismounting,* Extension Guide 2876, Columbia, Missouri, University of Missouri, 1971.

Davidson, D., and B. F. Yeates, *Showmanship at Halter*, Extension Service Technical Report Number 21, College Station, Texas, Texas A & M University, 1969.

Jordan, R. M., *Horse Care and Management*, Extension Bulletin 358 (Revised), St. Paul, Minnesota, University of Minn., 1972.

Livingston, P., *The Art of Saddlery*, Extension Service Technical Report Number 21, College Station, Texas, Texas A & M University, 1969.

Yeates, B. F., *Pre-bit Hackamore Training*, Extension Service Technical Report Number 25, College Station, Texas, Texas A & M University, 1970.

SHEEP PRODUCTION

29 The Sheep Industry

Many opportunities exist in the various phases of the sheep-producing industry for those who like livestock and are willing to put forth the effort to acquire the knowledge needed for success.

While world production continues to increase, the sheep population in the United States is at its lowest figure since the beginning of this century. The increased demand for synthetic fibers and the lack of efficiency in producing sheep and wool are the two main causes for the reduction. New research and technology within the industry shows renewed interest and economic strength for the future.

ADVANTAGES OF SHEEP

Sheep have many advantages over some other classes of livestock and are particularly well adapted to many areas. Sheep will produce two different kinds of crops each year, wool and lambs, providing the flock owner with two sources of income a year. Since the crops are entirely different, the price of one will not necessarily have a bearing on the other. Wool may be stored and held for higher prices or sold at shearing time, whichever seems advisable. A crop of lambs may be marketed from three to six months after they are born, bringing in rather quick returns.

Sheep will eat more different kinds of plants than any other kind of livestock. This makes them excellent weed destroyers, a class of livestock that can turn waste into profit, and, at the same time, improve the appearance of many farms. They do, however, respond very favorably—economically—to high-quality grain and roughages.

Since roughage is usually cheaper than grain and sheep have the ability to produce prime carcasses on roughage alone, they are especially well adapted to many areas unable to produce grain profitably. Studies show that sheep utilize an average of 13.6 acres of pasture and forages to each acre of grain to provide choice to prime carcasses; whereas beef cattle use 4.8 acres of pasture and forage to each acre of grain, and hogs only 0.2 acre of pasture and forage to an acre of grain.

Sheep do not require expensive buildings and equipment. Lambs, if born during cold weather, require warm housing at lambing time, but after that only protection from wind and storms and a dry place to lie down are needed. Natural protection furnished by hills or trees is all that many range flocks have.

Since sheep prefer to graze on hilltops and high land, the droppings are left where they are usually most needed. Farms that are low in fertility may be improved considerably with sheep.

CLASSES OF SHEEP PRODUCERS

Sheep producers may be divided into four general classes: (1) the rancher, (2) the farm flock owner, (3) the lamb feeder, and (4) the producer of purebreds. Some individuals combine two or more of the above classes of production.

Sheep Production on the Range

In the northern plains region, where the amount and quality of feed are usually good, most of the lambs are sold for slaughter at weaning time as milk-fat lambs. Where grazing conditions are less favorable, many lambs are sold into the grain-producing areas for finishing. There are also many commercial feed yards located in the sheep-producing areas, where the operators make a business out of buying feed and finishing lambs purchased from the ranchers. Recent trends indicate a reduction in range production and an increase in specialized farm flock units.

The Farm Flock

Farm flock owners usually have enough feed to produce finished lambs. Very few lambs are sold from farm flocks before they are ready for slaughter. Some owners produce fall and winter lambs for the off-season

FIGURE 29-1. Flocks of sheep can be established with inexpensive equipment. (Carr photo. *Courtesy* Hawkeye Institute of Technology)

markets, while others follow much the same procedures common on the northern range. Markets, labor, available feed, and the breed of sheep the flock owner raises are to be considered in planning a program. The number of farms specializing in sheep is on the increase. Flocks of 500 to 1,500 ewes are becoming more common.

FIGURE 29-2. (*Above*) An efficient flock must be comprised of good-quality breeding sheep. Note the modern type exhibited here by Heggemeier Sheep Farm, Kirkland, Illinois. FIGURE 29-3. (*Below*) Note the length of body and the amount of bone structure in this ram's legs. (Figures 29-2 and 29-3 *courtesy* American Hampshire Sheep Association)

The Lamb Feeder

Lamb feeders are generally located in the grain-producing areas. They buy feeder lambs, usually from the range, in the late summer or fall. Lamb feeders depend upon their profits from the increase in value per 100 pounds as a result of finishing, known as *margin,* and from the value of the gain over the feed and other costs.

The Purebred Breeder

The breeder usually produces purebred animals. The principal markets are other breeders, ranchers, and farm flock owners who buy rams or ewes for breeding purposes. The purebred breeder requires more skill than other types of producers because one needs to know the demands of consumers and flock owners, in addition to understanding feeding and management. The sheep industry depends largely upon the breeders for improvement in type, growing ability, reproductive and feeding efficiency, and qualities of fleece. Purebred sheep breeders are located wherever sheep are produced. The feeding of large quantities of grain is not essential in the purebred industry, except where the breeder is fitting animals for show. The size of the flock should be based on average feed conditions on the farm.

Combination Enterprise

Many producers prefer to combine two or more forms of sheep enterprises. Often purebred breeders start with a small flock of registered animals, depending upon their commercial flock to provide the profits until they can establish the purebred enterprise on sufficient scale to provide the expected income. The size of the flock should be based on average feed conditions on the farm.

SUMMARY

Many opportunities exist in one or more of the several different phases of sheep production.

Sheep provide two different crops per year, wool and lambs. These crops usually come at different seasons, which makes a good distribution of income. Sheep are excellent gleaners, making use of much waste feed and improving the appearance of the farmstead. They consume large quantities of roughage, converting a relatively cheap feed into a good cash product. Sheep help to improve soil fertility considerably because of their grazing habits.

Equipment for sheep need not be elaborate or expensive, unless lambing takes place during cold weather.

Sheep production in the range sections is characterized by large flocks and migratory enterprise, and usually represents the principal, if not the only, farm income to the owner. Many of the farm flock producers have small flocks which constitute only a part of the total income. These producers have livestock programs which are usually diversified. The number of farm flocks of 500 to 1,500 ewes is becoming more common.

Sheep producers may be divided into four distinct classes: rancher, farm flock owner, lamb feeder, and purebred breeder.

QUESTIONS

1 What are the classes of sheep producers? Explain.

2 What type of sheep production best fits your farm? your community?

3 What are the advantages of sheep production?

4 What are the two sheep-producing areas and how do they differ in methods of production?

REFERENCES

Ensminger, M. E., *Sheep and Wool Science,* Fourth Edition, Danville, Ill., The Interstate Printers and Publishers, Inc., 1970.

Research Reports and Sheep Management Programs, Brookings, South Dakota, South Dakota State University, 1965.

Sheep Management and Marketing in Wisconsin, Extension Reserve Bulletin, Madison, Wisconsin, University of Wisconsin, 1965.

30 Selection of Breeding and Feeding Stock

When entering the sheep business, the prospective shepherd must not only decide which phase of the industry to enter, but also determine which class and breed of sheep to produce.

CLASSES OF SHEEP

Sheep are classified depending on the factors one wishes to emphasize. Shepherds may talk about wool classification, western or native sheep, wool or mutton type, white face or black face, horned or polled. Sometimes sheep are classified according to the area in which they originate, such as mountains, uplands, lowlands, south, or north.

Wool Classification

The most common classification of sheep is based on types of wool. These classes are: (1) fine-wool type, (2) medium-wool type, (3) long-wool type, (4) crossbred-wool type, (5) carpet-wool type, and (6) fur sheep.

Fine-wool Type. The fine-wool breeds consist of those that produce a fine, wavy fiber of wool. The fleece is dense and contains a large amount of yolk or oil. The fine-wool breeds produce a heavy fleece of good quality. These breeds were originally developed primarily for their wool production, but modern breeders have been improving their mutton qualities so that lambs from the fine-wool class produce very acceptable carcasses when fed out and slaughtered. The fine-wool breeds combine a strong banding instinct with the ability to graze on poor-quality range. These characteristics make the fine-wool breeds especially adaptable for many Western range areas. Approximately 35 to 40 percent of the sheep in the United States are of the fine-wool type.

Medium-wool Type. The medium-wool breeds were developed primarily for meat production, but increasing emphasis has been placed on wool production during the last few years. They are called medium-wool breeds because their fleece is medium in fineness and length. The medium-wool sheep are used extensively in the western and mountain range areas. With the exception of the Tunis and Montadale, they were developed in England, with the major emphasis placed on mutton or meat qualities, and until quite recently little attention was given to wool production. Medium-wool breeds are considered the best when judged strictly as meat animals. Thirty-five to forty percent of the United States sheep population is of the medium-wool type.

FIGURE 30-1. This Columbia yearling ewe, a crossbred-wool type, owned by Verl Anderson, Fielding, Utah, was the 1972 Chicago International Champion. (*Courtesy* Columbia Sheep Breeders Association of America)

Long-wool Type. The long-wool sheep, as a class, are larger than the other classes and are so called because of the long wool fibers they produce. Fibers 12 inches long are not uncommon for these breeds. The wool is coarser than that of either the medium- or fine-wool sheep. They were developed when large, fat, and rugged sheep were popular in England; however, at the present time, they are too slow in maturing for profitable lamb production and the carcasses are of relatively poor quality overlaid with a layer of fat which makes them unpopular with both consumers and packers. Commercial purebred or straight-bred flocks of the long-wool breeds are seldom found. The population is limited to the coastal high-rainfall areas where feed is abundant. However, they have been used extensively and successfully for crossing purposes.

Crossbred-wool Type. The crossbred-wool breeds resulted from an infusion of long-wool and fine-wool breeds. As a result of selection, many breeds have recently been established from those crosses. The general objectives in developing the breeds comprising the crossbred-wool class were to improve the meat quality and length of wool fiber or staple, and to retain the banding instinct and general hardiness of the fine-wool breeds.

Breeds resulting from these crosses have become popular in many areas, especially in parts of the Western range country and among many Midwestern flock owners. The crossbred breeds are often classified as medium-wool rather than placed in a separate class.

Carpet-wool Type. Carpet wool requires a coarse, wiry, tough fleece which is not produced by the breeds of sheep that are popular in this country. Most of the wool used in the manufacture of carpets and rugs has been imported from Argentina, Pakistan, India, New Zealand, Syria, Iraq, and other countries.

Fur Sheep Type. New developments in processing wool for fur have made it possible to use the pelts from several breeds for fur purposes. However, the Karakul is the only breed raised primarily for its fur in this country.

The production of fur sheep is a comparatively new sheep industry in the United States. The value of the mature animals for mutton or wool is very low. Profits, other than the sale of breeding stock, result from the selling of the lamb pelts. The pelts are classified as follows: (1) *Broadtail* is produced from stillborn or premature lambs or lambs killed shortly after birth. The premature deaths are not forced but are the result of accidents or abortions. The hair is undeveloped and reflects light in such a way as to give the fur a watery appearance. Broadtails are the most valuable pelts, although production is small. (2) *Persian Lamb* is produced by killing the lamb when from three to ten days old, after the hair has formed a tight, lustrous curl. It is important to take the pelt when the curl is tight; therefore carefully watching for the right time to take the pelt is essential if the highest quality is to be obtained. Persian Lamb ranks next to Broadtail in value. (3) *Karakul* is the third type of pelt and is taken from the lambs after the curls have opened, which is usually when the lambs are two weeks or more of age. The Karakul is the least valuable of these three classes.

There are several grades of pelts within each class, depending upon color, luster, quality, and general appearance.

BREEDS AND BREED CHARACTERISTICS

Fine-wool Breeds

Two fine-wool breeds are popular in the United States. They are the Merinos and the Rambouillets.

Merinos. There are three types of Merino sheep, all originating from the same parent stock. The types are known as the A, B, and C, or Delaine, Merinos. The A-type sheep have extremely wrinkled skins from the head to the dock, making them hard to shear. The A-type Merino has lost its popularity in the United States and is almost extinct in this country.

The B-type have fewer wrinkles. The wrinkles are confined mostly to the neck, whereas the C-type, or Delaine, Merinos are comparatively smooth. The Delaines are the largest of the breed, with mature rams weighing from 180 to 220 pounds and ewes

TABLE 30-1 CLASSES AND COMMON BREEDS OF SHEEP

Fine-Wool Type	Medium-Wool Type	Long-Wool Type	Crossbred-Wool Type	Carpet-Wool Type	Fur Type
American Merino	Cheviot	Cotswold	Columbia	Black-Faced Highland	Karakul
Debouillet	Dorset	Leicester	Corriedale		
Delaine Merino	Hampshire	Lincoln	Panama		
Rambouillet	Montadale	Romney	Romeldale		
	Oxford		Targhee		
	Shropshire				
	Southdown				
	Suffolk				
	Tunis				

ranging from 120 to 150 pounds. The Delaines have the best mutton qualities of the Merino breed and the highest quality of fine-wool fleeces.

In the trade, the smooth-bodied Merinos are designated as Delaines, and the wrinkled type as American Merinos. All are purebred Merinos and may be registered in the same breed association because they are similar in type and ancestry.

The Merinos originated in Spain, and were first imported into the United States in 1793. Because of their banding instinct (desire to remain at all times in close contact with each other) they are easy to herd, enabling one shepherd to control large numbers of sheep. They are good grazers, foraging over large areas of poor grasslands, which enables them to survive where many less hardy breeds would fail. Where grazing conditions are not favorable for the production of grass-finished lambs, the wool-producing ability of Merinos has made them a favorite.

The Merino is a white-faced sheep with white feet. Most rams have horns, whereas the ewes are hornless. Most of the head and legs are covered by wool. Merinos have long been bred for wool production and do not carry the straight lines and compactness of the mutton breeds, but many have produced excellent carcasses. The average lamb crop is about 110 percent, which means that one ewe in ten will have more than a single lamb. Merinos are extremely hardy, and are able to survive under adverse weather as well as poor grazing conditions. The ewes live and produce longer than almost any other breed.

Rambouillet. The Rambouillet is a descendent of the Spanish Merino, and was developed as a breed in France. Approximately 50 percent of all sheep in the United States carry some Rambouillet breeding. Although Rambouillets are close relatives of the Merino, they have been bred to produce a better carcass. The carcass quality of the Rambouillets does not compare favorably with the best mutton breeds, but it does make a fair market lamb and produces an excellent fine-wool fleece. The rams may have horns or be polled. The ewes are polled. Rambouillets have large heads with white hair around the nose and ears. The body is not as smooth as that of the mutton types, but the lines are fairly straight, and they carry considerable depth. The breed is large, with mature rams averaging about 250 pounds and some weighing as much as 275 pounds. The ewes will weigh 150 pounds on the average with the larger ewes weighing up to 200 pounds. The fleece is heavy, close, and compact, covering most of the body including face and legs.

Rambouillets are good mothers, quite prolific, and unequalled for range qualities. A very large percentage of the range sheep carry some Rambouillet blood. When Rambouillets are crossed on medium- or long-wool breeds, the resulting lambs are good feeders and will produce top-quality fat lambs.

Only the B-type and C-type Rambouillets are found in the United States. The types are determined by the skin folds. The B-type is the more wrinkled variety, whereas the C-type is comparatively smooth. The B-type has lost much of its popularity in the United States and has largely disappeared. The C-type has consistently been improved, both in carcass and in fleece, until today it is enjoying its greatest popularity. The American Rambouillet Association has developed a ram testing program and recognizes certified and register of merit rams. Daily weight gain, clean wool production, staple length, face covering and skin folds are selection factors considered in the program. Rambouillets were first introduced into the United States in 1840 by Mr. D. C. Collins of Hartford, Connecticut.

Debouillet. The Debouillet breed was developed in New Mexico from crosses of Delaine-Merino and Rambouillet sheep. Development of the breed was begun in 1920. The Debouillet is well adapted for range production in the New Mexico-Western Texas area.

The Debouillet is medium in size with mature rams weighing from 180 to 230 pounds and mature ewes averaging about 130 pounds. The rams can be either horned or polled with the ewes being polled. Selection within the breed has been for adaptability to arid range conditions, open face, staple length, uniformity of fine-wool fleece, fleece quality, wool production, and body size.

Medium-wool Breeds

Among the medium-wool breeds are Cheviot, Dorset, Hampshire, Montadale, Oxford, Shropshire, Southdown, Suffolk, and Tunis.

Cheviot. The Cheviot, developed primarily in Scotland, is known for its hardiness and vitality. It is a beautiful breed with erect ears, a clean white face, and white legs. The face and legs are covered with short white hair. The nose, lips, and feet are black.

The breed is small, with rams weighing on an average of 175 pounds when mature and ewes averaging 125 pounds. They are blocky, compact, and well muscled over the back, loin, and leg of mutton. The fleece is light, averaging from 6 to 8 pounds, but contains a small amount of yolk and is light-shrinking. The Cheviot ewes are good mothers and quite prolific, averaging about a 125 percent lamb crop.

The first importation of Cheviot sheep into the United States was in 1838.

Dorset. There are two strains of Dorsets. One is polled and the other has horns. The ewes will breed during most of the year, which is uncommon among sheep breeds and makes the breed popular when late fall lambs are desired.

The Dorsets are medium-sized, with rams weighing from 200 to 250 pounds and

FIGURE 30-2. (*Left*) **Cheviot ewes lamb easily and are excellent mothers.** (*Courtesy* **American Cheviot Sheep Society, Inc.**). **FIGURE 30-3.** (*Right*) **A Polled Dorset yearling lamb.** (*Courtesy* **Continental Dorset Club**)

FIGURE 30-4. Horned Dorset yearling rams. (*Courtesy* Continental Dorset Club)

ewes from 150 to 175 pounds. The ewes are prolific, averaging nearly a 150 percent lamb crop. Dorsets will shear from 8 to 10 pounds of fleece.

Dorsets were developed in England and the first importation to this country was in 1885.

Hampshire. The Hampshires have most of the qualities desired in mutton sheep. They are large and grow rapidly. Mature rams will average 250 pounds in weight and ewes will average about 180 pounds. The ewes are prolific and good mothers.

The face, ears, and legs of Hampshires are dark brown or black. They are among the largest of the medium-wool breeds, being exceeded only by the Suffolk and Oxford. The fleece is not especially heavy, averaging from 7 to 8 pounds, and is of medium quality. Hampshires have been used extensively in crossbreeding with excellent results.

Hampshire sheep originated in England. Several importations were made into the United States prior to 1860. They grew rapidly in popularity and are one of the more popular breeds in the United States today.

Montadale. The Montadale breed was developed by Mr. E. H. Mattingly of St. Louis, Missouri; Columbia and Cheviots were used in the foundation breeding. The original cross was a Columbia ram on Cheviot ewes, but later the cross was reversed, using a Cheviot ram on Columbia ewes. After fourteen years of selection the breed has been well established and appears to be breeding true to type. The Montadale is of good mutton conformation and the wool clip averages from 10 to 12 pounds. The emphasis in selection has been for hardiness, prolificacy, total pounds of lamb, and pounds of wool.

Oxford. The Oxford is the largest of the medium-wool breeds. Mature rams will weigh up to 350 pounds, with an average weight of 300 pounds, and the ewes will range from 175 to 250 pounds. The head and ears are small compared to the body size. The face, ears, and legs will vary from gray to brown in color. The breed is polled.

Oxfords originated in England. Clayton Reynold, of Delaware City, Delaware, imported the first Oxfords into the United States in 1846. Desirable qualities which are useful

FIGURE 30-5. (*Left*) The 1972 International Livestock Show Champion Hampshire ram owned and exhibited by Robert Fritz, Albert Lea, Minnesota. **FIGURE 30-6.** (*Right*) The 1972 International Livestock Show Champion ewe owned and exhibited by C. M. Hubbard and Son, Junction City, Oregon. (Figures 30-5 and 30-6 *courtesy* American Hampshire Sheep Association)

to commercial breeders are growth rate, prolificacy, and milking ability.

Shropshires. The Shropshire is one of the smallest of the medium-wool breeds. Mature rams weigh from 175 to 200 pounds and ewes from 135 to 150 pounds. They are a good dual-purpose type, producing a very desirable carcass and a wool clip that averages 9 pounds.

The face, ears, and legs are a deep brown color. Because the face and legs are covered with wool, the Shropshire has lost much of its popularity. A heavy face covering of wool is conducive to wool blindness, to which many ranchers and other flock owners object.

The Shropshire breed was developed in England and first brought to the United States in 1885.

Southdown. The Southdown is a small sheep. Rams weigh about 175 pounds and ewes 125 pounds at maturity. The head is broad, with a wool cap that comes just below the eyes. The face is mouse-colored or light brown. The Southdowns are unexcelled for mutton conformation. They are low-set, compact, wide, and deep, with legs set wide apart. The breed matures early and has been used extensively in the production of hothouse lambs (see Chapter 32). Southdowns will produce a fleece weighing from 5 to 7 pounds. Because of the small size of Southdowns, they have never been popular in the West or Midwest, but they are quite popular in the East and Southeast.

The Southdowns are one of the oldest of the English breeds and have contributed to the development of many other breeds of sheep. The emphasis in selection in recent years has been on improvement of size and length. Selection has continued for muscling and carcass quality.

Suffolk. The Suffolk has enjoyed a recent popularity in the United States that is unequalled by any other breed. It is being used in many Western flocks for crossbreeding and has increased considerably in the Midwest. Its sudden increase in popularity is probably due to its open face, preventing wool blindness, and to its excellent conformation and comparatively large size. The head is small, resulting in less lambing difficulty.

The Suffolk is distinguished by its black face, legs, and ears. Rams average about 250 pounds and ewes 180 pounds when mature. The ewes are very prolific, producing a 150 percent or more lamb crop. The breed will shear a fleece ranging from 8 to 10 pounds in weight.

The Suffolk, like most of the medium-wool breeds, was developed in England.

Tunis. The Tunis is one of the oldest breeds of sheep known, among the smaller breeds, but it has never been popular in the United States. The breed is polled and open-faced, with a tan or red face. Rams average about 150 pounds and ewes 110 to 125 pounds when mature. The breed is of Asiatic origin.

Long-wool Breeds

Among the long-wool breeds which have been imported to the United States are the Cotswold, Leicester, Lincoln, and Romney.

FIGURE 30-7. (*Above*) A Shropshire ewe. (*Courtesy* Iowa Sheep Breeders Association) **FIGURE 30-8.** (*Below left*) Some large Suffolk ewes. (*Courtesy* Hawkeye Institute of Technology) **FIGURE 30-9.** (*Below right*) A Romney ewe. (*Courtesy* American Romney Breeders Association)

Cotswold. The most distinguishing characteristic of the Cotswold breed is the way the wool hangs in curls, with a tuft of wool hanging from the forehead down to the eyes. The wool is coarse and the fibers long. Cotswolds are large sheep, with rams weighing 300 pounds and ewes up to 225 pounds. The breed has never been popular in the United States.

Leicester. There are two types of Leicester sheep, the English and the Border. While they are regarded as two separate breeds in England, in the United States they are considered two types of the same breed and are registered in the same flock books.

Leicesters are medium in size, with clean faces and legs. The ewes are not very prolific. Leicesters have been used mainly for crossing purposes in the United States. The body form and wool are typical of the long-wool breeds. Leicester sheep were developed in England.

Lincoln. Native to England, the Lincoln is a large sheep with a broad head and large thick ears. It is very rugged and heavily fleshed. The wool is long and of good weight and quality. The ewes will produce from 12 to 16 pounds of wool annually. The ewes are fairly prolific, but are not recognized for their milking ability. While few purebred flocks exist in this country, Lincolns have been very successfully used in crossbreeding and in developing new breeds.

Romney. The Romney, sometimes called Romney Marsh or Kent sheep, seems to be better adapted to wet, swampy areas than most other breeds. Breeders of these sheep also claim they are less susceptible to foot rot and liver flukes (both common to wet areas), which probably gives them their reputation for being better adapted to countries of heavy rainfall. Like the other long-wool breeds, they are slow-maturing and coarse-wooled. Romney sheep were developed in the Romney Marsh region of southwestern England.

Crossbred-wool Breeds

The crossbred-wool breeds were developed by using crosses of long- and fine-wool sheep and by selection until the breeds were developed to the extent that they would breed true to form. As a class they are similar to the medium-wool class, but they are more hardy and better adapted to the Western range than are the medium-wool breeds. They produce a heavier fleece and better carcass than do the fine-wool breeds, without losing too much of their handling instinct and hardiness.

The Columbia, Corriedale, Panama, Romeldale, and Targhee constitute these breeds.

Columbia. This breed resulted from first crossing Lincoln rams on Rambouillet ewes. Selection was based on utility value without consideration to breed points which have little or no productive value. The Columbia, the first of the strictly American breeds to be developed, was started by the Bureau of Animal Industry in 1917.

The breed is the largest of the crossbred-wool class and well adapted to bitter range conditions. The mature rams will weigh from 225 to 275 pounds and ewes will weigh from 125 to 190 pounds in breeding condition. They will produce from 10 to 13 pounds of wool yearly. The lambs are of good market type, but they are not equal to that of the medium-wool breeds. The breed is somewhat rangy, a characteristic necessary under most range conditions where large areas must be covered for feed and water. The breed is open-faced and not susceptible to wool blindness. Columbias represent a successful breeding experiment since they produce a good market lamb and a good wool clip, are adapted to range conditions, and retain much of the herding instinct of their Rambouillet ancestors.

Corriedale. The Corriedale is blockier and smaller than the Columbia. Mature rams range from 185 to 250 pounds and ewes range from 125 to 185 pounds. The annual wool clip will average from 9 to 12 pounds. The quality of the wool is good, as is the carcass value of the lambs. Corriedales have less appeal to many range sheepmen than Columbias because of their smaller size and finer bone.

The Corriedale resulted from a combination of the Merino, Lincoln, and Leicester breeds. The oldest of all the crossbred-wool breeds, they were developed in New Zealand about 1880. The Bureau of Animal Industry imported the first Corriedales into the United States in 1914. They were tested under Western range conditions and were reported to be well adapted to the bitter range areas.

Panama. The Panama and Columbia originated from the same cross, the difference being that the cross was reversed. That is, Rambouillet rams were mated to Lincoln ewes to produce the Panama. The breed is an American breed originated by Laidlaw and Brockie of Mildoon, Idaho. The Panama closely resembles the Columbia, but it is a little smaller and shows somewhat better mutton qualities.

Romeldale. The Romeldale is a new breed developed by A. T. Spencer, Woodland, California. New Zealand Romney Marsh rams were used on Rambouillet ewes and, as a result of inbreeding and rigid selection, the Romeldale produces a fleece ranging from 10 to 13 pounds and is a compact, good-quality meat animal.

Targhee. Since 1926, the United States Department of Agriculture Sheep Experiment Station at Dubois, Idaho, has been working with a combination of Rambouillets, Corriedales, and Lincolns, breeding them into a strain which they have called Targhee. They are white-faced, hornless, and of medium size. Rams weigh about 200 pounds and ewes about 130. They are open-faced, produce a good-quality wool clip weighing from 10 to 12 pounds, and are a good mutton-type lamb. The breed gives promise of being well adapted to much of the range area.

Carpet-wool Breeds

Carpet-wool sheep are of minor importance in the United States. Only the Black-Faced Highland exists in any quantity in this country and there is only a small number of them.

Black-Faced Highland. Both sexes of this breed have horns. It is a typical carpet-wool type with a long coarse outer wool coat which resembles hair more than wool. They have a finer undercoat which helps to give protection against the inclement weather of their native home in the Scottish highlands. The breed is small, rugged, and able to graze on areas of sparse vegetation and rough terrain.

Fur Sheep Breeds

The fur sheep industry is new in the United States and many problems must be overcome before it succeeds. Since the primary purpose for raising fur sheep is to sell the pelts, a good market for pelts must exist if profits are to be made. Coats and other garments require matching pelts. To be able to market, at one time, a sufficient number of matching pelts for a coat has been difficult for the producer. Single pelts command much less in price than several matching pelts.

The feeding value of the lambs is very poor, as is the wool clip of mature animals. It is questionable whether it would be wise for a beginner to enter into the fur sheep business.

Karakul. The characteristics of the various types of pelts were discussed earlier in the chapter. The rams have horns, but the ewes are polled. Rams weigh about 200 pounds and ewes weigh about 145 pounds at maturity.

SELECTING AND ESTABLISHING THE BREEDING STOCK

Whether a small farm flock or a large range operation is to be established, success will depend upon the economical production of lambs and wool. Each individual breeding animal in the flock should be selected on the basis of efficiency in producing these products.

SELECTING A BREED

In selecting a breed of sheep the more important considerations are: (1) environmental conditions under which the animals will be produced, (2) market price and demand for the principal product to be produced, (3) cost and availability of breeding stock, and (4) personal likes.

Environmental Conditions

Certain breeds have the ability to forage and survive over range areas that have sparse vegetation, producing a wool clip and a lamb where other breeds would fail. Also, some breeds have what is known as a banding instinct, or the tendency to graze in close, compact groups. When large bands of sheep are grazed over unfenced areas and where predatory animals are prevalent, it is important that the sheep maintain a close formation. One shepherd is usually charged with the care of a flock. Even with the help of trained sheep dogs, the task is almost impossible under many conditions with breeds that tend to spread out as they graze. Certain breeds have a wool covering over their face. Some individual sheep of these breeds are subject to wool blindness, especially where snow, ice, and the awns of certain grasses may get into the fleece. Unless individual attention is given each animal, such breeds may not be a wise choice.

In areas where feed is plentiful and where the flock may graze on fenced pastures, mutton or slaughter lamb production will probably be the best source of income. The medium-wool or crossbred-wool breeds generally produce a better quality of lamb than either the fine-wool or long-wool breeds.

Some breeders of fur sheep have found the enterprise to be profitable. The number of fur sheep is not large in the United States and most breeders have been selling surplus animals for breeding purposes rather than for fur. The problem of having enough matching pelts at one time to produce a coat has made the marketing of lamb skins a problem for the breeders.

Market Price and Demand

Market outlets and the price paid for wool and lambs are a consideration in selecting a breed. The purebred breeder must rely upon commercial producers and other breeders for his sales. The number of breeding animals in proportion to the demand for such animals determines to a large extent the demand and price a breeder will receive for his stock.

Cost and Availability of Breeding Stock

The commercial producers must depend on the purebred breeders for improved breeding animals, especially rams. How far one will have to go to get these animals and what the cost will be on the farm and ranch are two questions to be answered before choosing a breed. It is usually cheaper and easier to find breeding animals if the breed is common to the community. However, if a new breed has been developed, or if the establishment of another breed in a community will be a decided improvement, the first producers to get a start in the breed will probably harvest some handsome profits. Such breeders will have a good demand for breeding stock until a general change-over of breeds has been accomplished in the community.

Personal Likes

Success depends largely upon one's interest in one's work. To choose a breed that does not appeal to the producer makes it difficult to develop the interest necessary for success.

SELECTING FOUNDATION STOCK

Having decided upon the breed of sheep to raise, the breeder is next confronted with the problem of selecting foundation stock. While some breeds are considered superior for the amount and quality of their wool, and other breeds for carcass qualities, the general type desired is very much the same, regardless of the breed.

The breeder of fine-wool sheep may be especially concerned about the quantity and quality of wool they produce, but he also depends upon the sale of lambs for a good share of his income. The meat qualities of the lambs largely determine the price they will bring, as either feeder or slaughter lambs. It is also true that the breeder of meat-type sheep depends on the wool clip as a secondary source of income. The chief concern is the carcass qualities of the animals, but if a heavy fleece of good quality is produced, the chances of a profit from the enterprise are considerably better.

Each breed has its own characteristics which are desirable and in many cases essential for registration, but are not important from a production standpoint. Color, wooled or clean legs, and horned or hornless heads are examples of breed characteristics. Purebred breeders are more concerned about individual breed characteristics than are commercial producers. The breed associations will furnish the needed information on the breeds.

Selection Based on Health, Soundness, and Uniformity

These three points should be given prime consideration when choosing animals for breeding purposes.

Health. Health of the stock is of the utmost importance. Animals that are listless or show dark skins or paleness in the lining of the nose and eyelids are not a good choice. Such animals may not live to produce a lamb and even if they do are not likely to produce enough milk for the rapid growth of the lamb.

Soundness. Old ewes with broken mouths are a poor choice, as they will have difficulty in eating and therefore will not be in condition for lamb production.

Udders on breeding ewes should be carefully checked for soundness. Abscesses, ruptures, blind teats, or missing teats make ewes useless for breeding purposes. Some ewes have hard meaty udders or abnormally large teats. These should be rejected.

The upper and lower jaws should be even in length, with the incisors of the lower jaw fitting against the hard pad or cartilage of the upper jaw. A sheep with either jaw protruding beyond the other will have difficulty grazing and will be a poor forager.

Any ewe that is exceptionally heavy or enlarged directly below the rear flank undoubtedly has a weak abdominal wall or a breach and should not be kept as a lamb producer.

Uniformity. To produce a lamb crop uniform in size and type, a uniform ewe flock of the same breed or cross is needed. Breeds of sheep differ considerably in size and type. Mixing the small, fast-maturing breeds with the larger breeds would result in a lamb crop lacking in uniformity and market value.

Selection Based on Body Conformation

Sheep, unlike other livestock, must be judged largely with the hands rather than the eye, because of the wool covering. Wool many times covers defects that cannot be seen. Going over the animal from the neck to the leg of mutton with the hands enables the trained judge to uncover any possible defects.

If the animals are shorn, more may be seen by the eye than when they are in fleece. While one must use both the eye and the hand to judge, a general idea of the conformation may be gained by sight alone. As in judging other livestock, the first step is to survey the animal from a distance of about 10 to 15 feet. From a side view the animal should appear large in scale with a straight top and underlines. The legs should be strong, with short, reasonably straight pasterns. From a front view the muzzle should be broad and square, the nostrils large and wide, and the eyes alert. The head should show masculinity in the ram and femininity in the ewe, and be free of coarseness. The chest should be deep, with good width between the front legs. The shoulders should blend smoothly with the body.

From the rear of the animal observe the width and uniformity of width from shoulders to dock. Remember the back and loin are where the lamb chops and choice roasts come from. They should be wide and uniform. The twist should be deep, the dock wide, and the legs full and plump. The animal should move with ease.

After gaining a general idea by looking over the animal from at least three positions, stop immediately behind the sheep, placing both hands around the neck. Observe whether the neck is short and thick, well blended into the shoulder. Bring the hands back on either side at the shoulders. Is the heart girth full, or is there a depression behind the shoulder down along the side? Next bring the hands up just below the topline on either side and work back to the dock. The animal should be well muscled, uniform, and firm.

FIGURE 30-10. Note the excellent body conformation of this ewe. (*Courtesy* American Hampshire Sheep Association)

The desired conformation is wide from front to rear, especially in the back and loin, and uniform in width to the dock. It should be heavily muscled, having a thick loin, large, plump, firm, thick leg, and deep twist. Good breeding stock will be straight and strong in the topline, well balanced, and will stand straight on the legs. The sheep should be large for their breed with good body length.

Selection Based on Age

The lamb has eight milk teeth in the lower jaw. Between the ages of 12 and 18 months, the center pair is replaced by permanent teeth which are much longer and wider. At two years of age, the second pair of permanent teeth appear, and at three years of age the third set is present. When the sheep reaches its fourth year, the last or fourth pair of permanent teeth has replaced the milk teeth. The animal is then said to have a full mouth.

After four years the teeth begin to slant forward or, in the case of range sheep, they may wear down short. As the sheep ages, the teeth either spread and eventually break off or wear completely away. This condition is referred to as *broken mouth,* and when all teeth have disappeared the sheep are called *gummers.*

Selection Based on Production Records

The surest way of selecting animals for breeding purposes is on a production test. Production testing includes: (1) the type and finish at weaning time, (2) prolificacy, (3) the rate of gain based on weight of

FIGURE 30-11. (*A*) Lamb mouth. (*B*) Yearling—first set of permanent teeth. (*C*) Two-year-old mouth. (*D*) Four-year-old—four pairs of permanent teeth, termed a "full mouth." (Campbell photos. *Courtesy* Hawkeye Institute of Technology)

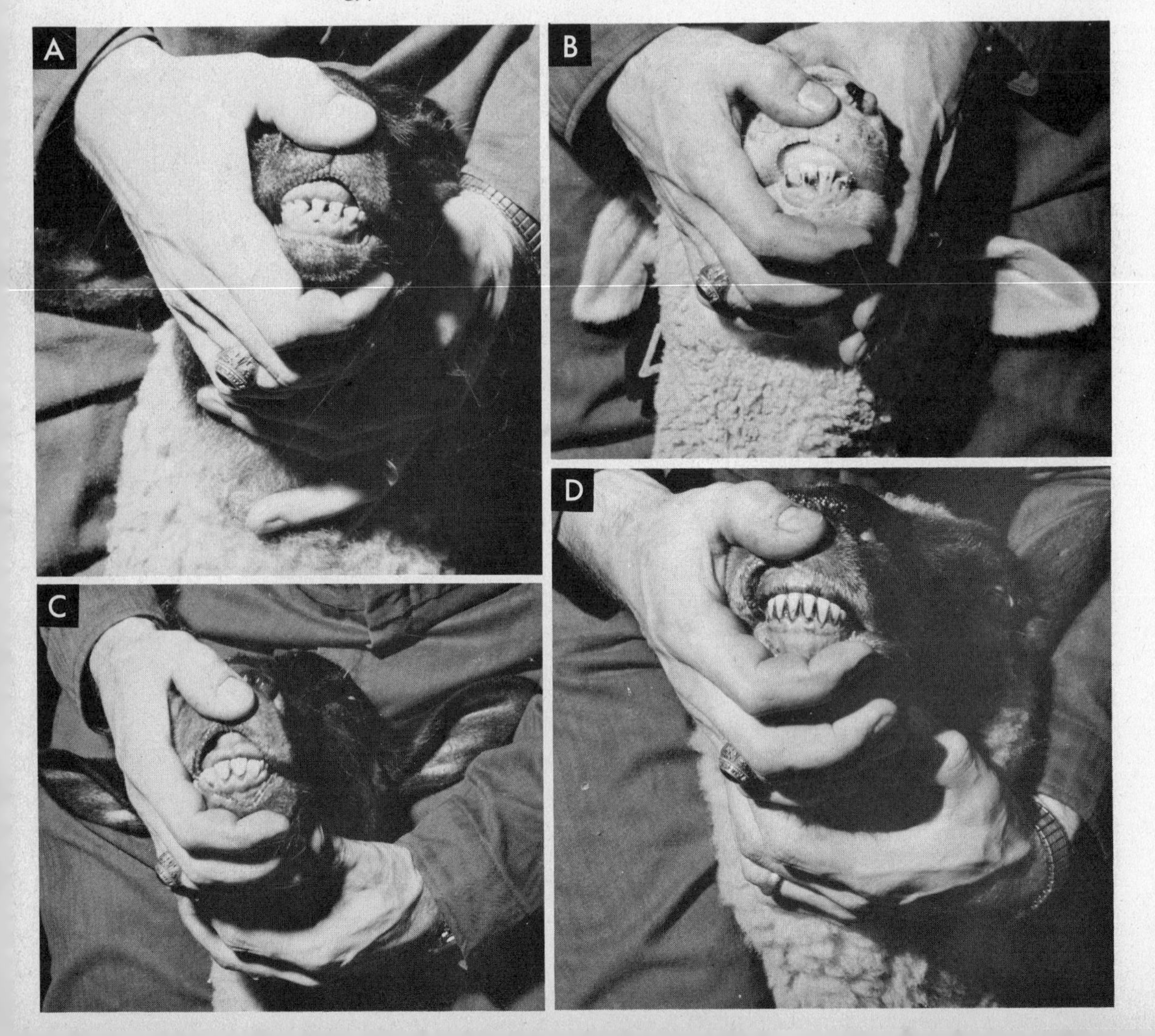

lambs at weaning time, and (4) the weight and quality of fleece. A ram whose offspring have shown up well in production tests is known as a desirably proven sire. Proven sires are not plentiful and breeding stock sired by a proven ram is less difficult to find and purchase than the proven sire himself. Animals sired by a ram with a number of offspring that have shown up well under production testing and have cut good carcasses are usually quite certain to be satisfactory.

Selection Based on the Heritability of Production Traits

The heritability for a given trait may vary between breeds or sheep populations of different genetic background. Heritability is the fraction of the total variation for a given trait within a population that is due to, or attributed to, the additive effect of the genes. Heritability is important to estimate the amount of improvement that might be expected in one year or one generation.

In general, heritabilities of less than 20 percent are considered low; between 20 and 40 percent are considered medium; and over 40 percent are considered high. Carcass traits are usually high in heritability, rate of gain and feed efficiency are medium, and multiple birth and weaning weight are low. Table 30-2 will show some of the production traits and the average heritability percentage expected.

SELECTING FEEDER LAMBS

Feeder sheep are sheep that are to be finished for slaughter. Most of the feeders consist of lambs, but a few ewes and some yearlings go into the feed yards.

Feeder Lambs

Feeder lambs are young animals, wethers or ewes under one year of age, that are not carrying sufficient finish for slaughter. The grade classification indicates the probable

TABLE 30-2 HERITABILITY OF PRODUCTION TRAITS

Trait	Average Heritability (%)	
Weaning weight	10	LOW
Multiple birth	15	LOW
Number of lambs raised	13	LOW
Carcass grade	12	LOW
Dressing percent	10	LOW
Milk production	26	MEDIUM
Birth weight	30	MEDIUM
Rate of gain	30	MEDIUM
Fleece grade	35	MEDIUM
Lean weight	39	MEDIUM
Mature body weight	40	HIGH
Face cover	56	HIGH
Clean fleece weight	40	HIGH
Loin eye area	53	HIGH
Fat weight	57	HIGH

length of the feeding period to produce a slaughter lamb, and the health, vigor, quality, and type of lamb. Fancy is the highest grade. It represents an excellent meat type, free from diseases and parasites and usually carrying enough finish so that only a short feeding period of 40 to 90 days is required to finish the sheep into prime slaughter lambs. Very few fancy grade feeder lambs are available, as the qualifications for this grade are very high and most of them are bought for exhibition purposes. Choice is recognized as the top commercial grade. The lower grades are usually thin and, in the case of common or inferior grades, may lack the health and vigor of the higher grades. They usually are shallow-bodied and upstanding, and show indications of being heavily infested with parasites.

Wethers

Wethers are male lambs castrated before reaching sexual maturity and before developing the characteristics of a ram.

Feeder Sheep

Sheep that are over a year old, but are best finished before marketing, are known as feeder sheep. A few ewes that have been culled are fed out and some ranchers, when the price is too low, may carry lambs over, take a wool clip, and market them as yearlings. Some of these yearlings are fed out on grass in the Corn Belt before going to slaughter.

Shearer Lambs

Shearer lambs carry nearly a full year's growth of wool. They are usually shorn before going into the finishing yard, especially during warm weather. Shearer lambs bring less money on the slaughter market per pound than lambs with fleeces. However, the value of the fleece plus the better gains during hot weather usually more than offset the lower market value.

SELECTING THE WEIGHT AND GRADE OF FEEDER LAMBS

The weight at which to buy lambs depends considerably upon the length of time they are to be fed. Lamb feeders usually plan a feeding period of from 30 to 120 days. Expected daily gains will range from ¼ to one pound, depending upon the kind of feed used and the method of feeding. The slaughter market prices are usually best for the finished lambs that weigh from 90 to 110 pounds. Heavier lambs are subject to a large price discount.

Lambs on a heavy grain ration may be expected to gain about ¾ pound per day. Those fed primarily on pasture or cured forage will average less daily gain.

The lamb feeder should calculate the gain his lambs will make, after considering the ration to be fed and the length of the anticipated feeding period. Ninety- to 105-pound finished lambs are usually bought by packers at top slaughter lamb prices. Therefore, 50- to 60-pounders will best fit a long feeding period of 120 days, while heavier lambs will be better suited for short feeding periods.

It should be pointed out that lambs of the small breeds will finish at weights of from 65 to 70 pounds. However, nearly all the feeder lambs available for sale are westerns. Most westerns are either of the large crossbred type, such as the Columbia or Corriedale, or fine-wool and the larger medium-wool crossbreds. These lambs are large and will not usually finish out at less than 90 pounds.

Grade

Generally speaking, choice feeder lambs are the best buy. Lambs grading lower than

TABLE 30-3 CLASSES AND GRADES OF FEEDER SHEEP

Sheep or Lambs	Sex	Age	Weight	Grades
Feeder sheep	Ewes and wethers	Yearlings	All weights	Fancy Choice Good Medium Common
	Ewes	2 years or older	All weights	Choice Good Medium Common Inferior
Lambs	Feeder lambs Ewes or wethers	All ages	All weights	Fancy Choice Good Medium Common Inferior
	Shearer lambs Ewes or wethers	All ages	All weights	Choice Good Medium

choice will require a longer feeding period and will have a heavier death rate. Most lamb feeders expect from 2 to 3 percent death loss during the feeding period. However, less thrifty lambs will run considerably above expected loss.

Price is an important consideration, but the lower-grade lambs will necessarily have to be bought at a figure considerably under the price of choice lambs if they are to be a good buy.

SUMMARY

Sheep are classified several different ways. However, wool-type classification is most common. Under the wool-type classification there exist six usual classifications. They are: fine-wool, medium-wool, long-wool, crossbred-wool, carpet-wool, and fur sheep wool.

The fine-wool breeds produce a fine fiber, but a heavy fleece containing a large amount of yolk. The American and Delaine Merinos and Rambouillets are the breeds of the fine-wool class most commonly found in the United States.

The medium-wool class consists of those breeds that have a fleece of medium fineness and length. They are unexcelled in mutton-type conformation. The breeds of this class commonly found in the United States are: Cheviot, Dorset, Hampshire, Montadale, Oxford, Shropshire, Southdown, Suffolk, and Tunis.

The long-wool class consists of those sheep with exceptionally long fibers of coarse wool. They are large and rugged and have been popular for crossbreeding purposes, especially on fine-wool breeds. The most common breeds of this class found in this country are the Cotswold, Leicester, Lincoln, and Romney.

The crossbred-wool type consists of a number of breeds developed by first crossing long-wool and fine-wool types. The main purpose for these breeds was to develop animals with a good fleece, good carcass qualities, and the herding instinct and hardiness essential for rugged range conditions. The common crossbred-wool breeds are: Columbia, Corriedale, Panama, Romeldale, and Targhee.

Carpet-wool sheep produce a long, coarse, tough wool suitable for carpet manufacture. The Black-Faced Highland is the most common breed, but very few are found in the United States.

Fur sheep, of which the Karakul is the chief breed, are produced for their lamb pelts. They have very poor carcass qualities and the fleece from the mature animal is of little value.

Environmental conditions, market price and demand, cost of available breeding stock, and personal likes are important factors in selecting a breed. Foundation stock should be selected on the basis of body conformation, production records, heritability of traits, age, mouth conditions, and soundness of udders. Good health is essential.

Feeder sheep and lambs are animals that need more finish before they are ready for the slaughter market. Sheep and lamb feeding is carried on by farmers and big commercial feeders. Profits come from one or more of the following sources: margin, gain, or wool clip.

Feeder lambs and sheep are classified and graded according to sex, age, weight, and grade. The weight and grade of lambs to buy for feeding will depend upon how long they are to be fed, the kind of ration to be used, and the prices of feeder and slaughter lambs. Choice grade lambs usually prove to be the best buy, as they are healthy, vigorous, and of high quality.

QUESTIONS

1. List the classes of sheep and the important breeds in each class.
2. What are the distinguishing characteristics of each class of sheep?
3. Which class is recognized chiefly for wool production?
4. Which class produces the best quality slaughter lambs?
5. For what purposes were the crossbred-wool types produced?
6. What breeds of sheep were used in the development of the crossbred-wool type?
7. What are carpet-wool sheep?
8. What are fur sheep?
9. What breeds of sheep best fit the conditions of your community? Why?
10. Explain what you would look for in selecting breeding stock.
11. How can you tell the age of sheep by looking in their mouths? Explain.
12. What factors would you consider in selecting a breed?
13. What information do production records furnish?
14. Under what conditions would you advise going into the purebred sheep business?
15. What is a feeder lamb?
16. What is the purpose of buying and feeding lambs?

17 Explain how lamb feeders may profit from their operations.

18 How are feeder lambs classified?

19 How are feeder lambs graded?

20 Explain the following terms: *wether* lamb, *sheep, shearer* lamb.

21 What are the factors to consider in determining the weight of feeder lambs to buy? Explain.

REFERENCES

ABC System of Selecting Replacement Ewes, Agricultural Extension Service, Publication AXT-178, Davis, California, University of California, 1965.

Alexander, M. A., W. W. Dorrick, and K. C. Fouts, *Farm Sheep Facts,* Agricultural Extension Service, Bulletin E. C. 255, Lincoln, Nebraska, University of Nebraska.

Brown, George, *Selection and Care of the Farm Sheep Flock,* Agricultural Extension Service, Bulletin 242, East Lansing, Michigan, Michigan State University.

Cadmus, W. G., *Sheep Records for Greater Profits,* Agricultural Experiment Station, Circular 182, Corvallis, Oregon, Oregon State College.

Cox, R. F., Donald T. Bell, and H. G. Reed, *Sheep Production in Kansas,* Agricultural Experiment Station, Bulletin 348, Manhattan, Kansas, Kansas State College.

Francis, Eugene, *Buying and Feeding Lambs,* Pamphlet 220, Ames, Iowa, Iowa State University Extension Service, 1955.

Improvement of Sheep through the Selection of Performance-Tested and Progeny-Tested Breeding Animals, Miscellaneous Publication 125, Texas Agricultural Experiment Station.

Scott, George E., *The Sheepman's Production Handbook,* First Edition, Abegg Printing, Denver, Colorado, 1970.

Selecting Sheep, Bulletin 13-858, College Station, Texas, Texas A. & M. University.

Selective Breeding for Better Sheep, Agricultural Extension Service, Davis, California, University of California.

31 Feeding and Management of the Breeding Flock

If a healthy productive flock of sheep is to be maintained, the shepherd must give special attention to feeding and management practices.

Flushing

About two weeks before the rams are turned in with the ewes, alert producers will put ewes on a grain ration, or move them to fresh pasture areas where feed is more abundant. This process is known as "flushing." Flushing the ewes will start the heat periods earlier, which is an advantage when early lambs are desired. It also has the effect of bringing all of the ewes into heat at nearly the same time, resulting in more uniformity in the time ewes are bred.

When the ewe is gaining flesh, her reproductive organs usually begin to function normally. The ovaries produce more healthy reproductive cells, fertility is increased, and a higher conception rate and more twin lambs result. However, twins under rugged range conditions are sometimes a disadvantage. Twins are smaller at birth than a single lamb and ewes milk less under scanty feed conditions. A strong single is more apt to survive. Generally speaking, however, twins are an advantage.

Rations for Flushing. Ewes that have been on pastures sufficient to keep them healthy without abnormally increasing their body flesh are in the best condition for flushing and may be moved to an area with abundant pasture. Immediately, the breeding flock will considerably increase their weight gains and be ready for breeding ten days to two weeks later. The same results may be obtained by giving the ewes a full feed of good-quality legume hay or from ½ to ¾ pound of corn, barley, oats, or sorghum grain daily.

Overfat Ewes. If ewes are overfat at breeding time, many unbred ewes will result. If it is apparent that ewes are going to be too fat, they should be placed on a sparse pasture or in a dry lot where feed will be limited at least six weeks before the breeding season. Although the feeding of ewes to lose body weight when not suckling lambs is not recommended, overfat ewes should be thinned down so that they may be placed on a flushing ration prior to breeding.

FEEDING THE EWES DURING GESTATION

The need for proper nourishment of the ewe during gestation cannot be overemphasized. Unless the ewes receive a proper ration, the owner will harvest weak or dead lambs. The 1968 NRC-recommended levels for the non-lactating period and the first 15 weeks of gestation are calculated to slightly exceed maintenance requirements.

Proper feeding of the ewes during pregnancy will: (1) increase the number of live lambs born, (2) decrease the number of weak or crippled lambs, (3) lessen the danger of lambing paralysis, (4) prolong the productive lifetime of the ewes, (5) increase the ewe's milk flow, (6) improve the quality and quantity of the wool clip, and (7) decrease the danger of ewes not owning their lambs as a result of their weakened condition.

Feeds for the Pregnant Ewe

As the unborn lamb, or fetus, develops within the ewe, the greater are her needs for nourishment. During the first half of the pregnancy the fetus grows rather slowly. For this reason, the demands on the ewe for nourishing her unborn offspring are not too great. If only a limited amount of good forage is available, it should be saved until the latter part of the gestation period when the nutritional requirements of the ewe become greater. Since approximately 70 percent of fetal growth occurs during the last six weeks of gestation, this period becomes critical in ewe nutrition.

A large combination of roughages and pastures may be successfully used during the gestation period. Small-grain stubble fields and the grasses growing in them, either those that have been seeded or the wild weedy plants, will furnish a large part of the nutrient

FIGURE 31-1. Polled Dorset group of ewes and a ram. These sheep are in excellent breeding condition. (*Courtesy* Continental Dorset Club)

needs of the ewes. Cornfields that have been harvested (and gleaned by other livestock to remove excess corn left by the harvesting machinery) will provide forage, some grain, and often much weedy green material for the ewe flock.

It is a good practice to make use of as much crop residue, fall pasture, and waste feed as possible during the early gestation period of the ewes. When this feed consists mostly of dry, nonlegume weeds and grasses, it should be supplemented with feeds high in protein and vitamin A.

Legume hay or good-quality legume grass, corn, or sorghum silage will supply the necessary amount of vitamin A when used to supplement the dry feeds of low quality. When fresh, growing pastures are available, little additional feed is necessary.

Under most circumstances, ewes will get sufficient minerals, except for salt, from the natural feed that is provided. However, phosphorous and many of the trace minerals may be lacking in feeds grown in some sections of the country. Because we cannot always be sure whether the natural feeds are providing the necessary minerals, it is a good plan to furnish a mineral mixture free choice. Salt should be available at all times. This may be provided separately or as a part of the mineral mixture. Minerals recommended for cattle (Chapter 13) may be used for sheep.

The use of a two compartment mineral feeder with salt in one compartment and the salt-mineral mixture in the other is recommended for pregnant ewes.

Many feeds, rich in protein, are available for sheep feeding. Among the more common protein concentrates are cottonseed oil meal, cottonseed cake, linseed oil meal, and soybean oil meal.

Rations for the First Ten Weeks of the Gestation Period

The gestation period of ewes will range from 143 to 151 days. The first half of the gestation period is less critical from a nutritional standpoint than the last half. While it is important that the ewes be properly nourished during the entire pregnancy period, the early part of gestation poses fewer nutritional problems than the latter part.

Some suggested rations for the first half of the gestation period are:

Rations	Pounds Daily
1. Legume hay	Free choice
2. Corn or sorghum silage	2 to 4
Legume hay	1 to 2
3. Chopped corn or sorghum fodder	Free choice
Soybean, linseed, or cottonseed oil meal	1/8
4. Grass hay	Free choice
Protein supplement containing 5 pounds urea and 95 pounds soybean or linseed oil meal	1/10

These are only suggested rations and may be altered using other available feeds. Ewes may find adequate roughages by grazing winter range or gleaning harvested grain fields, and need only a protein supplement added to these feeds.

Feeding Ewes During the Last Half of the Gestation Period

During the last half of the gestation period, the best forages available should be fed. The ration given for ewes during the first part of the gestation period may be continued, but, starting with the eleventh week, 1/4 pound of grain should be added to the rations. Beginning with the sixteenth week, the grain ration should be increased 1/2 to 3/4 of a pound. Unless the roughage is entirely legume hay, an additional amount of protein supplement should be provided.

Ewes, especially those carrying twin lambs, are frequently affected with lambing paralysis shortly before lambing, usually resulting in a dead ewe and loss of her lambs. The cause is thought to be due to the inability of the liver to transform body fats into food. While ewes that have had ½ pound of grain in the daily ration and plenty of exercise during the last six weeks of pregnancy are seldom affected, some operators feed one pint of liquid molasses per head daily during this time or supply the equivalent of from ¼ to ½ pound of sugar, either as dry molasses or in some other form. The molasses or sugar product may replace most of the grain, but since these products have little or no protein, a slight increase in the protein supplement would be advisable when they are used to replace the grain. Molasses is quickly digested and highly regarded by many producers as a precaution against lambing paralysis.

Following are some suggested rations for ewes between the eleventh and sixteenth week of pregnancy.

Rations †	Pounds Daily
1. Legume hay	Free choice
Corn, sorghum grain, or barley	¼ to ½
or oats	½ to ¾
2. Legume and grass silage	2 to 4
Legume hay	1 to 2
Corn, sorghum grain, barley, or wheat	¼ to ½
or oats	½ to ¾
3. Corn silage	2 to 4
Legume hay	1 to 2
Oats	¼ to ½
4. Grass hay	2 to 4
Corn-oats mixture	¼ to ½
35% protein supplement	1/10 to ⅛

† Salt and mineral mixture should be fed free choice with these rations.

The fetus or unborn lamb is developing rapidly during the last few weeks of gestation. This is a critical time in the gestation period. Some suggested rations for the last five weeks of gestation follow.

Rations †	Pounds Daily
1. Oats—5 parts Shelled corn, or grain sorghums—3 parts Bran—1 part Soybean or linseed meal—1 part Dry molasses—1 part	½ to 1
Corn silage	2 to 3
Legume hay	2
2. Barley, sorghum grain, or shelled corn	½ to ¾
Corn silage	2 to 3
Legume hay	2 to 3
3. Oats and shelled corn	¼ to ½
Liquid molasses	½
Legume hay	2 to 3
4. Wheat and oats	½ to ¾
Cottonseed meal	1/10 to ⅛
Sorghum silage	6 to 8

† Salt and mineral mixture should be fed free choice with these rations.

FEEDING EWES THAT ARE SUCKLING LAMBS

Since lambs make their most economical gains while being nursed, the ewes should be fed liberally while suckling so that they can supply a sufficient amount of milk.

Feeding Ewes for the First Ten Days after Lambing

Immediately after lambing, most successful operators reduce the concentrates in the ration. Ewes are less likely to have swollen udders or other udder trouble if the concentrates are reduced for the first ten days. The amount of grain given the ewes may be increased as the lambs grow and are able to take all the milk. The amount to reduce the concentrates depends upon the condition of the ewe, whether she has a single lamb or twins, and the amount of milk she is producing. When large flocks are involved, individual attention given to each ewe's ration may not be practical. Therefore, it is recommended that ewes be given all the legume hay they will consume but little concentrates for the first ten days after lambing.

Ewes that lamb after the start of the pasture season will not need any supplementary feeding, if the pasture is of good quality and plentiful.

Feeding Ewes Ten Days after Lambing until Weaning of the Lambs

Ewes that lamb on the pasture will not require supplemental feeding, other than salt and a mineral mixture, as long as they have an adequate amount of good-quality pasture. When pastures are short or dry they may not supply enough total digestible nutrients or enough protein to maintain milk flow at a high level. Under these conditions additional feed in the form of high quality hay, silage, or grain must be provided. If the lambs are creep fed, less feed will be required for the ewes. When supplementary feeding on pasture is necessary, the amount of additional feed may be calculated on the basis of one pound of good hay, plus 3 pounds of silage or ¾ pound of grain, to replace about half of the average ewe's daily pasture requirements.

Ewes that lamb during seasons when pasture is not available present a more complicated feeding problem. Since milk is high in protein, ewes suckling lambs need slightly more protein than pregnant ewes. Ewes suckling lambs will usually increase their consumption of hay or silage by one to two pounds daily over the amount eaten during gestation. Roots and silage are excellent feeds to stimulate the flow of milk.

Following are some suggestions for dry lot rations for ewes suckling lambs:

Rations	Pounds Daily
1. Good legume hay	2 to 4
Oats—60 pounds Corn or sorghum grain—25 pounds Wheat bran—15 pounds	¾ to 1
2. Corn, or sorghum silage, or root crops	3 to 5
Barley or oats—46 pounds Corn or sorghum—34 pounds Soybean, cottonseed, or linseed meal—20 pounds	½ to 1
3. Grass hay	2 to 4
Oats—30 pounds Corn—30 pounds Bran—20 pounds Linseed meal—20 pounds	½ to 1

FEEDING THE BREEDING RAMS

Well developed rams should be selected in advance of the breeding season. If the ram is overfat, he should be thinned down by gradual reduction in feed and plenty of exercise. If the ram is in normal condition at breeding time, he will need some extra grain unless the pasture is excellent during breeding season. The amount of grain to feed will vary with the size of the ram. A ration consisting of three parts oats, one part corn, and one part bran, fed at the rate of one pound per day, is usually sufficient for the ram during breeding season. Barley or sorghum grain may be substituted for corn or oats.

Usually, the feeds available to the ewes will provide adequately for the ram. Unless the ram is penned up separately part of the day he cannot be given a special ration. Under small flock conditions, when the ram is penned away from the ewes during the day and then turned out at night, the ram may be given a special ration.

MANAGEMENT OF THE BREEDING FLOCK

Successful management practices must necessarily vary depending upon the conditions and type of sheep enterprise with which one is associated. Management of a farm flock, which is usually considerably smaller than a range flock, involves many practices that would not be practical under range conditions. It is also true that many recommended range management practices would have little practical value under farm flock conditions.

The conditions under which the various practices are recommended will be described so that the student, beginner, or the established producer may select those that will be of the greatest value.

HEAT PERIODS, BREEDING SEASON, AND GESTATION PERIOD

Heat Periods

The heat period, or oestrus, is the period in which the ewes will permit the ram to mate. The duration of the heat period will range from 3 to 73 hours. Three fourths of the ewes will remain in heat from 21 to 39 hours. The heat periods will occur every 13 to 19 days, averaging 16½ days from one heat period to the next, during the season when the ewe will mate. Unlike other farm animals, ewes in general do not come in heat at regular intervals throughout the year but come in heat seasonally.

Mating Seasons

Most breeds come in heat during the periods of shorter daylight hours and cooler nights. Dorsets and Tunis come in heat quite regularly throughout the year, while the mating seasons for the other breeds fall during the months of August through November. The reason for this difference in the seasonal mating pattern of sheep is much disputed. Some authorities believe the length of daylight may be a factor, since the mating season changes with areas. Regardless of the reasons, it is well known that mating is seasonal except with the Dorset and Tunis, and that the seasonal pattern varies with breeds. Therefore, it is not wise to have a ewe flock of mixed breeds, if a short term drop of fall or early winter lambs is desired.

Gestation Periods. The gestation period will vary from 144 to 152 days.

PREPARING THE EWE AND RAM FOR BREEDING

Tagging

Before the breeding season starts the wool should be removed from around the dock; this is referred to as "tagging." Ewes sometimes are not bred because wool or tags prevent the ram from making satisfactory connection. Tagging also prevents the ewes from befouling themselves, especially when they are first turned on green pasture, which loosens the bowels.

Eyeing

Wool blindness may result from sharp grass seed awns and other materials collecting in the wool around the eyes. Open-faced breeds of sheep (those that do not have wool growing around the eyes) will not be troubled. However, to prevent wool blindness in those breeds with covered faces, the wool should be clipped away from around the eyes. This process is referred to as "eyeing."

Shearing and Ringing the Ram

Following these two practices, as described in the next paragraphs, will increase

FIGURE 31-2. These Horned Dorset ewes are an example of an open-faced breed of sheep. They are seldom troubled with wool blindness. (*Courtesy* Continental Dorset Club)

FIGURE 31-3. Shearing the ram before the breeding season helps increase his activeness. (*Courtesy* Hawkeye Institute of Technology)

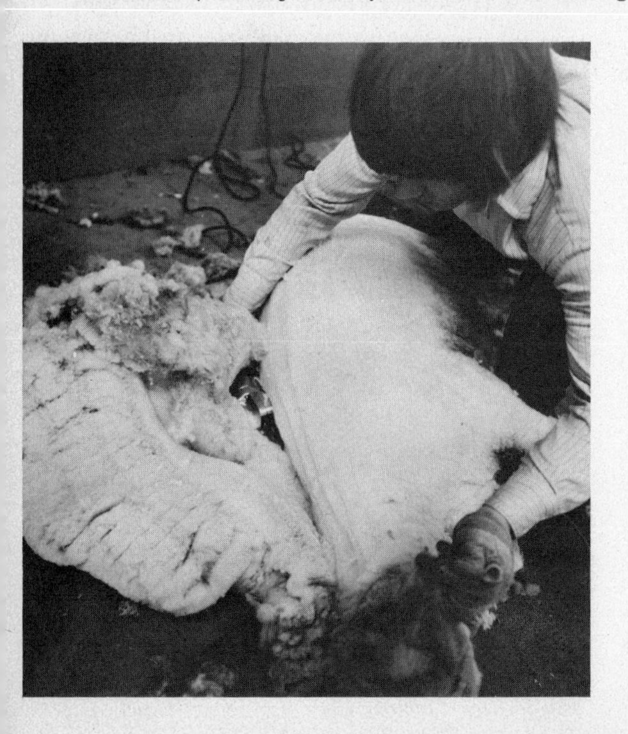

the physical comfort and effectiveness of the ram during the breeding season.

Shearing. Many rams that are fertile and good breeders during cool weather will become impotent during the heat of summer. Higher body temperatures, particularly that of the testicles, is probably the cause of this infertility. When the breeding season starts during warm weather, shearing the ram just prior to turning him in with the ewes will make him more active and, in many cases, will improve his fertility. Many operators shear the breeding rams twice a year and this practice is highly recommended when ewes are to be bred during the warm weather.

Ringing. If the ram is not completely shorn, he should at least be clipped from the neck and from the belly in the region of the

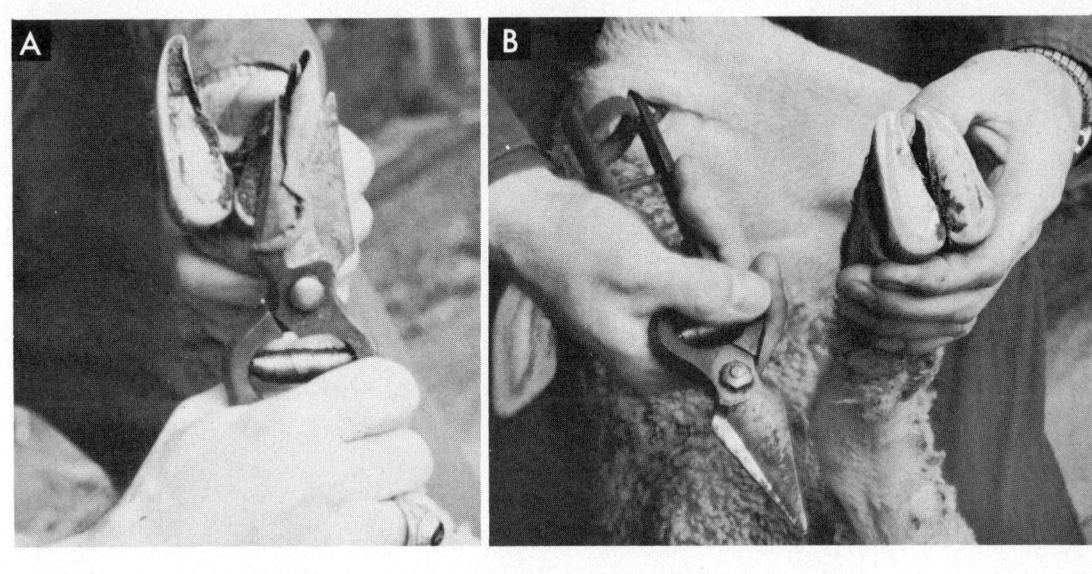

FIGURE 31-4. **(*A*) Long, grown-out hoofs accumulate filth and increase the danger of foot infection. This animal's foot needs trimming. (*B*) The same foot after trimming. (*Courtesy* Hawkeye Institute of Technology)**

penis (ringing). This will make it easier for the ram to make proper contact with the ewes during the act of mating.

Care of the Feet

Both ewes and rams should have their feet well trimmed so they can walk naturally. Sheep raised under range conditions usually keep their feet worn down smooth. Sheep under average farm conditions, where less walking for feed is necessary and when they are kept confined in small lots, do not wear down the hoof. Filth accumulates under the overgrown hoof, causing the feet to become sore. This filth may also harbor organisms responsible for infection—foot rot—which is difficult to control once it is established. A sharp knife or pruning shears may be used to trim the feet.

Marking the Ram

It is very difficult to tell, when a ram is breeding, whether or not he is sterile and whether ewes have been bred, unless the ram is properly marked and notes as to the breeding date of the ewes are kept. Marking the ram may be done either by painting the breast of the ram or using a marking harness.

A good breast paint may be made by mixing lamp black or venetian red with raw linseed oil into a thick paste. One of these mixtures should be applied to the brisket with a paddle at least once a week. When the ram mounts the ewe in the act of breeding, she will be marked on the rump. It is important to identify the ewe by ear tag number and make a note of her identity and the date she was bred. The owner will now have a record of the date the ewe may be expected to lamb. If several ewes return in heat, the ram is probably sterile and a replacement may be made before it is too late. If the color of the paint is changed every 16 to 18 days, detection of unbred ewes and breeding dates will be easier.

Devices consisting of an apron-like pad with straps for attaching them to the breast of the ram are available commercially. Some are fitted with a metal slot fastened between the ram's front legs, where a special type of crayon may be placed for marking the ewes. In one type, known as a marking harness, a colored chalk is used.

EWE-RAM RATIO AND AGE OF THE RAM

While a nine- to ten-month-old well-developed ram lamb may be used to breed ten or twelve ewes, his use is not recommended unless the supply of good yearling or older rams has been used. Good rams over five years of age may be used successfully, but they should be checked carefully as they may be infertile and may also lack the vigor and ability to get around under range conditions.

Yearlings and rams up to five years old may be relied upon breed up to 45 ewes during the breeding season. Under rough range conditions the rates should be one ram to 30 ewes.

Many farm flock owners have too many ewes for one ram and not enough for two. These owners can increase the breeding capacity of their rams by removing them from the flock during the day and feeding them from 2 to 3 pounds of grain. The rams will soon learn to come out of the flock for their feed. Since most of the breeding takes place at night, no breeding time is likely to be lost.

Age to Breed and Average Productive Life of Ewes

The recommended age to breed ewes has been the first breeding season after they are one year of age, or so that they will produce their first lamb at about 24 months of age. However, many experiment stations and flock owners have reported success in breeding ewe lambs.

The productive life of ewes will vary considerably depending upon both the breed and environment. Under average range conditions, ewes may be expected to produce about five crops of lambs if bred for the first time as yearlings. This calls for eliminating them from the flock when they are about seven years of age. Under farm flock conditions, which are less rigorous, the life span will average a little longer.

Should Ewe Lambs Be Bred? A summary of the experimental results and the experience of practical producers seems to indicate the following: (1) Ewe lambs must be well grown out if they are to be bred. (2) They should not be bred until they are nine months of age. (3) Ewe lambs bred at nine months of age gain faster than those not bred, but they weigh less when the lambs are weaned. (4) Lambs will be born later than the average for the rest of the flock; therefore, they must be weaned earlier if the ewes are to conform to the rest of the flock for the next season lamb crop. (5) First lambs will be smaller at birth than those from older ewes, but subsequent lambs will be equal in size. (6) Ewe lambs will need a better ration during pregnancy than older ewes to prevent underdevelopment. (7) More lambing difficulty (ewes needing help at lambing time) will be evident among ewe lambs. (8) Ewe lambs of the larger breeds should be mated the first time to small type rams such as the Southdown. (9) Under good feeding and management, ewe lambs bred at nine months of age produced a wool clip equal to those bred the first time as yearlings, and averaged about three fourths lamb per head more during their life span. (10) More broken mouths occur at six to seven years of age among ewes bred as lambs than among those bred as yearlings.

How Long the Rams Should Be Allowed to Run with the Band. A uniform lamb crop requires that the rams be allowed to run with the ewes for not more than eight weeks but preferably for only six. If several rams are used, the ewes have been properly conditioned for breeding, and each band or flock consists of the same breed or crossbreed, very few unbred ewes will result.

Pregnancy Tests

Early detection of pregnancy would enable the sheep producer to correctly feed and manage a ewe flock. Pregnant ewes would be fed recommended rations for the different stages of gestation, while open ewes could be rebred, fed less, or sold.

Several methods have been used in pregnancy testing sheep. Laparotomy, which requires surgery, and radiology equipment are both accurate methods but are also expensive. A less-expensive method is done by the use of ultrasonic Doppler equipment similar to that used in diagnostic human medicine. The use of ultrasound can detect pregnancy in sheep within 60 days of conception with an accuracy of 90 to 95 percent. Continued research is being done with ultrasound to determine if the ewe will deliver a single or multiple birth.

A rectal-abdominal palpation technique employs the use of a plastic hollow rod. This technique has proved to be the most rapid and accurate method for diagnosing pregnancy. The rod is inserted in the rectum by gentle forward and backward motion until it has penetrated the abdominal cavity approximately 35 centimeters. While the plastic rod is manipulated with one hand, the other hand is placed on the posterior abdomen. The rod is drawn from side to side across the internal surface of the ventral abdominal wall until a significant obstruction is encountered. The size, shape, location, and attachment of the obstruction, as detected by both hands, are then mentally evaluated and a diagnosis made. If no obstruction is observed, the ewe is classified as non-pregnant. The palpation technique can very quickly and easily be taught to a technician, suggesting that it can be widely used in sheep production.

MANAGING THE EWES DURING GESTATION

Lack of exercise during pregnancy contributes to lambing difficulties, including lambing paralysis. Range sheep are much less likely to suffer from lack of exercise than are the farm flocks.

Dry winter pastures and cornfields, when not covered with snow, provide good exercise areas as ewes will forage over them all winter long. Loading the hay or other forage onto a wagon and scattering it at a considerable distance from the bedding area will induce ewes to exercise when snow covers the fields.

Ranchers who supplement the winter range with cottonseed cake can force the ewes to get more exercise by feeding the cake at considerable distance from the bed grounds.

Shade

Ewes, during extremely warm weather, should have protection from the hot sunshine. Under farm flock conditions this can be accomplished either by buildings, trees, or temporary shades.

Protection from Other Livestock

It is not a good practice at any time to let ewes run in the same area with cattle or hogs. This is especially true of ewes heavy with lamb. Cattle may kick or bunt ewes, injuring them and causing abortions. Hogs, especially old sows, may attack a ewe, causing her to bleed. After getting a taste of the blood, they proceed to kill and eat her.

MANAGEMENT DURING THE LAMBING SEASON

Dividing the Flock Before Lambing

When the flock is large, much time and labor may be saved by separating the ewes that are nearest to lambing. If breeding dates have not been recorded, the next best method is by general observation and noticing bagging. Ewes about to lamb will sink away on either side of the rump in front of the hips, the vulva will enlarge, wax will form on the ends of the teats, there will be distention of the udder, and the teats will be tight and show signs of filling.

Marking Ewes

In making the examination and marking ewes for separation into drop bands, a chute with a swinging gate at one end (dodge gate) and a recommended solvable branding paint consisting of two colors are recommended equipment. The chute may be filled and udder examinations made. Those about to lamb can be marked with one color, those about two weeks away marked with another color, and those more than two weeks away left unmarked.

FIGURE 31-5. A temporary lambing pen with an outlet for an electric brooder or heat lamp, if necessary. (Campbell photo. *Courtesy* Hawkeye Institute of Technology)

Handling the Ewes About to Lamb

The ewes about to lamb may be successfully handled by several different methods, depending upon conditions. When flocks are large, lambing on pasture or in the open is generally more practical than attempting to furnish lambing barns or sheds. Naturally the weather must be reasonably warm for pasture lambing. Many successful range operators will confine the heavy ewes to a corral at night, fitting the area with temporary pens where the ewes may be confined for a day or so with their lambs. Individual lambing tents may be used to give the ewe and her newborn lambs temporary protection from inclement weather.

Lambing Pens or Jails

If the flock is not too large, it will pay to provide individual lambing pens for the ewes. The use of lambing pens (often referred to as jails) helps to prevent losses of lambs. Very often, ewes producing twins will wander away with the stronger lamb and leave the weaker one to chill or starve. Ewes having twins sometimes move, after having the first lamb, to a different area where the second is born. The first lamb may not be recognized and owned by the mother. It is difficult to identify lambs with their mothers when the flock is running together and several ewes are lambing at the same time. By numbering the ewes and giving the lambs the same identification marks, poor milking ewes, or those not producing strong lambs, may be more easily identified and later removed from the breeding flocks.

Lambing pens may be temporary and made from panels 36 to 40 inches high and 4 feet long, hinged or wired together. Burlap sacks may be placed around and over the pens to make them warmer, or they may be of a permanent type fitted with feed and watering facilities. The ewes should be left in the pens until the lambs are at least three to four days old.

Brooders

If ewes are lambing during cold weather, brooders may be provided in the lambing pens or in the lambing barn as a means of preventing chilled lambs. These brooders are of two types: those that provide for the lambs from several ewes, known as *colony brooders;* and those which provide for only two to three lambs, known as *individual brooders,* which are used in connection with lambing pens. The colony type is used when ewes lamb in a barn or shed but are not confined to lambing pens. They are enclosures heated with electric heat bulbs (usually 250 watt size) and will accommodate several lambs. They have openings that will admit the lambs but exclude the ewes. The individual type is used in the lambing pen and is built to accommodate only one ewe's lambs at a time.

Preparing the Ewe for Lambing

Just prior to lambing, the shepherd should tag his ewes. This is the process of cutting off any locks of wool that have accumulated dung and filth. He also clips the wool from around the teats so the lambs will be able to find them and will not suck on a lock of wool. Wool-faced breeds may also have the wool clipped away from around the eyes at the same time.

Delivering the Lamb

Ewes that are vigorous, well fed, and well managed seldom have difficulty in lambing. Do not disturb the ewe during the first stages of labor. However, if it becomes evident that the ewe will need help, have the assistant wash and disinfect his hand and arm, and apply a coating of Vaseline to his hand and arm. Lay the ewe on her right side, make entrance, and determine whether the lamb is in the proper position for delivery. If the lamb is in the normal position, it will be right side up, back toward the back of the ewe, front feet extended with the nose on or between the front feet. In a normal delivery the nose will show first with the feet on either side.

Unfortunately the position of the lamb inside the ewe is not always normal and the ewe will be unable to expel the lamb without assistance. The most common difficulty is when one or both front feet are doubled back instead of protruding outward. If the ewe has labored long there is likely to be considerable swelling of the lamb's head, adding to the difficulty. The procedure is to push the lamb back slowly and carefully. This will have to be done against the straining of the ewe. By sliding the hand down the neck and locating the front legs, they may be straightened out in a normal position. Lubricating the walls of the vagina by applying a light mineral oil to the vaginal walls will make it easier to maneuver the lamb into position.

Occasionally, the legs are properly presented but the head and neck are twisted back. If the head cannot be guided out with the hand, tie a heavy string around each leg leaving plenty of string protruding, and double the legs back into the womb. Next make a loop, just large enough to slip over the lamb's head, from a piece of clean unrusted wire. Slip the loop inside the uterus and over the lamb's head, then by pulling on the strings and the wire the lamb may be successfully delivered.

The position of the lamb may be backwards with the tail presented first. Usually the lamb can be delivered in this position by gently pulling on the tail and rear of the lamb.

Whenever it is necessary to go inside the ewe with the hands, it is advisable to

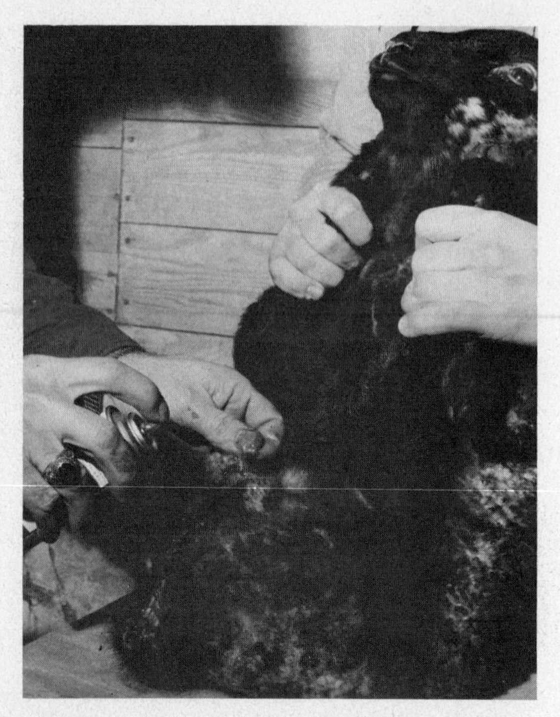

FIGURE 31-6. The navel of a newborn lamb, or the castration wound, should be disinfected with a tincture of iodine to prevent infection. (Campbell photo. *Courtesy* Hawkeye Institute of Technology)

give the ewe an injection of penicillin to help prevent infection.

Care of the Ewe Immediately After Lambing

The ewe should be kept under observation for several days after lambing. The shepherd should be sure the ewe expels the afterbirth and that her udder and bowels are in good condition. If the ewe is constipated, a drench consisting of ¼ pint of raw linseed oil or 3 ounces of epsom salts dissolved in warm water will start bowel movements. A rectal injection of one quart of warm soapy water brings quicker results than giving a laxative. The ewe should be given plenty of water. A light feed of bran and legume hay is recommended for ewes in pens or dry lot. Careful examination of the ewe's udder may detect any abnormality or infection in time to administer treatment before the udder is damaged.

MANAGEMENT OF THE NEWBORN LAMBS

The most critical period in the life of a lamb is during the first 48 hours. Most of the problems causing lamb losses, such as chilling, weak lambs, dry ewes, ewes with plugged teats, and ewes failing to claim their lambs, occur during this period.

When the lamb is born, pinch off the umbilical cord about 4 inches from the body. Be sure all mucus is removed from the nose and mouth. The navel should be disinfected with a tincture of iodine to prevent infection.

The lamb should be placed near the ewe's head. She may rest and clean the lamb. If the ewe will own the lamb, it is best to leave her alone. A strong lamb will usually nurse unassisted.

Chilled Lambs

Lambs born during cold weather or during storms often become chilled. Quick action on the part of the operator will result in reviving many lambs that appear to be nearly lifeless. The best and quickest method of reviving a chilled lamb is to immerse it in water as hot as one's elbow can bear. Place the entire body except the head in the water and hold it for a few minutes. Then after rubbing it dry, wrap it in a burlap sack or some heavy material along with another lamb that is warm. As quickly as possible, feed it some milk, as nothing revives a chilled lamb more quickly than some warm milk in its stomach.

Weak Lambs Will Need Help

Weak lambs will not usually survive unless given assistance in getting something to eat. The first step in assisting a weak lamb is to milk a few drops from the ewe to make sure the milk channels are open. Place the teat in the lamb's mouth and squeeze some milk into it.

Ewes that own and mother their lambs will lick the lamb under the tail while they are sucking. Weak lambs may be encouraged to suck by tickling the lamb under the base of the tail with the finger in imitation of the ewe.

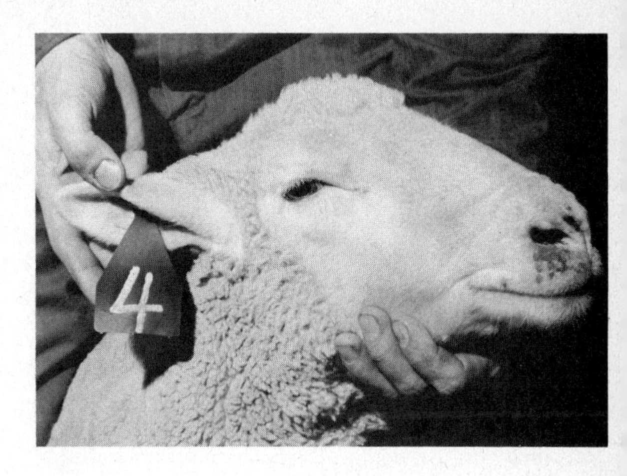

FIGURE 31-7. More complete lambing data can be recorded if individual ewes are identified. A plastic ear tag is shown here. (Campbell photo. *Courtesy* Hawkeye Institute of Technology)

How to Encourage Ewes to Own Their Lambs

Most ewes can be coaxed into owning their lambs. Ewes identify their lambs by smell. Sprinkling the ewe's milk on the lamb will often help. Tying the ewe in the lambing pen a few days so that the lamb may suck is often all that is necessary.

Yearling ewes or ewe lambs having their first lamb are most likely to disown them. For that reason, where a considerable number of ewes carrying their first lamb make up the flock, they should be penned separately and not allowed to lamb in the open with the rest of the flock. When the lambs are born, they should be placed near the ewe's head, where she will lick and clean them. They should then be placed at the teats and the ewe closely watched by the operator until it is sure that she will own them.

Grafting

Many times a ewe that has lost her lamb may be induced to take one or more lambs from another ewe if the proper procedure is followed. Occasionally, sprinkling the ewe's milk over the lamb will work and should be tried first as it is the easiest. Some ewes, if caught quickly after they have lambed, may be induced to take another lamb by rubbing kerosene or some similar agent over their nostrils. This temporarily disturbs their sense of smell and makes it difficult to detect the difference in lambs. Tying the ewe up in a pen or jail so that she cannot butt the lamb away and keeping her there for a couple of days will often work.

When a ewe has lost her lamb and it is desired to place another on her, the most successful method is to remove the hide of the dead lamb and place it over the new lamb. In removing the hide, cut the front legs off at the knees and slit the hide around the hind legs just below the hocks. Make a slit from hock to hock on the under side of the carcass and through the tailbone so the tail will stay on the hide. Start pulling the hide from the rear toward the front, down over the neck and cut it off. The hide is then pulled over the new lamb so its head sticks out of the neck hole in the hide, and the front legs are through the leg holes. This will hold the hide on the lamb and no further tying is necessary.

The ewe should be confined to the place where she lambed and the new lamb presented to her tail first and she will then get the odor of her own lamb from the hide.

Raising Orphans

Lambs that are left without a mother where no ewe is available for grafting, will have to be hand fed if they are to survive. The lamb should be fed two or three feeds of colostrum, or first milk, either from the mother or another ewe, by milking the ewe and using a bottle to feed the lamb. The milk should be warmed to 98° F. before feeding. Later, cow's milk may be substituted. The secret of using cow's milk is to feed often but only small amounts at a time. For the first day or so, one ounce fed at two-hour intervals is sufficient. Later the amount may be slowly increased and the feeding intervals spread further apart.

Commercial milk replacers have recently been developed which have proven effective in feeding young orphan lambs. These replacers can be purchased from your local feed dealer.

MANAGEMENT OF GROWING LAMBS

Pinning Passages

The first bowel passages of the lambs are sticky and often pin the tail to the bodies, preventing any further excretion. This is called pinning. The tail should be loosened and the excrement cleaned from the lamb.

Castrating

Castration of the lambs should be done when the lambs are from 7 to 14 days old. Lambs to be castrated should be placed in a clean, dry pen where they may be easily caught. Instruments used for the operation and the hands of the operator should be clean and disinfected. The ewe's pen should be clean and freshly bedded, unless they are on clean pasture, so that the lambs will be in a clean place when placed with their mothers after the operation. There are several methods of castrating. Each method has some advantages as well as disadvantages.

Proper Position to Hold the Lamb While Castrating. Holding a lamb in proper position for castrating requires that the holder place the lamb on its rump in a sitting position on the castration table. He then pulls each hind leg forward and upward to a position inside the front leg on the same side. Four fingers should be around the hind leg just above the hock, while the thumb is around the front leg at the knee. Thus he is holding a front and hind leg in each hand. This positions the lamb and renders it nearly motionless, unless the lamb is so large that he is well beyond the age recommended for castration.

Using a Knife. The most common method of castrating lambs is to cut off a third of the lower end of the scrotum. With the thumb and forefinger of the left hand, force out the testicles by squeezing the base of the scrotum next to the abdomen. Grasp the testicles, either with the teeth or the thumb and crooked forefinger of the right hand, and pull gently outward one at a time, until the cord breaks. There is some aversion to the use of the teeth, but it is the quickest method and where large numbers are to be castrated it will save considerable time. It is more sanitary than grasping with soiled fingers.

A mild disinfectant may be used to wash the wound, but if the operation has been done under sanitary conditions this is usually not necessary. The operation should be done on a clear warm day.

Castrating by removing the testicles is sure and does not require any special equipment. It does, however, create an open wound which may become infected if precautions are not taken to prevent infection. In screwworm-infested areas, considerable danger of screwworm infestation exists when animals have open wounds.

Castrating with Emasculator or Pincers. The emasculator is an instrument used to

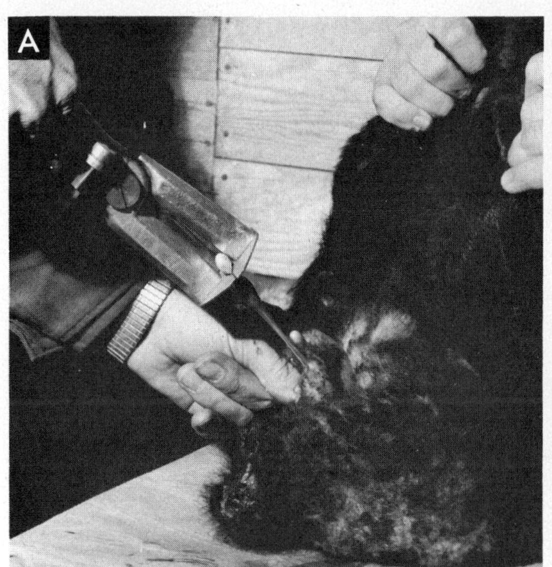

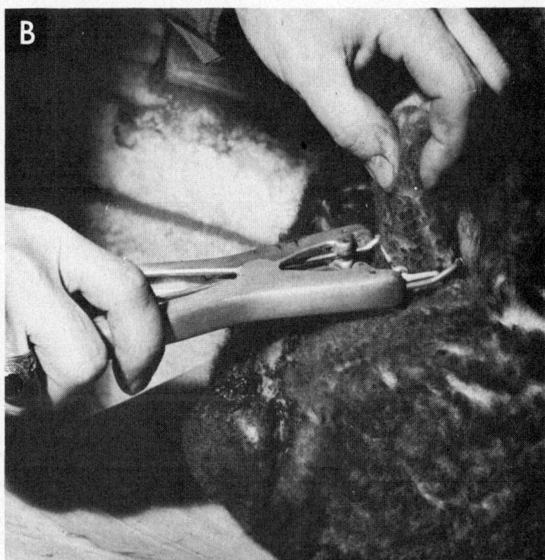

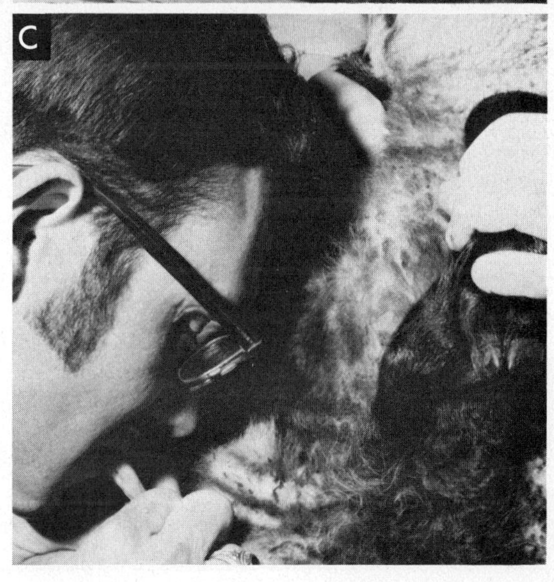

FIGURE 31-8. Castration can be accomplished by different methods. (*A*) A specially designed tool is used in cutting off one-third of the scrotum and, with the testicles protruding, pulling the testicle until the cord breaks. (*B*) An elastrator is used to place a rubber band around the scrotum between the testicles and the body. (*C*) After the lower one-third of the scrotum is cut off, one's teeth may be used to remove the testicles. (Campbell photos. *Courtesy* Hawkeye Institute of Technology)

crush the cord leading to the testicles. The cord is destroyed, leaving the testicle to gradually dry up or wither away. This method does not leave an open wound and is bloodless. While the emasculator, or "Burdizzo," gained considerable popularity for a time, careless use of the instrument resulted in so many partly castrated lambs, which become staggy as they grow older, that it has lost much of its former popularity. The operator must be skilled in its use if the operation is to be successful.

Elastrator. The elastrator is an instrument used to spread tight rubber bands which are slipped over the scrotum and released. The bands fit tightly around the cords at the base of the scrotum near the abdomen. The circulation is shut off and the scrotum and testicles gradually dry and drop off.

Docking

Lambs may be docked at the time they are castrated. Castrating should be the first operation, followed by docking. Several

FIGURE 31-9. An elastrator and rubber bands that are used to castrate and dock lambs. (Campbell photo. *Courtesy* Hawkeye Institute of Technology

FIGURE 31-10. This docking tool makes a clean cut with little or no bleeding. (Campbell photo. *Courtesy* Hawkeye Institute of Technology)

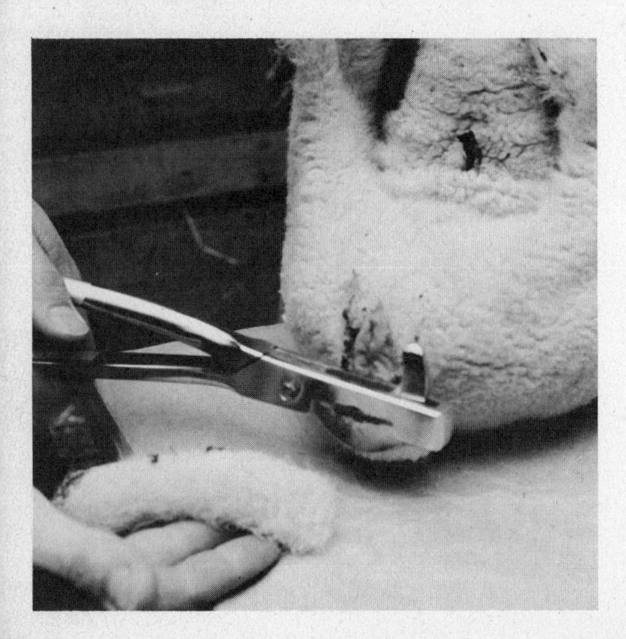

methods, all of them designed to remove the tail from the body, may be employed in docking lambs. Lambs not docked accumulate a great deal of filth around the tail, which often results in fly strike and maggot infestation.

Docking with a Knife. When a knife is used to dock lambs, locate the joint to be cut by feeling on the under side of the tail and push the skin toward the body so there will be a surplus of skin to cover the stub. After cutting off the tail, pinch the end to prevent excessive bleeding. By twisting the tail a quarter of a turn before cutting, a diagonal cut across the blood vessels is obtained and increases the possibility of the blood clotting quickly.

The use of a knife should be confined to lambs that are not more than a week or ten days old as older lambs can easily bleed to death. The knife is undesirable when lambs are confined to pens or corrals where the sanitation is not good or when docking is done during fly season. The bleeding makes it difficult to apply a disinfectant. Lambs should not be overheated just prior to docking with a knife as bleeding will be more severe.

Docking with a Hot Iron. While much slower than the knife method, the hot iron is much safer since it not only sterilizes but sears the wound and prevents bleeding. Docking irons may be homemade or purchased from livestock supply houses. Good homemade irons can be shaped from a piece of iron 2 feet long and the thickness of a heavy car or truck spring leaf. The cutting edge should not be knife sharp, but rather blunt so that when heated it will burn through rather than cut through the tail.

Heating the Iron. The iron should not be so hot it will go through the tail so rapidly that the large blood vessel will not be seared, nor so cold that it slows the cutting too much, causing a slow-healing sore. A black heat just turning red, rather than a red heat, is desirable. If bleeding occurs after the docking process, the operator should just touch the main blood vessel with the corner of the iron. By keeping several irons in the fire the process will not be slowed waiting for irons to heat. Thermostatically controlled electric irons are excellent if docking can be done where electricity is available.

Even when the hot iron is used, it is good practice to disinfect the stub with a swab soaked in iodine.

Rubber Band Method of Docking. The use of rubber bands fitted around the tail with the elastrator is a bloodless method of docking. The principle is the same as that for castrating in that the circulation is prevented and the tail drops off. Many object to it because the tail does not come off immediately, and may be held by the tailbone and putrify, creating a condition favorable to fly strike.

Docking with the Emasculator. The emasculator is sometimes used for docking. The tails of very young lambs may be clamped in the emasculator and pulled off, but older lambs usually require the use of a knife to sever the tail just outside the jaws of the emasculator. The pressure tends to squeeze the main artery and prevent bleeding.

The Point where the Tail Should Be Removed. Ewe lambs that have good prospects of becoming replacements in the breeding flock, based on the breeding, quality, and records of their ancestry, should have the tail removed about 2 inches from the body. Wethers and ewe lambs to be sold may have their tails cut longer for easier identification at market time.

Holding the Lamb for Docking. The lamb's rump should rest on a block of wood, or on a bench of convenient height for the operator. The lamb should be held in much the same position as for castrating with the tail resting topside down on the wooden cutting surface.

FIGURE 31-11. A lamb that has been paint-branded. (Campbell photo. *Courtesy* Hawkeye Institute of Technology)

Marking

Marking of the lambs for identification is an important sheep management practice. Purebred lambs are marked so that proper identification as to their sires and dams can be made for pedigrees. Range shepherds mark their lambs so that if one owner's sheep become mixed with another's, they can be sorted out and separated. Such mixing of bands happens frequently when grazing on unfenced range.

If ewe lambs are marked differently from the wethers they can be more easily sorted out and separated as they are run through a dodge chute. Since ewe lambs are often held back as replacements in the breeding flock, or sold separately from the wethers, quick identification provided by a good marking system saves time and labor. Well-marked lambs are a precaution against thieves, as the lambs may be more easily traced.

Marking may be done at the same time the lambs are docked and castrated.

Paint Brands. Paint brands are quickly applied and are quite reliable. However, they may wear off and be difficult to see. Ordinary lead-base paints should not be used as they cannot be removed with normal wool-scouring and may lower the market value of the fleece.

Commercial branding fluids that will remain on the sheep for a year, and may be removed from the fleece in the regular scouring process, are now available.

Metal Ear Tags. Purebred producers use metal or plastic ear tags as a means of identification. Many different colored plastic tags with any combination of letters and numbers may be purchased. Two tags, one in each ear, may be used. One tag carries the individual number and the other the flock number. This system not only provides identification of the owner, but also identifies the lamb with its sire and dam. The ear tag marking system is too slow and costly for use in large commercial flocks.

Ear Notches. For commercial flock identification ear-notching provides a quick means of marking and one that is easily recognized. It does disfigure the ears to some extent and, for that reason, is seldom used by purebred breeders. Ear-notching may be done with a knife or a commercial ear-notcher designed to cut a V or a U notch.

Catching Sheep

The proper method of catching sheep is to grab them under the chin with one hand and the rear flank with the other. Small sheep can usually be stopped by holding them under the chin and forcing the head upward. It is necessary to hold large animals by the rear flank as well as the chin.

Under no circumstance should sheep ever be caught by the wool. Grabbing them by the wool, especially on the back, opens the fleece so that rain may penetrate down to the skin. Grabbing the wool also damages the fleece, lowering its market value.

SHEARING

Shearing is more than just removing the wool from the sheep. A poor shearing lowers the quality and therefore the market value of the wool.

Shear in a Clean Place

Foreign materials such as dung, straw, water, and sand or soil should not be allowed to get mixed with the fleece. A wooden shearing floor, canvas, or an old rug may be used to shear sheep on and to help keep the fleece clean. The shearing floor should not be so close to the bedding or storage area as to permit chaff to be carried by the sheep or to be blown out to the shearing floor. Straw and other vegetable materials are much more difficult to clean from the fleece than sand or other soil particles. Hard bare ground is a better place to shear than a floor where straw and chaff accumulate.

Holding pens, where sheep are crowded into compact groups to await shearing, become saturated with droppings and urine. When this material becomes mixed with the fleece the quality and value drop considerably. Providing elevated slatted floors in the holding pens will allow the manure to drop through and will keep the fleeces free of such material.

When to Shear

The schedule will vary with various operators, and the shearing time will not always be the same. Whenever possible, shearing should not be done until there have been enough warm days to bring out the grease in the fleeces. Well-greased fibers are stronger. The natural oil in the fleece lubricates the shears and they cut more easily and more uniformly. Sheep shearers dislike shearing sheep with dry fleeces.

If the grease is out sufficiently, ewes should be shorn before they go on pasture. Pasture has a loosening effect on the bowels and the wool will be stained if sheep are shorn after they have been turned on pasture. Proper tagging before the ewes are turned on the pasture will reduce the amount of damaged wool.

When the wool is left on too late in the year, losses from natural shedding will reduce the weight of the fleece, and when the weather becomes warm, ewes in fleece suffer a great deal of discomfort.

Lambs not ready for market, or those intended as replacements to the breeding flock, should be shorn during July or August. Shorn lambs feed better, are more comfortable, and will gain more quickly during warm weather if the fleece is removed. Lambs intended for the early market should be left in fleece because of the added market value of the pelt.

Lambs that are to be marketed for slaughter and are in good condition at shearing time may grow a No. 1 fleece if shorn with handpieces with high runner combs. The fleece is not removed as close to the body and the length of time required to grow a fleece is reduced. Lambs on a good ration will grow a No. 1 pelt in 60 days when shorn close.

FIGURE 31-12. Shearing should be done in a clean place. (Campbell photo. *Courtesy* Hawkeye Institute of Technology)

FIGURE 31-13. An experienced shearer's job will look like this—uniform strokes with no cuts. (Campbell photo. *Courtesy* Hawkeye Institute of Technology)

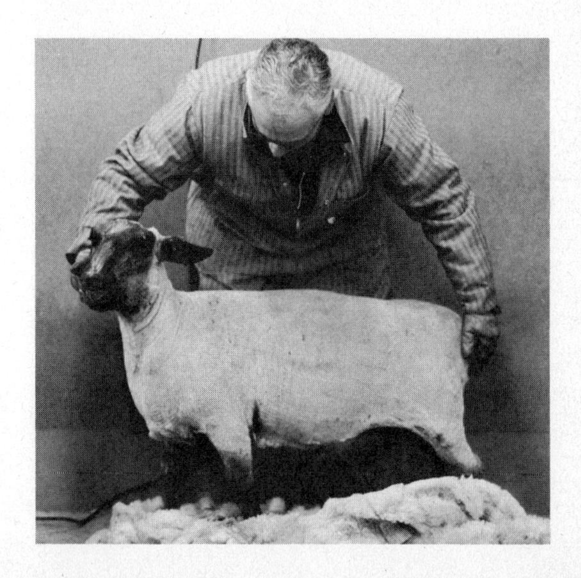

Removing the Fleece

Three fundamental principles should be kept in mind when removing the fleece from a sheep. They are: (1) Remove it in one piece; never break or tear the fleece apart. Shear close to the animal's body except in the case of finishing lambs, as previously mentioned. (2) Avoid second cuts. Second cuts simply add short fibers to the fleece, thereby lowering the value. Never attempt to smooth up the job by going over parts of the animal a second time. (3) Remove all dung locks and grease tags first and keep the fleece free of other foreign materials.

It is extremely difficult to describe how to shear sheep in writing or in words. Observing or working with an experienced shearer is the best way to learn.

Australian Shearing Table

An Australian invention known as the Barthwick power-driven shearing table, which works on the assembly-line principle, is expected to speed up the process of shearing sheep. The sheep walk up a ramp and are held in a specially designed loading pen until the rotary table comes around for loading. The loader is tilted by pulling a string and the sheep falls on its back on the table where the operator fastens its legs to specially designed holders. The shearing is accomplished without the operator having to hold the sheep.

FIGURE 31-14. A fleece being tied properly with twine. (Campbell photo. *Courtesy* Hawkeye Institute of Technology)

Sacking the Fleece for Market

Properly preparing a fleece will enhance its market value. Lamb's wool, black wool or fleeces containing black fibers, buck wool, and tags of all sorts should be packed separately.

Every fleece will contain several different kinds of wool referred to in the trade as *sorts*.

There may be 15 or more sorts in a single fleece depending upon the breeding of the sheep. Every fleece will contain the following sorts: (1) top knot, (2) head and neck wool, (3) shoulder, back and side wool, (4) leg wool, (5) belly wool, (6) breech wool, and (7) tags. The shoulders, back, and side wool make up about ¾ of the fleece and are the most valuable. The fleece should be tied so that the most valuable wool is on the outside.

Tying the Fleece. After the fleece is removed, spread it out flesh side down. Place the belly wool in the center. Fold in each side and roll the fleece from both the head and breech ends. The shoulder and side wool thus will be exposed. A fleece prepared in this manner may be spread out at the mill and sorted into the portions used for different yarns.

The fleece should be tied with a paper twine. Never use a sisal or a hemp twine, as the fibers from such a twine become mixed with the wool fiber and are difficult to separate. Go around the fleece once with the paper twine, cross, and go around once at right angles to the first, and tie. The fleeces should not be rolled or tied too tightly as springy fleeces are preferred.

Packing the Fleece

Wool may be packed in bags 6 or 7 feet in length. Burlap or commercial waterproof bags should be selected. The seams should be turned to the outside so the bags may be more easily opened for display at the warehouse. If tags are stuffed into the corners of the bag and tied off, they will form handles that will make handling the filled bags easier.

The sacking process is simplified if a sacking stand is used. The sacking stand holds the bag up and open so the fleeces may be properly packed.

Storage of the Packed Bags

The bag should not touch the ground after being removed from the sacking frame. Wool picks up a coating of dirt, which may lower its market value, if the bags are rolled on the ground. The wool must be stored in a dry, clean place. Moisture will cause staining and mildew. Trucks or freight cars should be thoroughly swept and cleaned before loading the wool. Wool loaded into open trucks should be covered with a tarpaulin to protect the fleeces from dirt and moisture.

RECORDS AND RECORD KEEPING

What the Record Should Reveal

Good records should identify the offspring with the parents. They should also give the birth date, sex, and the final disposal of each individual sheep. Most important, in selection for breeding flock replacement, or culling from the flock, are growth rates, fleece weight, quality, and prolificacy. This information together with desirable conformation is the basis upon which selection for flock improvement should be made.

On the following pages are examples of record forms which will give the needed information for wise selection and culling.

SHELTERS AND EQUIPMENT FOR SHEEP

Sheep do not require elaborate or expensive equipment. Many sheep flocks both large and small are maintained with little or no shelter even in the colder regions of the United States. However, newborn lambs must be protected against inclement weather and older sheep must be protected from cold, wet weather immediately after shearing.

Sheep Shelters

Single-story buildings of pole construction, which are relatively cheap to build and maintain, and are flexible in usage, are becoming popular among producers. When fitted with large doors on the side opposite the prevailing cold winds (which is the south side in most of the United States) it may

FIGURE 31-15. Packing the wool in burlap bags 6 to 7 feet in length. (Carr photo. *Courtesy* Hawkeye Institute of Technology)

EWE RECORD CARD

OWNER ____________ ADDRESS ____________ EWE NO. ________
COUNTY ________

BREED(S) ____________

Reg. P.B. ________ P.B. ________

Grade ________ XB ________

Date of Birth ____________

Birth Weight (lb.) ____________

Type of Birth S( ) Tw( ) Tr( )

Sire No. & Breed ____________

Ewe's Record as Lamb

Dam No. & Breed ____________

Adj. Wt. at 4 Mos. (lb.) ____________

Adj. Staple Length (cm.) ____________

Index as Lamb ____________

Ewe's Fleece Record			*Ewe*	*Notes about Ewe (Trouble at Lambing, Wouldn't Own Lambs, etc.)*
Date	*Wt. (lb.)*	*Grade*	*Index*	

Extension Service Circular 470, Madison, Wisconsin, University of Wisconsin, 1953.

INFORMATION ON EWE'S PROGENY

Lamb No.	*Date of Birth*	*Sex*	*Birth Wt. (Lbs.) (Optional)*	*Sire No.*	*Date and Age when Weighed*	*Wt.*	*Wt. Adj. to 120 Days of Age*	*Staple Length*	*Staple Length Adj. to 120 Days of Age*	*Lamb Index*	*Notes About Lamb (Born Dead, Date and Reason for Disposal, etc.)*

Extension Service Circular 470, Madison, Wisconsin, University of Wisconsin, 1953.

either be made tight for lambing or left as an open shed to give protection to breeding sheep or feeder lambs. Many such sheds are used as open shelters for breeding or feeder sheep until lambing time. The shelter is then fitted with lambing pens and the doors closed, and thus it is converted into a lambing shed.

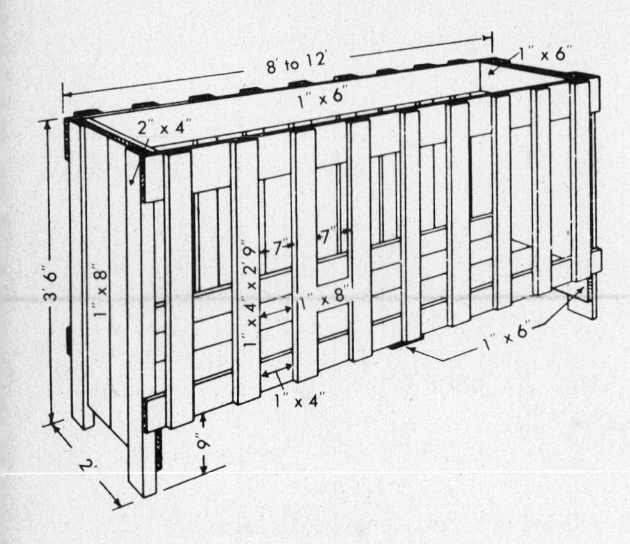

FIGURE 31-16. Combination hay and grain racks are convenient equipment for small flocks. This straight-sided rack will keep fleeces fairly free of chaff, except for the head and neck wool. Wool from the necks can be separated at shearing time.

Feeding Equipment

Feeding equipment varies with the kind of feed being fed.

Chopped hay or stacked long hay, may be self-fed; chopped self-feeding hay feeders may be constructed, or a portable manger may be built along one end of a loose haystack. As the sheep eat into the hay the manger is moved closer to the stack.

Grain feed bunks may be of two different types, depending upon which will better fit the sheep program. Stationary feed bunks may be constructed along one side of the lot. This type of feed bunk is best adapted to a lamb feeding enterprise where the lambs are confined to a feed yard.

Plans for several types of movable feed bunks are available. The important features of good feed bunks are ease of cleaning, ease of access, and durability.

A salt and phenothiazine feeder which provides the sheep with free access to salt and phenothiazine and which will keep the material protected from the rain and snow is

FIGURE 31-17. (*A*) The knock-down or panel feed rack is an economical piece of equipment for feeding roughage to finishing lambs. These racks can be set up as bunks, as shown in the drawing, or arranged as fences with the sheep feeding through the fence. They can be used for a breeding flock by placing them around a haystack with the hay being fed against the panels on the inside. Placing the stack away from the shed forces the ewes to exercise. (*B*) A single non-reversible trough. It is of cheaper construction and is not as satisfactory from a cleaning standpoint as the reversible trough.

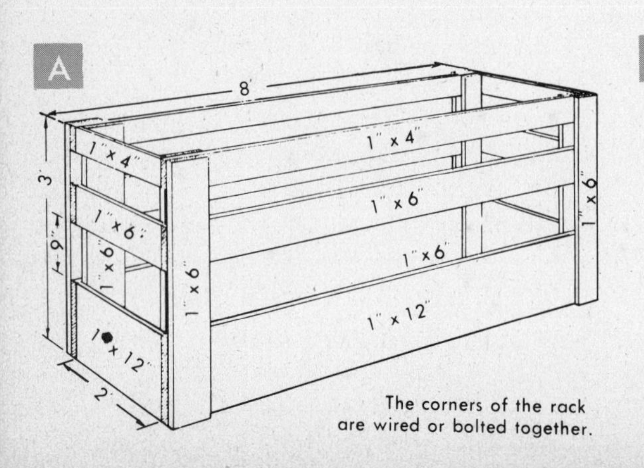

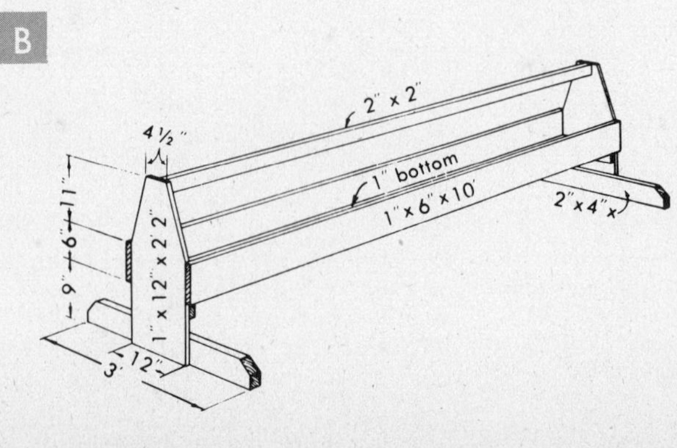

an important part of the feeding equipment. Phenothiazine keeps the flock worm-free.

Panels

A number of hinged panels, important for making temporary lambing jails or otherwise confining sheep that may need special attention, are handy equipment.

Loading Chutes

Loading chutes may be of a permanent or movable type. When sheep that are to be moved can easily be confined to a yard, the permanent loading chute works very well. However, if they are loaded from the pasture where they may be confined in a temporary corral, a movable loading chute is necessary.

Cutting Chute

Every operator needs a sorting device as sheep must be sorted one or more times a year. A properly constructed cutting chute with a dodge gate that will swing from one side to the other permits sorts to be made with comparative ease.

A good cutting chute may be constructed by using solid panel sides 30 inches high, 12 inches apart at the bottom and slanting so they are 18 inches apart at the top. The flare top makes it possible to sort both ewes and lambs in the same chute. The length may range from 12 to 14 feet.

The dodge gate should swing from one side to the other to permit cutting into different pens.

SUMMARY

Flushing is the term applied to the procedure of putting ewes in a rapid gaining condition just prior to breeding. Flushing helps to bring the ewes into heat and results in a more uniform lamb crop and a higher percentage of lambs.

Overfat ewes should be placed on a sparse pasture and given plenty of exercise in order to gradually reduce them in flesh prior to flushing. Breeding overfat ewes results in a smaller lamb crop.

The first half of gestation is the less critical, and little or no grain needs to be provided if plenty of good to fair quality roughages are available. Starting the eleventh week of gestation, grain or molasses should be included in the ration. From ¼ to ¾ pound of grain or the equivalent in molasses should be added to the daily feed.

Minerals are essential and can best be supplied by using a mineral mixture recommended for the area. Vitamins A, D, and E should be provided. Other vitamins are manufactured in the digestive system of ruminants and need not be supplied except in the case of young animals before rumination has started.

The ewe suckling lambs needs more feed and slightly more protein than the pregnant ewe. Good pastures will meet the nutritional requirements of ewes suckling lambs. However, ewes in dry lots need good-quality roughages and grain to maintain milk flow and grow out lambs economically. Ewes from weaning to flushing can make very good use of poor-quality pasture and roughages. This is the least critical time in the production cycle.

The breeding ram should not be too fat at breeding time. During the breeding season he will need about one pound of grain daily plus a full feed of good forage when confined to a dry lot. Good pasture will meet the nutritional needs of the ram.

Heat periods in ewes occur every 13 to 19 days and average 16½ days during the breeding season. Ewes remain in heat from three to 73 hours. Three-fourths of the ewes will stay in heat from 21 to 39 hours. Except for a few breeds, ewes do not come in heat

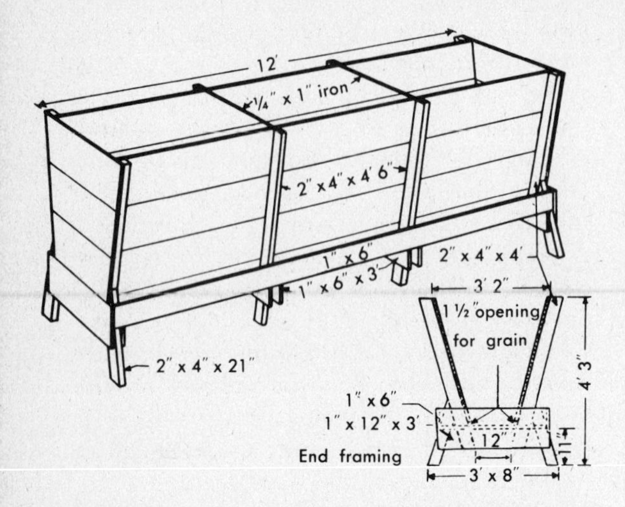

FIGURE 31-18. Where many lambs are being fed, self-feeding may be desirable, using ground hay and grain as the ration. An adjustable feeder permitting feeding down of the feed is necessary. This feeder is cheap and easy to construct. With hand feeding, 12 inches of space is needed for each lamb; with self-feeding, 12 inches for three lambs. Self-feeders need attention several times a day to keep the feed poked down so that it is always available.

regularly throughout the year, but are seasonal in this respect. Breeds differ as to their mating season. Mixed breeds of sheep running in one band are not likely to have a uniform lamb drop.

The gestation period of ewes will range from 144 to 152 days.

Ewes should be flushed, tagged, and eyed before breeding. Ringing or shearing the ram, in addition to tagging and eyeing, is important before the start of the breeding season. Both ewes and rams should have their feet checked and trimmed if necessary. Marking the ram either by breast paint or a marking harness helps the producer to determine the fertility of the rams, make up drop-bands, and determine ewe breeding dates.

Ram lambs, 9 to 10 months old and well grown out, may be used to breed 10 to 12 ewes. Rams over five years of age should be checked carefully for fertility. Yearling and up to five-year-old rams may breed up to 45 ewes per season. Whenever it is practical, supplemental feeding of the ram during the breeding season is recommended.

The generally accepted age for breeding ewes the first time is when they are yearlings. However, under proper conditions ewe lambs may be bred.

The ram should not be allowed to run with the flock for more than eight weeks during the mating season if a uniform lamb drop is desired.

Pregnancy testing ewes can be done by several methods. The most economical is the rectal-abdominal palpation technique. This method is rapid and accurate.

Exercise, shade, and protection from other livestock and predatory animals are the most important aspects of bred ewe management other than proper feeding.

Careful observation and management during the lambing season will save many lambs. Dividing the flock into drop-bands or separating ewes about to lamb, either by tagging or by the use of the marking system, is the first step. Placing the ewes in jails and giving them protection from inclement weather will prevent chilling of the lambs and decrease the number of disowned lambs. Under extreme cold weather conditions lamb brooders are recommended. Tagging the ewes and clipping the wool away from around the teats improves sanitation and enables the lambs to find the teats more quickly.

Ewes that cannot expel their lambs without help will be lost unless the shepherd gives them assistance. After lambing, the shepherd should be sure the ewe cleans properly and that her bowels are in good condition. Examine the ewe for possible udder infection.

Chilled lambs may be revived by immersing in warm water. They should be given some warm milk as quickly as possible. Weak lambs will need assistance in getting milk. Lambs from dry ewes or ewes that do not produce enough milk for their lambs will have to be hand fed or put onto other ewes. Succulent feeds will help to bring ewes to their milk. Lambs should be left with the ewes and given supplemental feed until the ewes have had plenty of time to come to their milk.

Ewes that disown their lambs may be encouraged to own their lambs by sprinkling the ewe's milk on the lamb or by tying the ewe in the lambing pen for a few days.

Lambs should be watched the first few days for pinning passages. If they occur loosen the tail and clean the lamb.

Ewes that have lost their lambs may be induced to take that of another by grafting the skin of the dead lamb over the lamb that is to take its place. Orphaned lambs may be raised on cow's milk. Success depends upon proper amounts, frequency of feeding, sanitation, and temperature of the milk.

Castrating may be done with a knife, emasculator, or elastrator. The knife is the most popular because it is sure. Lambs should be castrated when they are from 7 to 14 days of age.

Docking may be done at the time lambs are castrated. A knife, heated docking iron, rubber bands, or the emasculator can be used. The heated iron is the most sanitary and when properly used will prevent bleeding.

Marking for identification is important to good management and to the selection of breeding stock. There are several methods and combinations that may be used. Paint brands, ear tags, and ear notches are the commonly used methods.

The characteristics of a good shearing job are clean wool, allowing the grease to come out before shearing, removing the fleece in one piece, and avoiding second cuts.

The Australian shearing table is new and is expected to speed up the shearing process where large numbers are involved.

Proper sacking of the fleeces is the final step in preparing the wool for market. This requires keeping lambs' wool, black wool or fleeces containing black fibers, buck wool, and tags separate. Fleeces should be spread flesh side down, the edges turned in and rolled from both the head and tail ends into a fluffy, reasonably loose bundle and tied each way with a paper twine. Burlap bags or commercial waterproof bags with the corners tied off and seams to the outside should be placed on a sacking stand and the wool packed in layers. Either a round or a flat pack may be made.

Good records will reveal the following information: date of birth, sex, disposal, growth rates, fleece quality and weight, and prolificacy.

Sheep will need shelters for lambing in cold weather or for protection if shorn during inclement weather. Expensive or elaborate shelters are not necessary.

Hay racks and feed bunks are necessary equipment in many sheep enterprises, especially where lambs are confined to a feed yard or where the breeding flock is fed hay or grain in lots. Salt and phenothiazine feeders are essential.

Panels for making temporary jails, loading chutes, and sorters should be available.

QUESTIONS

1 What are the periods in the productive cycle of the ewe?

2 Discuss a feeding program for each of these periods.

3. What are the various roughages, grains, and protein supplements commonly fed to sheep?
4. What are the primary dangers of keeping ewes on too low a nutritional plane? Discuss.
5. Give some good rations using feeds common in your locality for ewes in various stages of pregnancy.
6. Discuss feeding the breeding ram.
7. Discuss the length and frequency of heat periods.
8. Discuss the mating seasons, breed differences in mating seasons, and how they affect the lamb drop.
9. What is the average length of the gestation period in sheep?
10. Explain the steps in preparing the ewes for breeding.
11. How would you prepare the ram for breeding?
12. Explain how ram marking is done for breeding purposes.
13. Why is ram marking at breeding time essential to good management?
14. Discuss the ewe-ram ratio under various breeding conditions.
15. Discuss ram feeding and management during the breeding season.
16. At what age should ewes be bred? Discuss breeding ewe lambs.
17. How long should the rams be allowed to run with the flock?
18. What ewes would you cull from the flock? Why?
19. Discuss the essentials of good ewe management during the gestation period.
20. How would you make up your drop-bands?
21. Explain how you would handle ewes at lambing time under various weather conditions.
22. Discuss the use of lambing jails and brooders.
23. How would you prepare the ewe for lambing?
24. What are some abnormal positions of an unborn lamb that will need to be corrected before delivery is possible?
25. How would you correct these positions?
26. How would you care for the ewe immediately after lambing?
27. Discuss methods of reviving chilled lambs.
28. How would you handle the lambs and ewes when the ewes fail to come to their milk?
29. How may ewes be encouraged to own their lambs?
30. What are pinning passages? How would you correct this condition?
31. Discuss grafting lambs on a ewe that is not their mother.
32. Give a plan for raising orphaned lambs.

33 Discuss the method of holding lambs for castration.

34 Explain the various methods of castration and give the advantages and disadvantages of each.

35 When should lambs be docked? Explain the methods used.

36 Why should a marking system be used? Discuss methods of marking.

37 Discuss the proper shearing and handling of the fleeces.

38 Why are records important to good management? What is included in a good record?

39 What type of shelters would be most practical on your farm or ranch? Why?

40 Discuss the type of feeding equipment that is most practical in your community.

41 What other kinds of equipment should producers in your community have available?

REFERENCES

Alexander, M. A., W. W. Dorrick, and D. C. Fouts, *Farm Sheep Facts,* Extension Service Bulletin E. C. 255, Lincoln, Nebraska, University of Nebraska.

Ensminger, M. E., *Sheep and Wool Science,* Fourth Edition, Danville, Ill., The Interstate Printers and Publishers, Inc., 1970.

Flushing Range Ewes. Progress Report 2252, College Station, Texas, Texas A. & M. University, 1962.

Mason, R. W., P. O. Stratton, N. W. Hilston, Irene Payne, L. C. Parker, and J. O. Tucker, *Animal Tallow as Part of the Wintering Ration for Pregnant Yearling Ewes,* Mimeograph Circular 52, Agricultural Experiment Station, Laramie, Wyoming, University of Wyoming, 1955.

Perry, T. W., W. M. Beesen, P. J. Reynolds, and Claude Harper, *Supplementing Grass Silage for Pregnant Ewes in Drylot,* Mimeo A. H. 140, Agricultural Experiment Station, Lafayette, Indiana, Purdue University, 1955.

Progress and Research in the Sheep Industry, Extension Service, Madison, Wisconsin, University of Wisconsin, 1965.

Research Reports, Brookings, South Dakota, South Dakota State University, 1965.

Scott, George E., *The Sheepman's Production Handbook,* First Edition, Abegg Printing, Denver, Colorado, 1970.

Sheep and Lamb Feeders Day, Morris, Minnesota, University of Minnesota, Western Central School, 1964, 1966.

Sheep Production in Kansas, Agricultural Experiment Station, B 348, Manhattan, Kansas, 1961.

Van Horn, J. L., *Milk Production of Ewes,* Research Seminar, Animal Industry, Range Management and Wool Laboratory, Montana Agricultural Experiment Station, Bozeman, Montana, 1954.

Van Horn, J. L., William H. Burkett, Gene F. Payne, Curtis G. Hughes, and Fred S. Wilson, *Nutritional Requirements of Ewes Wintered under Range Conditions,* Progress Report, Animal Industry, Range Management and Wool Laboratory, Montana Agricultural Experiment Station, Bozeman, Montana, 1950.

32 Feeding Lambs

Lamb production methods necessarily have to vary considerably to fit various feed, economic, and climatic conditions.

PRODUCING FINISHED LAMBS ON MILK AND GRASS

Lands capable of producing good grass, but otherwise unfit for tillable crops, may be profitably used for grass-finished lamb production. This method, producing finished lambs on grass and milk from ewes, is used throughout the United States and is the most common method in the better range areas. The milk production of the ewe largely determines how quickly the lambs will develop.

Lambs require short, tender, fresh green pasture made up of palatable plants. The grass should be abundant, allowing them to get a fill with a minimum of travel.

Creep Feeding

Creep feeding lambs is accomplished by putting feed in an enclosure that will permit lambs to enter but will exclude the ewes. Many types of creeps have been constructed and found to work successfully.

Why Creep Feed? Creep feeding of lambs has several advantages. The most important are: (1) It will speed the growth and finishing rate of lambs. (2) Ewes, especially those under range conditions, will not suckle down as thin and will maintain a milk flow over a longer period of time. (3) It tends to even up twin lambs and single lambs in size. (Twin lambs are smaller at birth and usually get less milk per lamb than singles. Creep feeding helps to make up for this difference.) (4) Lambs will be ready for an earlier market which often has a price advantage. (5) Creep feeding enables many farmers to produce finished lambs on range that is too poor to produce anything more than feeder lambs unless some extra feed is provided.

(6) Creep feeding prevents close grazing as would be harmful to the range.

When to Start Lambs on the Creep. If lambs are to be creep fed on the range, they should be started when they average about three weeks of age. If there is too long a spread in the dropping period, the flock should be separated into bands according to the age of the lambs. Lambs at three weeks of age have not developed a taste for grain to the extent that the danger of overeating exists. Older lambs may overeat, or they may have developed a taste for grass to the extent that it will be difficult to get them started on grains.

Under farm flock conditions, especially when lambs are born a month or more before the pasture season, they should be started on creeps as soon as they begin to nibble at feed. Lambs will usually start to eat when about ten days old.

How to Start Lambs on the Creep. When lambs are started on the creep, a sprinkle of grain may be placed in the troughs or feed bunks and left for the day. Each day it should be changed and fresh grain put in the creep. When the lambs have learned to eat the feed, they should be given the amount they can eat in 20 to 30 minutes.

Lambs will eat an average of about 1/4 pound of grain per head daily during the creep feeding period. Lambs dropped in late winter or early spring prior to the grass season will need to have both hay and grain fed in the creep. Choice hay, such as high-quality legumes, should be selected for the lambs. The best hay should be held in reserve for creep-fed lambs.

Some good mixed creep rations for lambs are:

FIGURE 32-1. (*Above*) **An easily constructed lamb creep.** (*Courtesy* **Hawkeye Institute of Technology**). **FIGURE 32-2.** (*Below*) **Hay and grain feeders for lambs need not be expensive.** (**Carr photo.** *Courtesy* **Robert Kimm**)

1. **300 pounds corn**
   300 pounds oats
   100 pounds bran
   100 pounds linseed, soybean, or cottonseed oil meal
2. **350 pounds sorghum grains**
   300 pounds barley or wheat
   150 pounds soybean oil meal
3. **60 pounds corn**
   20 pounds oats
   20 pounds soybean oil meal

When a mixed ration is fed, grinding will be necessary to properly mix and blend the ration.

PRODUCTION OF HOTHOUSE LAMBS

Hothouse lambs are those that are dropped in the fall or early winter, and finished for market when they are from 6 to 12 weeks old and weigh from 30 to 60 pounds. They should be marketed during the period from Christmas to Easter to bring a premium price. Boston and New York furnish the principal markets for hothouse lambs.

Dorset, Rambouillet, Merino, or Tunis ewes are selected to produce hothouse lambs since they will breed at almost any season and are excellent milkers. Southdown rams, because of their early maturity and high-quality carcasses, are used on these ewes. These crossbred lambs finish at a very early age when properly handled.

High-quality feeds for both the ewes and lambs are essential for hothouse lamb production. A large milk flow by the ewes is a must, and this requires liberal feeding of good forage, plus from one to 2 pounds of grain daily during the suckling period.

FEEDING REPLACEMENT EWE LAMBS

Ewe lambs selected for replacement ewes in the breeding flock should be fed for maximum growth. When good pastures and high-quality forages are available, grains are not essential to the growing out of ewe lambs. The best pastures and forage should be conserved for the replacement ewes. If the quality of the pasture or forage is not high, the supplemental feeding of grain is recommended.

FINISHING LAMBS

Lamb finishing is an enterprise usually carried out by a farmer in the grain-producing area or by large, commercial lamb feeders. The majority of lamb feeders purchase

FIGURE 32-3. Replacement ewe lambs. (Carr photo. *Courtesy* Robert Kimm)

lambs from the range. Most feeder lambs come from the range, where the large breeds are used. Sixty-pound lambs are generally considered a good weight. When a 60-pound lamb is fed during the average feeding period of 90 to 100 days, it should finish at 90 to 100 pounds, which is very acceptable market weight. The lamb feeder makes a profit from margin, gain, and wool clip. However, a fourth source of profit, manure value, should be considered.

Methods of Finishing Feeder Lambs

Feeder lambs are generally finished in dry lot on grains and preserved forages. However, they may be finished on pasture without the use of grains.

Dry-lot Feeding. Dry-lot feeding consists of confining the lambs in a lot devoid of green feed and using a combination of the various cured forage feeds, grains, and protein supplements to bring the lambs up to the desired weight and finish.

Pasture Finishing. When there is plenty of high-quality pasture available, lambs may be finished on pasture without grains. This method is common in the Southwest where wheat pastures may be used. Lambs or yearlings are brought in off the range and turned onto the wheat field. Pasture finishing may be speeded up by supplementing the pasture with grains.

Roughages for Finishing Lambs

Lamb feeds are many and varied. The rations used depend upon cost and availability of feeds. Roughages should make up a large part of the lamb ration. They are economical and, when properly fed, decrease the cost of gain.

High-quality legume hay is recognized as the most nutritional roughage. It is high in protein, minerals, and vitamins. Alfalfa, the clovers, lespedeza, soybeans, and cowpeas are all excellent common legumes that can be used for hay.

FIGURE 32-4. Dry-lot feeding of lambs. (Carr photo. *Courtesy* Robert Kimm)

Many of the grasses, such as brome, fescue, orchard grass, and western native grasses, are excellent roughages but low in protein, vitamins, and minerals compared to legumes.

Chopped corn or grain sorghums stover, when properly supplemented, will produce substantial gains on finishing lambs. They must be supplemented with some high-quality forage, such as alfalfa hay or a high-quality legume silage, to make up for the low protein and vitamin content of these low-quality forages.

Alfalfa, bromegrass, oats, and the other legumes or small grain crops all make good silage for lambs. These crops vary considerably in their nutrient content. Silage made from legumes will contain more protein than that made from grasses.

Corn and the sorghums make excellent silage. The silage from these crops will be lower in protein than legume, or legume and grass mixed silage, but higher in carbohydrates.

Grains and Grain Substitutes for Finishing Lambs

The grains are the most important concentrates used for feeding lambs. Corn is the standard finishing grain, and all other grains or grain substitutes are compared to corn in determining their value. When legume hay is fed, lambs will make very satisfactory gains on legume hay and corn as the entire ration.

Barley may be substituted for corn. Barley has about 87 percent of the value of corn for lamb feed.

Wheat is a very satisfactory feed, equal to corn for finishing lambs, if mixed with other grains. When fed alone, it is inferior to either corn or barley, having a value about 83 percent that of corn.

Oats are very satisfactory for starting lambs on feed. They are more bulky than other grains and are less likely to throw lambs off feed. However, when oats are continued as the only grain in the ration, they have a value of about 80 percent that of corn.

Rye has produced results about equal to barley in feeding tests for finishing lambs. However, a mixture of rye with corn or barley is probably better than rye alone.

Grain sorghums are about equal in value to corn for finishing lambs. There seems to be little difference in the feeding value of the various varieties of grain sorghums.

Recent Texas experiments show that animal fat may be used successfully as up to 10 percent of the entire ration for finishing lambs. The fats must be stabilized to prevent rancidity.

Molasses is well liked by lambs and is often used as a part of the finishing ration. It should not be used to replace grain entirely but has its highest value when fed at the rate of from ⅓ to ¾ pound per head daily. Molasses is very palatable and the increased rate of gain when molasses is fed may be due to the greater feed consumption of the lambs. Molasses has very little protein and is deficient in vitamins A and D.

Proteins and Protein Substitutes for Finishing Lambs

When a high protein roughage, such as good legume hay or legume silage, is fed as the principal roughage, protein concentrates are not essential for finishing lambs. Feeding trials have shown somewhat higher gains when protein supplements are fed to lambs receiving high protein roughages, but the extra cost in some cases offsets the value of the weight gains. One Iowa trial showed that lambs had a faster gain, a higher degree of finish, and sold higher on the market when a protein supplement was fed with corn and alfalfa. When low protein roughages are used, the feeding of protein supplements will be profitable. There are a large number of protein concentrates that may be successfully used in lamb feeding.

Soybean oil meal is equal to or higher in protein than other protein concentrates. Linseed oil meal has long been a favorite source of protein among lamb feeders, but it is being used less extensively now than in former years because of the greater availability of other protein supplements. Cottonseed meal is a widely used protein supplement. It is especially popular as a cheap source of protein among producers in the Cotton Belt.

Dehydrated alfalfa meal or pellets range from 17 to 20 percent protein and therefore closely approach what may be termed a protein concentrate. Dehydrated alfalfa may be used to replace all or part of the protein in lamb-finishing rations. It is also rich in carotene (vitamin A) and several essential minerals.

Peanut oil meal is about equal to cottonseed meal, soybean, or linseed oil meals for finishing lambs.

Soybeans provide a good source of protein but the price is generally too high to make them an economical feed. Corn gluten feed and meal give fairly good results when fed as the only protein to lambs. Urea, a nitrogen compound, may be used as a partial replacement in the rations of finishing lambs. Biuret, another nitrogen compound, is safer to feed than urea because the absorption of ammonia is less rapid. The rules for feeding urea and biuret to finishing lambs are the same as for cattle.

Mixed Supplements. There is a variety of mixed supplements designed for various types of feeding programs. Where good-quality forage is used as the primary roughage, any of the protein feeds that have been discussed will prove satisfactory. When low-quality roughages are used, a protein supplement reinforced with additional vitamins and minerals is recommended. If urea or biuret is to be used to replace part of the protein in a ration, it is usually fed in a mixed supplement. Following are some mixed supplements that are especially designed for lamb feeding with average to poor forage.

TABLE 32-1 COMPARATIVE FEED VALUE
A Per-Pound Basis of Some of the Commonly Used Protein Concentrates for Finishing Lambs

Feed	Value Compared to Soybean Oil Meal
Soybean oil meal	100
Linseed oil meal	95
Cottonseed meal	95
Peanut oil meal	95
Dehydrated alfalfa	70
Corn gluten feed	65–70
Corn gluten meal	95
Soybeans	95

1. **853 pounds linseed oil meal**
   900 pounds soybean oil meal
   100 pounds urea
   15 pounds trace mineral mixture for ruminants
   100 pounds feeding bone meal
   31 pounds sodium sulfate
   1 pound of irradiated 9F yeast

2. **1,344 pounds linseed oil meal**
   500 pounds soybean oil meal
   150 pounds feeding bone meal
   5 pounds trace mineral mixture for ruminants
   1 pound irradiated 9F yeast

Minerals

When high-quality legume roughages make up the major part of the forage fed to lambs, nothing more than salt and a mineral mixture fed free choice will be necessary to insure adequate minerals. However, when the forage contains no legumes, or only a small amount, a mixed supplement containing minerals would be advisable. Salt should also be fed free choice.

Antibiotics

Several feeding trials using antibiotics, mostly Aureomycin, have been conducted. The results of these trials have varied considerably but a large number have shown generally healthier lambs, greater gains, a reduction in the incidence of over-eating disease, and some savings in feed when 10 milligrams of Aureomycin chlortetracycline was fed per lamb daily.

Hormones

The feeding of hormones, or implanting them in pellet form under the skin, has given increased gains ranging up to 25 percent and reduced feed per 100 pounds of gain up to 22 percent. However, various complications such as lower carcass quality, prolapse of the rectum, and blockage of the urinary tract have been reported from the use of hormones.

Stilbestrol, progesterone, estradiol testosterone, and androgen are among the hormone or hormone-like substances that have been used in feeding trials. More research must be done before a recommendation as to their value can be made. Illinois experiments have indicated that the addition of sulphur to lamb rations increases gains when lambs are implanted with a hormone.

Preparation of Feed for Lambs

Feed may be either ground or pelleted, and these two preparations are discussed in the following paragraphs.

Grinding. Producers are quite well agreed that grinding grain is undesirable, except when a complete mixed ration is fed and grinding is necessary for proper blending of the ration. Some of the combine types of sorghums, such as milo, are extremely hard and cracking of these sorghums is recommended, but fine grinding is discouraged.

The grinding or chopping of roughage will induce greater consumption of poor-quality feed but otherwise has little value. When lambs are fed a complete mixed ration, including roughage, it then becomes necessary to grind the roughage and grains for mixing.

Pelleted Rations. Pelleting of feeds requires grinding them first, then running the ground material through a machine which forms the feed into pellets. These pellets vary in size but will average approximately the size of a pea. Pelleted grains, hays, and complete pelleted rations have been used in feeding trials. Lambs like the pelleted rations and the labor used in feeding is reduced. Most of the trials have shown somewhat faster gains when pelleted rations were fed.

Starting Lambs on Feed

Lambs will lose 4 to 7 pounds in shipment, depending upon the distance and length of time they are on the road. Upon arrival they will be tired, hungry, and thirsty. The best practice is to give the lambs access to a mixed hay (grass and legume), a grass hay, or a grass pasture for two hours before filling the water troughs to prevent water founder. Lambs that are hungry and thirsty will be inclined to over-drink, creating some serious complications. Salt and mineral mixture should be made available. Block salt should be used the first few days; then a change to loose salt, kept in a protected place, should be made.

From the third day, the lambs may be fed 1/4 pound of oats per head and 1/10 to 1/8 pound of linseed oil meal pellets daily. The linseed oil meal is a good conditioner and will help to regulate the bowels and digestive system. After the lambs have been on the oats and linseed oil meal pellets for four days, a gradual shift to the finishing ration may be made.

Methods of Finishing Lambs

There are three commonly used systems of finishing lambs. Hand feeding in a dry lot, self-feeding in a dry lot, and cornfield feeding with either hay or pasture.

Hand Feeding. Hand feeding consists of feeding the lambs concentrates, usually twice a day, and limiting the amount to what they will clean up in 20 to 30 minutes after they have been brought up to full feed. Hay is kept in the racks where the lambs will have access to it at the rate of 1 to 1½ pounds per head daily.

When the hand feeding method is to be used, a gradual shift from the starting ration to the finishing ration should be made. Good-quality legume hay is best, but since it tends to have a loosening effect on the bowels, the shift should be made over a period of about two weeks. The concentrate ration to be fed should be started the fourth day at the rate of ¼ pound per head daily and gradually increased over a two- to three-week period until the lambs are on full feed. Full feed will consist of the amount the lambs will clean up in from 20 to 30 minutes twice daily.

Following are some proven rations.

RATIONS FOR FINISHING LAMBS HAND FED IN DRY LOT †

Roughage	*Concentrate Mixture (Pounds)*	
1. Alfalfa, 1¼ lbs. daily	Corn or sorghum grains	90
	Linseed, soybean or cottonseed meal	10
2. Legume hay, 1¼ lbs. daily	Corn or sorghum grain	40
	Oats	30
	Barley	30
3. Corn or sorghum silage, 1¾ lbs. daily	Corn or sorghum grain	85
Legume hay, 1¼ lbs. daily	30 to 40% protein concentrate	15
4. Grass hay, 1¼ lbs. daily	Barley	40
	Corn or sorghum grain	40
	30 to 40% protein concentrate	20
5. Mixed hay, 1¼ lbs. daily	Wheat	20
	Corn or sorghum grain	40
	Barley	25
	30 to 40% protein concentrate	15
6. Corn or sorghum silage, 3 to 4 lbs. daily	Corn or sorghum grain	60
	Barley	20
	30 to 40% protein concentrate	20
7. Mixed hay, 1¼ lbs. daily	Molasses	30
	Corn or sorghum grain	50
	30 to 40% protein concentrate	20

† Salt and mineral mixture feed should be given free choice with the above rations.

SELF-FED RATIONS FOR FINISHING LAMBS ON FULL FEED †

Feeds	*Pounds*
1. Corn or sorghum grain	40
Molasses	15
Linseed, soybean, cottonseed meal	9
Legume hay	36
2. Corn or sorghum grain	20
Barley	20
Molasses	10
Linseed, soybean, cottonseed meal	10
Legume hay	40

† Salt and a mineral mixture feed should be given free choice with the above rations.

Self-feeding. Lambs may be placed on self-feeders after having become accustomed to grains, if the ration is made bulky by mixing roughages with the grains. A greater danger of overeating exists, but when large numbers are fed, the labor-saving factor is important. If self-feeding is to be practiced, the roughage and grain will have to be ground and mixed into one complete ration. The self-feeding of grain and hay separately usually results in overeating and heavier death losses.

The change from hand feeding to self-feeding should be done slowly. The lambs are first given their regular feed, and later the self-feeders are opened. The first self-fed mixture should contain not more than 30 pounds of grain and protein concentrate mixed with 70 pounds of chopped roughage. As the lambs become used to self-feeding, the mixture may be made with a larger percentage of grain to speed the finishing process. Molasses is valuable in mixed rations as a means of holding down the dust and providing a binder which holds the grain and protein in the mix.

Tests show that lambs may be finished with as little as 30 percent concentrates and 70 percent good-quality legume hay in the ration. They may be finished on 80 percent concentrates and 20 percent legume hay. However, the most economical ration is 50 to 60 percent concentrates and 40 to 50 percent legume hay.

Cornfield Feeding. Lambs may be successfully finished in the cornfield if properly handled. The death loss will be somewhat higher on the average than when they are fed in the lot, but labor will be considerably reduced. The lambs will harvest the corn, which not only reduces harvesting labor, but also reduces the labor required for the feeding operation.

It is very important that the roughage the lambs consume is sufficient to balance the corn and prevent overeating. This can best be accomplished by keeping the lambs in dry lot overnight and providing them with from 1¼ to 1½ pounds of hay per head before they go into the cornfield the next day. The hay should be at least half legume to insure adequate protein. If a good pasture is available adjoining the cornfield, the lambs will generally consume enough forage to prevent overeating. Salt and a mineral mixture should be made available free choice for lambs being finished in the cornfield.

It is important to condition the lambs to corn, before they are turned into the cornfield, by starting them on corn two weeks

COMPLETE RATIONS INCLUDING SALT AND MINERALS

Feeds	*Pounds*
1. Corn or sorghum	60
Soybean oil meal	4
Chopped legume hay	34.57
Salt	0.15
Limestone	0.70
Bone meal	0.50
Trace minerals	0.08
2. Corn or sorghum grain	50
Soybean, linseed, cottonseed meal	3
Legume hay	45.67
Salt	0.15
Limestone	0.50
Trace minerals	0.08
3. Corn	50
Molasses	10
Protein mineral supplement (see p. 474)	5
Legume hay	35

before and bringing them up to ½ to ¾ pound daily.

Expected Daily Gains

The rate of gain one may expect on finishing lambs will depend upon the size of lambs when started on feed, their thriftiness, and the ration fed. One-third pound gain daily on 60- to 70-pound feeders is considered a good average gain, but up to a half pound daily gain per lamb has been accomplished with good feeding and management.

Feed Requirements Per Hundredweight Gain

Lambs on full feed will consume from 1 to 1½ pounds of hay and from 1 to 1½ pounds of concentrates daily. To finish lambs from 90 to 100 days will require about 2 to 2½ bushels of corn (or the equivalent in other grains) and about 140 pounds of legume hay. The feed requirement for 100 pounds of gain will be five to six bushels of corn (280 to 340 pounds) and 420 pounds of alfalfa hay when fed a standard ration. The amount of grain and roughage will vary, if a high grain or high roughage ration is used.

FEED LOT TROUBLES

It takes a considerable amount of experience to feed lambs successfully. There is greater danger that lambs will have digestive troubles from overeating than any other kind of farm livestock.

Overeating Disease

Overeating disease causes heavy losses among finishing lambs unless the proper precautions are taken. The disease is thought by many to be caused by an organism found in the soil that is normally present in the lower bowel. It produces a toxin or form of poisoning. Lambs on full feed of grains are more susceptible to the disease than other sheep or lambs, since the heavy feeding causes rapid growth of the organism.

Veterinarians use perfringens type D bacterin, antitoxin, and toxoid for vaccinating against the disease. Antitoxin gives quick immunity and the toxoid gives a longer lasting immunity. Many veterinarians are using, with considerable success, a combination of the latter two as a prevention for overeating disease. Mixing Aureomycin in the ration to provide 10 milligrams daily per animal will help to control the disease.

Coccidiosis

Coccidiosis is caused by a protozoan parasite. The disease has become more prevalent among feeder lambs in recent years, causing heavy death losses in many instances. Outbreaks occur most frequently two to three weeks after an exhausting shipment. Prevention consists of keeping feed bunks and watering facilities clean, and isolating all animals showing symptoms. Veterinarians have had some success treating the disease with the sulfa drugs.

Shipping Fever

This disease is caused by one or more bacteria. Usually the outbreak in feeder lambs occurs shortly after shipping or exposure to bad weather. Vaccinating lambs two to three weeks before shipment with a mixed bacterin has helped to prevent the disease. Veterinarians have reported good success with the sulfa drugs as preventatives by administering them through the water fed to the lambs on arrival, and treating those showing visible symptoms with antibiotics.

SUMMARY

Finished lambs may be produced on grass and the milk from the ewes. This method requires plenty of good, fresh, tender grasses or legumes having a high feeding value.

Supplying high-quality forage, grain, or both to suckling lambs in an area where it is not available to the ewes is known as creep feeding, and is often desirable. Creep feeding speeds the growth rate of the lambs, prevents ewes from suckling down as much, and tends to even up the size of the lambs.

Under range conditions, if lambs are to be creep fed, they should be started on a creep when they are about three weeks of age. Under farm flock conditions lambs may be placed on creeps as soon as they will eat, which is at about ten days of age.

Grains may be placed in a creep and left for the day to get lambs started on feed. After they are eating, feeding them the amount of grain they will clean up in 20 minutes once or twice daily is recommended.

Hothouse lambs are lambs finished for market when 6 to 12 weeks old and weighing from 30 to 60 pounds.

Lamb finishing is the feeding out of lambs to a desirable market finish. Lambs that are under-finished are either put into the feed lot and fed grain or placed on a highly nutritious pasture with or without grain until they are ready for market.

The lamb feeder depends upon margin, value of gain over cost of grain, wool clip, and the fertilizing value of the manure for his profits.

Dry roughages, silage, root crops, animal fats, grains, molasses, grain by-products, and several protein supplements may be used in proper combination for finishing lambs. Minerals and salt should be provided either in mixed rations or separately. Urea or biuret may be successfully used as a partial protein substitute when properly fed in the correct combination with other feeds.

Antibiotics have shown beneficial results when fed at the rate of 10 milligrams per head daily. Hormones are in the experimental stage and no recommendations are made at this time.

Except for some of the combine type sorghums, such as milo, grinding feed for lambs is not recommended unless a complete mixed ration is fed and grinding is necessary for blending.

Lambs like pelleted feeds and experiments show some increased gains resulting from pelleted feeding.

Feeder lambs may be finished by hand feeding, self-feeding, or confield finishing.

Lambs may be expected to gain from ⅓ to ⅔ pound daily and will consume from four to five bushels of corn, or its equivalent, and 420 pounds of legume hay per 100 pounds of gain.

Overeating disease, coccidiosis, and shipping fever are the most common ailments affecting feed lot lambs. Vaccination, antibiotics, sulfa drugs, and sanitation are the best prevention and treatment methods.

QUESTIONS

1. What methods are commonly used to produce finished lambs?
2. Explain the requirements for the production of grass-finished lambs.
3. What are feeder lambs?
4. Explain what is meant by creep feeding lambs.
5. Discuss the advantages of creep feeding lambs.
6. When and how should lambs be started on creeps?
7. What are some recommended rations for creep feeding lambs?
8. Explain what is meant by a hothouse lamb.
9. How do lamb feeders expect to make their profits?
10. Discuss the various roughages, grains, and grain substitutes fed to lambs.
11. Discuss the protein feeds and protein blends that may be used in lamb feeding.
12. Have antibiotics and hormones shown any beneficial effects on fininishing lamb? Discuss.
13. How should feed be prepared for lambs? Discuss.
14. Explain how you would start lambs on feed.
15. Describe the common methods of finishing lambs.
16. What advantages and disadvantages do each of the methods described in Question 15 have?
17. What are the normal weight gains finishing lambs may be expected to make?
18. How much feed will lambs consume under average conditions per hundred pounds of gain?
19. Discuss the cause and prevention of the common feed lot lamb ailments.

REFERENCES

Andrews, F. N., and W. M. Beesen, "The Effects of Various Methods of Estrogen Administration in the Growth and Fattening of Wether Lambs," *Journal of Animal Science*, Vol. XII, No. 1, February, 1953.

Botkin, M. P., and Leon Paules, *Effect of Aureomycin in Various Ratios of Roughage to Concentrates for Feeder Lambs,* Agricultural Experiment Station, Mimeograph Circular No. 44, Laramie, Wyoming, University of Wyoming, 1954.

Colorado Feeders' Day Report, General Series No. 635, Fort Collins, Colorado, Colorado A. & M., 1956.

Henneman, H. A., R. E. Rust, and J. Meites, *The Effect of Steroid Hormones on Fattening Lambs,* Michigan Agricultural Experiment Station Report, 1956.

Illinois Sheep Day Reports, Agricultural Experiment Station, Urbana, Illinois, University of Illinois, annually, 1961–1971.

Iowa Sheep Day Report, Agricultural Experiment Station, Ames, Iowa, Iowa State University, 1965, 1966, 1968, 1970.

Kansas Agricultural Experiment Station Progress Report, Manhattan, Kansas, Kansas State College, 1965, 1966.

Lamb Feeding Experiments, Kansas Agricultural Experiment Station, Garden City Branch, Garden City, Kansas, annually, 1955–1966.

Means, T. M., F. N. Andrews, and W. M. Beesen, "The Effect of Hormones on the Growth and Fattening of Lambs," *Journal of Animal Science,* Vol. XII, No. 1, February, 1953.

Pelleted Rations for Fattening Lambs Progress Report, Agricultural Experiment Station Report No. 19, Bozeman, Montana, Montana State College, 1954.

Rath Packing Company and Iowa State College, *Modification of a Standard Feed Mixture for Fattening Lambs,* Agricultural Experiment Station Leaflet 205, 1956; Leaflet 196, 1955; Leaflet 184, 1953; Ames, Iowa, Iowa State College.

Sheep Day Report, Lafayette, Indiana, Purdue University, 1965, 1966, 1967, 1970.

"Studies Concerning the Value of Cooked Gull Beans, Stilbestrol, Progesterone and Estradiol in Rations for Fattening Lambs," *Lamb Feeding Experiments,* Agricultural Experiment Report, Ithaca, New York, Cornell University, 1954–1955.

33 Disease and Parasite Control

Sheep diseases, parasites, and other ailments may be divided into four groups: (1) external parasites, (2) internal parasites, (3) contagious diseases, and (4) noncontagious ailments. The average producer is not a veterinarian; therefore, the chief preventive measures in maintaining a healthy flock are good sanitation, the proper use of sprays and disinfectants, and the vaccination of his flock against those diseases for which vaccine has been perfected, when recommended by a reliable veterinarian.

While the list of ailments that affect sheep is almost endless, few shepherds have experienced difficulties with more than two or three diseases and parasites. Therefore, the principal sheep troubles will be discussed, but those that rarely cause losses in the United States will not be given consideration in this book.

Most sheep infections and diseases are brought into the flock as a result of poor management practices. Parasites and diseases usually breed best under filthy conditions. Many diseases spread from infected animals to healthy ones.

A PROGRAM OF DISEASE AND PARASITE PREVENTION

The following steps are important if a disease- and parasite-free flock is to be maintained:

1. **Bring only clean animals into the flock.**
2. **Keep animals in well-drained lots or pastures free of stagnant water.**
3. **Isolate all animals that are known to have contagious infections. Animals that have been purchased or otherwise added to the flock should be isolated until it is reasonably certain they are free of parasites and disease.**
4. **Vaccinate for diseases for which a successful vaccine exists, when recommended by a reliable veterinarian.**
5. **Disinfect housing and equipment regularly.**

6. **Treat open wounds on all sheep and the navels of newborn lambs with a recommended disinfectant.**
7. **Provide plenty of exercise for the breeding flock.**
8. **Spray, dip, or dust for external parasites at regular intervals, and eliminate manure piles and filth accumulation.**
9. **Follow a rigid program of internal parasite control using the recommended materials.**
10. **Avoid overcrowding of the flock.**
11. **Provide clean warm quarters for newborn lambs.**
12. **Rotate pastures often.**
13. **Observe the animals for signs of parasites and infections.**
14. **Consult a veterinarian when sickness or death occurs. Prompt diagnosis and treatment can reduce losses.**

EXTERNAL PARASITES OF SHEEP

Blowfly

This group of flies includes several species. The black blowfly and the bluebottle fly are the most dangerous to sheep.

Life History. After warm weather starts in the spring, the fly eggs hatch from the dead and putrifying materials where they have been laid. The hatched flies mature rapidly and seek out places to deposit their eggs. Accumulations of filth attract them. Sheep with stains or dung on them provide attractive places for the blowflies to lay their eggs.

Symptoms. The first symptoms one will notice in sheep is twitching the dock, stamping the feet, running short distances and showing other forms of restlessness.

As the eggs hatch and the maggots enter the flesh, discolored patches of wool caused by serum leakage will begin to show. Longwool sheep must be checked carefully as entire maggot-infested areas may be completely hidden by the wool.

Prevention and Treatment. Prevention consists of docking all lambs and tagging all sheep, keeping them clean and free from dung. All wounds should be treated with a good fly repellent. Insecticide sprays, such as lindane, sprayed on the crotch and dock help to prevent attacks by the flies.

Treatment consists of removing the wool from around the affected area. To be sure the entire area is exposed, clip the wool back to where it is clean and dry. Scrape off all the larger maggots and force the others out with a mild sheep dip. Chloroform may be applied to the area to kill those lodged deeply in the flesh.

Aftert he maggots have been eliminated, let the area dry and apply EQ 335, a smear developed by the United States Department of Agriculture. The EQ 335 should be diluted in water at the rate of one part EQ 335 to seven or eight parts water.

Screwworm

These pests are more prevalent in the Southern and Southwestern states, but may infect sheep in other areas.

Life History. The screwworm fly is bluish green in color, with orange shading below the eyes. The three prominent, dark stripes along its back distinguish it from similar insects. The flies lay their eggs in masses along the edges of the open wounds. The eggs hatch into tiny maggots that burrow into the flesh, where they feed from four to seven days.

When the worms have reached their full growth, they drop to the ground and burrow into the soil. A few days later they emerge from the pupa, or dormant stage, as adult flies. The entire life cycle may be completed within 21 days under favorable conditions.

Symptoms. The symptoms of screwworm infestation are much the same as for the fly strike (blowfly maggots). The animals show a general restlessness due to the irritation caused by the maggots feeding on the flesh.

Prevention and Treatment. Prevention requires that operations such as castrating and docking be done during the season when screwworm flies are not active. All wounds should have a fly repellent applied to them. EQ 335 is an effective repellent and will also kill the screwworm maggots. When a dye is mixed in the EQ 335, treated animals are easily detected and can be readily separated for later inspections. When the wound is deep and the screwworms are present, the smear should be diluted (see blowflies). By diluting, the smear penetration is deeper. The worms come to the surface where they die and drop off. If the undiluted smear is applied over the deeply imbedded worms, infection may result from the putrifying worms even though they may be killed.

Lice

The louse is a flat, wingless insect. There are several species and two general types, biting lice and sucking lice. The sucking lice are the most injurious. Lice are more prevalent on those animals that are out of condition because of improper feeding or management. They are more abundant when animals have been confined to small areas, and during the winter.

Life History. Lice spend their entire life cycle on the sheep. They attach their eggs or "nits" to the wool, where they hatch in about two weeks. The females begin laying eggs about two weeks after hatching and die after reproduction.

Symptoms. Intense irritation and itching caused by the lice make the animals scratch, rub, and gnaw at the skin. The wool may take on a dead, dry appearance. Growth is retarded, and badly infected animals will go out of condition.

Prevention and Treatment. See Table 33-1.

Ticks or Keds

The sheep tick is not a true tick but a bloodsucking fly without wings. It is dark brown in color, has a hairy body with six legs, and averages about 1/4 inch in length when mature. Ticks may infest all sheep. However, the medium- and long-wool breeds are more susceptible to infestation than are the fine-wool breeds.

Life History. Ticks, like lice, spend their entire life on the sheep. The eggs develop into the larvae while still in the body of the female. Females will deposit 12 or more larvae, which develop into the pupa, or dormant stage, and remain attached to the wool fibers. The pupa stage lasts from 20 to 25 days, when the pupa shell is broken open and the young tick emerges. The young tick will start to deposit larvae 14 to 16 days later.

Symptoms. Because ticks feed by sucking blood from the skin, an intense irritation is set up causing the sheep to bite and scratch itself in an attempt to gain relief from the itching. The wool will be damaged and in cases of heavy infestation, the animal will become anemic.

Prevention and Treatment. See Table 33-1.

Sheep Scab

Sheep scab is a reportable disease, which means that state and federal authorities must be notified. The disease is caused by an insect-like parasite so small it is nearly invisible to the naked eye.

Life History. The scab mite spends its entire life on the body of the sheep. It lays from 10 to 30 eggs over a two-week period. These hatch and the mites reach maturity in another two weeks. Thus a new generation of mites is produced every 15 to 20 days.

Symptoms. The mites live on the blood serum that oozes from the skin punctures which the mites make. The serum becomes mixed with dirt which soon dries forming a crust or scab. The skin thickens and becomes hardened. The infected sheep become restless, and rub, scratch, and bite at their wool.

Prevention and Treatment. Refer to Table 33-1.

TABLE 33-1 CONTROLS FOR TICKS, LICE, SCABS, AND WOOL MAGGOTS

Pest & Chemicals	Spray †	Dip †	Smear	Days before Slaughter
Ticks (keds) and lice:				
Coumaphos (Co-Ral)	4 lbs. 25% w.p.			15[1]
Diazinon	1 lb. 50% w.p.[2]			14
Dioxathion (Delnav)	1 gal. 15% e.c., *or* 0.5 gal. 30% e.c., + 2 lbs. detergent [3]			0
Lindane [1]	1 qt. 20% *or* 3 pt. 12.4% e.c.	½ strength of spray		*Spray:* 30 *Dip:* 60
Malathion	3 pt. 57% e.c. + 1 lb. detergent in 50 gals. water [4]			
Ronnel (Korlan)	1 gal. 24% e.c., *or* 8 lbs. 25% w.p., + 2 lbs. detergent			84
Toxaphene [5]	As for cattle lice			28
Scab (scabies, wet mange): [6]				
Lindane		1 pt. 20% *or* 3 cups (24 oz.) e.c.[1]		60
Toxaphene		As for cattle mange [7]		28
Wool maggots (fleece worms):				
Coumaphos (Co-Ral)	4 lbs. 25% w.p. + 2 lbs. detergent [8]			15
Lindane			EQ 335 *or* other screw-worm smears [9]	
Ronnel (Korlan)	Aerosol bombs available; follow label instructions			

† All ingredients are mixed in 100 gals. of water, unless otherwise noted.

[1] Do not use on lambs less than 3 months old.

[2] Apply 1 qt. per animal with low-pressure spray.

[3] Do not apply oftener than every 2 weeks.

[4] Apply 2–4 qts. per animal after shearing, or treat thoroughly with 4 or 5% Malathion dust. Do not treat lambs less than one month old.

[5] Do not use on suckling lambs.

[6] Consult a veterinarian.

[7] Dip animals in vat for 2 minutes.

[8] Apply 2 qts. per animal.

[9] Shear infested areas and treat as often as necessary.

GENERAL RECOMMENDATIONS IN USING SPRAY

Sprays and dips may be harmful to both humans and animals when improperly used.

The authors have suggested lindane because of its all-around effectiveness against external parasites. All insecticides, especially those used on livestock, are undergoing severe scrutiny by the Food and Drug Administration. Some have been, and others may be, prohibited. The stockman should follow the recommendations of reliable agricultural teachers and veterinarians.

Read Labels and Directions

Manufacturers put the same products out under different concentrations, so it is important to follow directions when mixing a dip or spray. Precautions for safe use of the product are also included and should be observed.

INTERNAL PARASITES

Internal parasites are more numerous and heavy infestation is more likely to occur when sheep are confined to the same area year after year, without benefit of a lot or pasture rotation. Low, wet land and humid conditions are favorable to the development of internal parasites. Proper nutrition is of extreme importance in the control of internal parasites.

Common Stomach Roundworms

The common stomach roundworm is one of the most prevalent forms of internal parasites to cause losses in sheep.

Life History. The female lays eggs which pass out of the body with the feces. The eggs hatch in a few days, when temperature and moisture conditions are favorable, into tiny larvae. The larvae crawl up on the blades of grass where they are swallowed by sheep during grazing. The worms travel to the abomasum, where they grow, and the cycle is repeated.

FIGURE 33-1. This sheep is heavily infested with scab mite. (Carr photo. *Courtesy* Hawkeye Institute of Technology)

Symptoms. Infected sheep become listless, thin, weak, and unthrifty. When infestation is heavy, a swelling, known as bottle jaw, may develop under the jaw.

Prevention and Treatment. Since the tiny worms that appear on the blades of grass are dependent upon a host for survival, the rotation of pastures every two weeks will help to control the common stomach worm.

The most effective material for the removal of the common stomach worm from sheep is micronized phenothiazine. The drug may be given as a drench or in capsules or boluses. The recommended solution is one pound of powder mixed with three pints of water and the dosage should be four ounces for each ewe and ram. Do not dose lambs. Ewes and lambs should be treated with Thiabenzole when lambs are about three months of age.

A mixture of one part phenothiazine by weight and ten parts salt by weight should be provided free choice for the flock. When the flock has been properly dosed, the salt-phenothiazine mixture will help to keep them free of the worms.

Drenches may be given with either a bottle or a syringe. Care should be exercised not to get any of the mixture into the lungs. The head should be held in a natural position (not too high) and the drench given slowly, allowing time for the sheep to swallow.

A complete control program should include fall and spring treatment, pasture rotation, and free access to the salt-phenothiazine mixture.

Other Species of Stomach and Intestinal Worms

There are several other types of stomach and intestinal worms that attack sheep in addition to the common stomach roundworm that has been described. The more important ones are: the brown stomach worm, the small stomach worm, cooperias (four species), the nodular worm, and the hookworm. Except for the hookworm, the life cycle of these parasites is similar to that of the common stomach roundworm. The hookworm differs in that it may enter the body through the mouth or skin. The treatment is the same as that recommended for the common stomach roundworm.

Broad Tapeworm

The broad tapeworm in the sheep has a head which might be likened to an anchor. To the head are attached the segments of the tapeworm.

Life History. Each segment is much like an individual worm, as it is capable of producing eggs. The segments at the end break off, and may be seen in the manure as white flecks. The segments contain eggs which are ingested by soil mites in which they develop into the secondary stage. The mites crawl up on the blades of grass where the sheep swallow them while grazing, and thus become infested with tapeworm larvae which live on the partially digested food in the sheep's intestines.

Symptoms. Tapeworms seldom do damage that is particularly noticeable to sheep, but since they do use some of the food eaten by the sheep, some slowing down in growth rate is bound to result. Heavily infested sheep will show symptoms similar to those infested with stomach worms.

Prevention and Treatment. Prevention of broad tapeworm infestation is accomplished primarily by rotating pastures and generally good sanitation. A drug known as Teniatol is very effective in removing them from the digestive tract. The dosage is 15 cubic centimeters for the first 25 pounds of body weight and 5 cubic centimeters for each additional 15 pounds of body weight. For complete eradication the treatment will have to be repeated two or three times at two- to three-week intervals.

Gid Tapeworm

The eggs of gid tapeworms are found in the droppings of dogs, wolves, coyotes, and other carnivorous animals.

Life History. Dogs and similar animals harbor the adult tapeworms after eating the cysts in dead animals. After the dogs, wolves, or coyotes swallow the worms, the tapeworms' heads push out from the bladder and attach themselves to the intestinal lining, where they develop to maturity. The mature worm produces many eggs which leave the host in the feces. Sheep grazing on the area where these droppings have been distributed pick up the eggs and become infected with the worms.

Symptoms. As the blood distributes the parasites in various parts of the body, some reach the brain where they complete their development. The affected brain produces an incoordination in the sheep which causes it to circle or stumble. For this reason, the animal is said to be giddy.

Prevention and Treatment. Prevention lies in disposing of all dead carcasses by burning or deep burying. Properly disposing of the internal organs of slaughtered animals so that dogs and other carnivorous animals may not eat them is advised. Human beings should also take precautions, as the tapeworm may infect them. Meats should be properly cooked before eaten. There is no satisfactory treatment.

Grub in the Head (Sheep Bots)

The sheep bot deposits larvae in the nostrils of the sheep, causing irritation and catarrh.

Life History. The larvae, not eggs, are deposited around the sheep's nostrils. The larvae crawl up into the nasal passages to the frontal sinuses. The mature larvae return by way of the nasal passages, drop to the ground, and pupate after burrowing into the soil. The flies emerge from the pupa shell in about three to five weeks. The entire life cycle lasts from three to five months.

Symptoms. The common term used by sheepmen for the condition produced by the bot is "snotty nose." The nose runs, and the sheep keeps up a continuous snuffing. Infested sheep stand with their noses near the ground or lie down, sneeze, and seem very excited.

Prevention and Treatment. Tar the noses of the sheep, or apply some noninjurious repellent to the nose during the hot season, as a preventative. A 3 percent saponified cresol solution sprayed under a 38-pound pressure into the nasal cavity has proven quite effective in killing young larvae. A 21 percent emulsion of Ruelene used as a drench at the rate of 2 cc/10 pounds of body weight is a recommended control.

Coccidiosis

Coccidiosis is a parasite disease caused by tiny organisms that live in the intestinal lining. Infested animals pass hundreds of the organisms in the feces, which are picked up by other animals, thus spreading the infection from one to another.

Life History. There are two main phases in the life cycle of the coccidia. One is the free-living phase outside the body where they spread from one animal to the other. The second is the parasitic phase in the intestines. The oöcysts (free-living phase) have thick shells and are not easily destroyed by disinfectants. When the oöcysts are swallowed by the animal, they reach the upper part of the small intestine. The digestive juices destroy the outer membrane of the oöcysts, releasing the eight sporozoites contained in each. The sporozoites attack certain cells of the body which they destroy while at the same time growing to maturity and forming new oöcysts.

Symptoms. Lambs are more likely to be affected than older sheep. Severe diarrhea that is often bloody is the most common symptom. Animals become weak, lose their appetite, and often die. The stained, dirty wool resulting from the diarrhea is an invitation for fly strike.

Prevention and Treatment. Sanitation, keeping the quarters clean, clean feeding and watering facilities free from contamination, and avoiding low, wet places for grazing are preventive measures. All infected animals must be isolated. The treatment consists of using the sulfonamides under the directions of a veterinarian. Some veterinarians have reported good success with Teniatol as a treatment.

CONTAGIOUS DISEASES OF SHEEP

While the list of contagious diseases that may affect sheep is long, the number that has caused considerable loss in any one area is comparatively small.

Foot Rot

Foot rot is a highly contagious disease, and in severe outbreaks may affect over 70 percent of the flock. Foot rot is caused by specific organisms that enter the feet through a break in the skin at the skin-hoof junction. The germs are carried by infected animals and are spread from one to another. A lame ewe may infect her lamb within a few days after birth. Wet, muddy lots provide receptive places for infected sheep to plant the germs, where they can easily be picked up by other animals.

Sheep pastured on irrigated pastures or allowed to run in dry lots that become muddy when wet have a higher incidence of the disease than typical range sheep.

Symptoms. The most common symptom is extreme lameness in one or possibly all four feet. Upon examination, a break between the claws and a swelling will likely be found. As the conditions continues, a rupture will occur. The disease is characterized by a very offensive odor.

Prevention and Treatment. Prevention consists of keeping sheep away from muddy lots, isolating all infected animals, and providing a foot bath at regular intervals, especially in areas where the disease is a problem. The foot bath consists of bluestone (copper sulfate) or a formalin solution. The bluestone solution is made by dissolving a pound of fresh bluestone in a gallon of water.

If a formalin solution is used, it should consist of 2 percent formalin. The solution should be about 2 inches deep and placed in a long narrow trough. The sheep should be moved through the trough single file, allowing each animal to remain from one to three minutes in the solution. Sheep should be carefully watched and prevented from lying down in the copper sulfate solution, as it will stain the wool.

Treatment of infected animals consists of several steps outlined as follows:

1. **Clean the foot thoroughly and use a pocket knife to trim back the hoof until there is some oozing of blood. This is necessary in order to remove all of the horn that may harbor the germs. Remove all of the thin flap of the horn between the toes and scrape around the side of the hoof. All dead tissue must be uncovered and removed. Caution is necessary when trimming so as not to trim too deeply at the point of the toe, where a large blood vessel is located.**
2. **The second step is to thoroughly disinfect the trimmed foot. For this purpose, several good disinfectants are available. Butter of antimony, which is cheap, may be applied with a swab, working it into the tissue all around the foot.**
3. **The third step is to run sheep through the bluestone or formalin foot bath as a final disinfectant. The foot bath should be repeated once a week until complete recovery takes place. Animals that do not respond to the above treatment may be successfully treated with a sulfa drug administered under the direction of a veterinarian.**

Mastitis (Blue Bag)

Mastitis is a disease resulting from one or both of two organisms that enter the udder, probably as a result of injury or chilling, although the exact cause is not known. The disease is serious and may cause the death of the animal infected, but usually it does not cause heavy losses. Ewes that have the disease are no longer fit for breeding purposes; the udder is spoiled even though the animal recovers.

Symptoms. The first noticeable symptom is that the sheep will lag behind the flock with a spraddle-legged walk. Due to the soreness in the udder the ewe will not allow her lamb to suckle. Examination will reveal a caked or hard condition of the udder. While the entire udder may be affected usually only one half will show infection. As the disease advances, gangrene sets in, giving the udder a blue appearance.

Prevention and Treatment. Prevention consists primarily of providing a clean, dry place for the ewes to bed down and eliminating sources of udder injury such as sharp objects. Hot applications of a solution containing a tablespoonful of Epsom salts in a quart of water may help. One cubic centimeter of procaine penicillin in oil syringed into the udder is effective against one of the organisms, while dihydrostreptomycin is effective against the other.

Anthrax

Anthrax is a disease that will appear with little warning and death will occur within an hour. The disease is caused by a germ which can survive for years in the soil. The disease may strike any place, but it is more prevalent in certain areas.

Symptoms. Grinding of the teeth, hard and rapid breathing, pounding of the heart, and sudden death are the common indications of anthrax in the flock.

Prevention and Treatment. A vaccine has been perfected which gives immunity, and in areas where the disease is a problem, animal vaccination is recommended. Treatment is unsatisfactory. Since humans may contract the disease, extreme caution should be used when handling infected animals. Dead animals should be burned.

Shipping Fever

This disease is caused by a group of organisms that are most likely to attack lambs soon after birth, or older sheep that have lowered resistance because of shipping or exposure to inclement weather.

Symptoms. Discharge from the eyes and nostrils, coughing, rapid breathing, high temperature, and loss of appetite are the most noticeable symptoms.

Prevention and Treatment. The elimination of practices that cause fatigue, such as hard driving or long hauls without rest, regular feeding, and giving protection against inclement weather constitute the chief preventive measures. All animals that have been purchased, or otherwise obtained, should be isolated until the danger of shipping fever is past.

A bacterin is available that has given considerable protection, if administered before the disease develops. When sheep are to be shipped or moved over long distances, the bacterin should be administered one to two weeks before moving. Sick animals may best be treated with antibiotics, although serum has been beneficial in many cases. Animals that are fed regularly while being moved have much more resistance to the organisms that cause shipping fever than those that go hungry. Taking time to feed

FIGURE 33-2. Top management is a must in preventing disease or parasite outbreaks. (Campbell photo. *Courtesy* Hawkeye Institute of Technology)

regularly is very important in preventing shipping fever.

Sore Mouth

This disease is more common in lambs than older sheep. It is caused by a virus and is transmissible to humans.

Symptoms. The first symptoms are refusal to eat and a depressed appearance. Small sores may be found on the lips, gums, and tongue. The sores break and run, but they will usually heal by themselves without treatment. However, the animal loses weight and because of the sore mouth, lambs may not suckle the ewes, resulting in caked udders.

Prevention and Treatment. Sanitation, and a vaccine applied to a scratch made under the tail, will prevent the disease. Treatment consists of isolating the infected animals and treating the sores with iodine or a 2 percent potassium permanganate solution.

Lamb Dysentery

This disease may develop in very young lambs. The death rate is high and an outbreak may cause heavy losses.

Symptoms. Lambs two to five days old are most susceptible. The lambs scour considerably, become depressed, weak, and unable to suck. Death will usually occur in a few hours if the lambs are not treated.

Prevention and Treatment. Sanitation, preventing the lambs from chilling, and keeping them in a well-ventilated and well-lighted place are the best preventive measures. Antibiotics, streptomycin, Aureomycin, and Terramycin given in doses of 0.5 grams once or twice daily by mouth have all been relatively effective in treating the disease. A veterinarian should be called.

Foot-and-Mouth Disease

This is a highly contagious disease of sheep, swine, and cattle. So far, the disease has been kept out of the United States, but every precaution will be necessary if it is to remain outside.

Symptoms. The disease is characterized by blisters found on the tongue, lips, cheeks, and skin around the claws of the feet. The animals infected have a heavy flow of saliva which hangs from the lips in strings. Death losses are not generally high, but the economic value of infected diseases is considerably reduced.

Prevention and Treatment. Prevention consists of the elimination of all infected animals by slaughtering and burning or burying. If the disease is suspected, government authorities should be notified. The foot-and-mouth disease virus is quickly destroyed by a solution of lye, and the solution should also be used for cleaning the quarters where infected animals have been housed, if the area is to be properly and thoroughly disinfected. A new vaccine has been developed which is being recommended for areas outside of the United States. Unless the disease spreads into this country to a considerable extent, American farmers and ranchers are not likely to be advised to vaccinate. Vaccines have not been regarded as favorable to complete eradication of the disease.

NONCONTAGIOUS AILMENTS OF SHEEP

Overeating Disease (Enterotoxemia)

This disease is probably the greatest killer of feed lot lambs. It also attacks lambs that are on lush pastures and are suckling ewes that are milking heavily. The disease is caused by an organism that is usually found in the digestive tract and in the soil. When lambs are on concentrates or lush pastures, the growth rate of the organisms is increased and a very lethal poison is produced.

Symptoms. Very few symptoms are noted. The lambs go off feed, are depressed, and develop diarrhea. Some may stagger or develop convulsions.

Prevention and Treatment. A very effective vaccine has been developed and lambs going into the feed lot should be vaccinated. In ranges where there is a history of the disease, it will probably pay to vaccinate.

Stiff Lamb Diseases

There are a number of diseases that cause stiffness in lambs. Complete knowledge of their cause and prevention or treatment is lacking. A brief discussion of the more common ones will be given in this book.

White Muscle Disease. This disease is more common in farm flocks where early spring or winter lambing is practiced than in range flocks or lambs dropped on pasture.

The hind legs are especially affected and the lamb is unable to get around. Death usually results from starvation. A vitamin E deficiency is the most likely cause. Treating affected lambs with 10 cubic centimeters of cold-pressed wheat germ oil daily is recommended as both a preventative and treatment. Alpha tocopherol injected into the muscle at the rate of 5 cubic centimeters for a first treatment and 2½ cubic centimeters on alternate days is somewhat faster than the wheat germ oil.

Navel Ill and Joint Ill. These two types of diseases are caused by the entry of germs into the lamb through the navel or through wounds made when castrating and docking. Swelling of the knees and other joints are common symptoms. Lambs affected with navel ill will often be seen lying down and thrashing their legs.

Prevention consists of disinfecting the navel of newborn lambs with a tincture of iodine and castrating and docking under sanitary conditions.

The sulfa drugs, antibiotics, or injections of hemorrhagic septicemia bacterin may be used in the treatment.

Chronic Arthritis. This disease is probably caused by the entry of certain organisms into the body through wounds made during the docking and castrating operations. It affects the joints but seldom causes death, although few affected animals make a complete recovery. Prevention consists of sanitation during the castrating and docking operations.

Stiffness Due to Feed Changes. Lambs subjected to rapid, radical feed changes either by being separated from their mothers and moved to new pasture or rapidly changed to concentrate rations often develop stiffness.

Most of these lambs will recover in time. However, making gradual feed changes will reduce the incidence of this disease.

Lambing Paralysis or Pregnancy Disease

Pregnancy disease attacks ewes shortly before lambing, usually during the fourth month of gestation. Ewes carrying twins are more susceptible than those with single lambs. The disease is due to a carbohydrate deficiency, especially sugar, and lack of exercise.

Symptoms. Ewes will often develop a highly nervous condition followed by paralysis, coma, and death. In the early stages they tend to lag behind the flock and appear to be stiff. The head is held upwards; the neck is stiff; blindness may occur.

Prevention and Treatment. Ewes receiving good roughage and ½ pound of shelled corn daily (in the last six weeks of pregnancy) and getting plenty of exercise are seldom affected. Some shepherds feed one pint of molasses per head daily during the last six weeks of pregnancy. No successful treatment exists.

Bloat

This is a condition in which the rumen becomes filled with gas which the animal is unable to expel. The condition may be caused by a growth or other obstruction in the esophagus, but is more often caused by feeds that ferment rapidly in the rumen causing large amounts of gas to form. Feeding on legume pastures is one of the most common causes of bloat. Saponins (plant materials that produce a soapy lather) are probably the principal ingredient in legumes responsible for bloat.

Symptoms. The chief symptom is a great distention of the upper left side of the abdomen. Rapid breathing and uneasiness occur. If the pressure becomes great enough to prevent lung action, death will result from suffocation.

Prevention and Treatment. Experiment stations are carrying on extensive research in an attempt to find a method of bloat prevention. Antibiotics, detergents, and other materials have been fed with some success. If sheep can be fed a daily ration of grain or other roughage, in addition to grazing on pasture, poloxalene may be added to this ration to control bloat. The manufacturer's directions should be carefully followed in mixing poloxalene with any feed.

Since legume pastures are the main cause of bloat, some recommendations as to how to use them, if followed, will reduce the incidence of bloat.

Sheep going onto a legume pasture should, if possible, first be placed on a good lush grass pasture for a few days. This will condition the digestive system to lush green feed. Then, toward evening, after the sheep are full, place them on the legume pasture and leave them alone.

If no green grass pasture is available, give the sheep a fill of high-quality hay before turning them on the legumes. Never turn sheep into a legume pasture for the first time when the ground is wet.

Sheep that are moved about are more apt to die from bloat than those allowed to remain quiet. Some losses may result from bloat, but disturbing the flock to help one or two bloaters will probably result in more deaths.

POISONOUS PLANTS

Controlling losses of livestock from poisonous plants does not require that the producer become a plant specialist. The producer should, however, learn to identify the plants at all stages of their growth and recognize some of the general signs sheep will show when poisoned. There is no known treatment for many plant poisonings. Identification and elimination of the plant is the best preventive measure.

There are a number of poisonous plants that may affect sheep. Following is a list of the most common ones found in the United States: greasewood, horsebrush, death camass, sneezeweed, lupines, halogeton, rubberweed, locoweed or poison vetch, water hemlock, larkspur, copperweed, choke-cherries, arrow grass, corn cockle, dogbane, white snakeroot, nightshade, poison hemlock, jenson weed, laurels, castor bean, and sorghums.

SUMMARY

The use of proper sprays, dips, or dust at least once a year is essential for the control of external parasites. Internal parasites may be controlled through pasture rotation and the use of a recommended vermifuge.

Disinfecting the navels of newborn lambs and clean, dry, well lighted and well ventilated lambing areas will prevent most diseases of newborn lambs. Stiff lambs due to vitamin E deficiency may be prevented by the oral administration of wheat germ oil or by injecting alpha tocopherol into the muscle.

Feeding the proper amounts of sugar or other carbohydrates, together with adequate exercise, will prevent lambing paralysis.

Open wounds should be treated with a fly repellent to prevent fly strike and screw worm infestation. Docking and tagging will do much to prevent filth accumulations that attract blowflies.

QUESTIONS

1. Give the steps in a management program that will help to prevent sheep diseases and parasite infestation.
2. Which of these steps do you apply to your flock or band?
3. Give the treatments recommended for external parasites.
4. What are the recommended control measures for internal parasites?
5. List the internal parasites common to your area.
6. Which diseases may be prevented by vaccination?
7. What are the common ailments of very young lambs?
8. How can diseases in young lambs be prevented?
9. What is the most common ailment of fattening lambs and what is the prevention?
10. What are the cause and prevention of pregnancy disease?
11. How would you recognize and treat infectious foot rot?
12. What is bloat? Give some preventive measures.

REFERENCES

Elder, Cecil, and Donald E. Rodabaugh, *Internal Parasites of Sheep,* Agricultural Experiment Station Bulletin 527, Columbia, Missouri, University of Missouri.

Prier, J. C., *Disease Prevention in Young Livestock,* Agricultural Experiment Station Circular 47, Laramie, Wyoming, University of Wyoming.

Profitable Sheep Production, Extension Service Circular 470, Madison, Wisconsin, University of Wisconsin, 1966.

Ryff, J. F., and Ralph F. Honess, *Internal Parasites of Sheep,* Agricultural Experiment Station Circular 42, Laramie, Wyoming, University of Wyoming.

Scott, George E., *The Sheepman's Production Handbook,* First Edition, Abegg Printing, Denver, Colorado, 1970.

U.S. Department of Agriculture, *Animal Diseases,* The Yearbook of Agriculture, Washington, D.C., 1956.

34 Marketing Sheep and Wool

The sheep producer or lamb feeder is especially interested in the price of live sheep or lambs, and wool. We will attempt to discuss some of the factors that affect price and profit. The reader can determine when it is best to buy or sell, depending upon each particular situation.

MARKETING SLAUGHTER LAMBS

Most finished lambs and sheep are sold to the packers to be slaughtered. The producer may sell directly to the packer or the buyer, through a commission firm, at a terminal market, or to a private dealer. Regardless of how slaughter lambs or sheep sell, the price they bring at any one particular time will depend largely upon their grade and classification.

Grades and Classes of Slaughter Sheep

Slaughter sheep are those animals intended for immediate slaughter. The price per pound paid for them depends largely upon the value of the dressed animal. Young, well-finished lambs produce the best-flavored carcasses and generally sell for higher prices per pound than older or poorly finished animals. Excessive finish or fat is undesirable.

Sex

Following are definitions of the three sex classifications of sheep.

Ewes. Ewes are female sheep or lambs.

Wether. A wether is a male lamb that has been castrated before reaching sexual maturity.

Ram. A ram is an uncastrated male sheep of any age.

Age

The following paragraphs describe how slaughter lambs are classified.

Hothouse Lambs. Hothouse lambs are young lambs, usually born in the fall or early winter, that are milk fat and sold during the season between Christmas and Easter. They are usually sold by the time they are three months of age and weigh 60 pounds.

Spring Lambs. Spring lambs are those that reach the market in the spring as finished lambs weighing from 70 to 90 pounds. These lambs are usually born in the fall, but are older and heavier than typical hothouse lambs.

Lambs. Animals under a year old that do not classify as spring or hothouse lambs are considered lambs. When the term *lambs* is used by packers, it generally designates grass-finished lambs. Those receiving grains are referred to as *fed lambs.* However, the term lamb, when used in the broad sense by the packer, refers to all carcasses when the forefeet are removed at the break-joint. The break-joint is a temporary cartilage located just above the ankle. The break-joint of lambs is red and moist and has four points or ridges. In yearlings, the break-joint is more porous and does not have the red, moist appearance

FIGURE 34-1. Grand Champion Wether, International Livestock Show, 1972. Owned and exhibited by Warren Finder, Stoughton, Wisconsin. (*Courtesy* American Hampshire Sheep Association) FIGURE 34-2. These lambs should make desirable grade when they are ready for market. (Carr photo. *Courtesy* Robert Kimm)

of the lamb joint. In older sheep, this joint will no longer break, and the foot is taken off at the ankle leaving a round, or spool, joint. Carcasses with the spool-joint are sold as mutton, indicating older animals, rather than lamb.

Yearlings. Animals between one and two years of age are called yearlings.

Two-year-olds and Older. Sheep that are two years old and older are classified as mature in the slaughter market.

Weight

Slaughter sheep and lambs are classified as heavy, medium, or light. The weight divisions are determined by age.

Grades

There are two types of grading systems for sheep: (1) the on-foot quality grades, and (2) the yield grades. The on-foot quality grades are determined by the degree of finish and type of body conformation. Animals that are well developed in the areas of valuable cuts, such as the back, loin, and leg, and with the proper amount of finish, are graded prime or choice. Animals lacking in these qualities are placed in the lower grades. Thin animals with poor body conformation fall into the utility or cull grades.

Yield grade standards were established in March 1969 to provide a method of identifying the expected yield of retail cuts (cutability) which differ among animals. Leg conformation and the degree of finish are the two factors which the yield grade is based upon. There are five yield grades: 1, 2, 3, 4, and 5. Yield Grade 1 has the highest estimated yield of retail cuts, while Yield Grade 5 has the lowest.

FIGURE 34-3. Lambs sired by modern-type rams as pictured here should produce desirable carcasses. (*Courtesy* Columbia Sheep Breeders Association)

Types of Marketing Procedure

The producer who is selling slaughter sheep or lambs has a choice of markets and methods of selling similar to that described for beef cattle (see Chapter 15). A good job of selling is important if profit is to be made from hard labor and wise management.

Shrinkage

Shrinkage is the loss in body weight of the animals from the time they leave the feed yards or range until they arrive at their destination. Most of the shrinkage is the result of feed and water the animal eliminates that is not being replaced by regular feeding and watering. However, there may be a tissue shrink. The amount of shrinkage depends upon the fill, the time in transit, weather conditions, the length of time the animals are allowed to rest, feed, and drink before weighing, how the animals are loaded, and the age of the animals. The amount of shrinkage may be as high as 10 percent or more, but a normal shrink will range from 3 to 5 percent.

TABLE 34-1 MARKET CLASSES AND QUALITY GRADES OF SLAUGHTER SHEEP

Sheep or Lambs	Sex	Age	Weight	Pounds	Quality Grade
Sheep	Ewes	Yearlings	Light	90 down	Prime, choice, good, utility, cull
			Medium	90 to 100	
			Heavy	100 up	
		Mature	Light	120 down	Choice, good, utility, cull
			Medium	120–140	
			Heavy	140 up	
	Wethers	Yearlings	Light	100 down	Prime, choice, good, utility, cull
			Medium	100–110	
			Heavy	110 up	
		Mature	Light	115 down	Choice, good, utility, cull
			Medium	115–130	
			Heavy	130 up	
	Rams	Yearlings	All weights		Choice, good, utility, cull
		Mature	All weights		Choice, good, utility, cull
Lambs	Ewes, wethers, rams	Hothouse		60 down	Prime, choice, good, utility, cull
	Ewes, wethers, rams	Spring lambs	Light	70 down	Prime, choice, good, utility, cull
			Medium	70–90	
			Heavy	90 up	
	Ewes, wethers, rams	Lambs	Light	75 down	Prime, choice, good, utility, cull
			Medium	75–95	
			Heavy	95 up	

When to Sell Slaughter Lambs and Sheep

Western lamb producers make more money by selling weight than price per pound. This condition arises from the comparatively cheap price of grass and the fact that the grass crop will not be harvested unless eaten by sheep. However, weather conditions vary from year to year. Lambs on lush pasture will soon lose their bloom when pastures start to dry up. This makes it advisable to move the lambs earlier some years than others. Also many western range areas have grass varieties that produce sharp awns when the grass matures. Lambs will lose weight when irritated by awns that are caught in the wool. Lambs will need to be sold or shorn before the stickers develop. It is usually advisable to sell them.

Seasonal Prices. Lamb marketing is seasonal. During the periods of heavy marketing, prices usually decline. As the marketing tapers off, the price trend is upward. The heavy marketing of lambs starts in August and reaches a peak in October. The run of lambs begins to taper off in November and the price trend continues upward, reaching a peak in June. There can be several dollars difference between the high and the low

FIGURE 34-4. **(*A*) The parts of a sheep. (*B*) The shaded areas indicate the most valuable wholesale cuts; unshaded areas are the cuts of lower value. (Drawing by Tom Wickersham, Iowa State University Extension Service)**

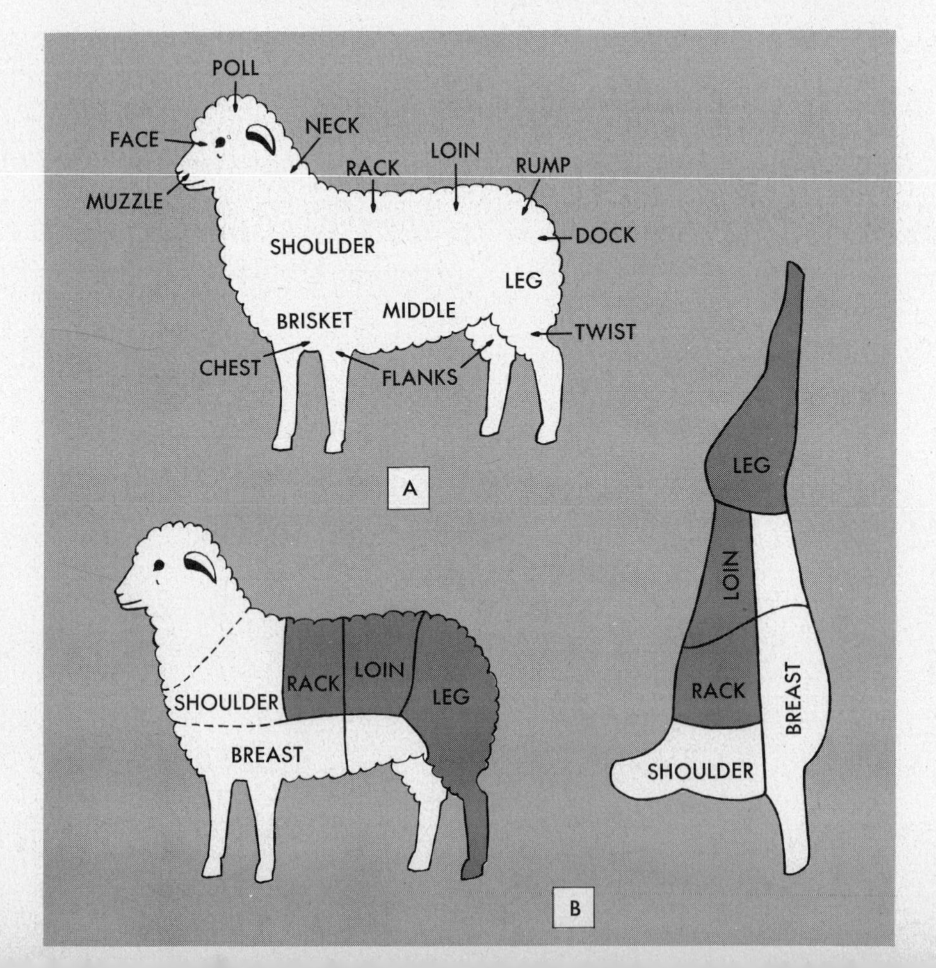

price paid during the year. Lambs and yearlings make up approximately 95 percent of the total sheep slaughter.

When to Sell Ewes. Ewe prices fluctuate more on the slaughter market than do lamb prices. The mid-summer low occurs just after shearing and is about 15 percent less than the year's average price. The March price of ewes is normally about 22 percent above the year's average level. The months of January, February, March, April, and May average considerably higher prices for slaughter ewes than the rest of the year.

MARKETING FEEDER LAMBS

Feeder lambs are usually marketed by the following methods: (1) private sale directly by producer to feeder, (2) through a dealer, (3) through a commission firm at a terminal market, and (4) at auction sales. (See the material on marketing feeder cattle in Chapter 15.)

When to Sell Feeder Lambs

The heavy period in shipments of feeder lambs to the Corn Belt occurs during the months of July, August, September, October, and November. Naturally the average seasonal price of feeder lambs is lower during these months. It is better to avoid selling during the summer and the fall months under average conditions. However, the producer must consider the amount of available feed and the additional weight he can get on his lambs by utilizing this feed.

MARKETING PUREBRED BREEDING STOCK

The marketing of purebred breeding stock is primarily limited to direct dealing between buyer and seller, or by auction sale. Unlike other classes of sheep, each transaction is limited to a single animal or to very small groups.

GRADING AND MARKETING WOOL

The producers of wool should have a working knowledge of the problems involved in wool marketing. They should be able to determine the value of the wool raised so as to look out for their own interests. Otherwise, they will find themselves at the mercy of the wool buyers at market time.

Wool is graded according to its *fineness*, the diameter of the individual fibers of wool within the fleece. Within limits of the grades, the finer wools are more valuable and the coarser wools less valuable. The *length* of the wool fiber determines the use that may

FIGURE 34-5. Feeling the lamb's loin and rib area helps to determine the fat cover on a live animal. (Carr photo. *Courtesy* Hawkeye Institute of Technology)

FIGURE 34-6. Grading wool. (Carr photo. *Courtesy* Hawkeye Institute of Technology)

be made of it in manufacture. Within limits of the length classifications, long wool is more valuable than short wool. A wool fleece, as it is shorn from the sheep, will contain, in addition to the wool, grease, dirt, various kinds of vegetable matter, and possibly other materials. The buyer looks at how much clean wool the grease fleece will *yield.* The difference in the amount of clean wool and the grease fleece weight is called *shrinkage.* Therefore, the value of a fleece of wool depends upon the following:

a. Fineness
b. Length
c. Shrinkage or yield

How Wool Is Graded According to Fineness

The fineness of wool refers to the diameter of the individual fibers of wool. Standard grades established by the U.S.D.A. are used in the wool trade. Prices are based on these standard grades.

TABLE 34-2 OFFICIAL U.S.D.A. STANDARDS FOR GRADES OF WOOL

Grade	Average Fiber Diameter (Microns)	Maximum Standard Deviation (Microns)
Less than 80's	17.69 or less	3.59
80's	17.70–19.14	4.09
70's	19.15–20.59	4.59
64's	20.60–22.04	5.19
62's	22.05–23.49	5.89
60's	23.50–24.94	6.49
58's	24.95–26.39	7.09
56's	26.40–27.84	7.59
54's	27.85–29.29	8.19
50's	29.30–30.99	8.69
48's	31.00–32.69	9.09
46's	32.70–34.39	9.59
44's	34.40–36.19	10.09
40's	36.20–38.09	10.69
36's	38.10–40.20	11.19
Coarser than 36's	40.21 or more	

Grades According to Fineness. The American or Blood System of grading wool refers to the amount of Merino breeding in the sheep from which the wool was shorn. The English woolen mills devised a very complicated system of grading wool, based on the number of hanks of yarn that one clean pound of wool would spin. This is known as the English or Spinning Count System of grading. In the years between 1920 and 1925, the two were combined into the present U.S.D.A. Standard Grades (effective January 1, 1966) which are based on average fiber diameter and variations of standard deviation in fiber diameter.

The Spinning Count System

When wool is graded according to the Spinning Count System, the grader actually indicates the number of hanks of yarn that can be spun from one pound of clean wool. A hank of yarn is 560 yards long. This being true, a clean pound of 64's quality wool will spin 64 × 560 yards or 35,840 yards of yarn. On the other hand, a clean pound of 50's quality wool will spin 28,000 yards of yarn. This is the reason why wool that grades fine is worth more per pound than wool grading ⅜ or ½ blood (see Table 34-3).

Fleece Grading. There is always some variation in the fineness of the wool grown on various parts of the sheep's body. The finest wool will usually be found on the head, while the coarsest wool is on the breech. For market purposes, wool is usually graded according to the fineness of the majority of the fleece.

The Relationship between Breed and Fineness. While there may be quite a lot of variation in fineness of fleece grown by individual sheep within a breed, the bulk of the individuals will grow a fleece of about the same fineness. The following table represents the fineness of fleece for various breeds.

How Wool Is Classified According to Length

Wool is generally classified according to the use that is made of it. Two broad classes are *carpet* wool, used in the manufacture of

TABLE 34-3 GRADE OF FLEECE PRODUCED BY THE COMMON BREEDS OF SHEEP

Breed	American Blood Grade	English Spinning Count
Rambouillet	Fine	64's & better
Delaine Merino	Fine	64's & better
Southdown	½ blood	58's & 60's
Hampshire	⅜ blood	56's
Shropshire	⅜ blood	56's
Suffolk	¼ blood	48's & 50's
Corriedale	⅜ blood	56's
Columbia	⅜ blood	56's
Horned Dorset	¼ blood	48's & 50's
Cheviot	¼ blood	48's & 50's
Targhee	½ blood	58's & 60's
Lincoln	Low ¼ blood	46's
Leicester	Braid	36's & 40's
Cotswold	Braid	36's & 40's

floor coverings, and *apparel* wool, used in the manufacture of clothing. Carpet wool is long and coarse and comes from the long-wool breeds. Apparel wool is further classified by length into *strict combing,* or *staple, French combing,* and *clothing wools.* The first is most valuable and the latter least valuable.

Length and Uses Made of Wool. Apparel wools may be used in either the *worsted* or the *woolen* process of manufacture. In the worsted process, the wool must be combed so that the fibers lie side by side when the wool is spun into yarn. The yarn made by the worsted process is a strong, relatively smooth, twisted yarn that may be woven into a very durable, well-finished fabric. The wool usually used in the woolen process is too short to comb and is therefore spun into yarn, which is relatively large, soft, and weak because the fibers do not lie parallel.

Length as Related to Fineness. Under the worsted and woolen systems of wool manufacture, the fineness of the wool determines the length it must be before it can be combed. The coarser the wool is, the longer it must be before it can be handled by the combing process. This being true, the coarser the grade, the longer the wool must be to stay in the strict combing class.

Classes of Wool According to Length. The following chart shows length requirements for each market grade of wool.

Putting the Grade and Class Together. In actual practice, wool is graded and classed at the same time. The fleece is inspected and the fineness grade is determined. Then the length of the fleece is studied and the length class determined. The fleece is then identified as to grade and class by using the name of each in describing that fleece. For example, fleeces will be called *Fine Staple, Fine French Combing, Fine Clothing, One-Half-Blood Staple, One-Half-Blood French Combing, One-Half-Blood Clothing,* etc. Each of these descriptions has a different market value.

Shrinkage—How It Is Determined and What Causes It to Vary. One of the first processes in the manufacture is to remove the grease, dirt, and vegetable matter. This process is called scouring. The remaining wool is called *clean wool.* The weight of the grease, dirt, and vegetable matter expressed as a percentage of the grease fleece weight is called *shrinkage.* The weight of the clean wool expressed as a percentage of the grease fleece weight is called *yield.* Example:

100 pounds of grease wool is scoured
40 pounds of clean wool remains
60 pounds of grease, dirt, and vegetable matter is lost in the scouring process

40 pounds clean wool = 40 percent *yield*
60 pounds lost in scouring = 60 percent *shrinkage*

Wool may vary in shrinkage from 35 to 80 percent. Fine-wool breeds produce more oil in their fleeces and therefore shrink more than medium- and long-wool breeds.

Defective and Offtype Wools

The several kinds of defective and offtype wools are less valuable to the mills because of the added expense of processing them or because of the limited uses that may be made of these wools. When these "off" wools are found at shearing time, they definitely should be packed and marketed separately from the sound wool. Some of the common defective and offtype wools are described in the following paragraphs.

Burry. These wools contain large amounts of the seed of bur clover, grassbur, threeawn or needlegrass, or other seeds or vegetable matter. These can be removed from the wool by a carbonizing or acid treatment process which is not necessary in the normal processing of wool. This increases the processing cost and therefore reduces the

TABLE 34-4 LENGTH REQUIREMENTS FOR THE MARKET CLASSES OF WOOL

Grade	Strict Combing, or Staple	French Combing	Clothing
64's and better	2½" or longer	1½" to 2½"	1½" or less
58's to 60's	2¾" or longer	1¾" to 2¾"	1¾" or less
56's	3" or longer	2" to 3"	2" or less
48's to 50's	3¼" or longer	2¼" to 3¼"	2¼" or less
46's	3½" or longer	None	3½" or less
40's	3¾" or longer	None	3¾" or less
36's or less	4" or longer	None	4" or less

TABLE 34-5 AVERAGE SHRINKAGE BY GRADES OF THE U.S. WOOL CLIP

Grade	Range in Shrinkage
64's and better	56–65%
58's to 60's	51–60%
56's	41–50%
46's	41–45%

market value. It is almost impossible to remove some of the seeds from the wool.

Cotted Fleeces. These are fleeces in which the fibers are badly tangled or matted. A special treatment not needed in processing sound wool is necessary to open cotted fleeces before they can be manufactured.

Black or Gray Wool. Black fleeces or fleeces that contain a large percentage of black or brown fibers cannot be dyed with all colors and are therefore less valuable.

Kempy Fleeces. These are fleeces that contain a large proportion of coarse hair called kemp. Kemp will not take a dye and is inferior to wool in quality. These fleeces can be used only in the manufacture of white fabrics of low quality.

Tags and Clippings. This wool consists of the parts of fleeces that are taken off when the ewes are tagged before the fleece is tied, or swept off the shearing floor.

HOW A GROWER MAY SELL WOOL

Sale to Independent Buyers

Independent buyers of one kind or another offer a market for wool in all localities in the areas where wool is produced. They offer to the grower an opportunity for immediate sale of the wool clip. The grower does not have to pay any marketing costs. He simply brings his wool to town, checks with one or more independent buyers, and sells the wool to the buyer making the best offer. The wool is weighed and he is paid immediately. The wool bought must later be sold on the going market. The buyers must, therefore, buy the wool at a price lower than its value if they are to make a profit.

Sale Direct to a Dealer or Mill Representative

Sales of this type are very unusual and are confined to growers who produce large clips of very desirable wool, and who have established a reputation for producing and putting up a top clip of wool over a period of years. Dealers and mill representatives know who they are and where they are located, and often seek them out to buy their clips. A grower, in this position, may contact several buyers and develop competition between them to his own advantage. Usually the grower must have a large clip of wool, for the buyers are not interested in one small clip. Often the grower must provide his own storage, carry his own insurance, and deliver the wool to the buyer's concentration point. These costs may offset any advantage that he might otherwise gain.

Consignment to a Commission Warehouse for Later Sale

This is the most common system of marketing wool in the range states. The grower delivers the wool clip to the warehouse. The warehouse will weigh and mark each bag and issue the grower a receipt for the wool. The warehouse provides storage for the grower, and the wool is insured immediately. The warehouse operator inspects the wool and selects sample bags to be shown to the buyer. The operator puts the grower's clip in one of several "lines" of wool that the warehouse will offer for sale. These "lines" of wool contain all the wool in the warehouse that grade the same for fineness and length.

The consignment system of marketing has some definite advantages:

1. **The price received for the wool will be nearer its real value than when marketed under any other system, except in rare cases.**
2. **The grower is in a better position in the market, since the system of "lining up" his wool with other similar type wool offers the buyer a large volume of uniform wool.**
3. **If the market is slow, the grower has his clip safely stored and insured and can easily hold it until the market is stronger.**
4. **The warehouse operator spends his entire time studying the wool market. He is in a better position to sell the wool than is the average individual grower.**
5. **Wool can be offered in a sealed bid sale to several buyers.**

Some of the disadvantages of this system of marketing are:

1. **The grower may need the income from his wool before the warehouse sells the "line" his wool is in.**
2. **A warehouseman can be mistaken in judging the wool market occasionally. When this happens, the grower loses money.**

Contracting for Future Delivery

Many times a mill will be in a position to close a contract for future delivery of a certain type fabric. The mill will contact warehouses and offer a specific grease price per pound for wool to be delivered by the grower at shearing time. The warehouse will contact the growers and give them this information. Those growers who are interested will contact the warehouseman and will sign a binding contract with the mill or dealer. The grower agrees to deliver a certain number of fleeces to the warehouse at shearing time. The buyer agrees to pay the grower $1 per head in advance, at the time of signing the contract, and the specified price per pound when the fleeces are delivered. The buyer usually reserves the right to discount the contract price if the fleeces delivered do not meet the description. In contracting, the grower stands to gain if the price of wool goes down at shearing time. On the other hand, if the price goes up, the grower loses.

Cooperative Marketing Agencies

Throughout the wool-producing area are located cooperatives which perform the same services as the commission warehouse. Some of these operate under the cooperative marketing laws of the various states, while others operate under the general corporation laws. The cooperative or grower-owned warehouses do not buy wool directly from the grower, but only handle the grower's wool on a consignment basis.

THE NATIONAL WOOL ACT

In 1954, Congress, recognizing wool as an essential product, passed the National Wool Act. This act created incentive payments to wool producers to encourage larger supplies of wool because the production for domestic use was insufficient. The incentive payments are financed from duties collected on wool imports from other countries. In the 1969 marketing year, the incentive price per pound was 69 cents.

The wool grower is advised to sell his wool for the highest possible price and obtain complete sales records. The incentive payment is paid on a percentage base of the average price received by all producers; thus, his total income is increased if he sold above the average price.

SUMMARY

Slaughter lambs and sheep are classified according to sex, age, weight, and grade. They may be sold directly to a packer or a packer-buyer on a live weight or grade-and-yield basis. Other methods of selling are through a commission company at a central market or directly to a private buyer.

Feeder lambs and sheep may be marketed by direct selling either on contract for future delivery or immediate transfer from buyer to seller. Other methods of selling feeder sheep are through auctions, dealers, and central markets.

The shipping cost, shrinkage, and prices are important considerations in determining the methods of marketing.

Generally speaking, lambs should be sold when they are ready for market. The highest seasonal prices for slaughter lambs usually occur in June and the lowest in October. Ewe prices average highest during March and lowest after the shearing season is over and during the late spring and summer months.

The heavy shipments of feeder lambs take place during July, August, September, October, and November. Price declines usually follow heavy shipments of feeder lambs.

Purebred animals are usually sold for breeding purposes, and auction sales or private sales offer the principal outlet for purebreds.

A progressive wool grower will learn to do the best job possible of preparing a wool clip for market. He will be in the shearing pen while the wool clip is being harvested. He will see that the tags and clippings are kept out of the wool bag. He will see that offwools are packed separately. He will pack fleeces of different grades, lengths, and yields separately. When he has done this, he will take to the warehouse a clip of wool that a commission agent can sell for him at a price that is fair.

A progressive grower should study the marketing facilities in his locality, look at other systems for marketing available in other localities, and determine which system of marketing will be the most profitable for him.

A grower can lose the profit he has made for 12 months of hard work in five minutes in the shearing pen or at the market place. This can be prevented by insuring that sheep are sheared properly and by understanding the requirements of the marketing system.

QUESTIONS

1 Give the grades and classes of slaughter sheep and lambs and describe each.

2 Explain the various methods of marketing both slaughter and feeder sheep and lambs.

3 Discuss the seasonal prices of both feeder and slaughter animals.

4 What determines the best time to sell?

5 What is shrinkage, and what are the principal factors that affect the amount of shrinkage?

6 Why is it important for the wool producer to understand how wool is classified and graded?

7 What determines the value of fleece? Explain.

8 How is wool graded?

9 Give the average grades of fleeces from the common United States breeds of sheep.

10 How is wool classified according to length?

11 What is the required length of the various market classes of wool?

12 How is shrinkage determined?

13 Describe the defective and offtype wools.

14 How does the wool buyer determine the value of wool?

15 How can the grower improve the quality of his wool?

16 What practices should the grower follow to improve the market value of his fleeces?

17 Discuss the various methods which may be used by a grower to sell his wool.

REFERENCES

Hamilton, Eugene, *Seasonal Market Variations and Their Importance to the Iowa Farmer,* Bulletin P 5, Agricultural Experiment Station and Extension Service, Ames, Iowa, Iowa State University.

Kammlade, W. G., *Sheep Science,* New York, J. B. Lippincott Co., 1957.

Marketing Feeder Cattle and Sheep in the North Central Region, Agricultural Experiment Bulletin 410, Lincoln, Nebraska, University of Nebraska.

Marketing Texas Wool on a Quality Basis, Bulletin No. 823, Texas Agricultural Experiment Station, 1955.

Potler, E. L., *The Marketing of Oregon Livestock,* Bulletin 514, Agricultural Experiment Station, Corvallis, Oregon, Oregon State College.

U.S. Department of Agriculture, *Official Standards of the United States For Wool Grades,* C & M No. 135, Washington, D.C.

Wool Marketing Problems in Texas, B-947, Texas Agricultural Experiment Station, 1961.

POULTRY PRODUCTION

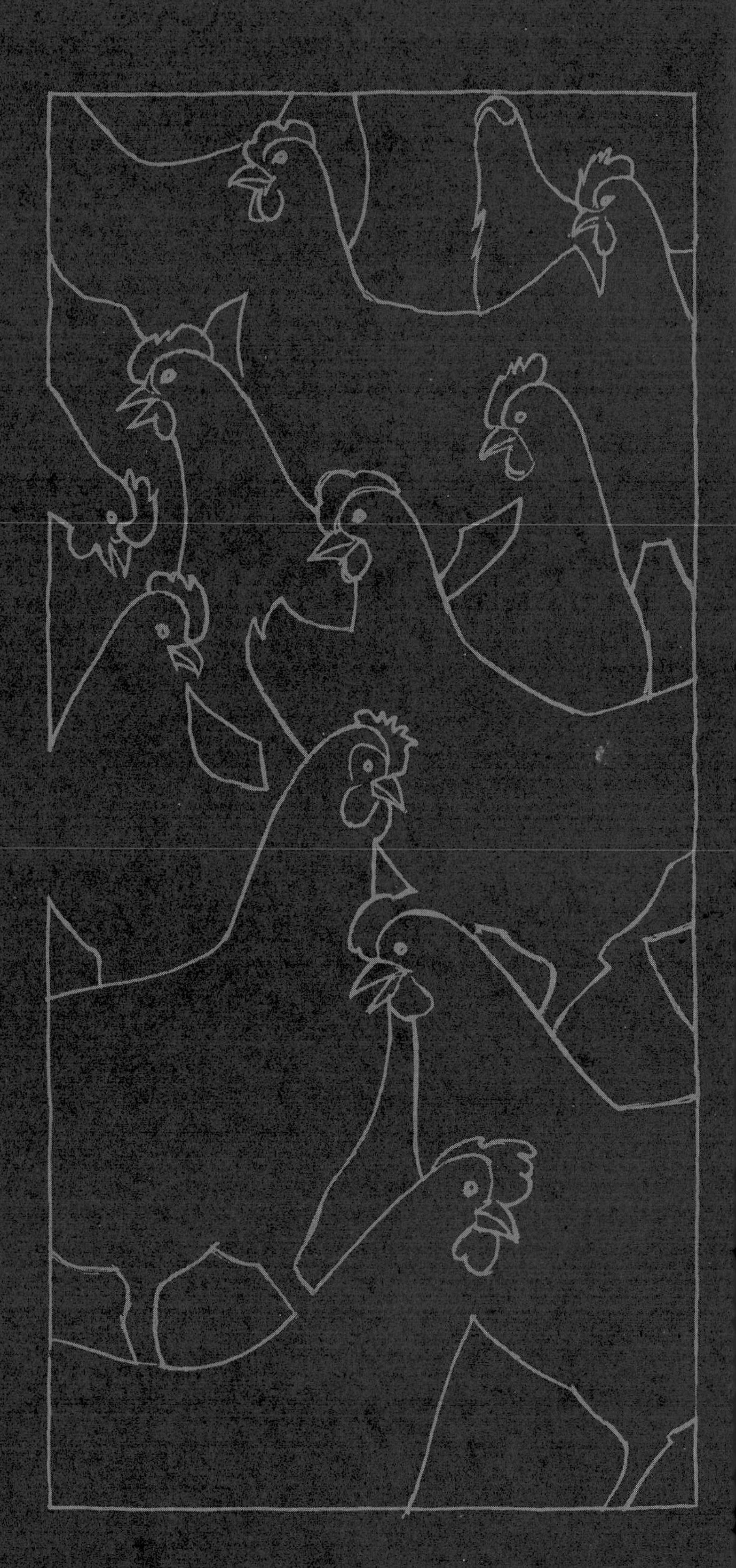

35 The Poultry Industry

Poultry and egg production was at one time a minor farm enterprise. Chickens or other fowl were kept either as a hobby or largely to produce meat and eggs for family consumption. The young couple beginning farming always had a small flock of hens and raised some young chickens to help pay the grocery bill until they had income from other enterprises.

Today, the poultry enterprise is big business. Flock owners consider the enterprise an economic unit which provides a substantial source of income. Chickens are to be found on about 40 percent of the 2.7 million farms in the United States. On January 1, 1972, there were 406 million chickens and 7.7 million turkeys (excluding broilers) on our farms.

INCOME FROM POULTRY AND EGGS

In 1970 poultry and eggs contributed 8.8 percent of our national farm income. Eggs were responsible for 4.4 percent, broilers and farm chickens for 3.2 percent, and turkeys for 1.2 percent of our 1970 farm income. The income from poultry during that year compared favorably with the income from other livestock enterprises: hogs produced 9.2 percent, dairy farming 13.3 percent, and beef cattle 27.7 percent.

The states with the highest percentages of farm income obtained from poultry in 1970 were Delaware, with 56.6 percent; Maine, with 42.6 percent; Georgia, with 34.6 percent; Alabama, with 34.5 percent; and Arkansas, with 31.3 percent. These states did not have the highest total farm production in terms of pounds or dollars.

Leading States in Gross Income from Broilers

The states having the highest gross income from the sale of broilers in 1972 were

FIGURE 35-1. Headquarters and research center of Arbor Acres Farm, Inc., international poultry-breeding firm, is located in Glastonbury, Connecticut. Arbor Acres maintains production farms and hatcheries throughout the U.S. and 23 nations abroad. (*Courtesy* Arbor Acres Farm, Inc.)

FIGURE 35-2. Poultry and egg producers received 54 cents of each consumer's retail food dollar in 1970 which was 7 cents less than in 1960. (*Courtesy* U.S.D.A. Economic Research Service)

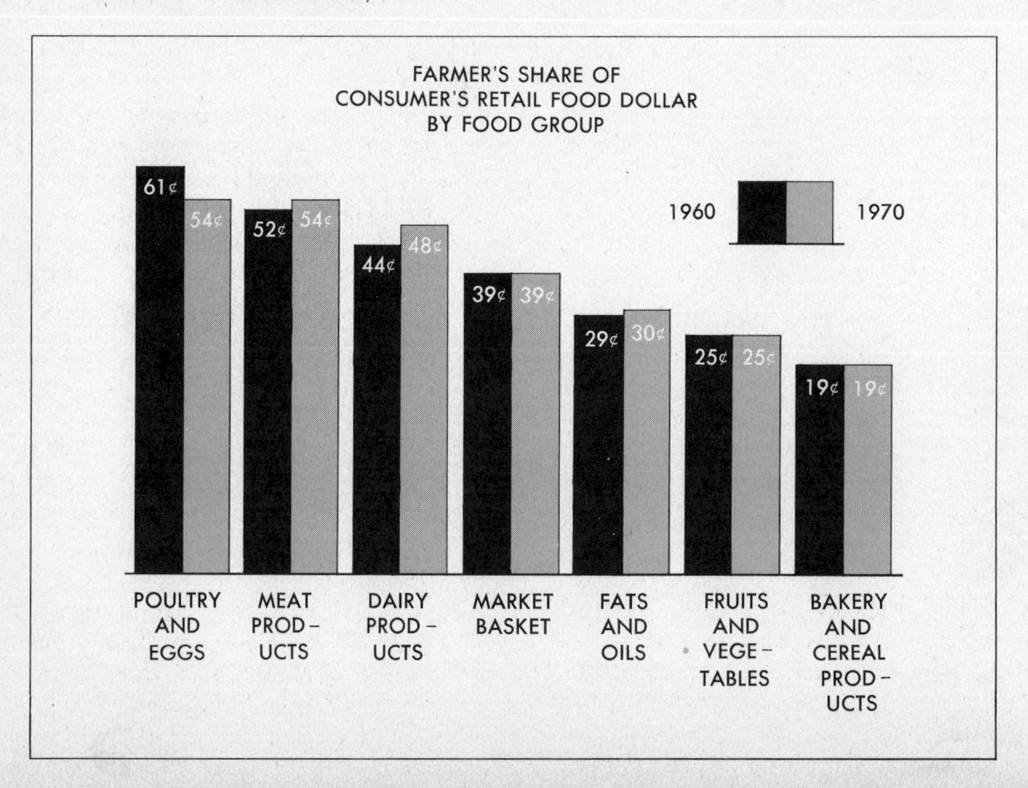

Arkansas, with $255 million; Georgia, with $214 million; Alabama, with $188 million; North Carolina, with $163 million; and Mississippi, with $125 million.

Leading States in Gross Income from Eggs

California led all states in 1971 in the sale of eggs, with $185 million. Georgia ranked second with $163 million. Next in order were North Carolina, with $117 million; Arkansas, with $101 million; and Pennsylvania, with $98 million.

Leading States in Gross Income from Turkeys

Minnesota led all states in the sale of turkeys in 1972 with $73.2 million. California had sales of $73 million; North Carolina had $44 million; Missouri had $42 million; and Arkansas had $35 million.

TRENDS IN POULTRY PRODUCTION

Chickens

In 1940 nearly 95 percent of the farms on which chickens were reported had flocks of 200 birds or less. Flocks of 1,000 or more chickens were reported on less than 1 percent of the farms. The percentage of farms on which chickens are raised is decreasing each year, but the size of flocks is increasing. A larger number of farmers are specializing in poultry farming.

The production of chickens in the United States, excluding broilers, in 1970 amounted to 330 million birds. This was about 363 million birds less than the 1945–1949 average.

California led all states in the number of chickens on farms on January 1, 1971, with 54.8 million; Georgia ranked second, with 40.0 million. The next states in order were Arkansas, with 24.0 million; North Carolina, 22.7 million; Alabama, 19.3 million; and Pennsylvania, 19.1 million.

California, Georgia, Arkansas, North Carolina, and Alabama led the states in the number of nonbroiler chickens raised in 1970. Iowa producers raised 11 million nonbroiler chickens in 1970.

Broilers

Broilers are young chickens, about eight to ten weeks of age, which have been raised for meat production. Broiler production is a comparatively new industry, yet more than 2.9 billion birds were produced in 1972. The production in 1950 was 616 million. Nearly 80 percent of all chicks hatched now become broilers—fried chicken is always popular at the dinner table.

The leading states in broiler production in 1972 were Arkansas, 532 million; Georgia, 443 million; Alabama, 399 million; North Carolina, 302 million; Mississippi, 256 million; and Texas, 179 million.

Turkeys

The production of turkeys in this country has increased rapidly since 1920. During 1920 there were about 4 million turkeys on farms. By 1930 there were 17 million, and by 1940 the number had increased to 33 million. The number of turkeys produced in this country decreased somewhat following World War II, but increased after 1948, and by 1972 approximately 129 million birds were produced.

Turkey is no longer just a Thanksgiving and holiday treat. It is served the year round and competes with beef, pork, mutton, and chicken at the consumer markets.

Minnesota led all states in the production of turkeys in 1972 with 20.8 million birds. California ranked second with 17.6 million, and North Carolina ranked third with 12.0

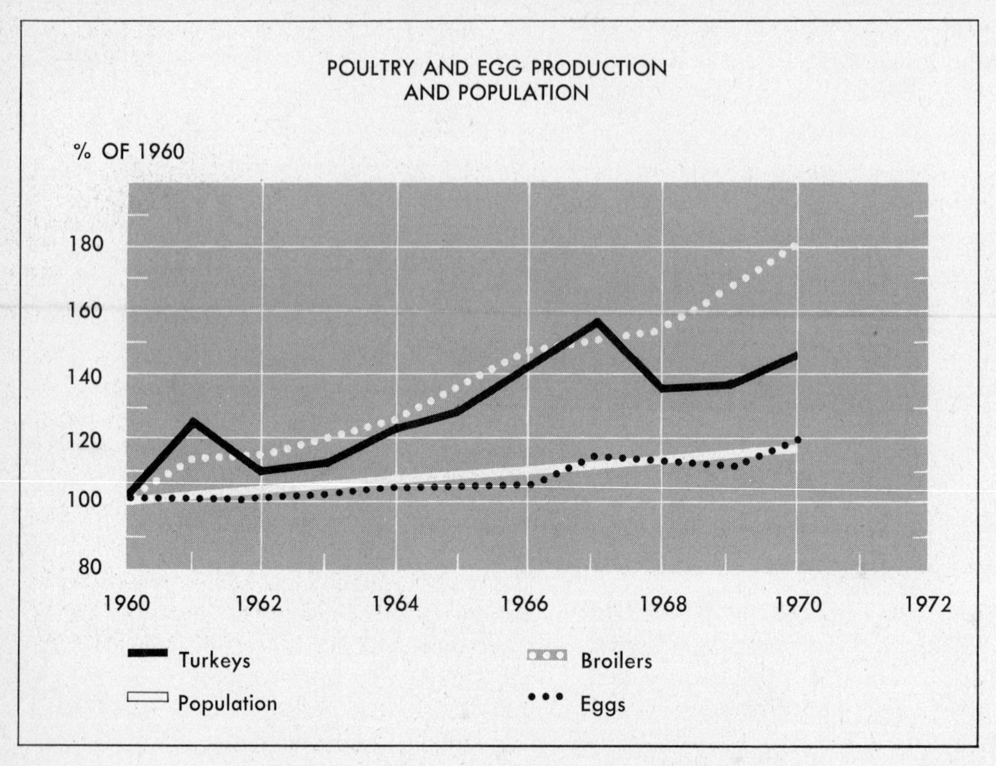
POULTRY AND EGG PRODUCTION
AND POPULATION
% OF 1960
180
160
140
120
100
80
1960
1962
1964
1966
1968
1970
1972
Turkeys
Broilers
Population
Eggs

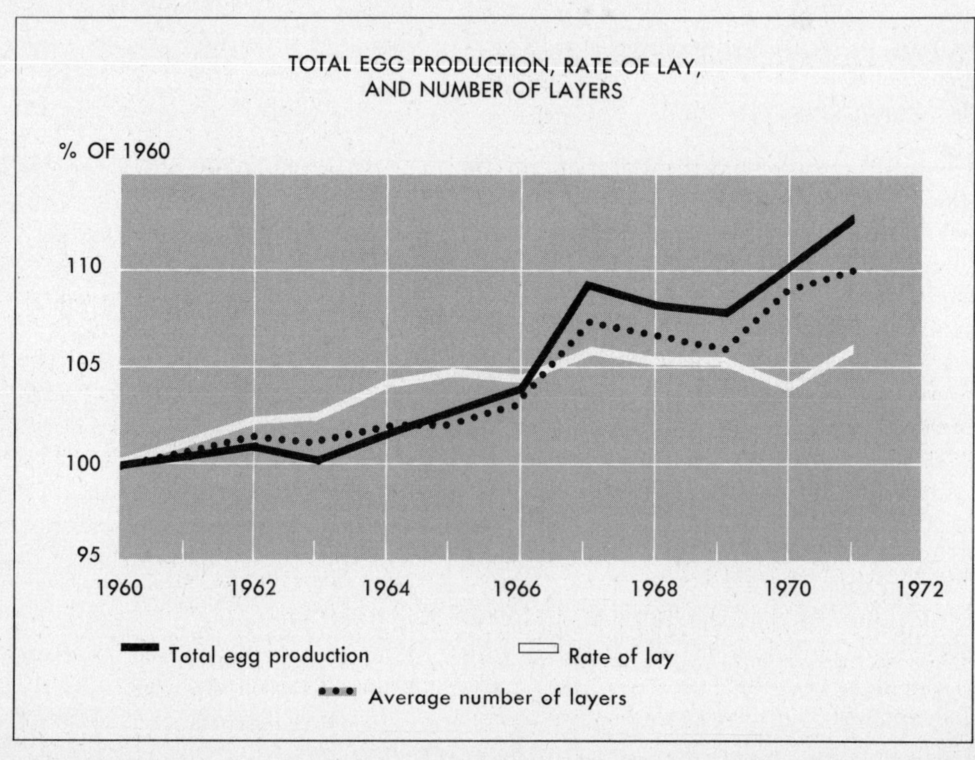
TOTAL EGG PRODUCTION, RATE OF LAY,
AND NUMBER OF LAYERS
% OF 1960
110
105
100
95
1960
1962
1964
1966
1968
1970
1972
Total egg production
Rate of lay
Average number of layers

million. Missouri produced 10.0 million; and Arkansas 8.2 million birds.

Eggs

Hens in this nation up to 1937 produced an average of 120 to 130 eggs per year. Since 1937, the average production per hen has increased annually, with an average production in 1972 of 227 eggs.

The high states in number of eggs produced in 1972 were California, 8.6 billion; Georgia, 5.5 billion; Arkansas, 3.8 billion; Pennsylvania, 3.6 billion; and North Carolina, 3.4 billion. Next high-ranking states were Indiana, Alabama, and Texas.

CONSUMPTION OF POULTRY PRODUCTS

Poultry and eggs rank with milk as the best balanced protein foods for the human diet. They have always been considered essential, but we have not always been able to provide them in the amounts needed by the people of this nation and the world. Refrigeration, improvements in methods of processing, and improved methods of distribution have increased the supply of poultry products.

Per Capita Consumption

Consumption of poultry products during the 1945-to-1958 period was near peak levels. Production of these products has been high and per capita income has also been high. Each year we have more people to feed as the population increases by about 2 million persons.

Chicken and Broiler Consumption. The average consumption of chickens and broilers in 1972 was estimated to be 42.9 pounds per person. Nearly 90 percent of this amount was consumed as broiler meat. During the 1935–1939 period, chicken consumption averaged 16.5 pounds, while only 1.4 pounds of broilers was consumed. Broiler consumption in 1972 was nearly double that in 1960.

Turkey Consumption. The average per capita consumption in the 1930–1934 period was about 2 pounds of turkey each year. In 1972 the average per capita consumption was 9.1 pounds. Turkey production and consumption are still increasing.

Egg Consumption. Egg consumption in the United States averaged 307 eggs per person in 1972. The per capita consumption between 1935 and 1939 was 298 eggs; during the 1947–1951 period, it averaged 389 eggs per person.

OPPORTUNITIES IN POULTRY PRODUCTION

Farmers must always be alert for methods of increasing net farm income. Poultry production on the average farm may be a small, secondary industry, but changes in production and management can bring about sizable increases in farm income. An increase in flock size from 500 to 5,000 hens will greatly increase the net income from the enterprise, providing other factors remain constant. By increasing the average egg production per hen, it is possible to increase further the income from the laying flock. The use of better methods in growing out larger

FIGURE 35-3. (*Above*) Egg production increased 13 percent; turkey production, 44 percent; and broiler production, 80 percent during the 1960 to 1970 period. FIGURE 35-4. (*Below*) The increase in egg production during the 1960 to 1971 period was due largely to the increased number of layers. Layer numbers increased 10 percent, whereas the rate of lay increased only 6 percent. (Figures 35-3 and 35-4 *courtesy* U.S.D.A. Economic Research Service)

numbers of broilers and roasters also presents opportunities for increasing farm income.

Specialized Poultry Farms

On some farms poultry production may be made the major enterprise. According to the 1964 *Census of the United States,* specialized poultry farms comprised 4.6 percent of the total farms of the nation, but they had on them about 33 percent of the nation's chickens, and produced nearly 70 percent of all poultry and poultry products sold, 45 percent of the eggs sold, and 90 percent of the broilers and turkeys raised. This trend has continued. Farms with housing facilities for 40,000 to 100,000 birds are not uncommon in both broiler and egg production programs. Young persons often find poultry raising stimulating and profitable, but a large percentage of our poultry farms are operated by men over 50 years of age. Specialized poultry production apparently attracts men who retire from general farming or from other employment.

Opportunities in Poultry for Youth on the Home Farm

The poultry enterprise on the average farm provides excellent farming program opportunities for 4-H Club and Future Farmers of America chapter members. In some cases it is a small enterprise and rather poorly managed. Parents may be very happy to turn the flock or enterprise over to a son or daughter on a partnership basis. Some excellent poultry enterprises have been developed in this way, and several of our best commercial poultrymen and poultrywomen began in this manner.

Advantages of Poultry Farming

Poultry production as a minor enterprise and as a specialized business has many advantages for the farmer who likes poultry and is industrious. For success a poultryman must be familiar with the best practices of production and marketing, and must use good judgment in managing the enterprise. The following are desirable features of poultry production.

Poultry and eggs are essential foods. The demand for poultry and eggs is permanent. Poultry is appetizing and popular as a meat. Eggs are highly digestable and nutritious. Future demand for high-quality poultry and eggs appears highly promising.

Poultry production provides quick returns. Three-pound broilers can be produced in eight or 10 weeks. Pullets begin laying in four months.

Poultry is efficient to produce. A three-pound broiler can be produced on 7½ to 8 pounds of feed. No other meat product can be produced as efficiently. A dozen eggs can be produced on 4 to 5 pounds of feed.

Poultry income is distributed throughout the year. Income from a laying flock and from broiler production is dependable, and does not come only once or twice during the year.

Poultry production is adapted to acreages and small farms. The average size of specialized poultry farms is less than 70 acres. Chickens, turkeys, and broilers may be produced equally well in backyards in cities and small towns, on acreages, and on large farms.

The poultry industry provides for the quick turnover of capital. Money invested in broiler production is returned in less than three months, and investments in a laying flock start coming back in about four months. Income from turkeys is somewhat slower, depending on the weight of the birds produced.

Poultry markets are standardized and well established. Eggs and poultry are sold according to grade, and many avenues are available for marketing poultry products.

Poultry nets high returns per $100 worth of feed used. According to Illinois Farm

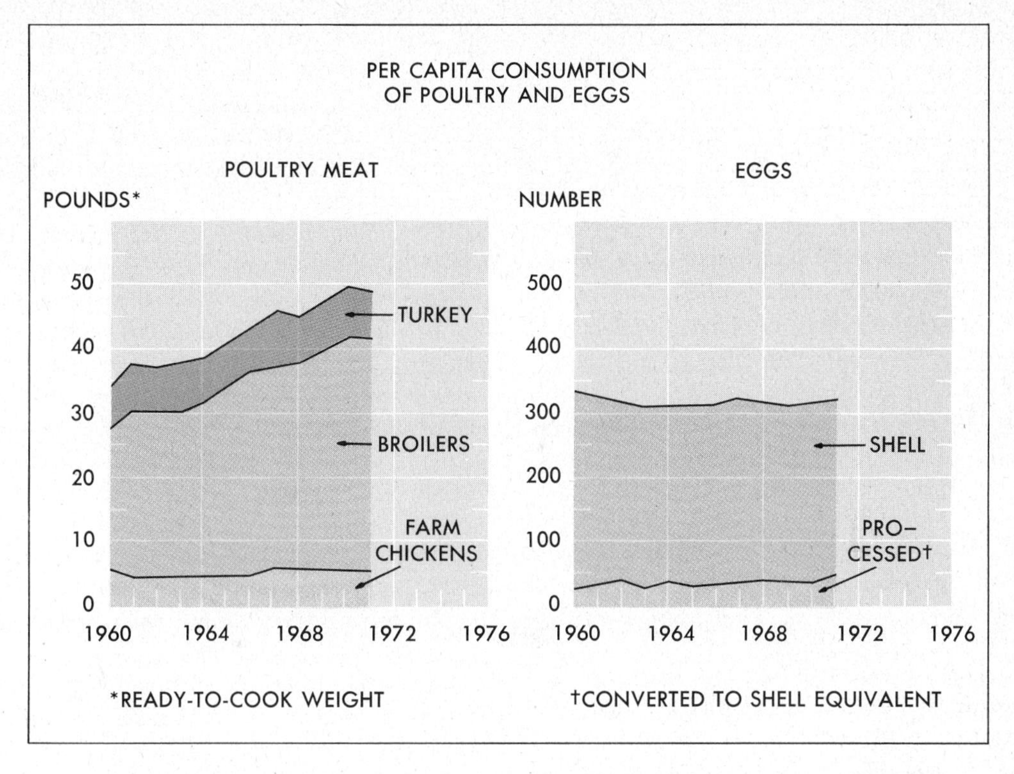

FIGURE 35-5. Per capita consumption of farm chickens and eggs decreased during 1960 to 1971. Broiler consumption increased from 28 to 42 pounds, and turkey consumption from 35 to 50 pounds. (*Courtesy* U.S.D.A. Economic Research Service)

Bureau Farm Management records, poultry enterprises on the Illinois farms studied returned $185 for every $100 invested in feed over a three-year period, 1968–1970. Only dairy cattle returned more income per $100 worth of feed. Dairy cattle returned $205.

Poultry production may be a full-time or part-time occupation, and on most farms and acreages, poultry is a part-time activity. On specialized farms, it may require all of the time of one or more persons. A sizeable income can be obtained from poultry, even though the individual producer also has regular full-time employment.

Producing poultry is a science which involves knowledge and proven practices. Much experimentation has taken place in poultry breeding, feeding, management, and marketing. An individual has information available to plan profitable production and management programs.

Young persons and elderly people can be successful in poultry production. The work is not strenuous and does not require great skill, but few farm enterprises can compete with poultry as a source of regular, substantial income.

Problems in Poultry Production

The production of broilers, chickens, turkeys, and eggs involves many problems. Certain limiting factors are apparent.

Diseases and parasites may cause problems. While chickens and turkeys are normally healthy, they are subject to many ailments, and losses may be heavy if proper methods of prevention and treatment have not been followed.

Problems may occur in feeding and management. Feed costs may become excessive and the enterprise unprofitable if the birds are not fed balanced rations in proper amounts. Housing, ventilation, and labor problems may also affect profits.

Improper selection of chicks, poults, and mature birds can ruin the enterprise. It takes good seed stock to have profitable production. Chicks and poults must be selected according to the use to be made of them, and they must be bought carefully. The laying flock must be carefully culled, and birds used in breeding pens must be carefully selected.

Other problems may arise in the marketing of poultry products. The profit from the poultry enterprise is influenced by demand and change in the market. Eggs, broilers, and turkeys bring more money during some seasons than others.

The size of the enterprise may have an undesirable effect on the poultry farmer. The fact that a chick, hen, or turkey is short-lived and requires only a small investment may cause the producer to become careless. The larger the unit, the easier it is to give the enterprise the care and management it should receive.

Finance is very important, as specialized poultry farming requires a large volume for efficient production.

Surplus production may cause marketing problems. Improvement in production efficiency may bring about surpluses, and producers must control production in order to maintain a satisfactory market.

Quality is, of course, a major consideration. Poultry products must compete with other products for the consumer food dollar, for there is no market for poorly produced or inferior products.

SUMMARY

Nearly 9 percent of our national farm income is derived from the sale of poultry products. This percentage compares favorably with the percentage of income from dairy farming and exceeds the income from hog production. In some New England states, poultry brings in as much as 40 to 60 percent of farm income. Georgia, California, Arkansas, and North Carolina led all states in total income from chickens, eggs, broilers, and turkeys in 1972.

In 1972 nearly 3 billion broilers and 330 million chickens were produced in this country. California and Georgia led all states in the number of chickens on farms on January 1, 1971. Broiler production has increased rapidly during the past few years, especially in Arkansas, Georgia, and Alabama.

Turkey production has increased from 4 million in 1920 to 129 million in 1972. Minnesota, California, and North Carolina were the high production states in 1972.

Egg production has increased from about 120 eggs per hen in 1937 to 227 per hen in 1972. California, Georgia, and Arkansas led the states in value of eggs produced in 1972.

About 42.9 pounds of chickens and broilers, 9.1 pounds of turkey, and 307 eggs per capita were consumed in the United States in 1972.

Poultry production is big business. Chickens are raised on about 40 percent of our farms. Poultry production offers excellent opportunities for ambitious young persons and for elderly people.

Poultry production has many advantages. Chickens and eggs are staple foods. Chickens are efficient converters of feeds into meat. The returns from the enterprise are rapid, as is the capital turnover. The income is distributed throughout the year, and the enterprise can be adapted to large or small farms. Poultry production may be a full- or part-time occupation.

Raising chickens and turkeys presents management problems related to stock selection, feeding, housing, disease control, brooding, management, and marketing. Knowledge is available, however, regarding proven practices. Poultry production can be a very profitable enterprise.

QUESTIONS

1 What percentage of your home farm income is derived from the sale of (a) chickens, (b) broilers, (c) turkeys, (d) eggs?

2 What states lead in the production of (a) chickens, (b) broilers, (c) turkeys, (d) eggs?

3 How much feed does it take to produce a gain of a pound in a broiler?

4 How much feed does it take to produce a dozen eggs?

5 What changes have come about in the production of poultry products during the past 20 years?

6 Is poultry production increasing or decreasing in your community? Why?

7 What are the advantages of poultry production over swine, sheep, dairy, and beef production?

8 What problems have you encountered in producing poultry products on your farm?

9 What should be the place of poultry production on your farm?

REFERENCES

Bundy, C. E. and R. V. Diggins, *Poultry Production,* Prentice-Hall, Inc., Englewood Cliffs, New Jersey, 1960.

Card, Leslie E. and M. C. Nesheim, *Poultry Production,* 10th Edition, Lea & Febiger, Philadelphia, Pennsylvania, 1966.

U. S. Department of Agriculture, *Livestock and Poultry Inventory,* Washington, D.C., February 1, 1971.

———, *Poultry and Egg Situation,* Washington, D.C., February, June, and September 1973.

———, *Selected Statistical Series for Poultry and Eggs Through 1968,* ERS 232 Revised, Washington, D.C., 1970.

36 Selection of Chicks and Birds for Production

The successful producer is careful in choosing chicks and birds for a flock, for these chicks and pullets are the foundation for the enterprise. Good management may compensate in part for poor stock, but for maximum profits you must begin with good stock.

SELECTING A BREED

There are some breeds of cattle which are kept primarily for milk production and some which are kept for the production of meat. Similarly, there are numerous breeds, varieties, and crosses of poultry. Some are kept primarily for egg production, while others are kept for meat.

Classes of Chickens

There are many classes of chickens, but most of the chickens produced in this country are produced from breeding stock of the following four: (1) American, (2) Mediterranean, (3) English, and (4) Asiatic.

American Class. The Plymouth Rock, Wyandotte, Rhode Island Red, New Hampshire, and Jersey Black Giant are the most popular of the American class. Other breeds in this class are the Rhode Island White, Java, Dominique, and Holland.

No birds in this class have feathered shanks. They all have yellow shanks and skin, and red earlobes. The breeds of the American class have been bred for both meat and egg production.

Mediterranean Class. The Leghorn, Minorca, and Ancona breeds are the most popular of the Mediterranean class in this country, but other breeds of this class, such as the Blue Andalusian, Buttercup, and Spanish, are also grown. The Mediterranean breeds are kept primarily for egg production. They are smaller than the American, English, and Asiatic breeds, and have white earlobes. They all lay white eggs.

English Class. The most popular breeds of this class are the Cornish, Australorp, and

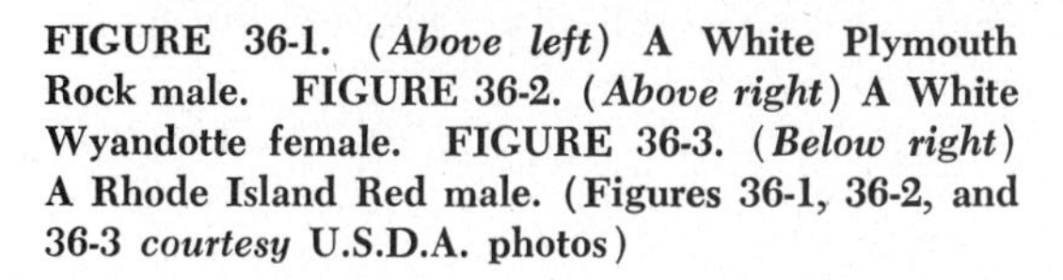

FIGURE 36-1. (*Above left*) **A White Plymouth Rock male. FIGURE 36-2.** (*Above right*) **A White Wyandotte female. FIGURE 36-3.** (*Below right*) **A Rhode Island Red male. (Figures 36-1, 36-2, and 36-3** *courtesy* **U.S.D.A. photos)**

Orpington. Three other breeds are grown—the Dorking, Sussex, and Red Cap. The English breeds have good meat. They are large and, with the exception of the Cornish, have white skin.

Asiatic Class. The three breeds of this class, the Brahma, Cochin, and Langshan, are not grown in large numbers in this country, but have been used in producing the American breeds. All three breeds have feathered shanks, are large and heavy-boned, and have red earlobes. With the exception of the Black Langshan, they have yellow skin. The Asiatic breeds are raised largely for meat production.

Breeds and Varieties of Chickens

The characteristics of the most common breeds of chickens are shown in Table 36-1. Some breeds are subdivided into varieties according to color of feathers and type of comb. There are seven varieties of Plymouth Rocks, the barred, white, buff, silver penciled, partridge, Columbian, and blue. Wyandottes have eight varieties classified according to color of feather. There are four varieties of Orpingtons, the buff, black, white, and blue. The Leghorn and Minorca breeds are subdivided into varieties according to both color and type of comb. White and buff are the most popular colors.

Nearly 200 varieties of chickens are listed in the *American Standard of Perfection*, but relatively few are of commercial importance.

Hybrid and Crossbred Chickens. Considerable work has been done in the production of crossbred or hybrid chickens. Agricultural experiment stations and com-

FIGURE 36-4. (*Left*) **A New Hampshire male. FIGURE 36-5.** (*Right*) **Barred Plymouth Rocks. (Figures 36-4 and 36-5** *courtesy* **U.S.D.A. photos)**

TABLE 36-1 CHARACTERISTICS OF COMMON BREEDS OF CHICKENS

Breed	Weight (Pounds)	Skin Color	Shank Color	Shanks Feathered	Type of Comb	Earlobe Color	Egg Color
American Breeds:							
Jersey White Giant	10 –13	Yellow	Yellow	No	Single	Red	Brown
New Hampshire	6½–8½	Yellow	Yellow	No	Single	Red	Brown
Plymouth Rock	7½–9½	Yellow	Yellow	No	Single and rose	Red	Brown
Rhode Island Red	6½–8½	Yellow	Yellow	No	Single and rose	Red	Brown
Wyandotte	6½–8½	Yellow	Yellow	No	Rose	Red	Brown
Mediterranean Breeds:							
Ancona	4½–6	Yellow	Yellow	No	Single and rose	White	White
Leghorn	4½–6	Yellow	Yellow	No	Single and rose	White	White
Minorca (white)	6½–8	White	White	No	Single	White	White
English Breeds:							
Australorp	6½–8½	White	Dark slate	No	Single	Red	Brown
Cornish (white)	8 –10	Yellow	Yellow	No	Pea	Red	Brown
Orpington (buff and white)	8 –10	White	White	No	Single	Red	Brown
Asiatic Breeds:							
Brahma (light)	9 –11	Yellow	Yellow	Yes	Pea	Red	Brown
Cochin	8½–11	Yellow	Yellow	Yes	Single	Red	Brown
Langshan (black)	7 –10	White	Bluish-black	Yes	Single	Red	Brown

mercial poultry breeders have developed inbred lines, and have made line crosses similar to those in the production of hybrid corn. The use of hybrid and crossbred chickens has been on the increase, in the production of both eggs and broilers.

Most farmers rely on the commercial hatcheryman to maintain the breeding flocks and make the crosses necessary to produce hybrid or crossbred chicks. The lines and crosses available vary with the community and the section of the country.

Eighty-eight percent of the hatchery supply flocks enrolled in the 1970 National Poultry Improvement Plan produced cross-mated or incross-mated birds. About 8 percent produced White Leghorn, 3 percent had White Rocks, and fewer than 1 percent produced New Hampshires, Barred Rocks, or Rhode Island Reds.

Eleven of the 12 entries in the 1970-1971 New Jersey Random Sample Egg Laying Test were White Leghorns. Most chickens used for production of eggs and meat are crosses. Single-comb White Leghorns are most popular in producing strain crosses and hybrids for eggs. The White Rock is widely used as the female parent of crossbred broilers. White Cornish males are generally used on White Rock, New Hampshire, or cross-mated females in broiler production. It is estimated that at least 90 percent of all hatchery flocks are cross-mated.

Popularity of the Breeds. Some breeds and varieties of chickens are more popular in certain areas than in others. An indication of the popularity of the breeds is shown in the summaries of entries in the official poultry production tests conducted in this nation, and in the list of breeds represented in the National Poultry Improvement Plan.

BUYING BABY CHICKS

Most poultrymen buy their chicks. Less than 1 percent of the chicks produced in the nation in 1965 was hatched on farms. Poultry-

FIGURE 36-6. Hubbard White Mountain male. (*Courtesy* Hubbard Farms)

FIGURE 36-7. Hubbard Meat Breeder pullet. (*Courtesy* Hubbard Farms)

FIGURE 36-8. (*Left*) A single-comb White Leghorn male. FIGURE 36-9. (*Right*) A White Minorca. (Figures 36-8 and 36-9 *courtesy* U.S.D.A. photos)

TABLE 36-2 DISTRIBUTION OF BIRDS IN RANDOM SAMPLE EGG PRODUCTION TESTS, U.S. AND CANADA, TWO-YEAR SUMMARY, 1969–1970, 1970–1971

Breeding	Number of Strains	Number of Pens
White Leghorn, strain cross	22	994
White Leghorn, incross	3	103
White Leghorn, pure strain	2	50
White Leghorn, incrossbred	2	20
White Leghorn × Black Australorp, crossbred	1	14
Rhode Island Red × Barred Plymouth Rock, crossbred	3	44
Rhode Island Red × White Plymouth Rock, crossbred	3	44
Rhode Island Red, strain cross	3	55
Rhode Island Red × Rhode Island White, crossbred	1	16
White Plymouth Rock × Rhode Island Red, crossbred	1	4
California Grey × White Leghorn, crossbred	2	59
Synthetic × New Hampshire, crossbred	1	40
Synthetic × White Leghorn, crossbred	1	85
Synthetic × White Leghorn, strain cross	1	23
Incrossbred	2	48
Strain cross	1	4
Total	49	1,603

Source: U.S.D.A., *1971 Report of Egg Production Tests, United States and Canada, March, 1972*

men rely on the commercial hatcheries for their chicks, for the hatcheries are able to maintain better breeding flocks and can produce healthier chicks at a lower cost than can be done by an individual producer.

The hatchery business has become a large industry. More than 980 hatcheries were in operation in the nation in 1973.

Arkansas, Georgia, Alabama, North Carolina, and Mississippi led the states in the number of chicks hatched in 1972. Of the 3,267 million broiler-type chicks hatched in the nation in 1972, 63 percent were hatched in these five states.

Chicks are raised each year to replace the laying flock, and for broiler production. While the use to be made of chickens may vary, much the same factors are involved in selecting and buying chicks. They are: (1) source, (2) time to order, (3) breeding and quality, (4) sex, (5) age, and (6) price.

Source

It is best to buy chicks from hatcheries close to home, where they can be picked up

FIGURE 36-10. (*Above*) Dekalb White Egg Layer. (*Courtesy* Dekalb Ag Research, Inc.) FIGURE 36-11. (*Below left*) Harco Sex-Link, a widely used brown egg commercial layer. (*Courtesy* Arbor Acres Farm, Inc.) FIGURE 36-12. (*Below right*) California Gray Cockerel. A heavy breed type, white egg-laying breed. (*Courtesy* Welp's Breeding Farm)

TABLE 36-3 DISTRIBUTION OF BIRDS IN NATIONAL POULTRY IMPROVEMENT PLAN HATCHERY FLOCKS, 1950–1970

Breed	1950 (Percent)	1955 (Percent)	1960 (Percent)	1965 (Percent)	1970 (Percent)
New Hampshire	38.8	16.0	1.7	0.5	0.1
White Leghorn	21.9	19.2	15.5	12.2	8.6
White Rock	10.4	29.5	7.2	5.0	2.9
Barred Rock	6.3	1.3	0.5	0.6	0.1
Rhode Island Red	4.3	2.7	1.0	0.7	0.3
Cross-mated	15.1	21.2	66.2	73.5	82.4
Others †	3.2	10.1	7.9	7.5	5.6

† Includes incross-mated.
Source: U.S.D.A., *Agricultural Statistics, 1971*

personally, if other factors are equal. Chicks can be transported long distances, but losses are usually lighter over short distances. It is easier to get an adjustment in case of loss when you buy close to home, and you are better able to determine the reputation of the hatchery.

Reputation and Reliability of Hatchery. The reputation of a hatchery spreads rapidly from one person to another. Before ordering chicks, check with other poultrymen concerning the reliability of the various hatcheries. Hatcheries which are supervised by the official state agency of the National Poultry Improvement Plan are usually good sources.

Shipping Chicks. Chicks should be placed in the brooder and fed within 24 to 48 hours after the hatching time. Government regulations limit to 72 hours the time chicks can be in transit. Choose a hatchery which can transport the chicks in as short a time as possible. Chicks should be shipped in ventilated boxes, with no more than 25 chicks in a section.

FIGURE 36-13. Approximately 100,000 eggs a week are gathered, processed, and incubated in this hatchery. (*Courtesy* U.S.D.A. photo)

Time to Order

The time to order chicks will vary with the hatchery, but it is usually best to place your order several weeks or months before you wish delivery. The best hatcheries have orders booked several weeks in advance.

Early Hatched Chicks Make the Most Money. Broilers hatched in November, December, March, April, and May can usually be marketed during February, March, May, June, and July, when broiler prices are high. Early hatched pullets can be brought into production before egg prices reach their peak in the fall. February and March chicks are best as laying flock replacements. Allow

FIGURE 36-14. (*Above*) An adult breeding flock in commercial broiler production. (*Courtesy* Pilch's Poultry Breeding Farms, Inc.) **FIGURE 36-15.** (*Below*) A pen of single-comb White Leghorns bred by J. A. Hanson and Sons of Corvallis, Oregon. This was the high pen in a Connecticut Standard Egg Laying Contest. (University of Connecticut photo. *Courtesy Poultry Tribune*)

a few more weeks for heavy breeds to mature than for lighter breeds.

Breeding and Quality

The most important factor in buying chicks is to get healthy chicks of desirable breeding.

Cheap Chicks Are Usually Expensive. The few cents difference between the price of a good chick and that of an inferior one may result in as much as a dollar or more increase in egg production per bird, or the difference between profit and loss in raising broilers.

Buy Chicks from Pullorum-free Flocks. Check with the hatcheryman concerning the health of the flocks. Pullorum may be communicated from the hen to the chick through the egg. Flocks supervised by the state agency of the National Poultry Improvement Plan are preferred.

Check the Production Records of the Parent Stock. Certain strains of the various breeds of chickens have been proven as egg producers or as good broiler producers. Find out everything possible about the birds used in the hatchery flocks. Visit the flocks if possible. Check broiler production and egg-laying test records. Buy chicks that are bred to produce.

Sex

The use to be made of the chicks and the price determine the sex to buy.

Straight-run Chicks. Chicks as they come from the incubator are straight-run chicks. The number of males is about equal to the number of females. Some farmers like to buy straight-run chicks. They feed out the cockerels as broilers and keep the pullets for the home flock. In raising straight-run chicks, about three chicks should be purchased for every pullet that is to be placed in the laying house.

Sexed Chicks. It is very often advantageous to buy sexed chicks. Leghorn cockerels are not usually profitable as broilers. Producers desiring pullet replacements often prefer to pay nearly twice as much per chick for sexed pullets than they would have to pay for straight-run chicks. The price spread between straight-run and sexed heavy breeds is not as wide as for Leghorns.

Methods of Sexing. Efficient chick sexers can determine the sex of chicks without injuring them. Three methods are in use. Some crosses may be sexed by the color of down on the newly hatched chicks, while the sex of others may be determined by the length of primary wing feathers or by the number of secondary wing feathers. The third method involves an examination of the rudimentary sex organs of the chick.

Age

Some purchasers like to buy started chicks which are from two to four weeks old. Chick losses usually occur during the first week or two. By buying started chicks, the poultryman may avoid these losses. The price of started chicks is usually high, but there are times when it is more practical, convenient, and profitable to buy them.

Price

When all other factors are equal, price may be the determining factor in buying chicks. Usually, however, other factors are not equal. Consider price in buying chicks, but remember that money invested in high-quality chicks brings in a better income than capital invested in poor-quality chicks.

CULLING THE FLOCK

Feed costs make up 50 percent or more of the egg-production cost. In a three-year study based on Missouri poultry records, hens that laid an average of 225 eggs each paid their owner a labor income of $2.86 each. Hens that laid an average of 124 eggs produced a labor income of only 22 cents each. When feed costs are high, the importance of culling is magnified. Every nonproductive and inefficient bird in a flock lowers the profit from the enterprise. Each hen in the average flock must lay 200 or more eggs for the producer to come out even. During some seasons, the production must be higher to avoid losses. One of the best means of increasing egg production per hen is to cull out the nonlaying boarders.

The amount of culling to be done will depend on the value of the birds to be culled.

It is sometimes more profitable to keep birds that lay fewer than 200 eggs per year than to cull them, since the loss in value of the birds may be greater than the cost of keeping them for the eggs produced. A 4½-pound layer may have cost $1.75, but be worth only 60 cents as a stewing hen. The owner would lose approximately $1.15 in culling the bird, an amount equal to about half of the cost of feeding the bird for one year.

Culling Chicks. Chicks and broilers need culling as well. There is no need to waste feed by keeping inferior birds. Crippled, weak, and runty chicks should be disposed of when chicks are placed in the brooder house. It does not pay to keep cull chicks. They seldom make a profit and they may become carriers of disease.

Culling Pullets. Pullets should be culled carefully as they are placed in the laying house in the fall. Only well-grown, healthy, mature birds should be kept. Birds which are undersized, deformed, or diseased should be culled out. Keep only pullets which have attractive heads, bright eyes, good beaks, and bright feathers.

Culling the Laying Flock. Most poultrymen try to maintain flocks with 60 to 70 percent average egg production for the year. They want 60 or 70 eggs from every 100 hens each day. To do this they must watch the flock closely and remove the nonlayers. Culling is not a practice to be done once a year when the pullets are placed in the laying house. It is a continuous process. Some culling should be done each week or month. The longer you leave the nonlayer in the flock, the greater the loss in feed.

Because the birds are easily caught when on the roost, some poultrymen do their culling at night with the aid of a flashlight. The culled birds are placed in a coop where they can be examined more carefully the following morning.

Culling Formula. Dr. N. R. Mehrhof of the University of Florida has developed a formula which gives the percentage of laying below which hens should be culled. The formula is as follows:

$$\frac{\text{Price of 100 lbs. of feed}}{\text{Price of 1 dozen eggs}} \times 3 = \text{Percent production necessary to pay feed bill}$$

Using the formula, if feed is $3.20 per 100 pounds and eggs are 29 cents per dozen, a producer would need a 33 percent rate of lay to pay the feed bill. Since feed costs represent about 50 percent of total production costs, another 33 percent rate of lay will be needed to pay other costs.

How to Select Good Layers

It is not difficult to select the layers from the nonlayers. Anyone familiar with chickens can become quite skilled in culling practices in a rather short time. Selection and culling are usually based on (1) condition of the comb and wattles, (2) brightness of the eyes, (3) body capacity, (4) handling quality, (5) condition of the vent, (6) amount of pigment, (7) stage of molt, and (8) health and vigor.

Condition of Comb and Wattles. The comb of a layer is large, red, and glossy. It is warm because of good blood circulation. The comb of a nonlayer is small, pale, dry, stiff, and scaly. The head of a nonlayer may be puffed or swollen. The wattles of layers are large, bright red, and warm. The wattles shrink and become pale when the hen goes out of production.

Brightness of Eye. A good layer has large, prominent, and bright eyes. Poor layers have small, sunken, or dull eyes. The eye is a good indication of health. Birds with gray eyes or irregularly shaped pupils should be culled. They may be infected with leukosis.

Body Capacity. To be a good layer, a bird must have a big body. Layers consume

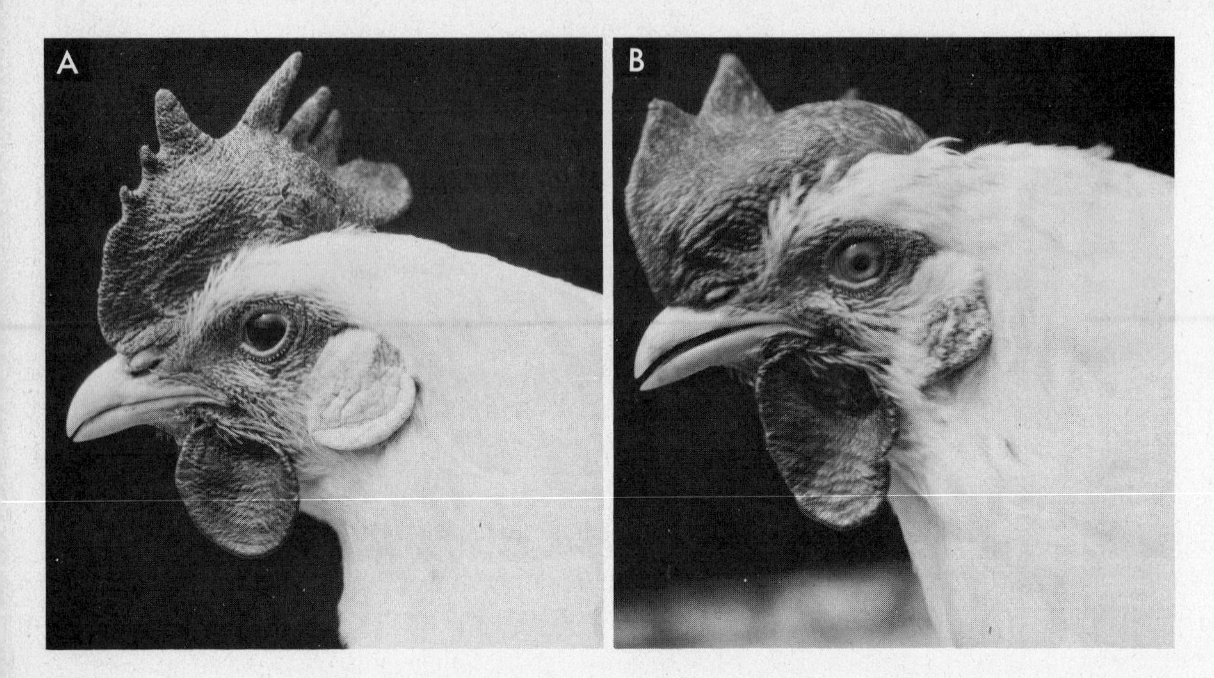

FIGURE 36-16. (*A*) The head of a good layer.(*B*) An undesirable head. (*Courtesy Poultry Tribune*)

FIGURE 36-17. (*Upper arrows*) Measuring the space between the pubic bones. There is room for three fingers between the pubic bones of the good layer on the left, whereas only two fingers can be placed between the pubic bones of the poor layer on the right. (*Lower arrows*) Measuring the depth of the abdomen. Four fingers can be placed between the pubic bones and the breastbone of the good layer on the left. There is room for only two fingers between these bones on the poor layer on the right. (*Courtesy Poultry Tribune*)

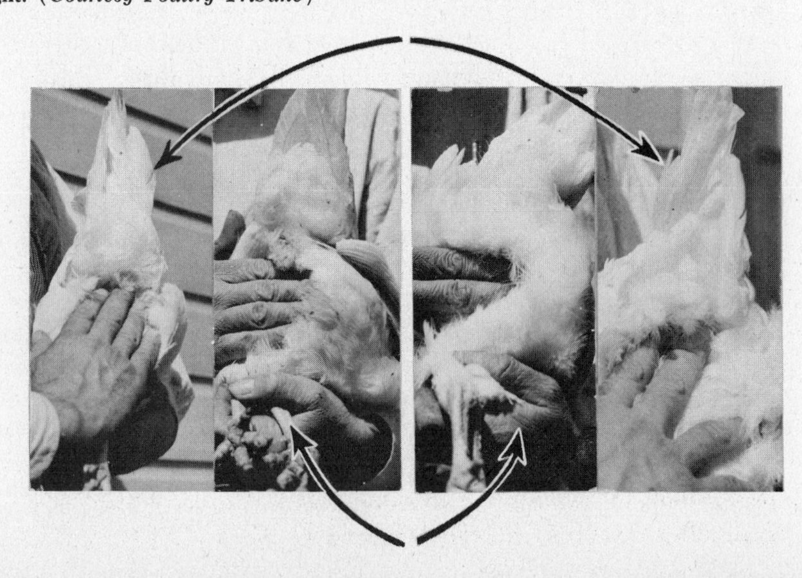

large quantities of feed and need space in the body cavity for the production of eggs. The abdomen of a layer is expanded. The width of the abdomen is determined by the width between the pubic bones. These bones are on either side of the vent. There should be room for two to four fingers between the pubic bones. The pubic bones in nonlayers are close together. There may be room for only one finger between them.

The depth of the abdomen is measured in terms of the number of fingers which can be placed between the pin, or pubic bones, and the breast bone, or the keel bone. There is room for three or four fingers between these bones on a good layer, while there is space for only one or two fingers on the nonlayer.

Handling Quality. The body of a layer is soft and pliable, while that of a nonlayer is hard and contracted. The layer has little fat on its abdomen, while that of the nonlayer may be fatty. Quite often the pubic bones of nonlayers are thickened because of fat.

Condition of the Vent. The best way to tell whether a hen is now laying is to examine the vent and the distance between the pubic bones. The vent of a layer is large, moist, and oblong. The vent of a nonlayer is small, dry, and round.

Amount of Pigment. Many of the common breeds of chickens have yellow skin due to yellow pigment. The color is also evident around the eyes, on the beak, around the vent, and on the shanks. As pullets go into production, this pigment and the pigment in feeds go to produce the yolks of the eggs. The heavier the production, the more pigment that is removed from the bird. An examination of the pigment is an excellent method of telling how long a hen has been laying, or how long she has been out of production.

When a hen begins to lay, the pigment leaves the vent rapidly. No pigment will remain after five or six eggs have been laid. A

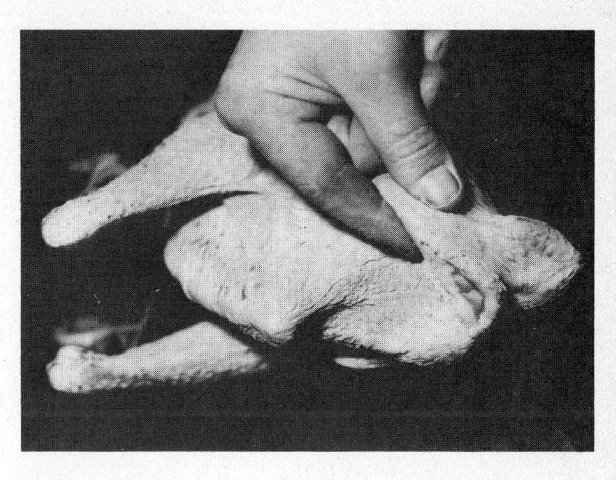

FIGURE 36-18. The abdomen of a layer is loose and pliable. The vent is large, moist, and oblong. (*Courtesy Poultry Tribune*)

yellow vent is a sign of a nonlayer. A white vent in a yellow-skinned breed indicates that the hen is a layer.

The pigment begins to leave the eye-ring shortly after it leaves the vent. After the second week of production, there usually is no pigment remaining in the eye-ring. The pigment in the earlobes of the Mediterranean breeds leaves a little later.

It takes a longer period of time for the pigment to leave the beak and shanks. The beak begins to fade at the face, and is fully bleached after five to seven weeks of production. The shanks lose their color more slowly than the beak, and are not completely bleached until the hen has been in production from four to six months.

Hens which continue to show yellow pigment should be examined carefully. Sometimes they can be culled on the basis of pigment alone, but it is usually best to consider other factors before marketing the bird.

Stage of Molt. Chickens grow new crops of feathers each year. Poor layers shed their feathers, or molt, slowly. Good layers molt rapidly. Poor layers usually begin to molt early in the year, while good layers do not

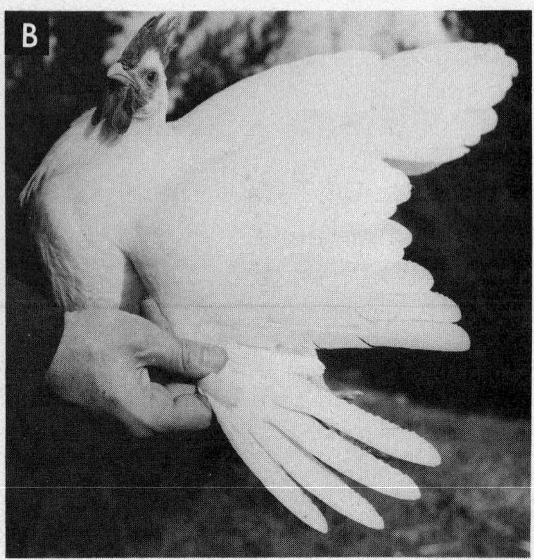

FIGURE 36-19. (*A*) The wing of a hen previous to the start of the molting period. (*B*) The wing of a hen in process of molting. (*Courtesy* Poultry Science Department, Oklahoma State University)

molt until late fall. Poultrymen who keep their pullet flocks for a second year of production can use molt as a means of selecting the good layers.

An examination of the feathers on a hen will indicate the stage of molt. Feathers are shed in the following order: (1) head, (2) neck, (3) breast, (4) body, and (5) wings and tail.

Pullets put into the laying house in the fall should not molt until late the following fall. Pullets which begin molting in May or June should be sold, since they go out of production during the molting period and it takes about three or four months to molt.

The length of time a hen has been out of production can be estimated by counting the number of new primary feathers in the wing. There are ten of these feathers in each wing. The feathers are shed from the axial feather, which separates the primary from the secondary wing feathers, at two-week intervals. Slow molters shed one feather at a time. Rapid molters may shed two or three feathers at a time. It takes about six weeks to grow a new feather. By adding the six weeks required to produce the first new feather and two weeks for each additional new feather, it is possible to calculate the length of time the bird has been out of production. A hen with three new feathers has been out of production for 10 weeks.

Producers with healthy, high-producing flocks may force them to molt and keep them for a second year's production. This can be done by reducing light, reducing feed, or a combination of the two. On January 1, 1971, 32 percent of the layers in California had been force-molted.

Poultrymen who maintain pullet flocks can make very little use of molt in culling the flock, but it is an important factor in selecting hens to hold over for a second year of egg production.

Health and Vigor. A good layer is well developed and has size and body capacity, as well as bright eyes and a clean-cut face. She is active, carries herself well, and appears

TABLE 36-4 GUIDE FOR SELECTION AND CULLING OF LAYERS

Characteristics	Select	Cull
Health and vitality	Vigorous, active, good capacity	Weak, sluggish, undersized, lacking capacity
Comb and wattles	Full, smooth, a glossy bright red	Shrunken, dry, dull, pale, scaly
Eyes	Prominent, keen, sparkling	Sunken, listless
Vent	Large, smooth, moist elliptical in shape	Small, puckered, dry, round
Pubic bones	Thin, flexible, well spread	Thick, hard, close together
Abdomen	Soft, pliable, expanded, covered with thin velvety skin	Contracted, firm, covered with thick, coarse skin
Pigmentation	Bleached vent, eye-ring, earlobe, beak, shanks	Yellow pigment in vent, eye-ring, earlobe, beak, shanks
Molt	Late, rapid	Early, slow

Source: Florida Bulletin 149.

vigorous. Small, weak hens with little body capacity or poor heads should be culled. Table 36-4 is a guide for the selection and culling of layers.

SUMMARY

Most chickens produced in the United States are of the American, English or Mediterranean breeds. The Asiatic breeds are grown largely for meat production. The Cornish is the most important English breed produced in this country. Only six breeds of chickens are of great commercial importance here. They are the White Leghorn, the New Hampshire, the White Plymouth Rock, the Barred Plymouth Rock, the Rhode Island Red, and the Cornish. The other breeds are grown in limited numbers and some are used in making crosses.

The production of hybrid and cross-mated chicks is increasing. Hybrid chicks are produced by crossing inbred lines of one breed with inbred lines of one or more other breeds. Crossbred chicks are produced by crossing noninbred birds of two or more breeds.

Most broilers are the result of crosses between White Rocks and White Cornish. Most white egg strains are of White Leghorn origin. Brown egg strains have been developed from New Hampshire, Rhode Island Red, and Barred Plymouth Rock parent stock.

Chicks should be ordered early and from a reliable hatchery. Only those chicks from pullorum-free and production-tested flocks should be purchased. It is best to buy chicks locally.

Culling is important in poultry production. Baby chicks should be culled when they are placed in the brooder house. Pullets should be carefully culled when they are placed in the laying house. Layers should be culled periodically throughout the year.

Keep only large, well-developed, healthy hens with large, bright red combs, prominent and bright eyes, and good body capacity. There should be room for two to three fingers between the pubic bones, and three or four fingers between pubic bones and breastbone.

Layers of the yellow skin breeds should have no pigment around the vent, eye-ring, beak, or shanks. Early and slow molters should be sold, as they go out of production during the molting period, which may last three or four months.

Buy only high-quality chicks. Keep only the best pullets for the laying flock, and cull them periodically. Don't waste feed, labor, and housing on inferior birds.

QUESTIONS

1. What are the major classes of chickens produced in this country?
2. How do these classes differ in conformation and use?
3. Which breeds of chickens are best for egg production?
4. Which breeds or crosses are best for broiler production?
5. What is the place of hybrid and crossbred chickens on our farms?
6. What factors must be considered in deciding where to obtain chicks?
7. Does it pay to buy started chicks?
8. Which is better to buy, straight-run or sexed chicks?
9. What factors should be considered in culling laying hens?
10. What type of head, comb, and wattles are desired in a laying hen?
11. How can pigment be used in culling a flock of chickens?
12. How can you measure the body capacity of a hen?
13. How can molt be used in culling hens?
14. Explain the formula used to determine the rate of culling based on rate of lay, price of feed, and price of eggs.

REFERENCES

Bundy, C. E. and R. V. Diggins, *Poultry Production,* Prentice-Hall, Inc., Englewood Cliffs, New Jersey, 1960.

Card, Leslie E. and M. C. Nesheim, *Poultry Production,* 10th Edition, Lea & Febiger, Philadelphia, Pennsylvania, 1966.

Harris, C. E., *Culling the Laying Flock,* Circular 557, University of Kentucky, Lexington, Kentucky, 1958.

Jaap, R. G., *Breeds and Breeding of Chickens,* MM-211, The Ohio State University, Columbus, Ohio, 1964.

U. S. Department of Agriculture, *Agricultural Statistics, 1972,* Washington, D.C., 1970.

———, *Poultry and Egg Situation,* Washington, D.C., November 1971, April 1973.

———, *Selected Statistical Series for Poultry and Eggs Through 1968,* ERS 232 Revised, Washington, D.C., 1970.

37 Feeding and Management of the Laying Flock

The spread between the price of eggs and the cost of production has gradually become smaller since about 1929. Poultrymen now must use better laying stock and methods of production in order to have profitable enterprises. A few years ago, the U.S.D.A. indicated that hens must produce at least 150 eggs per year to break even. A laying average of 200 to 240 eggs per hen per year appeared to be necessary in 1971.

A study conducted by the University of California in 1970 of the income and costs of 19 specialized egg ranches in San Diego County, California revealed that the average cost of producing one dozen eggs was 28.1 cents. The lowest cost was 25.7 cents, whereas the highest cost was 37.5 cents per dozen. The flock size varied among the 19 ranches from about 10,000 to 150,000 layers. The average production per hen was 230 eggs. The average price received per dozen eggs sold wholesale was 29.7 cents. These data emphasize the importance of good business management in commercial egg production.

METHODS OF INCREASING PROFITS

There are four ways of increasing the profit from laying flocks: (1) by increasing production, (2) by decreasing production costs, (3) by better marketing, or (4) by a combination of the three methods just named.

Feed and labor costs make up 60 to 70 percent of the cost of producing eggs. Feed costs per dozen eggs go down with increased production per hen. Labor costs go down with increases in flock size.

A profitable flock should average 65 to 75 percent production during the entire year. It takes good feeding and management to get this kind of production. Pullets must be of good quality and they must be properly housed. Adequate rations must be provided in proper amounts. Desirable roosting, feeding, watering, and laying equipment must be

NORTHCO
LAYING CAGES
NORTHCO
LAYING CAGES

provided. Adequate lighting and ventilation are essential, and health problems must be guarded against. Careful consideration of these factors may be the key to profitable production on your farm.

MOVING PULLETS TO THE LAYING HOUSE

Pullets of the light breeds will have their hen feathers and red combs and be ready to start laying when they are about five months old. Heavy pullets begin laying a few weeks later. February-hatched pullets should be moved to the laying house in August or early September.

Have the Laying House Clean

Clean the laying house and cages throughly a week or two before it is time to move the pullets. Spray the floor, walls, and equipment with disinfectant. Clean and disinfect the dropping pit, if used, and place new litter on the floor. Clean the feeders and waterers and have them in working condition. Clean and disinfect the nests and provide new nesting materials, if they are to be used. If you keep any old hens, have them in a separate place.

Care of Pullets at Moving Time

Each pullet should be carefully examined at the time it is moved to the laying house. All culls should be removed, and those which are to be kept should be treated for lice with powdered sulphur or sodium flouride. Try not to excite the birds at moving time. Watch the pullets the first night or two to see that they roost on the perches, rather than on the feeders or nests.

The laying house is an important part of the poultry enterprise. Hens are very responsive to their environment and they need a clean, dry, well–ventilated, quiet, and comfortable house. On some farms the laying house is inadequate, and the hens are permitted to have the run of the farmstead. It pays to keep the laying flock confined to the laying house at all times.

CAGE LAYERS

Nearly 90 percent of the layers in California are housed on wire. Some of them are in individual cages. Cages are also being used in the Southern and Central States. Practically all new layer houses involve from 5,000 to 100,000 birds, and cages are used.

Cage laying usually results in increased egg production, decreased mortality, control of cannibalism, ease in culling and record keeping, and decreased feed and labor costs per dozen eggs produced. The initial cost of equipment is high, and the cage-laying system of management requires more skill than does floor management. Flies may be a problem and overhead costs may be excessive, if equipment is not used to full capacity.

Cage birds require much the same housing conditions as birds on the floor. By using double-decked cages and putting two layers in each cage, it is possible to reduce the floor area per bird to less than 1 square foot. Cages vary in size, but are usually 8 to 10 inches wide and 18 inches deep. They are placed in rows, either single- or double-decked, and are

FIGURE 37-1. (*Above*) A high percentage of layers are housed in cages. These are Dekalb brown egg layers. (*Courtesy* Dekalb Ag Research, Inc.) FIGURE 37-2. (*Center*) Some poultry producers grow out pullets on range. (*Courtesy Poultry Tribune*) FIGURE 37-3. (*Below*) A cage layer installation in Kansas that houses 1,680 hens in standard double-deck cages. (*Courtesy* Northco Ventilating Co.)

hung from ceiling joists or roof rafters. The cages are usually placed back-to-back with an aisle between pairs of cages. The birds may be fed and watered by hand or automatically, and the eggs are gathered from the front.

In a Michigan study it was found that the building and equipment cost per bird housed in a floor laying house was $4.69. The cost was $5.98 per bird in 10-inch cages with two birds per cage, and $3.90 per bird in 8-inch cages with two birds per cage.

TRADITIONAL HOUSING

Space Required

It does not pay to crowd the hens in the laying house. One and one-half to 2 square feet of floor space should usually be provided for each bird of the light breeds and 2½ to 3 square feet or more for the heavy breeds. Hens do not produce well under crowded conditions, and the house becomes difficult to ventilate properly and to keep clean. Crowding may encourage outbreaks of disease. A house 40 by 100 feet will accommodate about 2,000 hens of the light breeds and 1,300 of the heavy breeds.

Poultrymen can reduce the floor space area to 1½ square feet per bird when the feeders and waterers are placed in the roost area, the hens are debeaked to control cannibalism, and the birds are provided with adequate feeder space and rations.

Roosts

From 7 to 8 inches of roost space should be provided for each bird. A flock of 2,000 hens will require 1,100 to 1,300 feet of roost space. Roosts usually are located at the back of the house, away from drafts. The roosts should be made of 2″ x 2″ or 2″ x 3″ material

FIGURE 37-4. A good arrangement of automatic feeders and waterers. Note the deep litter and specs on layers to prevent cannibalism. (Grant Heilman photo. *Courtesy* The Beacon Milling Co.)

and should be placed 13 to 16 inches apart. Low roosts are preferred.

Dropping Pits

Dropping pits should be constructed below the roosts. They are much more satisfactory than dropping boards. The pits may be made in sections for ease in cleaning. A pit 16 to 18 inches deep will allow for use of deep litter on the floor. The pits should be closed on all sides with boards or wire, and 1″ x 2″ welded wire should be placed under the roosts to keep the birds out of the pit. The pit will need to be cleaned during hot weather to avoid ammonia injury to the eyes of the hens.

Deep Litter System

Deep, or built-up, litter is the name applied to a system of litter management which involves placing 4 to 6 inches of fresh material on the floor of the house in August or September, and allowing it to build up with droppings and added litter with it is 8 to 12 inches deep. The deep layer of litter is easier to keep dry than a thin layer and has an insulating effect. The floor is warmer and drier. The addition of one pound of hydrated lime to 10 square feet of floor space aids the microorganism action in the litter, thus reducing dampness.

Litter Materials. Chopped straw, shavings, ground corncobs, or whole corncobs properly used make good litters. Some poultrymen make a litter of two-thirds ground cobs and one-third whole cobs. Cane pulp, peanut hulls, and peat moss may also be used when free from injurious materials.

Care of Litter. The surface of the deep litter should be stirred frequently to prevent matting. The finely pulverized bottom layer should not be disturbed. The litter should be spread evenly on the floor, and wet material around the waterer should be removed and replaced with new dry material. The new litter should be mixed with the old and kept level.

FIGURE 37-5. Layers in Ohio on a mesh floor. (*Courtesy* James Manufacturing Co.)

FIGURE 37-6. (*Above*) **A community rollaway nest. Wire bottom is slanted so that eggs roll to the front.** (*Courtesy Wallaces Farmer*) **FIGURE 37-7.** (*Below*) **Individual nests placed on the wall of the laying house.** (*Courtesy* **H. D. Hudson Manufacturing Co.**)

Slat Floors

While the combined use of litter and a dropping pit is recommended for poultrymen who give their layers proper care, the use of slat floors may have advantages for those who have difficulty providing a satisfactory dry litter. Slat floors require no litter and no cleaning during the laying year, but when they are used the humidity is higher, the birds appear more nervous, and there is more feather picking and egg breakage.

Nests

There should be 1 square foot of nesting space for every four or five hens, and the nests should be alike to keep the hens from crowding into certain ones. The number, location, and type of nests affects egg breakage, shell cleanliness, and the labor required in gathering eggs.

Community nests have become popular. They are 2 feet wide and from 4 to 12 feet long. Eight-inch openings are made every 4 or 5 feet. Eggs are gathered from the front when the openings are in the back. Some nests have hair screen bottoms which slope so that the eggs roll into a trough.

Regardless of the type of nest used, it should have the following features:

1. **The inside of the nest should be dark.**
2. **It should be easy to keep clean.**
3. **It should provide an easy way to gather the eggs.**
4. **It should be sturdy and easy for the hens to get into.**
5. **It should be able to be closed at night.**

Watering Equipment

A flock of 100 hens will drink from 5 to 7 tons of water in a year. Eggs are about 65 percent water, so the hens should have an ample supply at all times. Automatic waterers will provide a ready supply of water and will eliminate much of the labor required to carry water and fill fountains.

The waterer may be placed on a stand above the litter. A wire platform aids in keeping the water clean. A pendulum guard may be placed over the opening to keep chickens out of it. If water is not piped into the chicken house, air-pressure fountains may be used. Two 5-gallon fountains or a 4-foot auto-

FIGURE 37-8. (*Left*) Automatic feeding and watering equipment in use on the Cobb Breeding Corporation Farm. Note the type and arrangement of nests. (C. W. Holland photo. *Courtesy* Cobb Breeding Corporation) FIGURE 37-9. (*Right*) A commercial feeder and automatic waterer in use in an Iowa laying house. (*Courtesy Wallaces Farmer*)

matic watering trough should be available for every 100 birds. Four 8-foot trough-type mechanical (automatic) waterers should be available for every 1,000 hens. Additional drinking space may be needed during hot weather.

Housing space and litter may be conserved by locating the waterers in the roost section of the house over the dropping pit. Some farmers place containers below the waterers to catch splashed water.

Feeders

There should be about 5 inches of feeder space per hen, and the feeders should not be more than 10 inches above the litter level. It is usually best to have one end of the feeder toward a window or light so that the feed will not be shaded. All feeders should be at the same height and have reels or grills to keep the feed clean and to prevent waste.

A rack of lath or of 2-inch poultry netting can be made to feed hay and greens. Every 100 hens should have about 6 inches of space for grit and a foot of space for oyster shell in small feeders along the wall.

Ventilation

Adequate insulation and ventilation make it possible for poultrymen to control the temperature in the laying house during all seasons of the year. Ventilation is needed to provide fresh air for the hens and to remove the moisture given off in breathing and droppings. Three to 4 gallons of moisture will be given off by 100 hens in one day. Hens do best when the temperature is within a range of 55° to 75° F.

Slot-type Ventilation. An 8- to 10-inch slot ventilator located on the south wall just below the ceiling or eave and extending the length of the building is one of the best

TABLE 37-1 SCHEDULE FOR OPERATING SLOT VENTILATORS

Outside Temperature	House Temperature	Size of Opening
Below 0° F	25°–35° F	¼ to 1 inch opening
0°–30° F	35°–50° F	1 to 2 inch opening
30°–65° F	40°–65° F	2 inches to full opening
65° F and up	50° F and up	Full opening with windows out

Source: *Bulletin P108,* Iowa State University

systems of ventilating the average small-farm laying house. An adjustable board can be used to regulate the amount of air admitted. The amount of opening should vary with the extent to which the house is insulated, the outside temperature, and the number of birds in the house. Slot ventilators and windows are usually adequate, if properly managed, in ventilating houses from 20 to 25 feet wide.

Roof Ventilators. One 12-inch roof ventilator for every 125 birds is recommended in houses wider than 20 to 25 feet where forced ventilation is not used. The flue of the ventilator should extend to about 18 inches from the floor.

Forced Ventilation. Forced-air ventilation systems are usually recommended for commercial laying houses that are 25 feet wide or wider. Most systems use exhaust fans and air intakes. Poultry and electric-power specialists should be consulted in planning the installation of a fan system of ventilation. The system should provide 3 cubic feet of air per minute per bird for winter and 7 cubic feet per minute per bird for proper summer ventilation. In a 3,000-bird house, 4 fans, each with 2,250 cubic feet per minute capacity, may be used—one to run continuously and the others as needed.

FIGURE 37-10. Single-sire pedigree matings may be housed in cages or on the floor. Shown is one of 1,200 Arbor Acres pedigree mating units. (*Courtesy* Arbor Acres Farm, Inc.)

Insulation

Insulation conserves the heat generated by the birds in winter and keeps out the heat of the summer. In northern sections it is impossible to maintain desirable temperatures in the winter without the use of insulation. Accordion-type aluminum foil, blanket-type insulation, ground corncobs, and dry shavings make excellent insulations.

Lighting

Artificial lights may stimulate egg production during the winter months from October until March. A hen should have a day of at least 13 hours. When there are 10 hours of daylight, artificial light will be needed for three hours.

Morning Light Preferred. Providing artificial light in the morning avoids the problem of a dimming device to get the birds on the roosts at night. Time switches are available which turn the lights on and off automatically. One 100-watt bulb should be used for every 400 square feet of floor space, and each light should have a 16-inch reflector 4 inches deep. The lights should be placed in a row at least 7 feet above the floor and directly over the feeders.

Types of Laying Houses

Plans for many types of laying houses are available. Those pictured in Figures 37-10 and 37-11 are examples of satisfactory types.

Materials. Laying houses may be constructed of lumber, tile, concrete blocks, or metal.

Foundation and Floor. A 6-inch concrete foundation is needed, extending 18 to 24 inches into the ground. The floor should slope about 1 inch in 4 feet.

Roof. A shed or low gable roof is satisfactory.

FIGURE 37-11. (*Above*) A large multifloor laying house in New Jersey equipped with Thermopane windows. (*Courtesy* Libby-Owens-Ford Glass Co.) FIGURE 37-12. (*Below*) A modern caged laying house equipped with automatic feed-handling equipment. (*Courtesy* Dekalb Ag Research, Inc.)

Windows. One square foot of window space for every 20 to 25 square feet of floor space is ample. Windows may be placed on only one side of houses 25 feet wide or less. Cage houses usually have no windows; artificial lighting is provided.

Windowless houses are used successfully in large commercial enterprises. They are less expensive to build and provide better control of ventilation, temperature, and light. However, they require greater use of electricity and more fan capacity during summer months.

Insulation. Rigid board insulation made of wood or fiber is recommended because of its excellent insulation quality, ease in construction, and comparative low cost.

FEEDING LAYING HENS

Feed is the most important item of cost in producing eggs. Carefully culled pullets from high-production strains will usually produce profitably, if they are fed good rations and are properly housed. Good rations alone, or good laying stock alone, will not insure good production. Most flocks are capable of producing more eggs than they do produce. They have been fed inadequate rations. Hens use much the same nutrients as other farm animals, but they use them in different proportions and amounts. Hens consume very little roughage and bulky feeds. They need a large amount of protein of high quality, and they require certain vitamins and minerals.

AMOUNTS OF FEEDS NEEDED

The amount of feed needed by a hen depends upon the breed, her body size, her production of eggs, and her environment. A pullet producing an egg every other day uses nearly three-fourths of her feed in maintaining her body, in providing muscular activity, and in maintaining a body temperature of about 107.5° F. About one-fourth of the feed consumed is for egg production.

Heavy Producers Need More Feed

Heavy-producing hens need some additional feed for body maintenance and activity, but need considerable increases in feeds for egg production. After meeting the maintenance needs, it takes about one pound of dry feed to produce seven eggs.

The five high flocks of the Iowa Poultry Production Demonstration Flocks in 1965 consumed 85.3 pounds of feed to produce an average of 248 eggs per hen, while the five low flocks ate 91.4 pounds, but produced only 204 eggs per hen.

Small Hens Need Less Feed

It takes about 24 pounds more feed a year to maintain an 8-pound hen than is required to maintain a 5-pound hen. The lighter breeds such as the Leghorns and Minorcas are efficient converters of feed into eggs, and so are popular in commercial egg production.

Four-pound hens at 70 percent lay use about 4 pounds of feed to produce a dozen eggs. Hens of the same weight at 40 percent lay use about 6 pounds of feed.

ESSENTIAL NUTRIENTS

Hens in heavy production must receive adequate amounts of the following nutrients: (1) carbohydrates and fats, (2) proteins, (3) fiber, (4) minerals, (5) vitamins, and (6) water.

Carbohydrates and Fats

Chickens are active and have high body temperatures, so they require large amounts of heat- and energy-producing feeds. Corn, oats, barley, wheat, grain sorghums, and mill byproducts are the chief heat- and energy-producing feeds. They usually make up 75 to 80 percent of the ration.

Availability, price, and quality usually determine the grains to be fed. Yellow corn and oats are popular in much of the Corn Belt, but sorghums, barley, and wheat are fed in many areas. Hens like variety. The feeding of two or more grains is recommended.

Proteins

Eggs are rich in protein of high quality. If hens are to produce a large number of eggs, they must receive rations which are high in the amino acids essential in egg production. Since feeds vary in their amino acid content, it is usually desirable to include three or four different protein supplements in the ration. Protein should make up 15 to 17 percent of the total ration.

Animal Proteins. Approximately 25 percent of the protein in the ration should be of animal origin. Dried, condensed, or liquid skim milk and buttermilk are high in riboflavin and provide protein of high quality. Meat scraps and fish meal of good quality are also excellent sources of animal protein.

Plant Proteins. Soybean meal is an excellent source of protein. It is available in most areas, and is palatable and reasonable in price. Corn gluten, peanut, and cottenseed meals are used in some areas. Cottonseed meal, when fed in large amounts, may produce an objectionable yolk color.

The addition of vitamin B_{12} to a ration will permit a reduction in the percentage of animal proteins.

Fiber

Fiber is not usually considered to be a nutrient, but it has been found to be essential in the ration for maximum production. The feeding of grains and legume hays will provide adequate amounts of fiber.

Minerals

A hen has only about 2 ounces of mineral in her body when she begins to lay, but in producing 200 eggs in a year she puts into them nearly 40 ounces of mineral, mostly calcium.

Calcium. The shell of an egg is made up of calcium, which must be fed to the hen in the form of oyster shell, ground limestone, or steamed bone meal. Most poultrymen self-feed oyster shell or include mineral feeds in the mash ration.

Phorphorus. Chickens need only small amounts of phorphorus, which is available in steamed bone meal, meat scraps, fish meal, and milk, and in small amounts in farm grains.

Salt. Salt should make up ½ to 1 percent of the total ration.

Grit. Hens which are fed all-mash rations do not need grit. Those which are fed whole or cracked grains should have hard grit available free choice.

Vitamins

Vitamins in small quantities are essential in feeding hens for egg production.

Vitamin A. About 3,000 International Units of vitamin A should be included in each pound of ration. Yellow corn, fresh green forage, alfalfa meal, fortified fish oils, and vitamin A concentrates are the best sources.

Vitamin D. This vitamin is needed by the hens in order to utilize the minerals in the feeds. It should be provided at the rate of 400 International Units per pound of ration. Hens which have access to sunlight manufacture their own vitamin D, but hens confined to the laying house should be fed fortified vitamin A and D oils, or commercially mixed mashes which contain an adequate amount of vitamin D.

Riboflavin. Laying hens should have about 1 milligram of riboflavin per pound of ration, while breeding hens need about 1.3 milligrams. This vitamin is available in milk, yeast, liver, and alfalfa meal. Commercial premixed concentrates containing this vitamin are available. Riboflavin is essential in the production of hatching eggs.

Vitamin B_{12}. This vitamin, when fed with plant proteins, permits a reduction in the percentage of animal proteins. It is available in vitamin premixes and in commercially mixed feeds. From 1.5 to 2 micrograms of B_{12} should be provided in each pound of ration.

Water

Water is essential in poultry feeding. It permits normal body processes. A pint of water is contained in every dozen eggs. Keeping plenty of clean, fresh water before the hens at all times is a must.

Antibiotics

The feeding of antibiotics to healthy laying hens has not proven profitable. Egg production and feed efficiency, however, may be increased by feeding antibiotics to hens that are in poor health or housed under poor conditions.

SYSTEMS OF FEEDING

No combination of farm grains alone will provide the proper amounts of nutrients needed by laying hens. Grains must be supplemented with proteins and mineral and vitamin feeds. This means that both grain and mash feeds must be fed. Several systems are used in feeding laying hens.

Grain-and-Mash Feeding. With this method whole grains may be fed in hoppers every evening and mash fed in the morning, or both types of feeds may be self-fed. The mash should contain approximately 20 percent protein. In feeding whole grain, do not spread it in litter. Regular feed hoppers should be used.

Grain and Protein Concentrate. This system is much like the grain-mash system, but in this system a 26-percent protein concentrate is self-fed instead of the 20-percent mash.

All-Mash Ration. The grain is ground and mixed with the mash feed or with the protein concentrate to make up a complete ration. This system involves more expense in grinding, but a uniform ration results. All hens get the same feeds, providing the feeds have been well mixed.

Most commercial flocks receive complete mixed rations in the form of mash, crumbles, or pellets. Mash is least expensive, but more feed is wasted. Crumbles feed down in bulk bins better than mash and are more appetizings than either mash or pellets.

SUGGESTED RATIONS

The rations in Tables 37-2 and 37-3 may serve as guides in planning rations for your home flock. In formulating rations, remember that the content of the ration should supply the nutritional needs of the birds. The percentage of protein should be increased for (1) pullets just beginning to lay, (2) layers producing 80-percent lay, and (3) birds in heavy production but consuming less than ¼ pound of feed daily.

The rations presented are complete feeds, and no additional grain should be fed. Oyster shell, high-calcium limestone, or calcite grit should be fed free choice. When scratch grains are fed, a 20- or 26-percent protein mash should be fed as a supplement.

CANNIBALISM

The picking of one another by hens is referred to as *cannibalism.* Occasionally, when hens attempt to lay too large an egg, they will rupture the oviduct, or a part of the cloaca will protrude from the vent. Other

TABLE 37-2 COMPLETE LAYER RATIONS

These layer rations are a complete feed. No additional grain or calcium should be fed.

Ingredient	Layer (16%) (Pounds)	Layer (18%) (Pounds)
Ground yellow corn [1]	1334	1214
Animal fat	40	60
Dehydrated alfalfa meal (17%)	40	40
Soybean oil meal (50%)	320	420
Meat and bone scrap (50%)	90	90
Dried fish solubles [2]	—	—
Defluorinated phosphate (or equivalent)	30	30
Ground oyster shell (or equivalent)	120	120
Iodized salt	6	6
Manganese sulphate (or equivalent)	0.25	0.25
Methionine	1	1
Vitamin Additions		
Vitamin A (million units)	5	5
Vitamin D_3 (million units)	1.5	1.5
Niacin (grams)	10	10
Riboflavin (grams)	4	4
Pantothenic acid (grams)	4	4
Choline chloride (grams)	200	200
Vitamin B_{12} (milligrams) [2]	4	4
Total pounds (approximately)	2,200	2,200
Calculated Analysis		
Productive energy (Calories/pound)	950	950
Metabolizable energy (Calories/pound)	1300	1300
Protein (%)	16	18
Calcium (%)	3.2	3.2
Phosphorus (%)	0.7	0.7

[1] Sorghum grain can replace up to ⅓ of the corn by weight.

[2] Needed only if hens are being kept for production of hatching eggs. If producing hatching eggs, replace 40 pounds of soybean meal with 40 pounds of dried fish solubles.

Source: Iowa State University

hens may see this soft red membrane and pick at it. The intestines may be pulled out before the caretaker notices this practice. Cannibalism can also be caused by inheritance, by deficiencies in the ration, and by close confinement. Once hens get in the habit of picking at one another, it is difficult to stop, and heavy losses may result.

Prevention

The following are some of the best methods of preventing cannibalism:

1. **Cut off ½ to ⅔ of the upper beak.**
2. **Purchase metal appliances and attach them to the beaks as the pullets are placed in the laying house.**

3. **Allow plenty of feeder and floor space.**
4. **Provide good ventilation.**
5. **Feed fresh green feeds or dry green alfalfa hay.**
6. **Feed whole oats to growing pullets and laying hens.**
7. **Darken the windows and nests.**
8. **Delouse the birds and disinfect the quarters.**

RECORDS

A poultryman must keep records of his business and carefully analyze them. There is no way of telling how much profit is obtained from the laying flock if no records are kept. Neither is it possible to determine methods of cutting down production costs. Good business methods must be applied to egg production as in any other enterprise.

Kinds of Records

The following records will be valuable in analyzing the enterprise:

1. **Inventories of birds, equipment, and feeds at the beginning and at the close of the laying period. A monthly check of the number of hens is necessary.**
2. **Egg production records by days and by months.**

TABLE 37-3 COMPLETE LAYER DIETS (For both caged and floor layers)

Ingredients	Young laying pullets (lbs/ton)	Mature hens (lbs/ton)
Corn meal, No. 2, yellow, ground	1064	1299
Wheat shorts	—	100
Barley, pulverized	—	—
Hominy feed	200	—
Stabilized grease or vegetable oil	50	40
Soybean meal, low fiber, 50% protein	425	300
Fish meal, antioxidant-treated, 60% protein	—	—
Meat and bone scrap, 50% protein	—	50
Corn distillers dried solubles	50	—
Alfalfa meal, 17% protein (100,000 IU Vitamin A/lb)	30	50
Dicalcium phosphate	50	25
Limestone	40	45
Oyster shell, hen-size	80	90
Salt, iodized	5	5
DL-Methionine (or hydroxyanalog equivalent)	1	1
Layer vitamin and mineral premix	5	5
Calculated composition		
Protein, %	17.2	15.6
Metabolizable energy, Kcal/lb	1340	1350
Calcium, %	3.0	3.2
Available phosphorous, %	0.55	0.5
Linoleic acid, %	1.8	1.6

Source: Cornell University

3. A record of death loss by months.
4. A record of income from sales and products used at home.
5. Feed records by months.
6. Records of cash expenditures.

Use of Records

A carefully kept set of records is of no value unless it is used. Following is a partial summary of the records of flocks participating in the South Dakota Laying Flock Record Program in 1971. Note the factors explaining the differences in income per hen over feed costs.

FIGURE 37-13. These caged layers have been debeaked to prevent cannibalism. (*Courtesy* Nutrena Mills, Inc.)

SUMMARY

We can increase egg profits by increasing production per hen, by reducing produc-

TABLE 37-4 SOUTH DAKOTA LAYING FLOCK RECORD PROGRAM PROGRESS REPORT

	Flocks Closed Out August 1971		
	Largest Flock	Smallest Flock	Average
Days from 20 weeks	408	351	402
Number hens housed	18,240	1,700	6,917
Percent depletion	14.0	9.9	14.6
Percent hen day production	62.5	62.6	62.7
Eggs per hen	255.0	219.6	257.0
Percent hen housed production	58.1	59.5	58.1
Eggs per hen housed	237.1	208.8	238.3
Percent grade A eggs:			
Large	71.5	60.3	67.6
Medium	15.3	19.7	14.5
Small	3.0	2.8	2.4
Total grade A	89.8	82.8	84.5
Average income per dozen eggs sold (¢)	25.27	23.31	25.23
Feed cost per dozen (¢)	15.76	17.32	15.71
Income per hen housed over feed cost ($)	1.87	0.92	1.81
Pounds of feed per dozen eggs	4.24	4.53	4.4
Feed cost per ton ($)	74.32	76.46	71.38
Gross income for period ($)	90,972.00	6,679.00	36,144.00

Source: South Dakota State University

tion costs, and by using better marketing methods. Feed and labor costs make up 60 to 70 percent of production costs. Feed costs per dozen eggs go down with increased production per hen, and labor costs go down with increased flock size. A flock should average 220 to 230 eggs per hen each year to make a profit.

Pullets should be carefully culled and deloused before they are put in the laying house. The laying house, cages, and equipment should be thoroughly cleaned and disinfected. New litter should be placed on the floor and equipment prepared for use. Pullets should be moved into the laying house when they are about five months old.

Allow 1½ to 3 square feet of floor space, 7 to 8 inches of roost space, and 4 to 5 inches of feeder space per bird. Provide dropping pits and a deep litter. An 8-foot automatic waterer should be provided for every 250 hens. Individual nests may be used, but community nests are recommended, as they provide a square foot of nesting space per bird.

Insulate the house and provide slot or forced-air ventilation. Provide 1 square foot of window space for every 20 to 25 square feet of floor space. Supply artificial lights to lengthen the day to about 13 hours. Windowless houses should be considered, especially if cages are to be used.

Feed hens a ration containing 15 to 17 percent protein. The self-feeding of a complete mash ration is recommended. A hen will consume about 90 pounds of feed in a year. Grains should make up about 75 percent of the ration.

Include in the mash at least three protein feeds, of which one is of animal origin. Soybean meal, meat scraps, and dried milk is a good combination. Hens need vitamins A and D, which may best be supplied in a premix supplement. Vitamin B_{12} is especially needed when the ration is lacking in animal protein. It is available in vitamin premixes and commercially mixed feeds.

Provide hens with oyster shell, ground limestone, and salt. Riboflavin is especially needed in producing hatching eggs. Feeding antibiotics to healthy hens has not proven profitable; however, it may be profitable to feed them to unhealthy hens.

QUESTIONS

1 How many eggs does the average hen on your farm have to lay to produce a profit?

2 How many square feet of floor space should a Leghorn hen have in the laying house?

3 What rules do you need to follow in providing feeders, roosts, nests, and waterers?

4 What are the advantages and disadvantages of a dropping pit?

5 What are the advantages of deep litter?

6 What part of the total ration should be made up of protein?

7 Which of the farm grains make the best feed for laying hens? Why?

8 Is an all-mash ration better than a grain-mash ration? Why?

9 Should laying hens be self-fed or hand fed?

10 Plan a ration and feeding program for your flock.

11 What are the advantages of the use of cages in the management of layers?

12 Outline a program for improving the housing of your home flock.

REFERENCES

Balloun, S. L., and W. J. Owings, *Feeding Replacement Pullets and Laying Hens,* Pm 354, Iowa State University, Ames, Iowa, 1966.

Creek, R. D. *Feeding Laying Hens,* Fact Sheet 163, University of Maryland, College Park, Maryland, 1964.

Esmay, M. L. *et al., Poultry Housing for Layers,* Extension Bulletin 524, Michigan State University, East Lansing, Michigan, 1966.

Meyer, V. M. and L. Z. Eggleton, *Ventilate Your Poultry House,* Pamphlet 292, Revised, Iowa State University, Ames, Iowa, 1969.

Midwest Plan Service, *Poultry Equipment Plans,* Revised, Agricultural Engineering Department, Iowa State University, Ames, Iowa, 1973.

Naber, Edward C. and Sherman P. Touchburn, *Ohio Poultry Rations,* The Ohio State University, Columbus, Ohio, 1970.

National Academy of Sciences-National Research Council, *Nutrient Requirements of Poultry,* 6th Revised Edition, Washington, D.C., 1971.

38 Feeding and Management of Young Chickens

The methods used in feeding and managing young chickens vary somewhat according to the number being raised and the use to be made of them, but the principles are much the same. Chicken raising is profitable only under certain conditions. The chicks must be healthy and bred for production. They must be housed properly with adequate space, heat, and sanitation. The ration must provide the nutrients necessary for rapid and economical growth. Mortality must be kept to a minimum, and the chicks must be protected from diseases and parasites.

BROODER HOUSES

Some poultrymen use small, movable colony brooder houses and grow out fewer than 500 to 1,000 chicks each year. Specialized poultrymen and broiler growers usually grow out thousands of chicks in large permanent houses.

A satisfactory brooder house must be capable of maintaining the desired temperature under the hover regardless of weather conditions. It must provide space for the brooder and areas away from the brooder which will be cool. It should be tightly constructed, yet provide for ventilation. Windows may be provided, but an excess of sunlight should be avoided. It is difficult to maintain even temperatures in houses with large window areas on the south.

The floor and walls of the house should be such that they can be cleaned and disinfected easily.

Size of House

The size of the house determines the scope of the enterprise. Each chick should have ½ square foot of floor space during the first month and 1 square foot of space thereafter. No more than 750 chicks should be placed under a 1,000-capacity brooder. Each chick should have 7 square inches of hover space.

Many brooder houses for broiler production are 30 to 50 feet wide and 50 to 300 feet long. One square foot of floor area is provided for each chick, and the house is divided into pens, if chicks of varying ages are being housed in one building.

Ventilation

There should be sufficient ventilation to keep the litter dry. Slot ventilators or windows may be adequate in small houses. Electric fans provide the best movement of air through large brooder houses.

Most broiler houses in the northern states rely upon mechanical systems of ventilation. In southern climates, the houses may be ventilated by windows, open doors, or mechanical methods.

Getting the House Ready for Chicks

The brooder house should be thoroughly cleaned several weeks before the chicks arrive. The walls, ceiling, and floor should be swept or scraped. The lower walls and floor should be scrubbed with hot water and disinfected with lye water (one 13-ounce can to 15 gallons of water) or with a 5-percent chlorine solution. The house should be given plenty of time to dry out before it is used. The feeders and fountains should be cleaned and disinfected at the same time. The house and equipment must be free of diseases and parasites.

Litter

A good litter on the floor serves as an insulator in maintaining uniform temperature and as a blotter in absorbing moisture. Litter should be fairly coarse to permit the droppings to sift through to the floor. Shavings, ground corncobs, peanut shells, and sugarcane fiber make excellent litters. Hay chaff is satisfactory, if it is not moldy or dusty.

FIGURE 38-1. (*Above*) Baby chicks nicely started in a colony-type brooder house. (*Courtesy* Dekalb Ag Research, Inc.) FIGURE 38-2. (*Below*) A large-scale broiler setup for baby chicks. (*Courtesy* Buckeye Incubator Co.)

A layer of 2 to 4 inches of litter should be placed on the floor a day or two before the chicks arrive, and it should be piled up in the corners to prevent the chicks from crowding into the corners, where they may be injured or smothered. Litter should be stirred up two or three times each week to permit drying and to avoid matting. When new litter is needed, it should be placed on top of the old. It may be built up to a depth of 6 to 8 inches.

Wire or Slat Floors

Wire or slat floors may be used. They keep the chicks from walking over manure or damp litter. More heat must be provided to houses with this type of floor and they are somewhat unhandy in managing the chicks.

BROODERS AND MANAGEMENT

The brooder should be set up and put into operation a few days before the chicks arrive. It must be properly regulated. The temperature 2 inches above the litter near the edge of the hover should be from 90° to 95° F. during the first week.

Kinds of Brooders

Coal or oil brooders heat the houses as well as the areas under the hovers. The temperature under the hover can be a little lower with these brooders than with gas or electric brooders.

Gas or electric brooders heat the hover area but not the room. The temperature under the hover should be maintained at 93° to 95° F. at the start. Many poultrymen like these brooders because of their ease of operation.

Brooder Management

Ideal brooding conditions provide ample heat for the chicks under the brooder, but the rest of the house is comparatively cool.

Hover Guard. During the first few days a guard made of cardboard, wood, or sheet metal 15 to 18 inches high should be placed around the hover at a distance of about 2 or 3 feet to keep the chicks from getting away from the brooder and becoming chilled. The guard also helps to keep out drafts and cold air.

Decrease Temperature. The temperature can be reduced gradually about 5 degress each week until it is down to about 75° or 80°.

STARTING THE CHICKS

Feed and water should be available for the chicks when they are placed in the brooder house. The first feed for pullorum-free chicks may be placed on paper, on new egg-case flats, or in small chick feeders. Chicks must learn to eat and drink. Ordinarily they are ready to eat 18 to 24 hours after hatching and it is important that they receive feed within 36 hours after being hatched.

Feeders and Waterers

Small lath troughs may be used as feeders during the first week or ten days. The troughs should be kept about ⅔ filled. Two inches of trough space should be provided for each baby chick for the first five weeks. Several small water fountains should be used during the first few days. Four half-gallon fountains will provide adequate water for 100 chicks.

After a week or ten days the lath troughs should be replaced with 3″ × 6″ troughs, or with automatic equipment. Three inches of feeder space per chick is recommended after the fifth or sixth week.

Large feeders, or automatic feeders which are elevated, should be provided when the chicks are eight to ten weeks of age. Three-gallon waterers or automatic waterers should be used, and they should be placed on wire-covered platforms 6 inches high. Some brooding chicks will be out on range by the time they are eight to ten weeks of age, and can be fed in protected outdoor feeders. At least one feeder should be left indoors for use during rainy weather. Most breeding chicks are now reared in confinement.

NUTRITIONAL NEEDS OF CHICKS

Chicks require rations containing 22 to 24 percent protein during the first few weeks, and the protein must be of high quality. Baby chicks have little capacity and cannot consume coarse or bulky feeds. Most poultrymen use a commercial starter feed. Large commercial producers may require such a large volume of starter feeds that mixing their own rations would be justified. Starter and growing mashes which meet the nutritional needs of the chicks have been formulated. Proteins, carbohydrates and fats, minerals, vitamins, and antibiotics are essential for rapid and economical chick growth. Growing mashes for breeding chicks usually contain about 15 to 17 percent protein if no scratch grain is fed.

Proteins

The muscular tissue produced in chick and broiler production is largely made up of protein. The body of a five-pound Leghorn cockerel is about 25 percent protein. Starter and growing mashes must be high in proteins containing at least 13 of the amino acids. Sufficient arginine, lysine, methionine, tryptophane, and cystine may be critical in some rations. To guard against protein deficiency, three or four different protein supplements should be included in the ration. About 25 percent of the protein should be of animal origin. Milk products, meat scraps, fish meal, soybean meal, and corn gluten meal are good feeds to include in the chick mash.

Carbohydrates and Fats

The cereal grains, corn, wheat, oats, barley, and grain sorghums are usually used to provide carbohydrates. The use of two or three grains in the mash is recommended.

FIGURE 38-3. Interior of a large commercial house. Note the automatic feeders that are adjusted low enough for these 4- to 5-day-old chicks. (*Courtesy* Automatic Poultry Feeder Co.)

After the chicks are from five to six weeks old, they may be fed cracked corn, pellet feeds, or whole small grains in feeders in addition to the mash. Eight-week-old chicks should receive about 90 percent mash feeds and 10 percent scratch grains or pellet feeds. By the time breeding chickens are three months old, the grain ration should make up 30 percent of the total ration. Four-month-old birds should get about equal amounts of scratch and mash feeds.

Minerals

Common farm feeds contain most of the minerals needed by chicks with the exception of calcium, phosphorus, sodium, chlorine, and manganese. Calcium is easily supplied in ground limestone or marine shells. Phosphorus may be supplied in the form of steamed bone meal or defluorinated rock phosphate, or as meat and bone meal. Salt provides sodium and chlorine. Manganese must be purchased in a drugstore as technical anhydrous manganous sulfate. It is usually supplied to the feed mixer in the form of a mineral premix as the mash is manufactured.

Starter mashes should contain about 1 percent calcium, 0.6 percent phosphorus, 0.5 percent salt, 0.2 percent potassium, 25 milligrams of manganese per pound, and 0.5 milligrams of iodine per pound.

Vitamins

Vitamins are essential for rapid and economical chick growth. Vitamins A and D, riboflavin, niacin, pantothenic acid, and choline may be most critical in the ration.

Vitamin A. Chicks should receive rations containing 1,000 U.S.P. Units of vitamin A per pound of ration. This vitamin promotes growth and resistance to infection in chicks. Green feeds, yellow corn, and fish oils are the best sources.

Vitamin D. This vitamin promotes efficient use of calcium and phosphorus and prevents rickets in chicks. About 90 I.C.U.'s per pound of ration is recommended. Late-hatched chicks grown in the sunlight can produce their own vitamin D.

Riboflavin. Riboflavin is needed for chick growth and should be provided at the rate of 1.7 milligrams per pound of ration. Liver, yeast, milk products, and fermentation byproducts from the manufacture of lentyl alcohol are the best sources.

Niacin. Chicks need niacin for normal growth and development, and they lose their appetities and become nervous when it is lacking. Niacin may be obtained from liver, yeast, and fermentation byproducts. Wheat bran and middlings contain niacin, and it can be purchased in crystalline form. For chicks, 13 milligrams of niacin is recommended per pound of ration.

Pantothenic Acid. There should be 4.6 milligrams of pantothenic acid in each pound of chick ration. This vitamin prevents dermatosis, aids feathering, promotes normal pigmentation, and is essential for healthy nerves. Succulent green feed, alfalfa meal, milk products, yeast, and liver meal are good sources.

Choline. Chicks should receive about 600 milligrams of choline per pound of ration. It prevents perosis or slipped tendon conditions. Fish and animal products, soybean meal, and distillers' solubles are main sources of this vitamin.

Other Vitamins. Several other vitamins are essential, but they are usually available in the feeds included in the ration. Vitamin K (anti-hemorrhagic), biotin (to prevent dermatosis), and pyridoxine (to stimulate appetite and growth) are examples.

Antibiotics

Most poultrymen prefer starter mash that contains 4 to 10 grams of antibiotics per ton of feed. The maximum growth stimulation due to the feeding of antibiotics occurs

FIGURE 38-4. A houseful of young breeders. Note the automatic feeders. (*Courtesy* Brown's Led Brest, Inc.)

during the first four or five weeks. Rate of gain may be increased 10 to 15 percent and feed efficiency 5 percent by feeding antibiotics. Aureomycin, Terramycin, or bacitracin (each at 10 grams per ton), or procaine penicillin at the rate of 4 grams per ton are the most widely used antibiotics in chick rations. Antibiotics, like vitamins and minerals, are usually provided in the form of premixes and used in ration formulation.

Arsenicals

Tests indicate that the feeding of 3-nitro and other arsenicals to chicks in starter feeds, alone and with an antibiotic, increases the rate of gain and feed conversion, and reduces mortality. The value of arsenicals is dependent upon the quality of the ration, the health of the chicks, and the feeding methods being followed.

Arsenicals are used by most broiler producers. The directions of the manufacturer should be closely followed.

RATIONS FOR CHICKS

Most poultrymen rely upon commercial feed concerns for their chick rations. The materials needed can be purchased in large quantities by mixing firms, and they can put out a high-quality product at a reasonable price if they care to do so. The contents of the various starter and grower mashes must be studied, and selection should be based upon the extent to which the feeds meet the nutritional needs of the chicks and upon the cost.

MASH FORMULAS

Poultrymen may have homegrown grains available, and have their chick feeds mixed at the local elevator using homegrown grains, vitamin-mineral-antibiotic premixes, and other commercial feeds. Presented in Table 38-1 are suggested mash rations for chicks to be used as flock replacement birds and as broilers.

TABLE 38-1 STARTING AND GROWING RATIONS FOR CHICKS

Feedstuff or Ingredient	For Broilers			For Replacement Pullets	
	Broiler Starter 0–5 Weeks	Broiler Grower 5–7 Weeks	Broiler Finisher 7 Weeks-Market	Pullet Starter 0–8 Weeks	Pullet Grower 8–20 Weeks
Major Ingredients					
Ground yellow corn	480	610	700	530	450
Pulverized oats	0	0	0	0	200
Soybean meal (44% protein)	340	235	190	180	85
Wheat middlings or ground wheat	0	0	0	150	150
Meat and bone scrap (50% protein)	30	30	20	50	50
Whole fish meal (60% protein)	30	30	20	30	0
Dried whey or distillers' dried solubles	20	20	10	0	0
Dehydrated alfalfa meal (17% protein)	20	20	40	50	50
Defluorinated rock phosphate (13% phos.)	12	6	13	0	0
Feeding limestone (34 to 38% calcium)	0	4	2	5	10
Iodized salt	5	5	5	5	5
Stabilized feed grade fat	63	40	0	0	0
Total	1000	1000	1000	1000	1000
Minor Ingredient Additives (per 1000 lbs. of ration)					
Vitamin D_3, I. C. units	300,000	240,000	240,000	240,000	240,000
Vitamin E, I. units	5,000	2,700	0	0	0
Vitamin K, gm.	0.27	0.27	0.05	0	0
Riboflavin, gm.	1.8	0.9	0.4	1.9	0.5
Niacin, gm.	9.1	4.0	0	3.0	0
Pyridoxine, gm.	0	0	0	0	0
Pantothenic acid, gm.	3.3	3.7	3.8	2.4	2.1
Folacin, gm.	0.16	0.12	0.03	0	0
Biotin, mg.	0	0	0	0	0
Cobalamin (B_{12}), mg.	4.2	2.5	3.4	1.8	0.8
Choline gm.	64	163	120	162	145
Methionine or hydroxy analog, lbs.	2.4	2.2	2.0	1.8	0
Lysine, lbs.	0	0.7	1.4	0	0
Manganous oxide, gm.	41	41	41	34	32
Zinc oxide, gm.	32	32	32	20	23
BHT or Ethoxyquin	+	+	+	+	+
Antibiotic, gm.	2–5	2–5	?	2–5	—
Organic arsenical	+	+	—	—	—
Coccidiostat	+	+	—	+	?

Source: The Ohio State University

FIGURE 38-5. (*Above*) A poultry range shelter. Sturdy construction with braces glued and nailed. Midwest Plan Service plan No. 77773. FIGURE 38-6. (*Below*) A poultry range feeder built of plywood and fastened with casein glue. Midwest Plan Service plan No. 87410. (Figures 38-5 and 38-6 *courtesy* V. J. Morford)

RANGE FOR CHICKENS

Flock replacement chicks may be moved to range at eight weeks of age. Be certain to provide clean ground on which no chickens have been grown for at least two years.

Ladino clover, lespedeza, alfalfa, and red clover make excellent pasture crops for young chickens. From 10 to 30 percent of the feed can be saved by providing good range. Provide movable equipment and move it from time to time to maintain sanitary quarters. Provide plenty of fresh water and keep the feeders filled.

Range Shelters

Portable range shelters should be provided. Shelters enclosed with wire netting will permit the confinement of chickens at night to prevent losses to predators.

The range shelter may also be used as a sun porch, and some poultrymen grow their chicks for the first eight weeks in the brooder house and use the range shelter in this way.

ROOSTS

Chicks will begin to roost at about six weeks of age. Roosts should be placed along the back of the brooder house. Perches should be placed low when starting and raised after the chicks have learned to roost. Placing a frame covered with wire netting under the roosts aids in getting the birds on the roosts. Placing roosts level with the floor avoids the tendency for chickens to get to the top roosts.

Wide, flat roosts are recommended. They prevent chicks from developing crooked breastbones. At least 4 inches of perch space should be provided for each bird.

FIGURE 38-7. Many pullets are grown out in cages. (*Courtesy* Dekalb Ag Research, Inc.)

SEPARATE COCKERELS FROM PULLETS

Many poultrymen separate the cockerels from the replacement pullets when six to eight weeks of age. The pullets and cockerels are put on separate ranges or the cockerels are fed broiler ration and sold. Leghorn cockerels will use as much feed, putting on a pound of weight after 2½ pounds, as was required to put on the first 2½ pounds. It pays to separate the cockerels of the light breeds and dispose of them at 2½ pounds as broilers.

CANNIBALISM

Cannibalism in chicks may cause serious losses and the grown pullets may continue picking each other after being placed in the laying house. It can be prevented in the same way as was discussed in connection with the laying flock, by avoiding crowding, overheating, and insufficient feeder and water space. Use a sun porch to increase floor area. Debeaking is recommended.

Feeding finely ground oats to young chicks, or coarse-ground or whole oats to older chickens may help. Using salt as 4 percent of the mash for a few days may give good results. Injured and crippled chicks should be removed as soon as they are discovered.

BROILER PRODUCTION

The methods of feeding and managing broilers are quite similar to the methods used in starting and growing out chicks, with the exception that the broilers are kept in confinement until they are sold.

During the winter months, ¾ to ⅘ of a square foot of floor space per bird is recommended. One square foot of space per bird is desired in the summer. Roosts are not used in broiler production.

The feeders, waterers, and deep litter recommendations for chick production apply also in the production of broilers.

Broiler Rations

Most broiler producers feed starting mashes for the first eight weeks and continue feeding an all-mash ration until the birds are sold. Others start feeding pellet feeds or cracked corn with mash at about seven weeks and continue until the broilers are ready for market. Grain feeding slows up rate of gain but permits the feeding of homegrown grains. Most broilers are marketed when they are 8 to 10 weeks of age and weigh about 3 pounds.

Presented in Table 38-2 are suggested broiler rations. Arsenilic acid and a coccidiostat should be supplied as needed, according to the supplier's recommendations.

A pound of broiler should be produced on 2½ pounds or less of feed and the mortality loss should be less than 5 percent. One person with adequate facilities and equipment can care for 40 thousand to 100 thousand broilers in a year.

CAPONETTES AND CAPONS

The implanting of synthetic female hormones in young cockerels produces caponettes. The removal of the male organs produces capons.

Caponettes

Stilbestrol pellets implanted under the skin of the neck at the back of the head cause the cockerels to become quiet, lose their male characteristics, and take on the appearance of pullets. Cockerels treated when eight to ten weeks old could, at one time, be sold as caponettes when 12 to 16 weeks old. The treatment lasts only four to six weeks, after which they take on male characteristics again.

The U.S. Food and Drug Administration in 1959 prohibited further use of stilbestrol or diethylstilbestrol implants in poultry because of a possible relationship of these hor-

FIGURE 38-8. A commercial broiler setup. (Bob Streeter photo. *Courtesy* Shaver Poultry Breeding Farms Limited)

TABLE 38-2 BROILER RATIONS

Ingredient	Starter [1]	Finisher	Single All-Mash
Ground yellow corn [2]	1125	1440	1180
Soybean oil meal (50%)	550	350	550
Fish meal (65%)	40	40	40
Meat and bone scrap (50%)	100	50	50
Dehydrated alfalfa meal (17%)	40	30	40
Dried whole whey	40	—	40
Ground oyster shell (or equivalent)	—	5	5
Dicalcium phosphate (or equivalent)	20	30	30
Stabilized animal fat	60	30	40
Iodized salt	5	5	5
Manganese sulphate (or equivalent)	0.5	0.5	0.5
Methionine	1	0.5	1
Vitamin Additions			
Vitamin A (million units)	3	2	3
Vitamin D_3 (million units)	1	0.6	1
Niacin, grams	20	10	20
Riboflavin, grams	4	3	4
Pantothenic acid, grams	6	3	6
Choline chloride, grams	200	200	200
Vitamin E, IU	10	5	10
Vitamin B_{12}, milligrams	12	12	12
Menadione (vitamin K), grams	2	2	2
Antibiotic and coccidiostat [3]	+	+	+
Total pounds (approximately)	2000	2000	2000
Calculated Analysis			
Productive energy (Calories/pound)	1000	1050	1000
Metabolizable energy (Calories/pound)	1350	1400	1340
Protein (%)	23	17.7	22
Calcium (%)	1	0.9	1
Phosphorus, total (%)	0.8	0.7	0.8
Phosphorus, available (%)	0.5	0.45	0.45

[1] The starter diet is formulated to be fed from day-old through 5 weeks. The finisher diet should be fed the remainder of the period to market. The single all-mash can be fed from day-old to market.

[2] Up to 50 percent of the corn can be replaced by grain sorghum.

[3] Use at manufacturer's recommended levels.

mones to cancer in humans. The regulation is still in effect.

Capons

A capon must be kept six to nine months before it is ready for market, and 6 or 7 pounds of feed are required to produce a pound of meat. On the market capons compete with turkey. The capon business has not been promising in most areas.

Cockerels are caponized when they weigh 1½ to 2 pounds. The caponizing is

TABLE 38-3 ANALYSIS OF POULTRY FLOCKS IN ILLINOIS, 1972

Items	Number of hens per farm 200–999	2,000 and over
Number of farms	30	14
Average per farm		
Pounds of poultry produced	878	10,158
Total returns from poultry	$1,989	$50,432.00
Total value of feed fed	$1,792	$37,635.00
Returns above feed cost per hen	$0.48	$1.05
Returns per $100 feed fed	$111.00	$134.00
Average number of hens	410	12,129
Eggs produced per hen	197	233
Percent production	54.0	64.0
Feed units [1]	7,340	242,846
Feed cost per unit	$0.24	$0.15
Pounds of concentrates per feed unit	7.0	4.3
Cost per 100 pounds of concentrates	$3.42	$3.52
Price per dozen eggs sold	$0.34	$0.27

[1] One dozen eggs or 1.5 pounds of weight produced.

Source: *1972 Summary of Illinois Farm Business Records*, University of Illinois

usually done by a specialist. While many farmers have done some caponizing, the job is best given to a person who does the work sufficiently often to become skilled.

The removal of the testicles of cockerels produces birds with the appearance of pullets. They grow larger and meatier, but it requires many months to produce good capons. Two pounds of broiler meat can be produced on the amount of feed needed to produce a pound of capon. The price of capon usually is not twice the price of a broiler.

RECORDS

A careful record of chicken and broiler production, income, expense, mortality, feed consumption, egg production, and rate of gain will assist in developing a profitable enterprise. Table 38-3 is a summary of the costs and returns of poultry flocks in the state of Illinois for 1972.

SUMMARY

Chicks may be brooded in a colony-type or in a permanent brooder house. Each chick should have 7 square inches of hover space and ½ square foot of floor space. For broiler production each chick should have 1 square foot of floor area. The glass area in brooder houses should be kept to a minimum. There should be sufficient ventilation to keep the litter dry.

The brooder house and equipment should be cleaned and disinfected several days before the chicks arrive. The heating unit should be operated and 2 to 4 inches of peanut hulls, ground corncobs, or other good litter should be on the floor for a day or two before chicks arrive.

Feed and water should be available when chicks arrive. A hover guard should be used the first few days. The temperature under the hover should be about 95° F. the first week. The best room temperature is 65°. The temperature under the hover may be reduced 5 degrees each week.

A 22 to 24 percent protein starter feed should be fed in adequate feeders. The starter mash should meet National Poultry Council recommendations. Usually it pays to feed antibiotics, arsenicals, and a coccidiostat. Some broiler growers feed a 21 to 22 percent protein ration during the entire growing period. Breeding chickens are usually fed mash and scratch grain rations after the seventh week. Grower rations contain about 16 percent protein.

Some farm chickens, breeders, and capons are reared on range. Alfalfa and ladino clover provide excellent range. Adequate shelters, feeders, and waterers are necessary in range feeding.

Nearly 80 percent of the chickens reared in this country are reared in confinement as broilers. Cannibalism is best controlled by debeaking. It usually does not pay to caponize cockerels.

QUESTIONS

1. How much of the following should be provided per baby chick: (a) floor space, (b) feeder space, (c) waterer space, and (d) hover space?
2. What temperature is desired under the hover during the first week?
3. Outline the essentials of a good starter feed.
4. What kind of litter is recommended for chicks and broilers?
5. What vitamins are most critical in formulating chick rations?
6. Does it pay to feed chicks arsenicals?
7. How can you best provide antibiotics to your chicks?
8. Outline a program for feeding and managing chickens on range.
9. In what respects are the methods of feeding and managing broilers different from those used in growing out breeding chickens or laying pullets?
10. What methods can you use to prevent cannibalism in young chickens?

REFERENCES

Bundy, C. E. and R. V. Diggins, *Poultry Production,* Prentice-Hall, Inc., Englewood Cliffs, New Jersey, 1960.

Card, Leslie E. and M. C. Nesheim, *Poultry Production,* 10th Edition, Lea & Febiger, Philadelphia, Pennsylvania, 1966.

Management Guide for Broiler Growers, Poultry Husbandry Department, University of Kentucky, Lexington, Kentucky, 1963.

Naber, Edward C. and Sherman P. Touchburn, *Ohio Poultry Rations,* The Ohio State University, Columbus, Ohio, 1970.

National Academy of Sciences-National Research Council, *Nutrient Requirements of Poultry,* 6th Revised Edition, Washington, D.C., 1971.

U. S. Department of Agriculture, *The Broiler Industry: Structure, Practices, and Costs,* Marketing Research Report No. 930, Washington, D.C., 1971.

39 Turkey Production and Management

Turkey production has become a highly specialized industry. More than half of the 128.8 million turkeys marketed in the nation in 1972 were produced in five states: Minnesota, California, North Carolina, Missouri, and Arkansas. Minnesota marketed 20.8 million birds in 1972, which equaled the combined production of Ohio, Indiana, Wisconsin, and Iowa. Nearly 75 percent of our turkeys are grown on farms which have more than 2,000 birds.

Few turkey raisers maintain breeding flocks. Turkey poults (young birds) are usually purchased from specialized producers who maintain high-quality breeding flocks ana hatchery facilities. The problems encountered by the average turkey grower have to do with the acquisition of high-quality poults, the management of the poults during the brooding and range feeding periods, the control of diseases and parasites, and the marketing of birds which have been produced. We will deal, in this chapter, with the problems in production and management. Diseases are considered in Chapter 40 and the marketing of turkeys is covered in Chapter 41.

BUY GOOD POULTS

The same factors need to be considered in buying poults that were considered in buying chicks, but their importance is magnified because of the higher prices paid for poults and received from the sale of market turkey.

Selection of Variety

The most popular varieties of turkeys are the Broad Breasted Bronze, the Broad Breasted White, the White Holland, and the Beltsville (U.S.D.A.) Small White. Other varieties grown in small numbers are the Bourbon Red, Narragansett, Black Slate, and Jersey Buff.

In setting up a turkey business, the grower should choose the variety that is best suited to his market, is available in the desired quality and quantity, and can be purchased at the best price.

Production Qualities

Poults should be purchased from flocks where pullorum-control programs have been followed, and where meat production has been stressed. Thrifty, disease-free poults are essential to success. When possible, poults should be bought nearby in order to avoid overheating and chilling in transporting them. Data from official random sample meat production tests are helpful in selecting a source of poults.

Time to Start Poults

It takes from 20 to 24 weeks to grow out a prime turkey. Poults should be started about six months before they are to be sold, unless you plan to raise broilers. Broilers may be marketed in 12 to 15 weeks.

BROODING YOUNG POULTS

Brooder House

The brooder house should be thoroughly cleaned and disinfected several weeks before it is to be used. All litter should be removed and the interior of the house disinfected or

FIGURE 39-1. (*Above*) A Bronze male turkey. (*Courtesy* U.S.D.A. photo) FIGURE 39-2. (*Below left*) Beltsville Small White turkey. (*Courtesy Wallaces Farmer*) FIGURE 39-3. (*Below right*) Broad White toms ready for market. (*Courtesy Turkey World*)

scrubbed with lye water. All equipment should be cleaned and disinfected.

Centralized, stationary brooder houses divided into pen units are commonly used. Large houses may be 30 feet wide with a southern exposure or about 40 feet wide with a 6-foot alley in the middle and a row of pens on each side facing east and west.

Floor. Movable houses are usually built on skids and have wooden floors. Permanent houses usually have concrete floors underlaid with gravel.

Litter. This is very important in turkey production. A litter of 2 to 4 inches aids in keeping the quarters dry and sanitary. Poults will eat litter for the first week or ten days, so care should be taken in selecting litter material. Many growers use clean, silt-free sand or litter paper the first week. If litter is used the first week, it should be coarse, with particles ¼ inch or more in diameter. Permanent litter of crushed corncobs, peanut hulls, cane pulp, or shavings may be used after the first week or ten days.

Wire or Slat Floors. Some growers use a fine mesh wire on the floor. Welded 14 gauge wire with ½″ × 2″ mesh is recommended. Slat floors, made in sections with the slats placed on edge about ¾ inch apart, are also used.

Brooder

The brooder should be set up and in operation a couple of days before getting the poults. Brood only 250 to 350 poults per hover.

BROODING TEMPERATURES

Week	Temperatures
First	90–95° F.
Second	85–90
Third	80–85
Fourth	75–80
Fifth	70–75
After fifth week	Maintain 70° until heat is no longer needed.

FIGURE 39-4. Young poults on good litter. The gas brooder is economical and provides constant temperature. (*Courtesy Turkey World*)

FIGURE 39-5. Notice the guard which keeps the poults from crowding in the corner. (*Courtesy* **James Manufacturing Co.**)

Brooder Guard. A guard of wood, cardboard, metal, or wire should be placed completely around the hover to keep the poults from getting lost or chilled. The guard should be 3 to 4 feet from the hover, and corners should be closed off.

Starting Poults

Poults should be placed under the hover as soon as possible after they leave the hatchery.

Feeding. The beak of every third poult should be dipped in water and feed to get them to start to eat. Putting in a few poults five or six days old will help younger birds learn to eat. Spreading oatmeal or finely ground yellow corn on mash will also help to start poults eating. A bright light should be kept over the feeders and waterers during the first week.

Each five poults need one foot of linear feeder space for the first three weeks. The feeder space should be increased gradually. At five weeks, an 8-foot feeder should be provided for each 60 poults. Poults should be debeaked at about 10 days, removing half of the upper beak.

During the first eight weeks, poults need rations containing 27 to 29 percent protein, 1.3 percent calcium, 0.8 percent phosphorus, 0.5 percent salt, and the following amounts of vitamins and minerals per pound of feed: 4,000 U.S.P. units of vitamin A, 900 I.C.U.'s of vitamin D, 3.6 milligrams of riboflavin, 7.5 milligrams of pantothenic acid, 950 milligrams of choline, 0.7 milligrams of folacin, and 40 milligrams of manganese.

Farmers who grow turkeys in small numbers usually buy commercial mixed feeds. Large producers may mix their own, but the trend is for them to feed commercial mixed rations too.

Table 39-1 shows the nutrient recommendations for turkey rations prepared by The Ohio State University. Rations recommended by Iowa State University for turkey poults during various stages of production and for breeding birds are given in Tables 39-2 through 39-5.

Feeder Space. The following amounts of feeder space should be provided for every 100 poults:

First two weeks	33 linear feet
Third and fourth weeks	50 linear feet
Fifth through eighth week	66 linear feet

Sun Porch. A sun porch about the same size as the house is needed to provide sunlight and exercise, and floor space as poults become older. The floor may be of 1″ × 1″ welded wire or of wood slats placed about an inch apart. Start using the sun porch the third week if weather permits.

Roosts. Poults will begin roosting at three to four weeks of age. Roosts 3 to 4 inches wide may be provided at the back of the brooder house about 12 inches from the floor. Four to 6 inches of roost space should be provided for each bird.

Waterers. One 4-foot automatic trough waterer or its equivalent is needed per 100

TABLE 39-1 RECOMMENDED NUTRIENT SPECIFICATIONS FOR TURKEY GROWER AND BREEDER FEEDS

Nutrient or Additive		Grower						
	Age Fed, Weeks	1	2	3	4	5	6	Breeder
	Toms	0-3	3-8	8-14	14-18	18-22	22-	—
	Hens	0-3	3-8	8-12	12-15	15-18	18-	—
Crude Protein, %		29.4	27.3	23.1	18.9	16.8	14.5	17.0
Metabolizable Energy, kilocalories/lb.		1260 to 1350	1280 to 1365	1315 to 1390	1320 to 1400	1340 to 1420	1360 to 1440	1260 to 1320
Productive Energy, kilocalories/lb.		910 to 970	920 to 980	945 to 1000	950 to 1010	960 to 1020	960 to 1040	900 to 950
Amino Acids, %								
Arginine		1.68	1.56	1.32	1.08	0.94	0.80	0.97
Lysine		1.58	1.47	1.24	1.02	0.88	0.75	0.91
Histidine		0.59	0.55	0.46	0.38	0.33	0.28	0.34
Methionine		0.55	0.51	0.43	0.35	0.31	0.26	0.32
Methionine + Cystine		0.92	0.85	0.72	0.59	0.51	0.44	0.53
Tryptophan		0.28	0.26	0.22	0.18	0.16	0.13	0.16
Glycine		1.05	0.98	0.82	0.68	0.59	0.50	0.60
Phenylalanine		1.00	0.93	0.79	0.64	0.56	0.48	0.58
Phenlyalanine + Tyrosine		1.90	1.76	1.49	1.22	1.06	0.90	1.10
Leucine		2.00	1.86	1.57	1.29	1.12	0.95	1.16
Isoleucine		0.88	0.82	0.69	0.57	0.49	0.42	0.51
Threonine		1.00	0.93	0.79	0.64	0.56	0.48	0.58
Valine		1.25	1.16	0.98	0.80	0.70	0.60	0.72
Additives								
Antioxidant		+	+	+	+	+	+	+
Antibiotic		+	+	+	+	—	—	—
Coccidiostat		+	+	—	—	—	—	—
Histomonastat		—	—	+	+	+	+	—
Vitamins								
Vitamin A, U.S.P. units/lb.		4500	4500	3600	3600	3600	3600	4500
Vitamin D_3, I.C. units/lb.		820	820	600	600	600	600	820
Vitamin E, I.U./lb.		9.0	9.0	7.0	7.0	5.0	5.0	23.0
Vitamin K, mg./lb.		1.0	1.0	0.8	0.8	0.5	0.5	1.0
Thiamine, mg./lb.		1.5	1.5	1.5	1.5	1.0	1.0	1.5
Riboflavin, mg./lb.		3.0	3.0	3.0	3.0	2.5	2.0	3.0
Pantothenic acid, mg./lb.		7.5	7.5	7.0	7.0	6.0	6.0	11.0
Niacin, mg./lb.		48.0	48.0	32.0	32.0	24.0	20.0	20.0
Pyridoxine, mg./lb.		2.2	2.2	1.6	1.6	1.4	1.4	2.0
Biotin, mg./lb.		0.20	0.20	0.10	0.10	0.05	0.05	0.10
Choline, mg./lb.		950	900	800	700	600	500	600
Folacin, mg./lb.		0.70	0.70	0.45	0.40	0.35	0.30	0.55
Cobalamin (B_{12}), mcg./lb.		7.0	7.0	6.0	6.0	5.0	5.0	5.5
Minerals								
Calcium, %		1.32	1.24	1.12	0.98	0.90	0.80	2.48
Phosphorus, %		0.88	0.86	0.82	0.74	0.68	0.64	0.83
Sodium, %		0.17	0.17	0.15	0.15	0.13	0.10	0.17
Potassium, %		0.44	0.44	0.40	0.35	0.30	0.25	0.44
Manganese, mg./lb.		40	40	35	35	20	20	25
Iodine, mg./lb.		0.25	0.25	0.20	0.20	0.15	0.15	0.4
Magnesium, mg./lb.		250	250	230	230	200	200	250
Iron, mg./lb.		40	40	28	28	20	20	40
Copper, mg./lb.		4	4	3	3	2	2	4
Zinc, mg./lb.		50	50	23	23	18	18	40
Molybdenum, mg./lb.		1.0	1.0	1.0	1.0	1.0	1.0	1.0
Selenium, mg./lb.		0.1	0.1	0.1	0.1	0.1	0.1	0.1

Source: The Ohio State University

birds. Waterers should be shaded.

Lights. Small lights over the feeders and waterers will assist in keeping the poults from piling up and will also encourage them to eat and drink.

RANGE MANAGEMENT

Poults may be reared in confinement or on range. A large percentage of turkeys raised in the Corn Belt are on range.

The grower may move the poults to range when they are 8 to 12 weeks of age. By that time the ration should have been gradually changed from a starter feed to a grower mash and grain.

Pastures for Turkeys

Turkeys on alfalfa in Kansas consumed 17 percent less feed than turkeys raised in confinement. Green pasture furnishes vitamins, protein, and minerals which supplement the grower mash and grain.

Pasture Crops. Alfalfa, lespedeza, ladino, and red clover make excellent pastures for poults. Sudan grass and fall-seeded rye may also be used.

Rotate Pasture

Clean ground is a must in turkey production. Land should not be used for turkey range which has been used in growing either chickens or turkeys during the preceding three or four years, because blackhead disease may be spread by cecal worms still living in the ground following chicken production.

Most poultrymen move their turkeys every week or ten days. The lower areas of the range should be used first so the new range will not be contaminated later in the season. Temporary fences may be used. In wet weather, the poults should be moved more often. One acre of good range per season for every 65 birds is recommended.

Range Shelters

Portable range shelters with roosts both over and under the roof may be used. Shelters made on skids are easily moved. Birds

FIGURE 39-6. A Wisconsin turkey flock on range and ready for market. (*Courtesy* James Manufacturing Co.)

TABLE 39-2 COMPLETE TURKEY PRESTARTER AND STARTER RATIONS

Ingredients	Up to 3 Weeks	3–8 Weeks	0–8 Weeks
Ground yellow corn [1]	674 lbs.	847 lbs.	797 lbs.
Soybean meal (50% protein)	920	775	825
Wheat middlings	50	50	50
Dehydrated alfalfa meal (17% protein)	40	50	50
Meat and bone scrap (50% protein) [2]	50	50	50
Dried fish solubles	40	40	40
Dried brewers' yeast	40	30	30
Dried whole whey, low lactose	40	40	40
Stabilized fat	40	30	30
Dicalcium phosphate (or equivalent phosphorus)	50	40	40
Oyster shell (or equivalent calcium)	30	20	20
Feed grade salt, iodized	5	5	6
Trace mineral mix [3]	1.25	1	1
Methionine	1.5	1	1
Additives			
Vitamin A (million I.U.)	6	6	6
Vitamin D_3 (million I.C.U.)	2	2	2
Vitamin E (I.U.)	10,000	10,000	10,000
Vitamin K (gms.)	2	2	2
Riboflavin (gms.)	5	4	4
Pantothenic acid (gms.)	6	6	6
Niacin (gms.)	50	40	40
Choline chloride (gms.)	400	400	400
Vitamin B_{12} (mg.)	10	8	8
Antibiotics and medicants [4]	√	√	√
Folic acid (mg.)	500	500	500
Total	2,000 lbs.	2,000 lbs.	2,000 lbs.
Calculated Analysis of Ration			
Protein (%)	30	27	28
Calcium (%)	1.6	1.3	1.3
Phosphorus (%)	1.0	0.9	0.9
Productive energy (cal./lb.)	825	900	900
Metabolizable energy (cal./lb.)	1,200	1,250	1,200

[1] Sorghum grain can be used to replace up to half the corn by weight.

[2] May be replaced by 50 lbs. of 50% protein soybean meal plus 15 lbs. of dicalcium phosphate (or equivalent phosphorus).

[3] Trace mineral mix to contain 10% each of manganese and zinc

[4] Use at manufacturer's recommended levels.

Source: Iowa State University

TABLE 39-3 COMPLETE TURKEY GROWER RATIONS

Ingredients	8–12 Weeks Male	8–12 Weeks Female	12–16 Weeks Male	12–16 Weeks Female
Ground yellow corn [1]	1,010 lbs.	1,140 lbs.	1,180 lbs.	1,290 lbs.
Soybean meal (50% protein)	600	500	500	400
Wheat middlings	50	50	50	50
Dehydrated alfalfa meal (17% protein)	50	50	50	50
Meat and bone scrap (50% protein) [2]	50	50	50	50
Dried fish solubles	40	40	40	40
Dried brewers' yeast	30	—	—	—
Dried whole whey, low lactose	40	40	—	—
Stabilized fat	40	40	40	40
Dicalcium phosphate (or equivalent phosphorus)	40	40	50	40
Oyster shell (or equivalent calcium)	20	20	10	10
Feed grade salt, iodized	9	9	9	9
Trace mineral mix [3]	1	1	1	1
Methionine	1	1	—	—
Additives				
Vitamin A (million I.U.)	4	4	4	4
Vitamin D^3 (million I.C.U.)	2	2	2	2
Vitamin E (I.U.)	5,000	5,000	5,000	5,000
Vitamin K (gms.)	2	2	2	2
Riboflavin (gms.)	4	4	4	4
Pantothenic acid (gms.)	6	6	6	6
Niacin (gms.)	30	30	20	20
Choline chloride (gms.)	400	400	400	400
Vitamin B_{12} (mg.)	5	5	5	5
Antibiotics and medicants [4]	√	√	√	√
Total	2,000 lbs.	2,000 lbs.	2,000 lbs.	2,000 lbs.
Calculated Analysis of Ration				
Protein (%)	23	20.5	20	18
Calcium (%)	1.3	1.3	1.1	1.0
Phosphorus (%)	0.9	0.85	0.9	0.8
Productive energy (cal./lb.)	920	960	970	1,000
Metabolizable energy (cal./lb.)	1,290	1,320	1,330	1,360

[1] Sorghum grain can be used to replace up to half the corn by weight.

[2] May be replaced by 50 lbs. soybean meal plus 15 lbs. dicalcium phosphate (or equivalent phosphorus).

[3] Trace mineral mix to contain 10% each of manganese and zinc.

[4] Use at manufacturer's recommended levels.

Source: Iowa State University

TABLE 39-4 COMPLETE TURKEY FINISHER RATIONS

Ingredients	16–20 Weeks, Male	16–18 Weeks, Female	After 20 Weeks, Male	After 18 Weeks, Female
Ground yellow corn [1]	1,430 lbs.	1,520 lbs.	1,520 lbs.	1,625 lbs.
Soybean meal (50% protein)	400	310	310	225
Dehydrated alfalfa meal (17% protein)	40	40	40	40
Stabilized fat	40	40	40	40
Dicalcium phosphate (or equivalent phosphorus)	40	40	40	30
Ground oyster shell (or equivalent calcium)	20	20	20	10
Feed grade salt, iodized	9	9	9	9
Trace mineral mix [2]	1	1	1	1
Lysine	0.5	—	1	—
Additives				
Vitamin A (million I.U.)	3	3	3	3
Vitamin D_3 (million I.C.U.)	1	1	1	1
Vitamin K (gms.)	1	1	1	1
Riboflavin (gms.)	3	3	3	3
Pantothenic acid (gms.)	4	4	4	4
Niacin (gms.)	20	15	15	10
Choline chloride (gms.)	300	200	200	200
Vitamin B_{12} (mg.)	5	5	5	5
Antibiotics and medicants [3]	√	√	√	√
Total	2,000 lbs.	2,000 lbs.	2,000 lbs.	2,000 lbs.
Calculated Analysis of Ration				
Protein (%)	16.5	14.5	14.5	13
Calcium (%)	1.0	1.0	1.0	0.6
Phosphorus (%)	0.7	0.6	0.6	0.5
Productive energy (cal./lb.)	1,000	1,040	1,040	1,050
Metabolizable energy (cal./lb.)	1,380	1,400	1,400	1,430

[1] Sorghum grain can be used to replace up to half the corn by weight.

[2] Trace mineral mix to contain 10% each of manganese and zinc.

[3] Use at manufacturer's recommended levels.

Source: Iowa State University

need protection from both the sun and the weather.

A 10′ × 16′ shelter will accommodate 200 to 250 poults. From 9 to 12 inches of roost space should be provided per bird. Roosts may be made of 2 × 4's or of poles and should be placed about 3 feet off the ground.

Feeders

One foot of feeder space should be provided for every three birds. Many types of

TABLE 39-5 TURKEY CONCENTRATES AND BREEDER RATIONS

Ingredients	Concentrates [1] 8–16 Weeks	Concentrates [1] 16 Weeks to Market	Breeding Ration
Ground yellow corn [2]	—	—	1,224 lbs.
Soybean meal (50% protein)	1,240 lbs.	1,240 lbs.	250
Wheat middlings (or heavy oats)	150	150	100
Dehydrated alfalfa meal (17% protein)	100	120	100
Dried fish solubles	80	—	50
Meat and bone scrap (50% protein)	100	200	50
Dried whole whey	60	—	40
Dried brewers' yeast	50	—	40
Stabilized fat	70	80	—
Dicalcium phosphate (or equivalent)	80	110	60
Ground oyster shell (or equivalent calcium)	30	50	60
Feed grade salt, iodized	18	25	6
Trace mineral mix [3]	2	2	1
Methionine	2	—	—
Lysine	—	3	—
Additives			
Vitamin A (million I.U.)	8	10	8
Vitamin D_3 (million I.C.U.)	4	4	2
Vitamin E (I.U.)	10,000	—	40,000
Vitamin K (gms.)	4	4	4
Riboflavin (gms.)	8	10	6
Pantothenic acid (gms.)	10	12	10
Niacin (gms.)	50	50	30
Choline (gms.)	800	700	200
Vitamin B_{12} (mgs)	10	10	10
Folic acid (mgs.)	—	—	500
Antibiotics and medicants [4]	√	√	√
Total	2,000 lbs.	2,000 lbs.	2,000 lbs.
Calculated Analysis of Ration			
Protein (%)	38	38	16
Calcium (%)	2.3	3.5	2.31
Phosphorus (%)	1.5	1.9	1.0
Productive energy (cal./lb.)	750	750	910
Metabolizable energy (cal./lb.)	1,100	1,050	1,225

[1] Concentrates to be fed with corn or other grain.

[2] Sorghum grain or heavy oats can be used to replace up to ¼ of the corn by weight.

[3] Trace mineral mix to contain 10% each of manganese and zinc.

[4] Antibiotics and medicants to be added to concentrates at manufacturers' recommended levels.

Source: Iowa State University

feeders may be made or purchased. The feeder should provide for continuous feeding, without waste, and provide protection of the feed in case of rain. Half of the feeders should be filled with grain and the other half filled with mash.

FEEDING GROWING TURKEYS

Growing Rations

Growing turkeys are usually fed rations made up of farm grains and mash supplements to which grit, antibiotics, and other materials have been added.

Grains. Cracked corn or a mixture of corn, oats, wheat, and sorghums are good feeds for young turkeys on range. After the birds can eat whole grain, it need not be ground. Oats help to feather the birds and may prevent feather picking.

Mineral and Grit. Turkeys consume much fibrous material, so they need grit at all times. Granite grit should be provided. Minerals are usually fed as part of the mash.

Mash Feeds. Most mash supplements for growing turkeys contain from 25 to 40 percent protein. These concentrates permit the grower to feed large quantities of farm grains. The contents of any purchased supplement should be checked carefully, and the cost per pound of protein calculated. The best way to keep turkeys healthy is to keep them eating.

Growing turkeys 8 to 16 weeks old need rations containing 19 to 23 percent protein, 1.2 percent calcium, 0.8 percent phosphorus, 0.5 percent salt, 3,600 U.S.P. units of vitamin A per pound, and 600 I.C.U.'s of vitamin D per pound. The amounts of riboflavin, pantothenic acid, choline, folacin, and manganese needed by growing turkeys are shown in Table 39-1.

Most poultrymen feed grains and mash concentrates. The amounts of grains fed vary with the protein content and amounts of mash feed consumed by the birds. The rations given in Tables 39-2 through 39-5 may be helpful in formulating feeding programs and in evaluating commercial feeds.

FIGURE 39-7. Young turkeys are heavy eaters but grow rapidly. (*Courtesy Turkey World*)

Feeds Required per Pound of Gain

Turkey growers estimate that it takes 50 to 60 pounds of feed to grow out a hen and 80 to 100 pounds to grow out a tom to market weight. Feed required per pound of gain runs from 3 to 4 pounds. A turkey broiler can gain a pound on 2½ to 2¾ pounds of feed.

Antibiotics

Antibiotics will stimulate growth in turkeys as they do in chickens. The greatest increases occur during the brooding stage, but 5 percent increases in growth at 24 weeks of age have resulted from antibiotic feeding.

Aureomycin, bacitracin, penicillin, and Terramycin have been fed with excellent results. The rate of feeding penicillin is from 2 to 4 grams per ton. The other antibiotics should be fed at the rate of 6 to 10 grams per ton.

Arsenicals

Arsenilic acid and 3-nitro-4-hydroxyphenylarsonic acid produce increased growth when fed to turkeys in about the same amounts as antibiotics. Since they are toxic, they should be fed with care and according to the manufacturers' recommendations.

Water

Turkeys need plenty of clean, fresh water. On range, it is better to place the waterer in the sun than in the shade. Provide waterers which will maintain a supply of water, but will not permit the birds to pollute it.

FIGURE 39-8. The effect of feeding Aureomycin to turkey poults. (*Courtesy Turkey World*)

Hormones

Feeding hormones to growing turkeys speeds rates of gain and fattening. Hormones should be fed according to manufacturers' recommendations. Implanting hormones is prohibited.

Cannibalism

Feather pecking and cannibalism can be controlled in turkeys in the same manner as with chickens. Feeding oats in the ration, providing adequate range, and debeaking are the best methods.

CONFINEMENT REARING IN POLE SHEDS

A recent development in the confinement rearing of turkeys is the use of pole sheds. Sheds may be 40 to 60 feet wide and several hundred feet long. The poles are set into the ground 3½ to 4 feet. The sheds usually are gable roofed with 5- to 7-foot side walls, and have a 2- to 3-foot roof overhang. The walls and ends may be enclosed with strong wire

FIGURE 39-9. (*Above*) Turkeys in a large poultry house in Maryland. (Photo by Murray Berman. *Courtesy* U.S.D.A.) FIGURE 39-10. (*Below*) A long, low pole building and feeding equipment used in a large-scale turkey enterprise. (*Courtesy* Aluminum Company of America)

or a combination of siding and wire. In northern states, provision is made to enclose the sheds during bad weather with wood or steel panels, or with canvas.

The turkeys are kept in the shed at all times. Four to five square feet of floor space is provided for each bird. Dirt floors and deep litters are used, and new litter is added as necessary.

Wide partitions are used to separate the turkeys into groups of 300 to 400 birds, and most growers find it necessary to debeak the birds. Feeders and waterers are placed along the sides. The birds roost on the litter, so no roosts are necessary.

FINISHING TURKEYS FOR MARKET

As long as a turkey is growing, it is impossible to get it to put on a thick covering of flesh. Turkeys must be mature to be properly finished. Operations must be planned so that the poults are started in time to be finished for the desired market. The toms of larger varieties will finish out weighing 24 to 28 pounds, and the hens will finish out at 14 to 18 pounds.

Turkeys can be finished on grass and mash, but they should also receive a large amount of whole yellow corn in the ration. The birds should be examined before marketing to see that the breast is broad and full. The back and hip bones should be well covered, and the skin should appear white because of the layer of fat underneath. The bird should have a good growth of feathers. Pinfeathers indicate immaturity and a general lack of condition.

TURKEY BROILER PRODUCTION

The production of turkey broilers has increased with the development of the Beltsville Small White turkey. Twelve percent of the 1972 turkey crop was of the Beltsville Small White breed. Turkey broilers are fed and managed in much the same manner as chicken broilers. They dress out with more breast meat and less leg meat than the chicken broiler, and there is less loss in the dressing process. Turkey broilers dress out at 5 to 7 pounds.

Since day-old turkeys sell for 60 to 80 cents while chicks sell for 10 to 18 cents, the selling price of turkey broilers must be high to show a profit. Turkey broilers are very efficient consumers of feed, but the original cost of the poults limits the profit from the enterprise.

There are three limiting factors in turkey broiler production: (1) the high price of day-old poults, (2) the difficulty of getting a high finish on the young turkey, and (3) finding a sufficiently high market for the birds to yield a profit.

SUMMARY

Turkey growers should buy poults from pullorum-free flocks which have been bred for broad breasts and meat production. Broad Breasted Bronze, Broad Breasted White, White Holland, and Beltsville Small White are the most popular varieties.

It takes about 12 to 15 weeks to produce a turkey broiler and about 20 to 24 weeks to finish out a mature bird.

Turkeys are brooded in much the same manner as chicks, but not more than 250 to 350 poults should be brooded under one hover. Turkeys are more difficult to get to eat than chickens. A starter ration containing 27 to 29 percent protein and fortified with penicillin or other antibiotics should be fed.

Deep, coarse litter should be provided for the poults. The brooder house should be kept dry. A sun porch should be used to increase floor space. A foot of feeder space for every five poults should be provided during the first three weeks. The feeder space should be doubled after four weeks. One 4-foot waterer is required for every 75 poults.

Small lights will help keep the poults from piling up while feeding.

The poults may be moved to a good legume range at eight to ten weeks of age. Clean ground should be used. Feeders and shelters should be moved to new range every week or ten days. Good pasture may save 15 to 20 percent of the feed, depending upon the area. One foot of feeder space should be provided for every two or three birds. Birds on range should be fed a 25 percent protein mash and grains, or a complete mash ration. Provide self-feeders. It pays to feed antibiotics up to market time at the rate of 2 to 10 grams per ton.

It takes 60 to 100 pounds of feed to grow out a turkey, or 3 to 4 pounds of feed per pound of turkey. Turkey broilers will put on a pound of gain on less than 3 pounds of feed.

Sanitation, good rations, and careful management are essential in turkey growing.

QUESTIONS

1. What is the place of turkey production on your farm?
2. Outline a plan for successful brooding of turkey poults.
3. Formulate a starting mash for turkeys.
4. What are the essentials of a good growing mash for turkeys on range?
5. Outline a program of management of turkeys while on range.
6. Of what value is range to turkeys?
7. How can you tell when a turkey is ready for market?
8. What place do antibiotics and arsenicals have in turkey production?
9. What is the future of turkey broiler production in your state?
10. What factors should be considered in buying poults?
11. Under what conditions can turkeys be grown in confinement?
12. What are the advantages of rearing sexed poults?

REFERENCES

Biellier, Harold and Walter Russell, *Artificial Insemination of Turkeys,* University of Missouri, Columbia, Missouri, 1966.

Naber, Edward C. and Sherman P. Touchburn, *Ohio Turkey Rations,* The Ohio State University, Columbus, Ohio, 1970.

National Academy of Sciences-National Research Council, *Nutrient Requirements of Poultry,* 6th Revised Edition, Washington, D.C., 1971.

Owings, W. J. and S. L. Balloun, *Feeding and Managing Turkeys,* Iowa State University, Ames, Iowa, 1966.

Schroeder, Price, *California Turkey Meat Production Costs,* University of California, Davis, California, 1967.

U. S. Department of Agriculture, *Poultry and Egg Situation,* Washington, D.C., April and September 1973.

40 Disease and Parasite Control

It has been estimated that of the 15 to 25 percent of the chickens and turkeys produced in this country which die each year, 60 to 75 percent die because of disease. Poultrymen need to be familiar with the causes and symptoms of the diseases and parasites most likely to cause trouble in order to plan programs of prevention and cure. Carefully selecting chicks and laying stock, providing adequate housing, feeding satisfactory rations, and practicing sanitation in management help prevent disease and parasite losses, but disease outbreaks occur even when these practices are carried out.

DISEASES

The most serious diseases of chicks are pullorum and coccidiosis. Infectious synovitis and coccidiosis cause heavy losses in broiler production. Newcastle disease, Marek's disease, CRD (chronic respiratory, or air sac disease), lymphomatosis, bronchitis, fowl pox, and blue comb cause heavy losses in laying flocks. Many of these diseases and blackhead cause heavy losses in turkeys.

Pullorum

This is an acute, highly infectious bacterial disease which occurs between the second and third day after hatching and when the chicks are about three weeks old.

Symptoms. Infected baby chicks or poults become weak and unthrifty, and show a white diarrhea or "pasted-up behind." Chicks which have been shipped or chilled are most susceptible to the disease. Losses may exceed 30 to 40 percent.

The disease occurs in the ovary of infected hens and passes into the eggs, infecting them before they hatch. Most states have control programs under the National Poultry Improvement Program which encourage hatcheries to use only pullorum-free flocks as sources of hatching eggs.

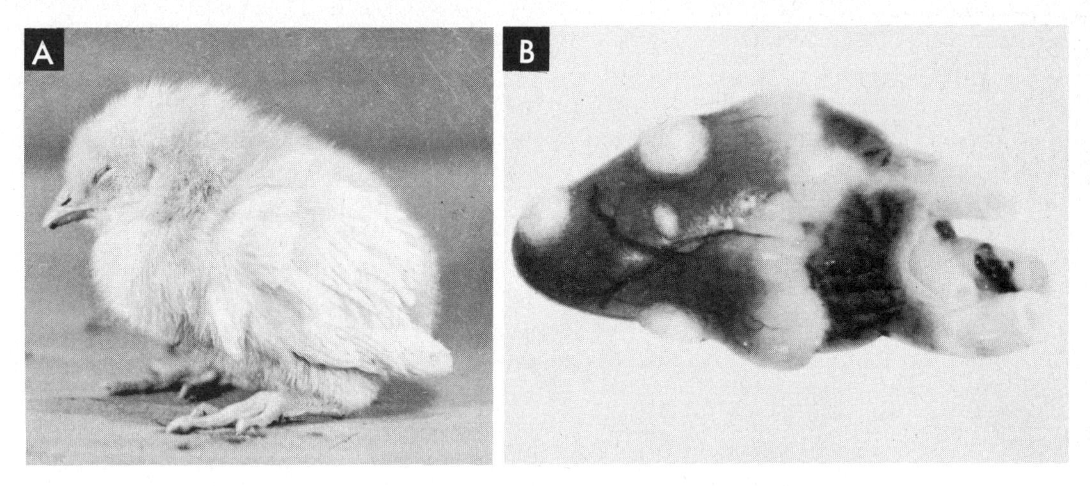

FIGURE 40-1. (*A*) A baby chick with pullorum disease. (*B*) Grayish bumps on the heart of a pullorum-infected bird. **FIGURE 40-2.** (*Below*) Typical symptoms of cecal coccidiosis. (Figures 40-1 and 40-2 *courtesy* Salsbury Laboratories)

Prevention. The best preventative is to buy chicks from hatcheries cooperating in the state pullorum-control programs. These programs require the blood-testing of all birds used in hatchery flocks. A new fast and accurate test has been developed.

Treatment. Infected chicks should be given nf-180 (Furazolidone), sulfamethazine, or other sulfa drugs in water. Follow the instructions of the manufacturer. Keep chicks warm and dry, and their quarters sanitary.

Coccidiosis

Coccidiosis is a protozoan disease which affects the intestinal tract of chicks, usually when they are from three to five weeks of age. Outbreaks occur most often during warm, humid weather in early summer. Cecal coccidiosis is considered the most serious chick disease in this country. It is estimated that from 15 to 20 percent of all chicks hatched in this nation die from this disease before they are a month old.

Symptoms. Chicks infected with the cecal type show bloody diarrhea, become listless, and have ruffled feathers. The birds eat little feed, and the comb and wattles become pale. Chicks with intestinal coccidiosis show the same symptoms as those with the cecal form, but there is less blood in the droppings.

Prevention. The litter must be kept loose and dry. Adequate rations containing Aureomycin or other antibiotics have proven helpful in preventing and treating coccidiosis outbreaks. Chicks develop an immunity to the disease if they recover from a mild case.

Treatment. When coccidiosis strikes, it moves rapidly. Diseased birds or the entire flock should be treated immediately with sulfa drugs according to the directions of the manufacturer or the veterinarian. Sulfaquinoxaline may be fed in mash at the rate of 4 ounces per ton as a control, or at the rate of one pound per ton of chick feed as a treatment. About 3½ ounces of Aureomycin and 2½ pounds of sulfamethazine mixed in each ton of feed is a very effective treatment. Amprol and Sulmet are also recommended.

FIGURE 40-3. (*A*) Symptoms of Newcastle disease. (*Courtesy* Salsbury Laboratories) (*B*) Vaccinating birds for Newcastle disease by using a spray method. (*Courtesy* American Scientific Laboratories, Inc.)

Infectious Synovitis

This disease has become serious during the past twenty years. It may affect 75 percent of the birds. It is widespread in broiler-producing areas, but may also affect layers and turkeys.

Symptoms. Infected birds lose their appetites and flesh, become listless, and have pale combs. They walk with a stiff, halting gait, and sit on their haunches much of the time. The hock joints of the legs are swollen and hot to the touch.

Prevention. Good management, consisting of sanitation, plenty of floor space, good ventilation, humidity control, adequate rations, and careful culling of suspicious birds, appears to be the best preventive measure. Medication may be used, but it is costly if used for long periods of time at high levels.

Treatment. In Delaware tests, the feeding of 200 grams of nf-180 (a nitrofuran derived from oat hulls and corncobs) or 100 grams of Aureomycin or Terramycin per ton of feed gave the best results.

Newcastle Disease

Newcastle disease has spread rapidly since about 1942, when it was discovered in California. It is a filterable virus which enters the birds through the respiratory tract. It affects chickens, turkeys, and sometimes other species of birds. It is highly infectious and causes death losses amounting to 25 to 80 percent in both young chickens and pullet flocks.

Symptoms. The usual symptoms are similar to those of bronchitis, tracheitis, and colds. Chicks sneeze, gasp, and have difficulty in breathing. They are weak and have ruffled feathers. Within a few days, nervous symptoms begin. They may twitch their heads and necks, and the head may become twisted and be drawn between the legs or back over the shoulders. Birds may walk in circles or backwards.

Laying hens have breathing trouble, and they gasp, wheeze, and rattle. Although they do not have serious nervous symptoms, egg production completely stops within a few days, and soft-shelled floor eggs are found.

Prevention. If Newcastle disease has not been present on the farm or in the community, practice sanitation in production and management. Isolate all new birds and keep visitors out of the poultry houses and yards. Vaccinate chicks at four weeks of age, using B_1 type vaccine.

On farms where Newcastle is known to exist, all chicks, regardless of age, should be vaccinated with Newcastle disease B_1 type vaccine.

Newcastle vaccines may be used alone, or with bronchitis vaccine in dust form, or in drinking water. Commercial vaccines are available and should be used according to manufacturers' directions. Breeders should be revaccinated in 90 days.

Treatment. There is no medical cure. Palatable rations should be fed and desirable housing quarters maintained. Birds that recover may be kept in the flock.

Bronchitis

This respiratory disease may affect chickens at any age, but death loss is highest among chicks. Infected layers may be out of production for several weeks, but few birds die.

Symptoms. Infected birds have a raspy cough and gasp, and usually there is a discharge from the nose and eyes. The disease may spread through the entire flock in a day's time. Bronchitis resembles Newcastle disease, laryngotracheitis, and coryza.

Prevention. Chicks, poults, or mature birds may be immunized with a live-virus vaccine. It can be administered as a dust, in drinking water, or by the nose-drop method; day-old birds can be vaccinated by the latter method.

Treatment. Although there is no cure, affected birds can be made comfortable, and when they begin to eat again, they should be fed good rations.

Chronic Respiratory Disease

This disease, which is commonly called CRD or air sac infection, was discovered in

FIGURE 40-4. (*A*) Typical signs of CRD infection. (*B*) Lesions of CRD. The heart sac is thick and white. The liver is covered with a slimy false membrane. (*Courtesy* Salsbury Laboratories)

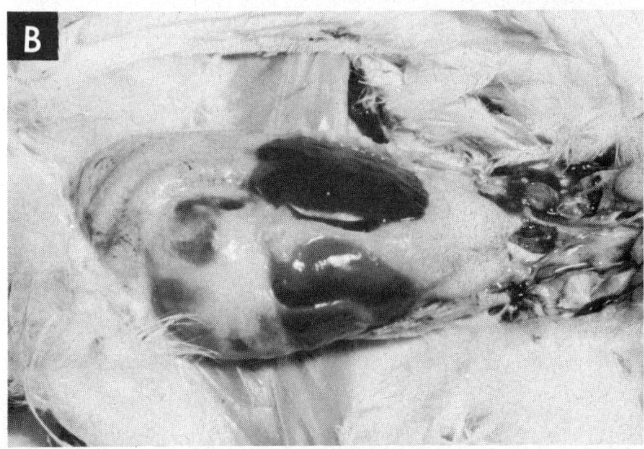

FIGURE 40-5. A bird infected with laryngotracheitis. Stretching the neck and inhaling is characteristic. (*Courtesy* Salsbury Laboratories)

1943. It is now considered the most serious disease of the broiler industry.

The infection is caused by a pleuropneumonialike organism, which can be transmitted by direct contact or to chicks through the eggs of infected hens.

Symptoms. CRD causes young chickens to sneeze, and they lose weight even though they continue to eat. The sinuses become swollen and breathing is accompanied by respiratory rales. Adult birds appear to have colds and their production decreases. Inflammation of the sinuses and trachea and thickening of the air sacs is common in chicks. Breathing may become difficult.

Prevention. Adequate rations, proper sanitation, good ventilation, and careful timing of vaccination to avoid stresses (periods when birds are physically weak) are musts. Chicks should be obtained from hatcheries that produce lines resistant to the disease, or from lines which by test have proven to be free from the disease. Feeding antibiotics is desirable as a preventative, as it retards the growth of the organism.

Treatment. There is no vaccine. Treat with nf-180 and feed antibiotics. Use good management practices.

Laryngotracheitis

Fowl laryngotracheitis is a highly communicable and fatal virus disease. It attacks the larynx and windpipe and is most serious in pullets from five to ten months old. Birds that recover may be carriers and spread the infection to others.

Symptoms. The infection strikes rapidly. Dead birds may be found before symptoms are noticed. The chief symptoms are labored, gasping, and rattling breathing, with the bird's head extended and the beak open. Watery discharge from the eyes and nostrils may also be present. Birds may cough violently and expel bloody mucus, and they often die of suffocation.

Prevention. Birds should be vaccinated when six weeks of age or shortly thereafter. The vaccine is applied in the vent with a brush. Immunity develops in six to nine days. All vaccinated birds should be checked after four or five days to see whether the vaccination took effect. A reddening of the vent or a cheesy mucus indicates a "take."

Treatment. Vaccinate all birds. Rely on a veterinarian, feed a good ration, and practice sanitation. Burn all dead birds immediately.

The vaccine may be administered in the drinking water. The birds should be vaccinated twice, first at four to six weeks and later at 14 to 16 weeks. Heavy doses of vaccine are recommended.

Marek's Disease

California poultrymen rate this as one of their major problems in egg production. The disease is very contagious and causes high mortality. It does most damage to broilers and layers.

Symptoms. The disease strikes fast. Many birds die. Diagnosis must be based on the presence of tumorlike accumulations of lymphoid cells in the nerves, skin, and viscera. The disease is caused by a herpes virus which spreads by contact and by air. It may live in litter and be present in dust. Birds 8 to 20 weeks of age are most susceptible.

Prevention. A vaccine was approved in 1971 which, when properly used, will protect chicks and birds against the disease. It is comparatively high in cost and it has been used largely with breeding and laying stock. It is recommended that chicks be vaccinated at the day-old stage. Protection lasts up to 18 months. Five hundred units of vaccine per chick is recommended. Thorough cleaning of quarters and use of chemical disinfectants are also recommended.

Treatment. There is no treatment other than providing sanitary quarters and adequate rations.

Infectious Sinusitis

This is a serious respiratory disease of turkeys and is caused by the same microorganism that causes CRD in chickens. Airsac infection, a form of infectious sinusitis, may affect birds of all ages and cause high mortality in poults. The more common form of infectious sinusitis does not result in high mortality, but heavy losses often result from retarded growth and poor condition.

Symptoms. One or both sides of the face become swollen. The eyes may be closed, and the birds have difficulty in breathing.

Prevention. There is no vaccine. Poults should be purchased from reliable hatcheries, and good management practices should be followed.

Treatment. Two treatments are used: (1) A mixture of streptomycin and penicillin may be injected into the muscle of infected birds, and the birds permanently removed from the flock. This mix of antibiotics may also be fed in dry mash. (2) Remove the fluid from the swollen sinuses with a sterile syringe and a 1½-inch 18- or 19-gauge needle. Inject 4 cubic centimeters of 1 percent silver nitrate solution into the sinus. Do not put the bird with the flock.

FIGURE 40-6. Leukosis. The ocular form causes birds to have gray eyes with irregular pupils. (*Courtesy* Salsbury Laboratories)

Lymphomatosis or Leukosis

Poultrymen may know this disease as fowl paralysis, big-liver disease, gray-eye, or thick-leg disease. All are forms of leukosis. The group of diseases is one of the greatest causes of mortality in half-grown and adult chickens.

Symptoms. The symptoms vary with the form of the disease. Range paralysis affects young birds from 6 to 12 weeks old. Infected birds show partial paralysis of one or both legs or wings. With the gray-eye form, the iris loses its normal color and becomes gray and distorted. Birds affected with visceral lymphomatosis develop pale combs and lose appetites and flesh. The liver and other organs become enlarged. Birds with the big-bone or thick-leg type develop abnormally thick legs.

Prevention. A virus which can be transmitted from hen to chick in the egg or from bird to bird causes the disease. Chicks or poults should be purchased from reliable hatcheries and reared in sanitary quarters, in isolation from all mature birds. Isolated, clean ground is most important. All questionable birds should be removed from the flock as soon as they are identified.

Treatment. The disease is cancerous in nature and there is no treatment. Isolation of infected birds and proper feeding and management of the flock should be practiced.

Fowl Pox

Chicken pox or sorehead is caused by a virus. It is quite infectious and shows up in two different forms. One form affects the skin and comb, the other affects the throat.

Symptoms. Small blisterlike spots appear on the face, comb, and wattles. Light infection results in a few scattered spots. Heavy infection may cover the face, comb, and wattles. The diphtheritic form affects the mouth and throat, and yellow puslike patches appear. The eyes may water, the birds will lose their appetites, and production will diminish.

Prevention. Almost complete protection will result from the vaccination of chickens while they are from 6 to 16 weeks of age. Veterinarians may use the stick method. If the disease has been present on the farm, vaccination each year is recommended. Vaccinate with pigeon pox if birds are laying and with fowl pox if not laying.

Treatment. There is no satisfactory treatment.

FIGURE 40-7. Fowl pox. (*Courtesy* Salsbury Laboratories)

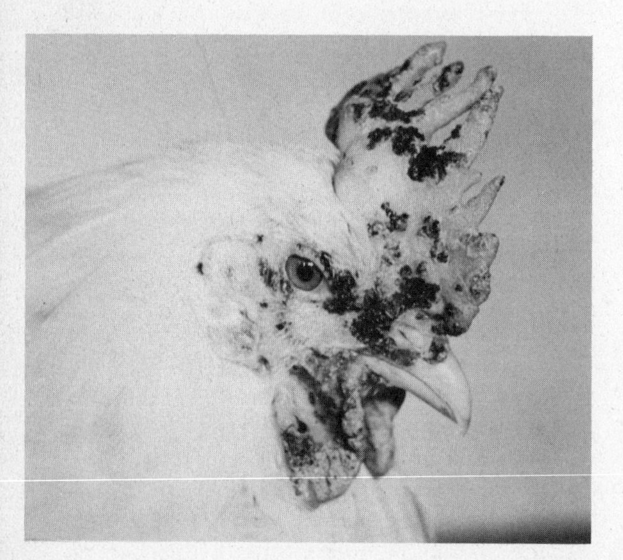

Roup

This infection is also known as coryza or cold. It is an infectious inflammation of the sinuses and upper respiratory tract and is caused by a bacterium. Death losses are low, but financial losses result from decreased growth and production of diseased birds.

Symptoms. A thin, watery discharge from the nostrils during the early stages of infection changes to a thick, sticky discharge which has an offensive odor. Roup is usually detectable by this odor. Yellow masses of mucus form around the nostrils and eyes, and the birds may cough and sneeze.

Prevention. The isolation of birds brought onto the farm, and the use of sanitation in feeding and management are the best preventive methods.

Treatment. Sulfa drugs are effective in controlling roup. A half pound of sulfathiazine to every 100 pounds of mash is recommended, or sulfamethazine may be fed in water. The directions of the manufacturer must be followed. Injecting 50 milligrams of streptomycin per pound of body weight is also effective.

Typhoid

Fowl typhoid is a bacterial disease which affects both chickens and turkeys. Losses average about 25 percent, but 80 percent losses have been reported. The disease occurs mostly in mature birds.

Symptoms. Birds with typhoid are droopy, have little appetite, and crave water. The livers are infected, and birds show a profuse, watery, green diarrhea. Chickens may suffer from a few days to about two weeks. Turkeys die within two or three days.

Prevention. All birds in the laying or breeding flock should be blood-tested and carriers eliminated. Birds may be vaccinated for typhoid with avisepticus-gallinarum bacterin, but a veterinarian or a poultry specialist should do it.

Treatment. Sulfa-drug treatments, with or without repeated injections of bacterins, will aid in controlling the disease. Sodium sulfamerazine fed in the drinking water is recommended. The directions of the manufacturer or veterinarian must be followed.

Fowl Cholera

This disease affects all species of poultry and causes heavy losses. It is very contagious and spreads rapidly, and, at times, birds begin to die before the disease has been discovered. It is caused by a bacterium.

Symptoms. The infected birds are weak, droopy, and inactive. They have ruffled feathers, a loss of appetite, and a fever. The heads, combs, and wattles become dark colored, and a greenish or yellowish diarrhea is evidenced. Birds rarely recover and birds which appear healthy may die suddenly.

Prevention. All healthy birds should be vaccinated if the disease is prevalent in the community or on the farm. The vaccine is a bacterin that controls cholera. A dual-purpose typhoid-cholera vaccine is less effective.

Treatment. Sulfamethazine in the drinking water will aid in preventing deaths. Sanitation is an absolute necessity in controlling cholera. Albamix has been approved as a treatment.

Tuberculosis

Tuberculosis is caused by microbacterin tuberculosis avium. It is infectious and occurs in many types of poultry.

Symptoms. The disease rarely affects birds less than 18 months old because it takes many months to take effect. Infected birds lose weight, are inactive, and stop laying. The combs and wattles shrink and become dry and pale in color. The birds finally die.

Prevention. There is no treatment for tuberculosis, and the best control method is to sell all hens at the end of the first laying season. The hens are sold as stewing hens, and the processors can identify any diseased birds and remove them.

FIGURE 40-8. Blackhead. A typically affected turkey. (*Courtesy* Salsbury Laboratories)

Pullets which have been raised under sanitary conditions and are housed in a well disinfected house rarely become infected.

Since the infection can spread from chickens to hogs, it is necessary that all diseased or dead birds be kept away from the hogs.

Blackhead

This is one of the very infectious protozoan diseases. It affects chickens, turkeys, and other fowl, but is particularly fatal to turkeys.

Symptoms. The common symptoms of the disease in turkeys are lowered head, drooping wings, ruffled feathers, drowsiness, and a yellowish diarrhea. Young turkeys may die without showing signs of the infection.

The disease is mild in chickens but spreads from chickens to turkeys. It lives over in the cecal worms, or worm eggs, of the chickens.

Prevention. Grow poults and chicks on clean ground and in separate quarters. Worm flocks with phenothiazine at the rate of one pound per 100 pounds of mash for every 1,000 birds.

Treatment. When an outbreak occurs, healthy turkeys should be removed to a clean range, and the birds treated with phenothiazine mash to remove cecal worms. Enheptin and other materials which are helpful in treating infected birds are available from veterinarians.

Many turkey producers use 4-nitro phenylarsonic acid in blackhead control. The drug may be given in feed or water throughout the growing period to prevent the establishment of the disease.

Blue Comb

This disease in chickens is similar to blue comb or mud fever in turkeys. Although the causal organism has not been found for either disease, it is known that they are not caused by the same organism. Losses may be heavy, especially in turkeys.

Symptoms. In chickens, the disease usually shows up in young pullets just coming into production. Birds lose their appetites, become listless, and develop diarrhea. The comb and face become dark or bluish in color. Egg production drops and birds lose weight. Dehydration takes place. The disease appears to be infectious, and the entire flock may be affected. The disease runs its course in 10 to 14 days. Similar symptoms are present in turkeys. High mortality results.

Prevention. Good sanitation, housing, feeding, and management are recommended. Vaccination and housing of pullets should be timed to avoid stresses which may make the birds susceptible to blue comb infection.

Treatment. The use of Aureomycin and other antibiotics in feed or drinking water is recommended. Potassium chloride may be fed in drinking water at the rate of one tablespoonful per gallon. The feeding of nf-180, or 3-nitro or 4-nitro phenylarsonic acid is also recommended. Manufacturers' directions must be followed.

Turkey poults may be treated by adding 25 pounds of calf milk replacer and 3 pounds of muriate of potash to every 100 gallons of drinking water. A broad-spectrum antibiotic should be used.

Erysipelas in Turkeys

This disease is caused by the same organism that causes swine erysipelas. Outbreaks usually occur in September and October. Young poults are not usually affected. Turkey toms are more susceptible to the disease than hens. Most outbreaks occur in flocks three to six months of age.

Symptoms. Birds may be listless and weak, with wings and tails drooped. There may be some diarrhea. The snood of the toms appears reddish-purple. There may be mot-

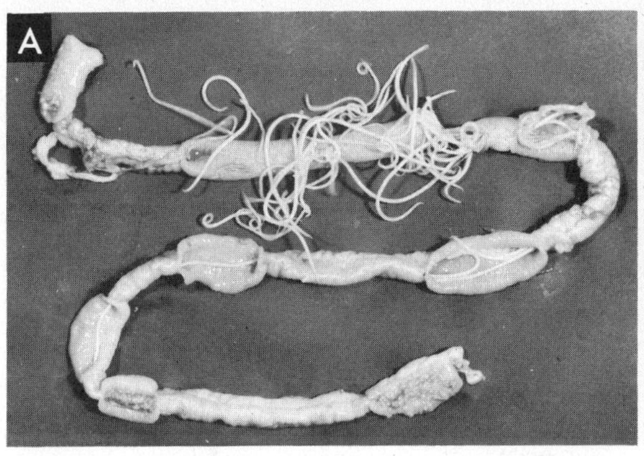

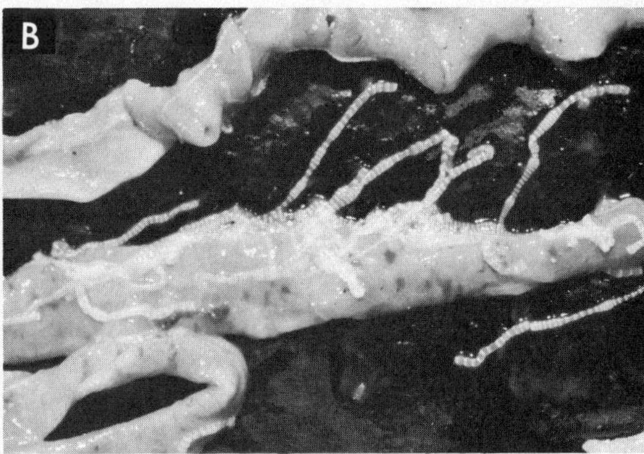

FIGURE 40-9. (*A*) Large roundworms in intestine. (*B*) Tapeworms. These flat, segmented parasites attach themselves to the intestinal lining. (*Courtesy* Salsbury Laboratories)

tling of the face color. Some birds die without showing outward signs of the disease.

Prevention. Birds should be vaccinated with erysipelas bacterin when 8 to 12 weeks old if erysipelas has been prevelant.

Treatment. Penicillin and streptomycin may be used in treating infected birds.

PARASITES

Large roundworms and cecal worms are the most serious internal parasites of poultry. Lice and mites are the serious external parasites.

Large Roundworms

These worms live in the small intestines, and the eggs are passed with the droppings. The eggs, which mature in about two months, are picked up by other chickens.

Symptoms. When heavily infested, young chickens become unthrifty and stunted. Some may die. Mature birds are droopy and have pale combs and wattles and ruffled feathers. Their appetites are poor, and they stop laying. The large worms live on food which has already been digested, thus using food needed by the bird in growth and egg production.

Prevention. Sanitation and rotated ranges are the best preventatives. Thorough cleaning and disinfecting of the houses and equipment is a must. Good litter, well managed, and the use of dropping pits are good aids. Young pullets ought to be grown on a clean range.

Treatment. One ounce of phenothiazine in one feeding for 60 medium weight mature birds, or one ounce per 100 three-pound young chickens is recommended. Hygromycin B may be used.

Piperazine compounds which can be fed in water are now available. They can be fed to young birds and layers without harmful results. Dibutyltin dilaurate may be fed to control tapeworms.

Cecal Worms

These worms are usually about half an inch long and grayish white in color. They live in the ceca, or blind pouches. The eggs are expelled with droppings, and the life cycle of the parasite is similar to that of the roundworm. Eliminating the worm is important because it serves as a host for blackhead disease.

Symptoms. The worms cause inflammation of the ceca but there are usually no outward symptoms.

Prevention. These worms are prevented in the same manner as roundworms. Sanitation practices and rotated ranges are necessary. Chickens and turkeys should be kept on separate, clean ranges.

Treatment. Phenothiazine may be fed in the mash at the rate of one pound to 100 pounds of mash for 1,000 birds. A compound containing 15 grams of 40 percent nicotine sulfate, 151 grams of phenothiazine, 287 grams of bentonite, and 44 pounds of chicken mash has produced excellent results.

Lice

Feeding on feathers and skin, head, body, and shaft lice infest poultry and cause losses in egg production and growth.

Symptoms. Birds are unthrifty, feathers may be ruffled, and production decreases.

Treatment. Birds should be dusted with malathion, Sevin or Co-Ral. These pesticides should not be used during the 10-day period previous to marketing the birds. Apply malathion in nests and on litter and roosts. One part of 50 percent malathion concentrate to 49 parts water or two tablespoonfuls of 50 percent Sevin (Carbaryl) wettable powder in one gallon of water may be used in spraying the inside of the house and the equipment. Follow Pure Food and Drug Administration recommendations.

Mites

Chicken mites live on the roosts or on other parts of the house during the day and attack the birds at night. They are blood-sucking insects.

Symptoms. Birds may appear droopy or listless. Masses of mites may be found during the day on roosts and in cracks.

Treatment. Roosts, nests, boards, and the lower part of the hen house should be sprayed with an insecticide, creosote oil, or Carbolineum mixed with kerosene. Scaly leg mite can be controlled by dipping the shanks in kerosene. The chemicals recommended for control of lice will also control chicken mites. One percent malathion or a 0.5 percent Sevin spray is recommended.

SUMMARY

Pullorum in chicks and poults can be controlled by buying chicks from hatcheries cooperating in the National Poultry Improvement Program. Coccidiosis is the most serious chick disease. It can be prevented by buying early chicks, by keeping litter dry, and by feeding antibiotics. Amprol, Sulmet, sulfaquinoxaline, and sulfamethazine are the best treatments. Nf-180 or other nitrofurans provide the best treatments for infectious synovitis.

A veterinarian should be called when an outbreak of the respiratory diseases takes place. There is no cure for typhoid. Birds may be vaccinated for laryngotracheitis when they are six weeks of age or older. On farms where Newcastle has not been present, sanitary methods should be used, and new birds must be isolated. Where Newcastle is known to exist, all birds should be vaccinated.

Buy chicks from strains known to be free from range paralysis because there is no cure. Roup can be identified by the offensive smell of the discharge from the nostrils. Use sulfa drugs and good management as controls.

For typhoid, give the birds blood tests. Vaccinate birds for typhoid and cholera, and burn dead birds. Practice sanitation, and feed sulfa drugs as controls.

Tuberculosis affects older birds. Pullets should be grown on clean ground, and all old hens should be sold off after one year's production. Any old birds held over should be given blood tests. Clean and disinfect the house and premises.

Blackhead is best controlled by using clean range and by getting rid of cecal worms. Enheptin and other medicines are available and should be fed according to directions. Infected birds should be moved periodically to new range.

Vaccination can be used for infectious bronchitis, Newcastle disease, infectious laryngotracheitis, fowl pox, fowl cholera, and typhoid.

Worm chickens with piperazine compound, phenothiazine, or nicotine-bentonite compounds. Follow directions. Lice and mites can be controlled by spraying facilities and equipment with malathion or Sevin solutions.

QUESTIONS

1 What practical method do you have for control of pullorum on your farm?
2 What methods would you use to control an outbreak of coccidiosis on your farm?
3 How can you tell the difference between Newcastle disease, bronchitis, roup, and laryngotracheitis? How would you treat these diseases?
4 Which of the diseases of poultry are transmitted through the egg from the hen to the chick?
5 How can you control range paralysis?
6 What are the means of controlling cholera and typhoid?
7 What are the best methods of controlling CRD and Marek's disease?
8 Describe the various forms of leukosis and indicate how this disease may be controlled on your farm.
9 Which is preferred in vaccinating broilers for fowl pox—pigeon pox or fowl pox vaccine? Why?
10 List treatments which will control both chicken lice and mites.
11 How can you control round and cecal worms in chickens?
12 Outline a program for maintaining a healthy poultry enterprise on your home farm.

REFERENCES

Bell, R. R., *et al.*, *A Manual of Poultry Diseases,* Texas A & M University, College Station, Texas, 1966.

Bundy, C. E. and R. V. Diggins, *Poultry Production,* Prentice-Hall, Inc., Englewood Cliffs, New Jersey, 1960.

Card, Leslie E. and M. C. Nesheim, *Poultry Production,* 10th Edition, Lea & Febiger, Philadelphia, Pennsylvania, 1966.
1969.

Dr. Salsbury's Turkey Disease Manual, Salsbury Laboratories, Charles City, Iowa, 1969.

Kingsbury, F. W., *Rutgers Vaccination Schedule for the Laying Flock,* Rutgers, The State University, New Brunswick, New Jersey, 1965.

Manual of Poultry Diseases, Salsbury Laboratories, Charles City, Iowa, 1966.

Price, Manning A., *et al., External Parasites of Poultry,* B-1088, Texas A & M University, College Station, Texas, 1969.

41 Marketing Poultry Products

Producers are confronted with four major problems in marketing their products: (1) how to keep the volume of production in line with consumer demand, (2) how to plan production so as to market the products at the peak seasonal price, (3) how to produce the quality of products desired by the processor and consumer, and (4) how to locate the best market.

As an enterprise, poultry contributes less than 25 percent of farm income in many states. The poultry producers in those states, however, are big producers. The small producer with a home flock contributes very little to the poultry income of the state. The enterprise may contribute to family living, especially in marginal, low-income areas.

A high percentage of poultry income comes from the enterprises of large-scale specialized producers. Production and marketing procedures are carefully analyzed and coordinated in order to obtain the highest possible return on capital invested and products produced.

MARKETING EGGS

It has been estimated that egg income per hen can be increased from 50 cents to $1.00 per year by having maximum production during the periods of high prices, by producing high-quality eggs, and by selecting markets carefully.

Seasonal Price Trends

Egg prices fluctuate during the year in much the same manner as hog prices.

Egg Production Trends. Production has been high during March, April, May, and June, and the price of eggs during these months is low. Egg production during August, September, October, and November is comparatively low, and prices are high. With improved methods of production, flock owners are now producing a higher percent-

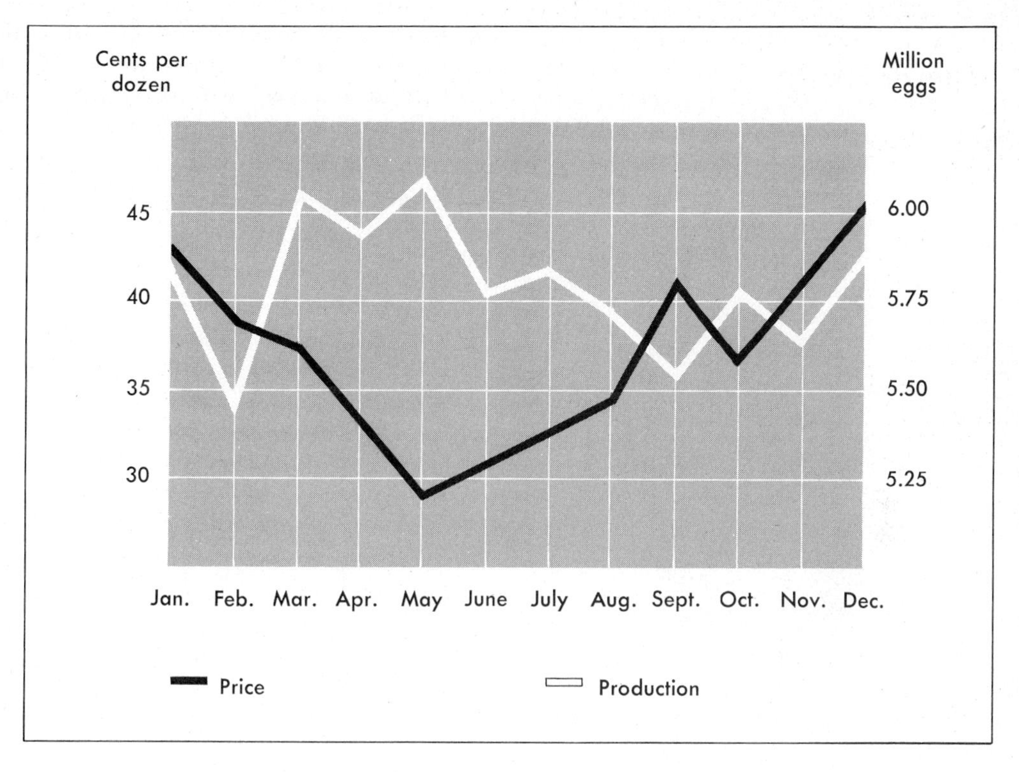

FIGURE 41-1. Egg production and average monthly prices (1968–1970). Note that when egg production is greatest, prices are lower. (U.S.D.A. *Agricultural Statistics, 1971*)

age of eggs during the months when prices are high than they were 15 or 20 years ago. Presented in Figure 41-1 are the average monthly egg production and the average price received by farmers for eggs each month during the three-year period 1968 through 1970. Note that fewer than 5.75 million eggs were produced during the months of August and September. Peak production averaging 6.03 million eggs was produced during March, April, and May.

Price Trends. As shown in Figure 41-1, egg prices during the months of peak production (March, April, and May) averaged 33.4 cents per dozen. The average price during September, October, November, and December, when production was low, was 39.6 cents per dozen. One of the keys to egg profits is to get high production during the months when prices are high.

Classes and Grades of Eggs

Care in the production and handling of eggs may net from 3 to 7 cents more per dozen if they are sold by grade. If a flock of hens averages 240 eggs per hen each year, from 60 cents to $1.40 profit per hen can be cleared by properly caring for the eggs and selling them by grade.

Eggs are classed in grading according to color, weight, and quality.

Consumer Grades of Shell Eggs. The main factors in determining the quality of an egg are the albumen quality, the absence of defects, and the shell quality. Commercially, egg quality is determined by candling. The four United States consumer grades of eggs are illustrated in Figure 41-3. The characteristics of these grades of eggs are as follows:

AA Quality: A clean, unbroken, practically normal shell. A practically regular air cell, ⅛ inch or less in depth. Clear, firm white. Yolk outline slightly defined; practically free from defects.

A Quality: A clean, unbroken, practically normal shell. A practically regular air cell, 3/16 inch or less in depth. Clear, reasonably firm white. Yolk outline may be fairly well defined; practically free from defects.

B Quality: Clean to slightly stained, unbroken shell, but may be slightly abnormal. Air cell ⅜ inch or less in depth. May be free or bubbly. Clear, slightly weak white. Yolk outline may be well defined; may be slightly enlarged or flattened; may show definite but not serious defects.

C Quality: Clean to moderately stained, unbroken shell, may be abnormal. Air cell may be over ⅜ inch in depth; may be free or bubbly. White may be weak and watery; small blood clots or spots may be present. Yolk may be enlarged, off-center, and flattened; outline plainly visible; may

FIGURE 41-2. (*Above*) Candling is necessary to determine egg quality. (*Courtesy Wallaces Farmer*) FIGURE 41-3. (*Below*) U.S. Standards for quality of eggs. (*Courtesy* U.S.D.A. Agricultural Marketing Service)

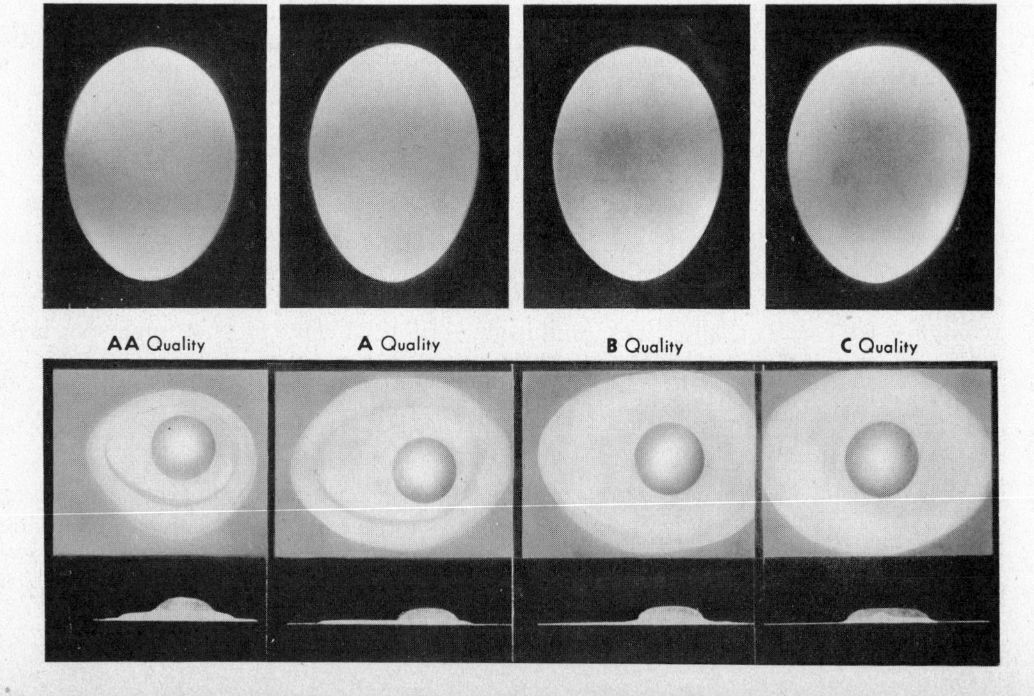

show visible germ development but no blood; may show other serious defects.

Weight Classes of Eggs. The United States consumer weight classes of eggs are shown in Figure 41-4. Wholesale eggs are graded according to weight as follows:

Extra large	Minimum weight, 26 ounces per dozen
Large	Minimum weight, 23 ounces per dozen
Medium	Minimum weight, 20 ounces per dozen
Small	Minimum weight, 17 ounces per dozen

Commercial Grades of Eggs. About 20 million cases of the eggs produced and available for market in the United States in 1955 were graded under Federal or Federal-State grading programs. In 1962, 34 million cases were graded under these programs. All eggs sold commercially for human consumption in 1972 were graded.

State Egg Laws. As of January 1972, all states had some type of egg law. Most of the egg laws were enacted: (1) to improve the quality of eggs on the consumer market, and (2) to insure the producer of high-quality eggs a higher price than that received for ungraded eggs or eggs of inferior quality.

Egg laws have improved the market for eggs. Improvement in egg quality is desirable. In most states, egg laws have tended to reduce the number of small flock owners and increase the number of large, specialized egg producers.

Practices in Maintaining Egg Quality

The marketing of eggs begins as soon as they are laid. Fresh eggs have high quality, but a few hours of heat and dry air will lower the quality of an egg to second grade. The following practices will aid in maintaining the high quality of fresh eggs. No practice should be disregarded. It takes all of them to maintain high quality in the eggs produced.

1. **Use the cage system with the mechanical egg-gathering system.**
2. **Gather eggs three times a day in winter and five times a day in summer.**
3. **Use wire baskets for gathering and cooling eggs.**
4. **Take the eggs to a cool place immediately when gathered, and cool them overnight.**
5. **Case the eggs in precooled cases, after cooling.**

FIGURE 41-4. U.S. weight classes of eggs (minimum weight per dozen). (*Courtesy* U.S.D.A. Agricultural Marketing Service)

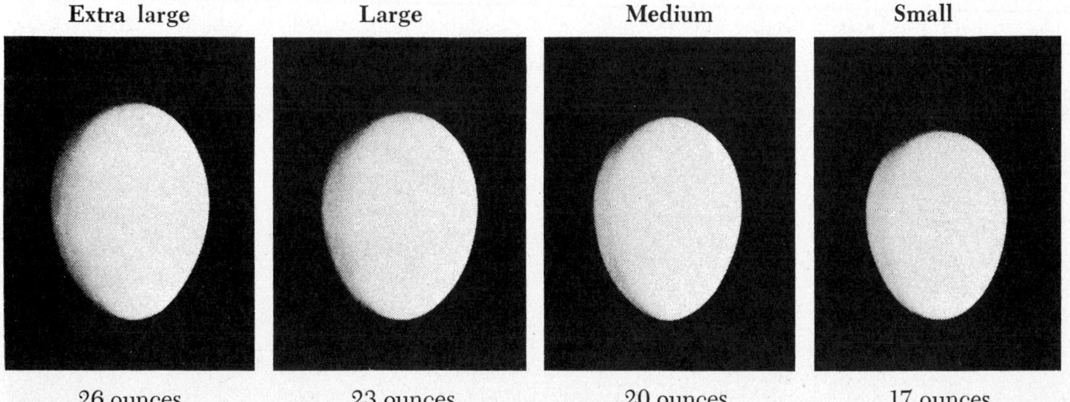

FIGURE 41-5. (*Above*) Eggs should be gathered and cooled in wire baskets. (*Courtesy* Nutrena Mills, Inc.) **FIGURE 41-6.** (*Right*) A mechanical egg cooler. (*Courtesy* Buckeye Incubator Co.)

6. Pack eggs with the small end down.
7. Store eggs in a cave, basement, or mechanical cooler where the temperature is 50° to 60° F. and the air is moist.
8. Sort out the small, irregular, and dirty eggs.
9. Keep male birds away from the laying flock. Produce infertile eggs.
10. Feed complete rations in adequate amounts.
11. Confine layers.
12. Provide at least one nest for every five hens, when nests are used.
13. Keep clean shavings, excelsior, or Chick Bed in the nests, if used.
14. Do not permit birds to roost in nests at night.
15. Keep dropping boards or pits covered with wire.
16. Provide ample floor space and keep the litter dry.
17. Keep broody hens out of the nests.
18. Keep eggs away from onions, kerosene, potatoes, and other products from which eggs may absorb odors.
19. Remove cracked eggs and wash all sound eggs that are soiled.
20. Oil-treat eggs in the basket by spraying them when they are gathered.

MARKETING POULTRY

The rapid increase in turkey and broiler production has changed the poultry marketing picture. At one time, practically all the poultry marketed in this country came from farm flocks and, in the main, was a byproduct of egg production.

In 1972 more than nine times as many pounds of commercial broilers as of farm-raised chickens were marketed in this country. According to U.S.D.A. data, 11,158 million pounds of commercial broilers, 2,424 million pounds of turkeys, and 1,195 million pounds of farm-raised chickens were marketed in this country in 1972.

The production of chicken and turkey broilers and improved methods of refrigeration and storage have helped to distribute marketing throughout the year and minimize price variations.

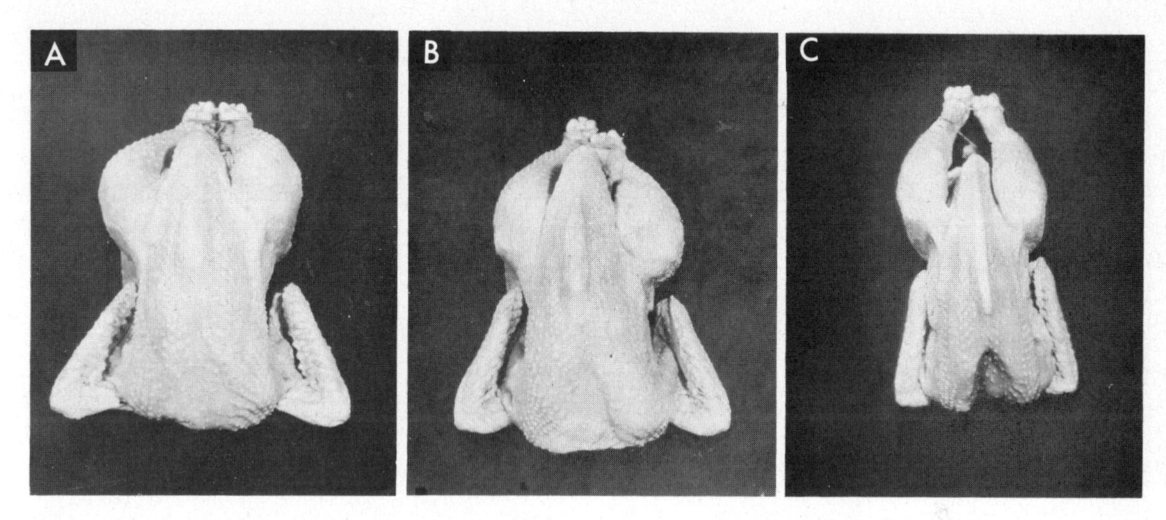

FIGURE 41-7. (*A*) Young chicken carcass—A quality. (*B*) Young chicken carcass —B quality. (*C*) Young chicken carcass—C quality. (*Courtesy* U.S.D.A. Agricultural Marketing Service)

MARKET CLASSES OF POULTRY

Market classes are provided for chickens, turkeys, ducks, geese, guineas, and pigeons. Three classes are provided—live poultry, dressed poultry, and read-to-cook poultry. Most poultry sold by farmers is sold as live poultry, although the marketing of dressed and read-to-cook birds is done on a small scale by some farmers. Commercial producers may process the birds or sell them as live poultry.

Chickens

The following classes of chickens are recognized by the Poultry Branch, Production and Marketing Administration of the United States Department of Agriculture.

Rock Cornish Game Hen or Cornish Game Hen. A young immature chicken (usually 5 to 7 weeks of age) weighing not more than 2 pounds read-to-cook weight, which has Cornish hen or Cornish hen crossed ancestry.

Broiler or Fryer. A broiler or fryer is a young chicken (usually 9 to 12 weeks of age) of either sex that is tender-meated and has soft, pliable, smooth-textured skin and flexible breastbone cartilage.

Roaster. A roaster is a young chicken (usually 3 to 5 months of age) of either sex with the same characteristics as a broiler or fryer, except that the breastbone cartilage is somewhat less flexible.

Capon. A capon is a surgically de-sexed male chicken (usually under 8 months of age) with the meat and skin characteristics of a broiler or fryer.

Stag. A stag is a male chicken (usually under 10 months of age) with a coarse skin, somewhat toughened and darkened flesh, and considerable hardening of the breastbone cartilage. Stags show a condition of fleshing and a degree of maturity intermediate between those of a roaster and a cock.

Hen or Stewing Chicken or Fowl. These birds are mature female chickens (usually more than 10 months old) with meat less tender than that of a roaster and nonflexible breastbones.

Cock or Old Rooster. A cock or old rooster is a mature male chicken with coarse

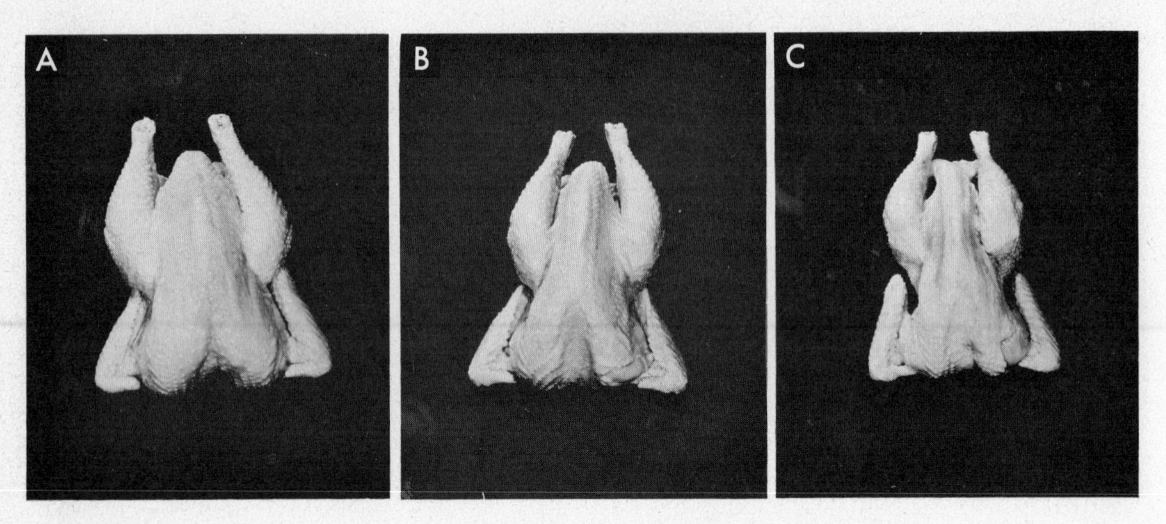

FIGURE 41-8. (*A*) Hen, stewing chicken—A quality. (*B*) Hen, stewing chicken —B quality. (*C*) Hen, stewing chicken—C quality. (*Courtesy* U.S.D.A. Agricultural Marketing Service)

skin, toughened and darkened meat, and a hardened breastbone.

Turkeys

The six official classes of turkeys are listed on the next page.

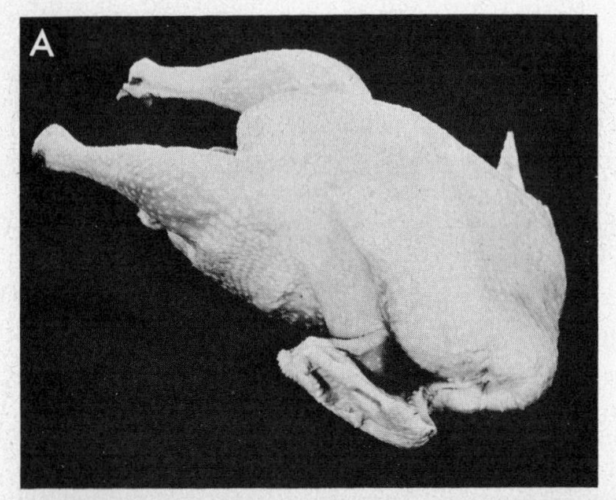

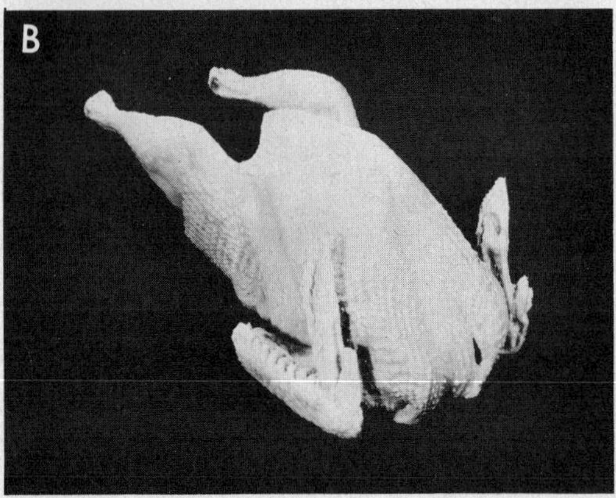

FIGURE 41-9. (*A*) Young tom turkey—A quality. (*B*) Young tom turkey—B quality. (*C*) Young tom turkey—C quality. (*Courtesy* U.S.D.A. Agricultural Marketing Service)

(**Minimum Requirements and Maximum Defects Permitted**)

Factor	A or No. 1 Quality	B or No. 2 Quality	C or No. 3 Quality
Health and vigor	Alert, bright eyes, healthy, vigorous	Good health and vigor	Lacking in vigor
Feathering	Well covered with feathers. Slight scattering of pinfeathers.	Fairly well covered with feathers. Moderate number of pinfeathers.	Complete lack of plumage feathers on back. Large number of pinfeathers.
Conformation:	*Normal*	*Practically normal*	*Abnormal*
Breastbone	Slight curve, ⅛″ dent (chickens), ¼″ dent (turkeys)	Slightly crooked	Crooked
Back	Normal (slight curve)	Moderately crooked	Crooked or hunched back
Legs and wings	Normal	Slightly misshapen	Misshapen
Fleshing	Well fleshed, moderately broad and long breast	Fairly well fleshed	Poorly developed, narrow breast; thin covering of flesh.
Fat covering	Well covered, some fat under skin over entire carcass.	Enough fat on breast and legs to prevent a distinct appearance of flesh through skin.	Lacking in fat covering on back and thighs, small amount in feather tracks.
	Chicken fryers, turkey fryers, and young toms only moderate covering. No excess abdominal fat.	Hens or fowl may have excessive abdominal fat.	
Defects:	*Slight*	*Moderate*	*Serious*
Tears and broken bones	Free	Free	Free
Bruises, scratches, and calluses	Slight skin bruises, scratches, and calluses	Moderate (only slight flesh bruises)	Unlimited to extent no part unfit for food.
Shanks	Slightly scaly	Moderately scaly	Seriously scaly.

Source: U.S.D.A., *Poultry Grading Manual* (Revised), 1971

Fryer-Roaster. A fryer-roaster is a young, immature turkey (usually under 16 weeks of age) of either sex that is tender-meated and has soft, pliable, smooth-textured skin and flexible breastbone cartilage.

Young Hen Turkey. A young hen turkey is a young female turkey (usually 5 to 7 months of age) that is tender-meated with soft, pliable, smooth-textured skin, and breastbone cartilage that is somewhat less flexible that that of a fryer-roaster turkey.

Young Tom Turkey. A young tom turkey is a young male turkey (usually 5 to 7 months of age) that has the same characteristics as a young hen turkey.

Yearling Hen Turkey. A yearling hen turkey is a fully matured female turkey (usually under 15 months of age) that is reasonably tender-meated and has reasonably smooth-textured skin.

Yearling Tom Turkey. A yearling tom turkey is a fully matured male turkey (usually under 15 months of age) that is reasonably tender-meated and has reasonably smooth-textured skin.

Mature Turkey or Old Turkey (Hen or Tom). A mature or old turkey is a turkey of either sex (usually in excess of 15 months of age) with coarse skin and toughened flesh.

Ducks and Geese

Three classes of ducks are specified: (1) broiler duckling or fryer duckling, (2) roaster duckling, and (3) mature or old duck.

There are two classes of geese: (1) young goose and (2) mature or old goose.

Guineas and Pigeons

There are two classes of guineas: (1) young guineas and (2) mature or old guineas.

There are two classes of pigeons: (1) squab and (2) pigeon.

MARKETING LIVE POULTRY

Much of the live poultry sold by producers goes directly or through dealers to commercial dressing plants. A comn tice up to about 1930 was to ship liv in cars by rail to processing plan large cities, but this was expensive of transportation costs and shrinka essing and dressing plants are no within trucking distance of prac producers and most live poultry ported by truck. Many large produ their own dressing plants.

Finishing Poultry for Market

Some poultry dressing plants practice of putting fowl, purch farmers and commercial produce teries and feeding the birds heav feeds for two to four days before sl them. The added finish improves quality and the market grade of t Usually the weight increase is v able.

It is a good idea to confine days the birds that are to be m satisfactory facilities are availab feed them heavily on a wet mas adequate and sanitary quarters provided, the birds can be hand- mash rations while on range. This ing period will improve the mark the bird, and the added gain in further increase the returns whe are sold.

Standards of Quality for Live Poultry

Official United States standa ity of individual live birds have for the classes described earlier i ter. The following factors are c determining the quality of an ind (1) health and vigor, (2) fea conformation, (4) fleshing, (5) the number of defects.

Three standards of quality a (1) A or No. 1 Quality, (2) B or ity, and (3) C or No. 3 Quality.

TABLE 41-1 SUMMARY OF STANDARDS OF QUALITY FOR LIVE POULTRY ON AN INDIVIDUAL BIRD BA

of the minimum requirements and maximum defects permitted in each of the quality standards is shown in Table 41-1.

MARKETING DRESSED AND READY-TO-COOK POULTRY

Consumers buy very little live poultry. Most of the poultry sold to consumers is sold as dressed or ready-to-cook fowl.

Dressed Fowl

The term "dressed fowl" refers to birds which have been slaughtered and bled, and have had the feathers removed. Dressed birds are usually sold as fresh-killed poultry. The dressed carcass weighs from 9 to 12 percent less than the live weight of the bird.

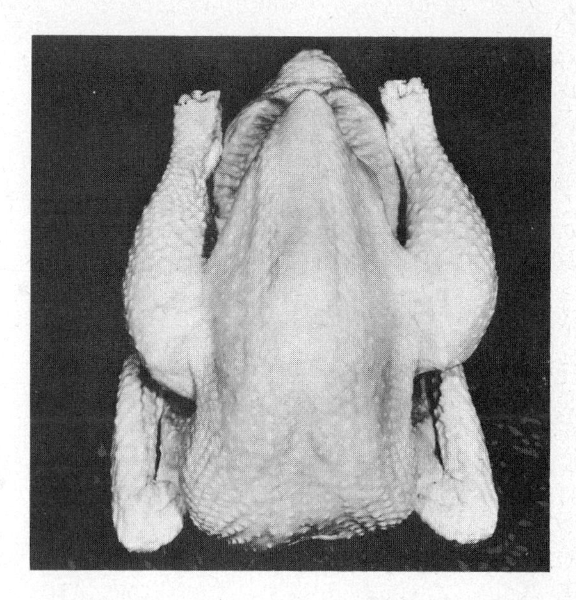

FIGURE 41-10. This type of dressed broiler is very much in demand by the consumer. (*Courtesy* Pilch-Dekalb, Inc.)

Ready-to-Cook Poultry

Much of the poultry on the consumer market is ready-to-cook poultry. The birds have been slaughtered and bled, and the feathers, entrails, head, feet, and shanks have been removed. The gizzard, liver, heart, and neck are usually wrapped in waxed paper and placed inside the bird. The carcass may or may not be cut up for frying.

A three-pound broiler will produce a dressed carcass weighing from about 2.6 to 2.7 pounds, and an eviscerated or ready-to-cook carcass (with giblets) weighing about 2.1 pounds. A six-pound roaster will yield a 5.3- to 5.4-pound dressed carcass, and a 4.2- to 4.4-pound read-to-cook carcass. The loss in dressing and drawing usually amounts to 25 to 35 percent.

Standards of Quality for Dressed and Ready-to-Cook Poultry

Official standards for quality of individual dressed and ready-to-cook poultry have been outlined by the Poultry Branch of the Production and Marketing Administration. The factors considered in determining the quality of an individual carcass are as follows: (1) conformation, (2) fleshing, (3) fat, (4) absence of pinfeathers, (5) lack of exposed flesh resulting from cuts, tears, and broken bones, (6) absence of skin discolorations, flesh blemishes, and bruises, and (7) lack of freezing defects.

The minimum requirements and maximum defects permitted in each of the quality or grade standards for dressed and ready-to-cook chickens and turkeys are shown in Table 41-2.

Weight Specifications

Dressed and ready-to-cook poultry is also sold according to weight specifications. The carcasses are weighed and grouped. Broilers, roasters, and hens are classified at half-pound intervals, while capon, stag, and cock classifications are based on one-pound differences in weight. The classifications of

TABLE 41-2 SUMMARY OF SPECIFICATIONS FOR STANDARDS OF QUALITY FOR INDIVIDUAL CARCASSES OF READY-TO-COOK POULTRY AND PARTS THEREFROM

(Minimum Requirements and Maximum Defects Permitted)

SEPTEMBER 1, 1965

FACTOR		A QUALITY			B QUALITY			C QUALITY
CONFORMATION:		Normal			Moderate deformities			Abnormal
Breastbone		Slight curve or dent			Moderately dented, curved or crooked			Seriously curved or crooked
Back		Normal (except slight curve)			Moderately crooked			Seriously crooked
Legs and Wings		Normal			Moderately misshapen			Misshapen
FLESHING:		Well fleshed, moderately long, deep and rounded breast			Moderately fleshed, considering kind, class and part			Poorly fleshed
FAT COVERING:		Well covered—especially between heavy feather tracts on breast and considering kind, class and part			Sufficient fat on breast and legs to prevent distinct appearance of flesh through the skin			Lacking in fat covering over all parts of carcass
PINFEATHERS:								
Nonprotruding pins and hair		Free			Few scattered			Scattering
Protruding pins		Free			Free			Free
EXPOSED FLESH: 1 Carcass Weight								
Minimum	Maximum	Breast and Legs	Elsewhere	Part	Breast and Legs 2	Elsewhere 2	Part	
None	1½ lbs.	None	¾″	(Slight	¾″	1½″	(Moderate	
Over 1½ lbs.	6 lbs.	None	1½″	trim	1½″	3″	amount of the	No
Over 6 lbs.	16 lbs.	None	2″	on edge)	2″	4″	flesh normally	Limit
Over 16 lbs.	None	None	3″		3″	5″	covered)	
DISCOLORATIONS: 3								
None	1½ lbs.	½″	1″	¼″	1″	2″	½″	
Over 1½ lbs.	6 lbs.	1″	2″	¼″	2″	3″	1″	No
Over 6 lbs.	16 lbs.	1½″	2½″	½″	2½″	4″	1½″	Limit 4
Over 16 lbs.	None	2″	3″	½″	3″	5″	1½″	
Disjointed bones		1			2 disjointed and no broken or			No limit
Broken bones		None			1 disjointed and 1 nonprotruding broken			No limit
Missing parts		Wing tips and tail 5			Wing tips, 2nd wing joint and tail			Wing tips, wings and tail
FREEZING DEFECTS: (When consumer packaged)		Slight darkening over the back and drumsticks. Few small ⅛″ pockmarks for poultry weighing 6 lbs. or less and ¼″ pockmarks for poultry weighing more than 6 lbs. Occasional small areas showing layer of clear or pinkish ice.			Moderate dried areas not in excess of ½″ in diameter. May lack brightness. Moderate areas showing layer of clear, pinkish or reddish colored ice.			Numerous pockmarks and large dried areas.

1 Total aggregate area of flesh exposed by all cuts and tears and missing skin.

2 A carcass meeting the requirements of A quality for fleshing may be trimmed to remove skin and flesh defects, provided that no more than one-third of the flesh is exposed on any part and the meat yield is not appreciably affected.

3 Flesh bruises and discolorations such as "blue back" are not permitted on breast and legs of A quality birds. Not more than one-half of total aggregate area of discolorations may be due to flesh bruises or "blue back" (when permitted), and skin bruises in any combination.

4 No limit on size and number of areas of discoloration and flesh bruises if such areas do not render any part of the carcass unfit for food.

5 In geese, the parts of the wing beyond the second joint may be removed, if removed at the joint and both wings are so treated.

Source: U.S.D.A., *Poultry Grading Manual* (Revised), 1971.

turkeys and geese are based on two-pound differences.

Turkey hens and lightweight birds bring higher prices than toms and heavy birds. The discount for heavy birds is greatest in the spring. The average weight of all turkeys marketed in 1972 was 18.8 pounds. Small turkeys made up about 11.6 percent of the 1972 turkey crop in number.

Seasonal Price Trends

Nearly 50 percent of the turkeys produced in this country are marketed in the months of September, October, and November. The increase in broiler production has helped spread production throughout the year and decrease seasonal price differences. The freezing of poultry has also helped minimize seasonal price differences.

As shown in Figure 41-11, turkey prices were lowest during the period from May to August and highest in October, November, and December. The spread in price between the low in July and the high in December was only 2 cents per pound. Broilers sold best in June, July, and August. October, November, and December were the poorest broiler marketing months. There was a spread of 2.4 cents per pound between the low market in December and the high market in July. It pays to plan production and marketing so that the birds will be sold during the periods of high prices.

It should be noted from data in Figure 41-11 that the large volume of broilers re-

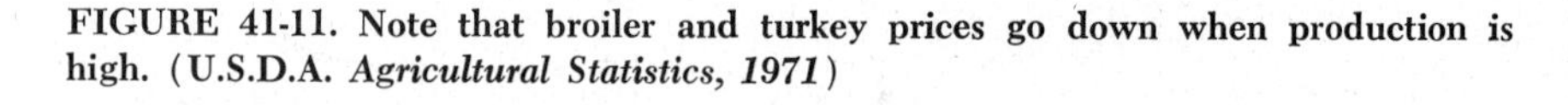

FIGURE 41-11. Note that broiler and turkey prices go down when production is high. (U.S.D.A. *Agricultural Statistics, 1971*)

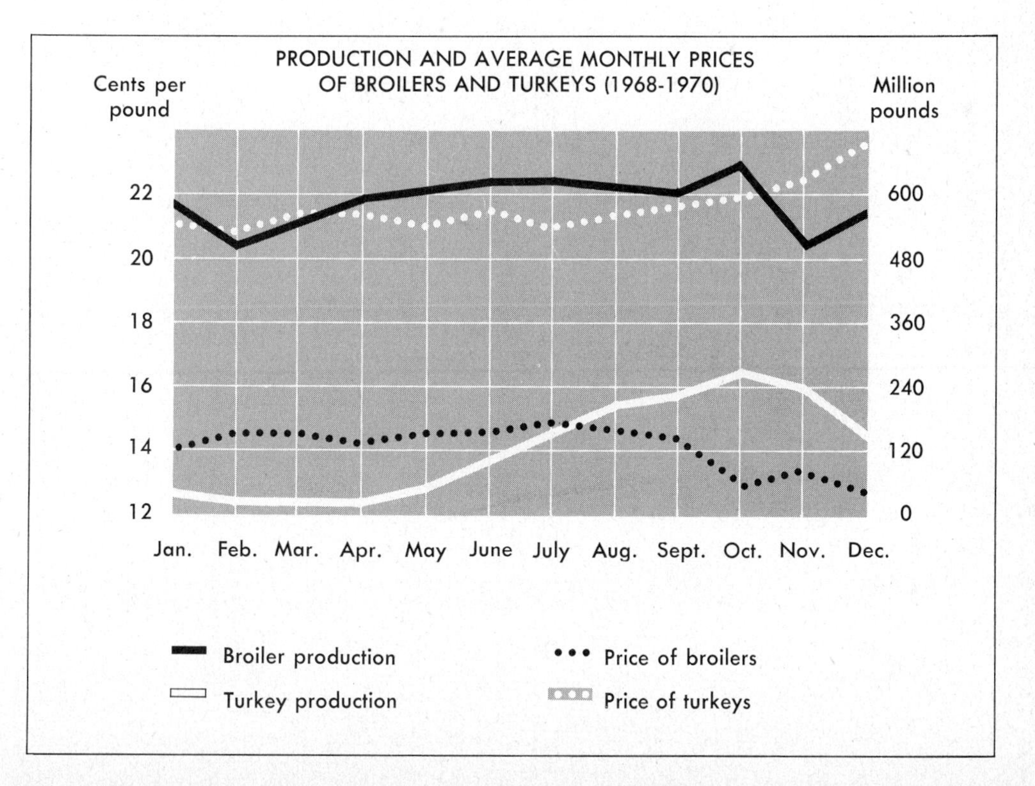

sulted in a comparatively low selling price. In comparison to broiler production, turkey production was low and the selling price high. The difference in selling price does not necessarily indicate greater profit in turkey production, since the cost of production is greater in producing turkeys than in producing broilers.

SUMMARY

Increased returns from the poultry enterprise may be obtained by (1) marketing poultry and eggs during the periods of high prices, (2) improving the quality of the poultry products, and (3) careful selection of the method of marketing. Eggs sold in November, December, and January sell for several cents more per dozen than eggs sold in April, May, and June.

The four United States consumer grades of eggs are AA Quality, A Quality, B Quality, and C Quality. The quality is determined by candling the egg. Albumen quality, freedom from defects, shell quality, and age are important in determining overall egg quality.

Eggs are classed according to size as jumbo, extra large, large, medium, small, and peewee. Most commercial eggs are now sold as graded eggs. Most states have egg grading laws.

To improve egg quality, eggs should be gathered several times a day, cooled in a wire basket in a cool, moist room or in a mechanical cooler, and cased the following morning, placing the large end up. Small, irregular, cracked, and dirty eggs should be sorted out, and only infertile eggs should be produced. Good nesting facilities should be provided. Eggs should be marketed two or three times a week and sold by grade.

Chickens are sold as Cornish game hens, broilers, roasters, capons, stags, hens, or cocks. Turkeys are sold as fryer-roasters, young hen turkeys, young tom turkeys, yearling hen turkeys, yearling tom turkeys, and old turkeys. Live birds are graded according to health and vigor, feathering, conformation, fleshing, fat, and degree of freedom from defects. There are three standards of quality: A or No. 1 Quality, B or No. 2 Quality, and C or No. 3 Quality.

Dressed and ready-to-cook poultry is graded on conformation, fleshing, fat, freedom from pinfeathers, amount of exposed flesh, and disjointed or broken bones, number of bruises, flesh blemishes, discolorations of skin, and freezing defects. Three quality grades are used—A, B, and C.

Toms and heavy turkeys sell for less than hens and light-weight finished turkeys. Turkey prices are highest in October, November, and December. Broilers should be sold in June, July, and August. Producers should have the desired quantity and quality of birds ready when the market is best.

QUESTIONS

1 During which months of the year are egg prices highest?

2 What are the characteristics of each of the U.S.D.A. consumer grades of eggs?

3 What are the advantages and disadvantages of selling eggs by grade?

4 What are the weight classes of eggs?

5 Does it pay to clean dirty eggs? Why?

6 Outline a program for improving the quality and the marketing of eggs on your farm.

7 What are the market classes of chickens and turkeys?

8 Describe the standards of quality and the grades for live poultry.

9 What are the standards of quality and the grades of dressed and ready-to-cook poultry?

10 What is the difference between dressed and ready-to-cook fowl?

11 During which months will broilers sell for the highest prices? turkeys?

12 Plan a feeding program for finishing broilers and turkeys.

13 Outline a program for improving the quality and methods of marketing poultry on your farm.

REFERENCES

Baker, Ralph L., *Marketing Poultry Meat,* MM-237, The Ohio State University, Columbus, Ohio, 1964.

Dawson, L. E., *Farm Handling Practices and Egg Quality,* FS 30.1, Michigan State University, East Lansing, Michigan, 1958.

Harper, W. W., *Marketing Georgia Turkeys,* Bulletin N.S. 79, University of Georgia, Athens, Georgia, 1960.

Stiles, P. G., *et al., Quality Control of Poultry Products,* 65–69, University of Connecticut, Storrs, Connecticut, 1965.

U. S. Department of Agriculture, *Egg Grading Manual,* Agriculture Handbook No. 75, Revised, Washington, D.C., 1969.

———, *Poultry and Egg Situation,* Economic Research Service, Washington, D.C., June 1972, September 1972, April 1973.

———, *Poultry Grading Manual,* Agriculture Handbook No. 31, Revised, Washington, D.C., 1971.

———, *Poultry Market Statistics, 1970,* Statistical Bulletin 466, Washington, D.C., 1971.

———, *Selected Statistical Series for Poultry and Eggs Through 1968,* ERS 232 Revised, Washington, D.C., 1970.

INHERITANCE AND REPRODUCTION

42 Reproduction, Inheritance, and Pedigrees in Animal Breeding

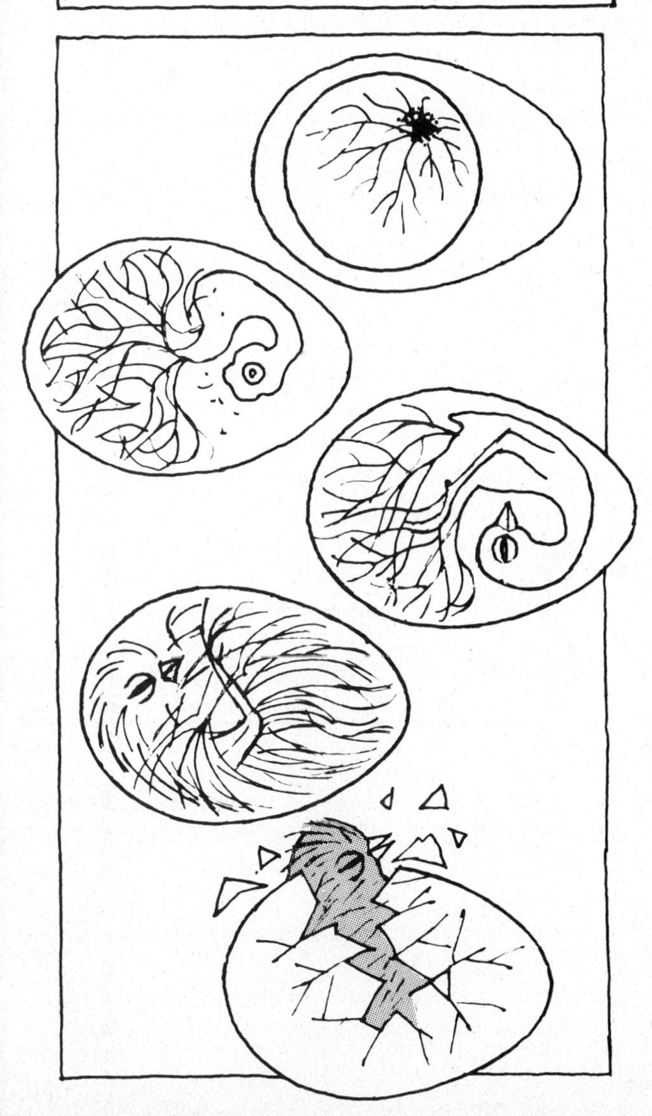

Much of the improvement in the conformation and type of livestock has come about through carefully planned breeding programs. An understanding of the basic principles in reproduction and inheritance is essential in planning livestock-breeding programs. These principles are much the same, regardless of the type of animal, but there may be variations in application. We present in this chapter a brief discussion of (1) reproduction, (2) the reproductive organs of male and female animals, (3) the laws of inheritance which affect reproduction, and (4) the value of pedigree information in animal breeding.

REPRODUCTION

Reproduction is the process by which new individuals are produced in plant and animal life. Life begins when the female reproductive cell is fertilized by the male reproductive cell. Fertilization occurs after the animal has been bred. The female cell is called the *ovum*, or egg, and the male cell is called the *sperm*.

The egg contains the hereditary materials of the mother (dam) while the sperm contains the materials of inheritance contributed by the father (sire). The contributions of the sire and dam, so far as inheritance is concerned, are equal. Fertilization of the egg takes place within the body of the mother and the offspring is nourished and protected by the mother until birth.

Female Reproductive Organs

As shown in Figure 42-1, the reproductive system of the female consists of the ovaries, the oviducts, the uterus, the vagina, and the vulva.

Ovaries. These two glandular organs are located in the sublumbar region and produce eggs. Each ovary contains many follicles in

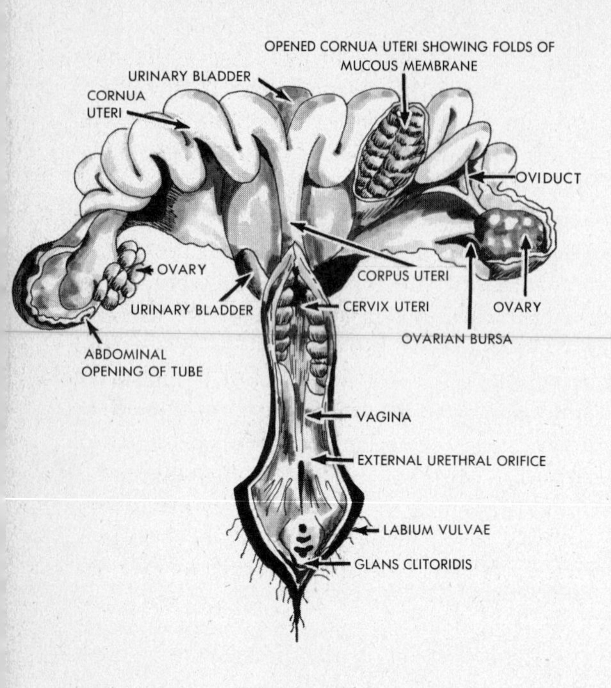

FIGURE 42-1. Reproductive organs of a sow. (Drawing by David C. Opheim)

which the eggs are produced. As the eggs mature, they are dropped into the oviduct. The process is called ovulation and takes place during or shortly after the heat period.

Oviducts. These tubes lead from the ovaries to the horns of the uterus. The fertilization of the egg usually takes place near the upper end of the oviduct.

Uterus. This is a hollow organ containing two horns, which are connected to the oviducts. The fertilized egg moves from the oviduct into the uterus, becomes attached to the wall, and develops.

Cervix. This is the neck of the uterus and separates it from the vagina.

Vagina. This tube connects the vulva and the cervix.

Vulva. Both the urinary and reproductive organs of the female terminate in the vulva.

Male Reproductive Organs

The reproductive system of the male consists of the testicles, the sperm ducts, the seminal vesicles, the prostate, Cowper's glands, the urethra, and the penis.

Testicles. Sperm cells are produced in the two testicles, which are suspended in the scrotum.

Sperm Ducts. These tubes connect the testicles with the urethra. Sperm passes through and may be stored at the upper end of these tubes.

Seminal Vesicles. These glands open to the urethra and secrete a fluid.

Prostate. The prostate gland is located near the bladder and the urethra. It also produces a secretion that becomes a part of the seminal fluid.

Cowper's Glands. These glands secrete a fluid that precedes the passage of the sperm cells down the urethra. The sperm cells and the fluids make up the semen.

Urethra. This long tube extends from the bladder to the penis and carries both urine and semen.

Penis. This organ deposits the sperm cells within the female reproductive system.

Conception

Reproduction begins with the female's heat period, the time when she will be receptive to the male. The heat period varies with the type of animal, but in cows it usually occurs on an average of every 21 days. Within 6 to 20 hours after the heat period, the female reproductive cell or egg is released from the ovaries. If the female is bred and a sperm (male reproductive cell) comes in contact with the egg, fertilization takes place and the female becomes pregnant. It takes from four to nine hours for the sperm to travel from the vagina to the oviducts where fertilization normally takes place. Therefore, a female bred near the end of her heat period is more apt to conceive than if she were bred early in the heat period.

Growth of the Fetus

The cells within the fertilized egg multiply and accumulate rapidly after the egg enters the uterus. The outer layer of the fertilized egg breaks up in about ten days and the formation of the body organs and parts begins. The heart of a calf embryo may begin to beat 22 to 24 days after conception.

The developing embryo floats unattached in the uterus fluids for about 30 days. The membranes of the embryo then unite with the caruncles, or "buttons," on the inside of the uterus. The developing embryo at this stage becomes a *fetus* and continues growth by receiving food through the blood from the capillaries in the caruncles of the uterus.

Fetal Membranes. These consist of three separate structures which surround the fetus. One contains many blood vessels which lead into the placenta and unite the mother and her unborn offspring. Another membrane contains a fluid which protects the fetus from injury. The third membrane serves as a holding place for the urine from the fetal bladder.

Placenta. This vascular structure is the portion of the fetal membrane that unites the mother and the fetus. The blood vessels of each lie close together, which permits the interchange of food materials through their extremely thin walls.

Navel Cord. This cord connects the fetus and the placenta and serves as a passageway for blood to and from the fetus.

LAWS OF INHERITANCE

Chromosomes and Genes

Each reproductive cell (sperm or egg) contains chromosomes, which in turn carry genes. The genes determine the characteristics to be found in the individual. The color of hair, the conformation of the body, the type, and other characteristics are determined by the genes in the germ cells.

The number of chromosomes in the nucleus of a cell is constant. They occur in pairs. In swine there are 19 pairs, and in sheep, cattle, and horses there are 30 pairs. The members of each pair carry genes that affect the same characteristics of the animal's physical condition, but they may not affect them in the same way. For example, in breeding sheep, factor A on chromosome 1 may affect the color of the wool by producing black color, whereas factor A on the other chromosome of the pair may produce white color. If neither color is dominant, the progeny can be either color.

Dominant and Recessive Characteristics

Certain characteristics are dominant and others are recessive. When a pure polled ram is mated to a horned ewe, the offspring is polled. This is a dominant feature. When dominant and recessive characters are brought together, the progeny will possess the dominant characteristics, but will produce, in the third generation, some animals that will show the dominant characteristics and some that show the recessive. The best example of dominant and recessive factors and how the ratio works out is that of the color and polled characteristics in cattle.

Black color in cattle is dominant to red. When an Angus cow is mated to a Hereford bull, the calf produced will be black with a white face. Darker colors in cattle are usually dominant to lighter colors.

When a polled beef animal such as the Angus is mated to a horned animal such as the Hereford, the first generation or first cross will be polled. However, the crossbred offspring will carry the recessive factor for horns and if a crossbred from this parentage is mated to a similar crossbred, an average of one animal in four will be horned.

White hair color in hogs is dominant to black and to red. Many of the factors which represent good health in hogs—vigor, rate of gain, feed efficiency, and carcass quality—are thought to be dominant, while those that tend toward poor health are thought to be recessive. For example, navel hernias appear when two recessive genes are brought together.

Chromosome Segregation

The normal reproductive cell in sheep and cattle contains 30 pairs of chromosomes, yet when two reproductive cells are brought together there still are only 30 pairs of chromosomes. This is because of a reduction process in reproductive cell production which reduces the number of chromosomes by half, so that only one chromosome of each pair is included in the reproductive cell. The union of the two reproductive cells through fertilization restores the normal number of chromosomes.

Hybrid

An animal is considered a hybrid for any one character when it possesses one dominant and one recessive gene. When a hybrid is crossed with another hybrid, about 75 percent of the progeny show dominance, but only about one-third of the group are pure dominant for the one character. The other two-thirds have the appearance of the dominant but are hybrid. Those that show recessive characteristics are always pure recessives.

Hybrid Vigor

The crossing of two superior animals of different breeds usually results in increased growth rate, increased efficiency of fertilization, improvement of body conformation, and increased production. Crossbreeding in hogs, cattle, sheep, and chickens has been done for years, but not until rather recently has it been done on a truly scientific basis.

Heterosis, or hybrid vigor, is the term applied to the increase in vigor and performance resulting when two animals of unrelated breeds are crossed. The increase in vigor may be explained in terms of the genetic principles discussed in previous paragraphs. The genes producing vigor are dominant to those producing a lack of vigor. By crossing breeds, a larger number of dominant genes are brought together in the progeny than are involved in breeding animals of the same breed.

Upgrading and Purebreeding

These two systems of breeding involve the mating of animals of the same breed. Upgrading is the successive mating of purebred sires and grade females of the same breed. Purebreeding is the mating of purebred sires and purebred females of the same breed. In both cases, animals to be mated are selected according to their body conformation, vigor, growth rate, and production record. Purebred animals are quite *homozygous*—the genes are usually alike. The hybridization referred to previously does not exist. The animals tend to produce progeny like themselves—like begets like.

Inbreeding and Linebreeding

Inbreeding is the mating of closely related animals, such as (1) brother to sister, (2) son to dam, and (3) sire to daughter. It is done primarily to intensify the degree of homozygosity, or the similarity of the genes in the reproductive cells of the animals. The crossing of inbred lines of two or more breeds results in hybrid vigor greater than that produced when two or more noninbred lines are crossed.

Linebreeding is similar to inbreeding but involves the breeding of animals less closely related. The mating of cousins and of grandsire with granddaughters are examples of linebreeding. It is done to conserve and perpetuate the good traits of certain outstanding breeding animals. It ends to produce an homozygous genetic condition.

VALUE OF PEDIGREES IN ANIMAL BREEDING

Pedigree information is very valuable in livestock breeding. Most breeding programs use purebred sires for which pedigree information is available, and some breeding programs use females that are purebred. A careful analysis of the pedigree may indicate the prolificacy, productiveness, and conformation of the animal and of its progeny.

Pedigrees of dairy animals not only provide a list of the ancestors but may also give the production records of the animals. Shown on pages 612 and 613 are a photograph of a Guernsey bull and its pedigree. Note that the ancestry and production information are provided for the sire and dam, and for the grandsires and granddams. An extended pedigree provides information concerning the great-grandsires and great-granddams. This pedigree, while not extended, shows the type of information that may be available concerning breeding animals. This information, supplemented by an inspection of the animal and of the herd from which it came, provides the breeder with sufficient information to decide upon the desirability of using the animal or his progeny in his breeding program.

SUMMARY

The reproductive cell of the female is called an ovum. The male cell is called a sperm. Each parent contributes equally in the reproduction process.

The ovum is produced in a follicle of the ovary and is dropped into the oviduct. It is fertilized by a sperm in the upper end of the oviduct. From there it moves to the uterus, where it becomes attached to the wall and develops.

The sperm is produced in the testicle and moves through the sperm ducts to the urethra and the penis. The seminal vesicles, prostate, and Cowper's glands add secretions to the seminal fluid.

The heat period in cows occurs about every 21 days and ovulation takes place from 6 to 20 hours after the cow has gone out of heat. Breeding should be timed so that the sperm is in the oviduct when the ovulation occurs. It normally takes the sperm from four to nine hours to reach the ovum after being introduced.

The chromosomes in the reproductive cells carry genes which determine the characteristics to be transmitted to the progeny. Chromosomes occur in pairs, with one of each pair in the male cell uniting with one of each pair in the female cell to produce the new individual. The reproductive cell in swine has 19 pairs of chromosomes, while the reproductive cell of cattle, sheep, and horses has 30 pairs. The two members of a pair of chromosomes carry genes which affect the same physical conditions of the animal, but they may not affect them in the same way.

Some factors of inheritance are dominant, while others are recessive. A hybrid condition exists when the reproductive cell has one dominant and one recessive gene.

The most common systems of breeding are upgrading, purebreeding, inbreeding, linebreeding, and crossbreeding. Upgrading is the successive use of purebred sires on grade females of the same breed. Purebreeding is the mating of purebred males and females of the same breed. Inbreeding is the mating of closely related animals. Linebreed-

FIGURE 42-2. Burger's Place Billy, a proven Guernsey bull owned by American Breeders Service, Inc. This bull sired, through artificial insemination, 651 daughters that produced an average of 11,482 pounds of milk and 537 pounds of milkfat. Ten daughters were classified Excellent and 325 as Very Good. (Strohmeyer and Carpenter photo. *Courtesy* American Breeders Service, Inc.)

BURGER'S PLACE BILLY
511244 G★S, A.I. G★S Wt. 2050 lbs.
Born: October 24, 1953
Bred by: Albert Burger, Sebastopol, California
Purchased from: Arnold Kessler, Galt, California
Owned by: American Breeders Service, Inc., DeForest, Wisconsin

USDA Daughter-Herdmate Comparison, 5/67

2010 Dtrs., 2935 Recs., 586 Herds	10,333 M	4.52%	467 BF
All Herdmates Average	9,728 M	4.72%	459 BF
Difference	+605 M	−0.20%	+ 8 BF
EDS or Pred. Diff.	+646 M		+10 BF

A.G.C.C.

651 Dtrs., 1023 Recs., Average	11,482 M	4.7%	537 BF

653 Class. Dtrs. Average 83.7
10 EX, 325 VG, 257 D, 52 A, 9 F

ing is similar but uses animals less closely related. Crossbreeding is the crossing of two breeds.

Hybrid vigor is produced by crossing two unrelated breeds. The crossing of inbred lines of two or more breeds increases the heterosis, or hybrid vigor.

Pedigrees provide the ancestry of an animal and, in some cases, a record of their prolificacy and productiveness. The pedigree, supplemented by an inspection of the animal and of the herd from which it came, provides the breeder with information necessary in selecting animals for a breeding program.

CURTISS CANDY FABRON 454224

USDA Daughter-Herdmate Comparison, 5/66

845 A.I. Dtrs.,			
2119 Recs.	8199 M	4.78%	392 BF
All Herdmates Avg.	8655 M	4.76%	412 BF
Difference	−456 M	+.02%	−20 BF
EDS	−479 M		−21 BF

A.G.C.C.—Volume 19
209 Dtrs.,
564 Recs., Avg. 8818 M 4.84% 427 BF

165 Class. Dtrs. Avg. 81.5

ELMCREST DEMONSTRATOR'S PATSY 1278008 EX

AR Record:
Jr. 2 305c 2x 12071 M 4.95% 597 BF

CURTISS CANDY LEVITY HEIR 346065

USDA Daughter-Herdmate Comparison, 3/64

370 A.I. Dtrs.,			
851 Recs.	7779 M	4.62%	360 BF
All Herdmates Avg.	8386 M	4.78%	401 BF
Difference	−607 M	−.16%	−42 BF
EDS	−634 M		−42 BF

A.G.C.C.—Volume 18
161 Dtrs.,
386 Recs., Avg. 8348 M 4.64% 387 BF

96 Class. Dtrs. Avg. 80.5

MCDONALD FARMS HESTER 855759 EX

AR Records:

Jr. 2	363	3x	10597 M	5.62%	596 BF
6y	305c	3x	12093 M	4.96%	600 BF
12y5m	305	2x	10872 M	5.45%	593 BF

CAUMSETT DEMONSTRATOR 400912

A.G.C.C.—Volume 12
18 Dtrs.,
29 Recs., Avg. 10232 M 4.73% 484 BF

4 Class. Dtrs. Avg. 86.3

ELMCREST GRAY PAULINE 752822 VG

AR Records:

Jr. 2	365	3x	14260 M	5.08%	724 BF
Sr. 3	305c	3x	12260 M	4.69%	575 BF
Sr. 4	365c	3x	12010 M	4.66%	560 BF

2 AR Dtrs:
1. At left.
2. Elmcrest Prophet's Pauline, EX
Sr. 2 305c 2x 12870 M 5.13% 660 BF

2 PR Sons.

QUESTIONS

1. Name the reproductive organs of the male and indicate the function of each.
2. What are the reproductive organs of the female and what are their functions?
3. What is meant by ovulation and when does it take place in relationship to the heat period?
4. What is the difference between a gene and a choromosome?
5. Explain the process of producing and fertilizing reproductive cells.
6. Which contributes most to inheritance, the ovum or the sperm? Explain.
7. Explain hybrid vigor in terms of the gene composition of the chromosomes.
8. What differences exist in the gene composition of the chromosomes in purebred and crossbred animals?
9. Explain the following systems of breeding: upgrading, purebreeding, inbreeding, linebreeding, and crossbreeding.
10. Of what value are pedigrees in livestock breeding?

REFERENCES

Hafez, E. S., *Reproduction in Farm Animals,* 2nd Edition, Lea & Febiger, Philadelphia, Pennsylvania, 1968.

Hafs, H. D. and L. J. Boyd, *Dairy Cattle Sterility,* Hoard's Dairyman, Fort Atkinson, Wisconsin, 1964.

Lasley, John F., *Genetics of Livestock Improvement,* Prentice-Hall, Inc., Englewood Cliffs, New Jersey, 1963.

Lush, Jay L., *Animal Breeding Plans* (Revised), Collegiate Press, Inc., Ames, Iowa, 1962.

INDEX

C

D

H

T